# Biology

## *Laboratory Manual*

*Thirteenth Edition*

**Darrell S. Vodopich**
*Baylor University*

**Randy Moore**
*University of Minnesota*

BIOLOGY LABORATORY MANUAL, THIRTEENTH EDITION

Published by McGraw Hill LLC, 1325 Avenue of the Americas, New York, NY 10019. Copyright ©2023 by McGraw Hill LLC. All rights reserved. Printed in the United States of America. Previous editions ©2020, 2017, and 2014. No part of this publication may be reproduced or distributed in any form or by any means, or stored in a database or retrieval system, without the prior written consent of McGraw Hill LLC, including, but not limited to, in any network or other electronic storage or transmission, or broadcast for distance learning.

Some ancillaries, including electronic and print components, may not be available to customers outside the United States.

This book is printed on acid-free paper.

1 2 3 4 5 6 7 8 9 LMN 27 26 25 24 23 22

ISBN 978-1-264-13727-5 (bound edition)
MHID 1-264-13727-3 (bound edition)
ISBN 978-1-264-40598-5 (loose-leaf edition)
MHID 1-264-40598-7 (loose-leaf edition)

Senior Portfolio Manager: *Ian Townsend*
Product Developer: *Erin DeHeck*
Marketing Manager: *Kelly Brown*
Content Project Managers: *Maria McGreal, Tammy Juran*
Buyer: *Laura Fuller*
Designer: *MPS Limited*
Content Licensing Specialist: *Lorraine Buczek*
Cover Image: *©Darrell S. Vodopich*
Compositor: *MPS Limited*

The Internet addresses listed in the text were accurate at the time of publication. The inclusion of a website does not indicate an endorsement by the authors or McGraw Hill LLC, and McGraw Hill LLC does not guarantee the accuracy of the information presented at these sites.

mheducation.com/highered

# Contents

# Preface

We have designed this laboratory manual for an introductory biology course with a broad survey of basic laboratory techniques. The experiments and procedures are simple, safe, easy to perform, and especially appropriate for large classes. Few experiments require more than one class meeting to complete the procedure. Each exercise includes many photographs and illustrations, traditional topics, and experiments that help students *do* biology as they learn about life. Procedures within each exercise are numerous and discrete so that an exercise can be tailored to the needs of the students, the style of the instructor, and the facilities available.

## TO THE STUDENT

We hope this manual is an interesting guide to many areas of biology. As you read about these areas, you'll probably spend equal amounts of time observing and experimenting. Don't hesitate to go beyond the observations that we've outlined—your future success as a scientist and an informed citizen depends on your ability to seek and notice things that others may overlook. Now is the time to develop this ability with a mixture of hard work and relaxed observation. Have fun, and learning will come easily. Also, remember that this manual is designed with your instructors in mind as well. Go to them often with questions—their experience is a valuable tool that you should use as you work.

## TO THE INSTRUCTOR

This manual's simple, straightforward approach emphasizes experiments and activities that optimize students' investment of time and your investment of supplies, equipment, and preparation. Simple, safe, and straightforward experiments are most effective if you interpret the work in depth. Most experiments can be done easily by a student in 2 to 3 hours. Terminology, structures, photographs, and concepts are limited to those that the student can readily observe and understand. In each exercise we have included a few activities requiring a greater investment of effort if resources are available, but omitting them will not detract from the objectives.

This manual functions best with an instructor's guidance and is not an autotutorial system. We've provided background information for context and understanding, but the focus of each exercise remains on students doing interesting and meaningful activities to learn basic information about biology. We've tried to guide students from observations to conclusions, to help students make their own discoveries, and to make the transition from observation to understanding biological principles. But discussions and interactions between student and instructor are major components of a successful laboratory experience. Be sure to examine the "Questions for Further Study and Inquiry" in each exercise. We hope they will help you expand students' perceptions that each exercise has broad application to their world.

## DIGITAL INTEGRATION

Today's students are digital learners, and this lab manual integrates that learning with interesting activities that help students learn about biology. Virtually every exercise of this manual is accompanied by tailor-made digital resources, including assignable questions and a variety of high-definition videos, PowerPoint images, and other resources that demonstrate basic techniques, emphasize biological principles, test for understanding, and engage students as they learn biology in the laboratory.

Digital resources are available to instructors at **connect .mheducation.com.** Instructors will want to assign these resources to help students know what they'll be doing, what principles they'll be investigating, and what concepts they'll need to understand before coming to lab.

## WHAT'S NEW IN THIS EDITION

Throughout the manual, we have expanded and improved several of the most popular and effective features of previous editions, including

- **Learning Objectives** have been updated to provide an overview of what students will do and learn in the exercise.

- **Procedures** and **Doing Biology Yourself** require students to *do* biology as they apply skills they've learned to develop and study hypotheses they formulate about biology.

- **Questions** throughout each exercise encourage students to pause and think about their data and what they've learned.

- **Questions for Further Study and Inquiry** at the end of each exercise help students apply what they've learned to broader topics and issues in biology.

- **Writing to Learn Biology** encourages students to use writing to develop their ideas about what they learned in lab.

- **Caution** and **Safety First** icons make students aware of safety issues associated with the procedures they'll use in lab.

- **Boxed readings** titled **Inquiry-Based Learning** encourage students to apply what they've learned to independently answer questions about intriguing biological topics.

- Updated health-related exercises help students better understand how topics such as genetics, cell biology, blood pressure, atherosclerosis, and their risk of cardiovascular disease relate to our health.

- Several illustrations have been replaced with photographs to provide more realistic images to support the Exercise content.

- Approximately 90 illustrations and photos have been revised.

- Questions within procedures now include lines on which students can write their answers.

- An assignable, updated library of videos and Connect questions helps students prepare for lab and understand the instruments and techniques that will be important for their investigations. **Instructors may assign these videos before class time to help ensure that students arrive prepared for lab.**

## Exercise-Specific Changes

- Exercise 1—Edited text for improved readability and relevance (e.g., climate change, COVID-19); Improved questions to help students better understand what science is and how science is done

- Exercise 2—Improved the readability of the text and the presentation of metric units; Specified the differences in using a triple-beam balance and an electronic scale; Emphasized the importance of significant figures in measurements; Emphasized that in biology, the mean is usually preferred to the median when reporting descriptive statistics; Added a question about measurements of COVID-19

- Exercise 3—Improved the instructions for how to use a compound light microscope

- Exercise 4—Added an objective for understanding the relative sizes of cells and organelles; Added a boxed insert about surface-area-to-volume ratios in cells; Added a boxed insert about cellular structure and human disease

- Exercise 5—Reorganized and edited the text for increased understanding and readability

- Exercise 6—Replaced figure 6.9 with a better, more informative image; Added a table for students to summarize the biochemical tests they performed in the lab; Added a question to emphasize the significance of acid precipitation; Added a boxed insert about using the iodine test to detect counterfeit money; Added a boxed insert about dietary fats

- Exercise 7—Reorganized the procedures for better use of time in the lab

- Exercise 9—Revised the Introduction and Diffusion sections to emphasize the relevance of osmosis and diffusion to general physiology; Enhanced the safety notice to use appropriate PPE; Added question for problem-solving based on experimental data; Revised captions for figures 9.7 and 9.9 to emphasize the flow of water into and out of cells

- Exercise 10—Revised the Introduction to reinforce understanding of how membranes regulate the movement of materials into and out of cells

- Exercise 12—Replaced figure 12.1 (i.e., rising bread dough) to show the production of carbon dioxide; Edited questions for improved understanding; Updated the terminology for the citric acid cycle

- Exercise 13—Replaced figure 13.1 to emphasize the production of oxygen by photosynthesis; Edited the text for improved readability and understanding; Corrected figure 13.10 for improved entry of data by students

- Exercise 14—Enhanced the readability of the Introduction; Expanded the description of chromatids versus chromosomes; Added new figure 14.6 showing the metaphase plate and chromosomal alignment

- Exercise 15—Revised the Introduction to emphasize the value of genetic recombination for adaptation to changing environments; Revised labels of figure 15.1 to better distinguish maternal homologues from paternal homologues; Revised figure 15.2 to emphasize (1) the replication of chromosomes and (2) the formation of chromatids; Added new figure 15.6 of spermatogenesis to emphasize the steps of maturation from spermatogonium to spermatozoa

- Exercise 16—Updated the information about the use and yield of genetically modified crops; Edited questions to emphasize critical thinking about genetically modified crops

- Exercise 17—Edited the text for improved readability and understanding; Added updates about phenylketonuria, Huntington's disease, and familial hypercholesteremia; Added information and a new image to improve students' understanding of transposons

- Exercise 18—Added an example of calculating Hardy-Weinberg frequencies

- Exercise 19—Revised figure 19.2 to reflect recent discoveries about human evolution; Revised Procedure 19.2 to compare the sizes of brain cases in apes versus humans; Added new figure 19.10 comparing skeletons of humans and chimpanzees

- Exercise 20—Clarified the definitions of soil types; Revised Procedure 20.3 to clarify calculations

- Exercise 21—Edited the objectives for improved understanding

- Exercise 22—Plagues; Added a boxed insert about Population Growth and Our Carbon Footprint; Updated information in the text about population and population growth; Expanded table 22.1 to include 10 generations of bacterial growth; Emphasized and added a question about how population growth affects public health, economic stability, social structure, and the well-being of our environment

- Exercise 23—Edited text to improve readability and accuracy

- Exercise 24—Relabeled figure 24.6 to help students better understand the structure of bacterial cell walls; Replaced figure 24.7 to better show steps of the Gram stain procedure; Revised the description and interpretation of antibiotic effectiveness apparent on bacterial sensitivity plates

- Exercise 25—Enhanced explanations of autotrophic versus heterotrophic protistans; Added new figure 25.1 to distinguish between algae and protozoans; Replaced figure 25.5 to better explain *Chlamydomonas* life cycle; Expanded the explanation of asexual versus sexual reproduction in unicellular algae; Rearranged the descriptions of brown algae and red algae to adhere to current phylogeny based on molecular taxonomic techniques

- Exercise 26—Moved the coverage and procedures about slime molds forward to better reflect current phylogeny; Added new figure 26.8 showing a scanning electron micrograph that emphasizes the cell surface of a ciliate

- Exercise 27—Multiple clarifications of the structures and processes of asexual versus sexual reproduction in fungi; Revised figure 27.1 to highlight aseptate hyphae; Revised figure 27.2 to distinguish between sporangia and sporangiophores; Expanded the coverage of the major phyla of fungi to include phylum Glomeromycota; Added new figure 27.3b to show infection by chytrid fungi; Revised table 27.1 to include description and artwork of key reproductive features of Glomeromycota; Updated figure 27.4 to better illustrate stolons, spores, and sporangiophores

of Zygomycota; Expanded explanation of asexual versus sexual reproduction in Zygomycota; Revised figure 27.6b to emphasize distinctions between sexual reproduction and asexual reproduction in bread molds; Expanded descriptions in Procedure 27.3 to help students better interpret conjugation plates of Rhizopus; Revised figure 27.9 to better distinguish between a sporangium and conidiophore; Revised figure 27.13 to better distinguish asexual from sexual reproductive structures and processes; Revised figure 27.15 to emphasize sexual reproduction in mushrooms; Included coverage and new procedures for examining Glomeromycota and other mycorrhizae; Added descriptions and illustrations of mycorrhizae, including arbuscular and ectomycorrhizae forms; Added new figure 27.18e illustrating the structure of a lichen cross section

- Exercise 28—Updated classification information; Replaced figures 28.6 and 28.11 to help students better understand the information

- Exercise 29—Enhanced figures 29.1 and 29.11 for better understanding

- Exercise 30—Edited text for better readability and understanding; Added a question about the distinguishing features of the groups of plants that students examined in this lab

- Exercise 31—Improved table 31.1 and figure 31.5 for better understanding; Improved "Dichotomous Key to Major Types of Fruit"; Replaced figure 31.18 with better, more informative images and information; Added a question to emphasize the differences between monocots and eudicots

- Exercise 32—Edited text for improved readability and understanding; Improved the description of the endodermis and its function; Replaced figure 32.1 to better show the differences in tap versus fibrous root systems; Added scale-markers to figures; Edited the text to better emphasize the differences between gymnosperms and angiosperms; Enhanced figure 32.16 for better understanding; Added a question to emphasize the differences between stomata and lenticels

- Exercise 33—Edited the Introduction for improved understanding; Removed the redundant instruction in Procedure 33.2; Added an alternate procedure for making a leaf-impression for counting and visualizing stomata

- Exercise 34—Emphasized and added a question about how plants, unlike animals, have a small number of growth regulators that influence many traits; Added scale-markers to figures; Added information about the use of 2,4-D; Added information about how gibberellic acid is important for increasing yields and profits for grape growers

- Exercise 35—Added text to improve understanding about bioassays and standard curves; Added a more specific question to the "Inquiry-Based Learning" assignment; Added graph paper for reporting students' results

- Exercise 36—Clarified functional relationships among spicules, spongin fibers, porocytes, and amoebocytes; Expanded the description of water flow through a wall of a sponge as depicted in figure 36.4; Revised figure 36.12 to show the relative size of cnidarian medusae; Revised figure 36.16 to show the relative size of ephyrae; Expanded the description of corals to include information about coral bleaching and coral symbioses with algae

- Exercise 37—Significantly revised the sequence of coverage of invertebrate phyla to adhere to current phylogeny based on molecular taxonomic techniques; Included taxonomic classifications of lophophorazoa and ecdysozoa; Positioned coverage of nematodes to immediately precede coverage of arthropods, as both are now considered ecdysozoans; Mollusk coverage now immediately follows that of flatworms, as they are both considered lophophorazoans; Added new figure 37.3 to illustrate a trochophore larva; Revised table 37.1 to replace nematode descriptions with mollusk descriptions; Replaced figure 37.3 with new art illustrating flatworm anatomy; Replaced figure 38.5 with new art illustrating molluscan radula

- Exercise 38—Coverage of nematodes now follows that of annelids

- Exercise 39—Revised figure 39.16 to clarify position of retinula cells

- Exercise 40—Revised legend of figure 40.18 to better describe the evolution of jaws among fish ancestors; Changed common name of chordate class Actinopterygii from boney fish to ray-finned fish; Added new table 40.3 to provide space for students to organize classes of vertebrates and their major characteristics

- Exercise 41—Revised Procedure 41.1 to emphasize safety when using stains; Revised figure 41.5 to clearly label nuclei of simple columnar epithelial cells; Clarified the varied functions of connective tissues; Expanded Procedure 41.3 to describe the appearance of red blood cells and leukocytes on prepared slides; Included new terminology of central canals in place of Haversian systems of bones

- Exercise 42—Clarified the differences between tendons and ligaments; Added new figure 42.1 to illustrate the parts of the human skeleton; Revised figure 42.2 to include labels of the ileum, ischium, and pubis; Expanded the Questions for Further Study and Inquiry

- Exercise 43—Modified labels of figure 43.2 to show the origin and insertion of triceps brachii

- Exercise 44—Revised figure 44.4 to emphasize how changes of internal air pressure affect the mechanics of breathing; Emphasized the value of measuring lung capacity to understanding respiratory disease; Clarified Procedure 44.2 to better describe the use of a spirometer

- Exercise 45—Expanded the procedure for examining a cow heart to include the use of a heart model; Added a new question to describe heartbeat sounds heard with a stethoscope; Revised figure 45.2 to better show differences in the walls of arteries versus veins; Revised Procedure 45.2 to better describe the steps to measure blood pressure; Added new figure 45.7 to illustrate the anatomy of venous valves; Updated the table for scoring risk factors of cardiovascular disease; Questions for Further Thought and Inquiry now include library research to understand diseases of the heart and circulatory system

- Exercise 46—Quantified differences in retinal resolutions among humans and other animals; Described and distinguished sensorineural versus nerve deafness; Clarified the steps of Procedure 46.8 to better determine nerve deafness; Updated figure 46.6 to show the size of the ear drum; Modified Procedure 46.1 to include safety procedures

- Exercise 47—Expanded Questions for Further Study and Inquiry include an analysis of bipedalism

- Exercise 48—Added new figure 48.7 to include art and a photograph showing the structure of microvilli; Relabeled figure 48.6 to show the common bile duct

- Exercise 49—Added new figure 49.4 to illustrate kidney anatomy with sagittal section

- Exercise 50—Clarified the distinction between an embryo and a zygote; Expanded the description of gray crescent formation; Added new figure 50.5 to illustrate the formation of a gray crescent; Added new figure 50.8 to illustrate differences between the vegetal pole and animal pole; Relabeled figure 50.9 to clearly distinguish the endoderm and mesoderm; Quantified the egg sizes among birds to emphasize variety in egg anatomy; Relabeled figure 50.12 to show albumin

- Exercise 51—Added questions to encourage students to think about agonistic behaviors in humans and why it is important to try to integrate all aspects of an organism's behavior

- Appendix II Updated information about the metric system

# Teaching and Learning Tools

## McGraw Hill Connect®

**McGraw Hill Connect** provides online presentation, assignment, and assessment solutions. It connects your students with the tools and resources they'll need to succeed at **connect.mheducation.com**.

With Connect Biology, you can deliver assignments and quizzes online. A robust set of questions and activities is presented and aligned with this lab manual's learning outcomes. Pre-lab worksheets and Investigation worksheets are also included within Connect. As an instructor, you can edit existing questions and write entirely new questions. Track students' performance—by question, by assignment, or in relation to the class overall—with detailed grade reports. Integrate grade reports easily with Learning Management Systems (LMS), such as Blackboard—and much more.

## Virtual Labs and Lab Simulations *Virtual* Labs

While the biological sciences are hands-on disciplines, instructors are now often being asked to deliver some of their lab content online, as full online replacements, supplements to prepare for in-person labs, or make-up labs.

These simulations help each student learn the practical and conceptual skills needed, then check for understanding and provide feedback. With adaptive pre-lab and post-lab assessment available, instructors can customize each assignment.

From the instructor's perspective, these simulations may be used in the lecture environment to help students visualize complex scientific processes, such as DNA technology or Gram staining, while at the same time providing a valuable connection between the lecture and lab environments.

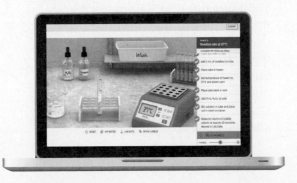

## McGraw Hill Create™

With **McGraw Hill Create**, you can easily rearrange exercises, combine material from other content sources, and quickly upload content you have written, such as your course syllabus or teaching notes. Find the content you need in Create by searching through thousands of leading McGraw Hill textbooks. Arrange your book to fit your teaching style. Create even allows you to personalize your book's appearance by selecting the cover and adding your name, school, and course information. Order a Create book and you'll receive a complimentary print review copy in 3–5 business days or a complimentary electronic review copy (eComp) via e-mail in minutes. Go to **create.mheducation.com** today and register to experience how McGraw Hill Create empowers you to teach *your* students *your* way.

## Laboratory Resource Guide

The *Laboratory Resource Guide* is essential for instructors and laboratory assistants and is available free to adopters of the Laboratory Manual within Connect under the Instructor Resources tab.

SUPPORT AT every step

# Students: Get Learning that Fits You

## Effective tools for efficient studying

Connect is designed to help you be more productive with simple, flexible, intuitive tools that maximize your study time and meet your individual learning needs. Get learning that works for you with Connect.

## Study anytime, anywhere

Download the free ReadAnywhere app and access your online eBook, SmartBook 2.0, or Adaptive Learning Assignments when it's convenient, even if you're offline. And since the app automatically syncs with your Connect account, all of your work is available every time you open it. Find out more at **www.mheducation.com/readanywhere**

*"I really liked this app—it made it easy to study when you don't have your textbook in front of you."*

- Jordan Cunningham,
  Eastern Washington University

## Everything you need in one place

Your Connect course has everything you need—whether reading on your digital eBook or completing assignments for class, Connect makes it easy to get your work done.

Calendar: owattaphotos/Getty Images

## Learning for everyone

McGraw Hill works directly with Accessibility Services Departments and faculty to meet the learning needs of all students. Please contact your Accessibility Services Office and ask them to email accessibility@mheducation.com, or visit **www.mheducation.com/about/accessibility** for more information.

# Welcome to the Biology Laboratory

**W**elcome to the biology laboratory! Although reading your textbook and attending lectures are important ways of learning about biology, nothing can replace the importance of the laboratory. In lab you'll get hands-on experience with what you've heard and read about biology—for example, you'll observe and manipulate organisms, do experiments, test ideas, collect and organize data, and make conclusions about what you've learned. You'll *do* biology.

You'll enjoy the exercises in this manual—they're interesting and informative and can be completed within the time limits of your laboratory period. We've provided questions to test your understanding of what you've done; in some of the exercises, we've also asked you to devise your own experiments to test hypotheses and answer questions that you've posed. To make these exercises most useful and enjoyable, follow these guidelines noted in the next sections.

## THE IMPORTANCE OF COMING TO CLASS

Biology labs are designed to help you experience biology firsthand. To do well in your biology course, you'll need to attend class and pay attention. **Remember this: Attending and being prepared for class are critical for learning about biology and earning a good grade in this course.** To appreciate this, examine figure 1, which is a graph

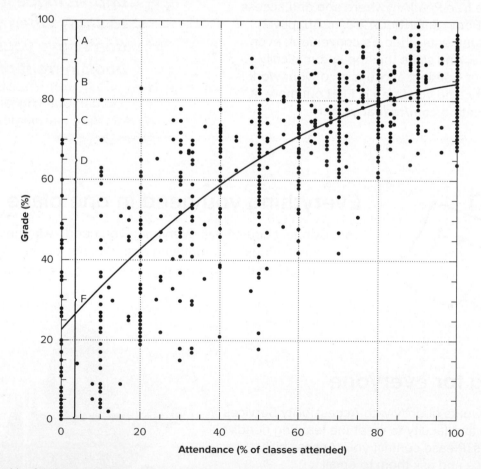

**Figure 1** Relationship of students' grades in an introductory biology course to their rates of class attendance.

showing how students' grades in an introductory biology course correlate to their rates of class attendance. Data are from a general biology class at the University of Minnesota. On page xv, write an analysis of the data shown in figure 1. What do these data mean?

## BEFORE COMING TO LAB

**Watch the lab video.** Videos are provided for several of the labs in this manual. Be sure to watch any assigned video associated with the lab you will be completing. These videos will help you know more about what you will be doing, what principles you will be investigating, and what concepts you need to understand before coming to lab.

**Read the exercise before coming to lab.** This will give you a general idea about what you're going to do, as well as why you're going to do it. Knowing this will not only save time, it will also help you finish the experiments and make you aware of any safety-related issues associated with the lab.

**Review any of the lab safety concerns.** Before doing any procedures, you'll encounter a section of each exercise titled "SAFETY FIRST" that is marked with its icon:

This icon will warn you of safety concerns (e.g., solvents, acids, bases, hotplates) associated with the work. If you have questions about these safety issues, contact your lab instructor before starting the lab work.

Notify your instructor if you are pregnant, are color-blind, are taking immunosuppressive drugs, have allergies, or have any other conditions that may require precautionary measures. Also, before coming to lab, cover any cuts or scrapes with a sterile, waterproof bandage.

## WHEN IN LAB

1. Know what you are going to do. Read and understand the lab before coming to lab.
2. Don't start the exercise until you've discussed the exercise with your laboratory instructor. She or he will give you specific instructions about the lab and tell you how the exercise may have been modified.
3. Work carefully and thoughtfully, and stay focused as you work. You'll be able to finish each exercise within the allotted time if you are prepared and stay on task.

4. Discuss your observations, results, and conclusions with your instructor and lab partners. Perhaps their comments and ideas will help you better understand what you've observed.
5. Always follow instructions and safety guidelines presented by your instructor. Speak up!
6. If you have questions, ask your instructor.

## SAFETY IN THE LABORATORY

Laboratory accidents can affect individuals, classes, or the entire campus. To avoid such accidents, the exercises in this manual were designed with safety as a top priority. You'll be warned about any potentially hazardous situations or chemicals with this image:

When you see this image, pay special attention to the instructions.

The laboratory safety rules listed in table 1 will help make lab a safe place for everyone to learn biology. Remember, it is much easier to prevent an accident than to deal with its consequences.

Read the laboratory safety rules listed in table 1. If you do not understand them, or if you have questions, ask your instructor for an explanation. Then complete table 1 and sign the statement at the bottom of page xv.

## BEFORE YOU LEAVE LAB

Put away all equipment and glassware, and wipe clean your work area.

## AFTER EACH LABORATORY

Soon after each lab, review what you did. What questions did you answer? What data did you gather? What conclusions did you make?

Also note any questions that remain. Try to answer these questions by using your textbook or visiting the library. If you can't answer the questions, discuss them with your instructor.

Welcome to the biology laboratory!

**Table 1**

**Laboratory Safety Rules**

| Rule | Why is this rule important?<br>What could happen if this rule is not followed? |
|---|---|
| Behave responsibly. No horseplay or fooling around while in lab. | |
| Do not bring any food or beverages into lab, and do not eat, drink, smoke, chew gum, chew tobacco, or apply cosmetics when in lab. Never taste anything in lab. Do not put anything in lab into your mouth. Avoid touching your face, chewing on pens, and other similar behaviors while in lab. Always wear shoes in lab. | |
| Unless you are told otherwise by your instructor, assume that all chemicals and solutions in lab are poisonous, and act accordingly. Never pipette by mouth. Always use a mechanical pipetting device (e.g., a suction bulb) to pipette solutions. Clean up all spills immediately, and report all spills to your instructor. | |
| Wear safety goggles when working with chemicals. Carefully read the labels on bottles and know the chemical you are dealing with. Do not use chemicals from an unlabeled container, and do not return excess chemicals back to their container. Report all spills to your instructor immediately. | |
| Unless your instructor tells you to do otherwise, do not pour any solutions down the drain. Dispose of all materials as per instructions from your instructor. | |
| If you have long hair, tie it back. Don't wear dangling jewelry. If you are using open flames, roll up loose sleeves. Wear contact lenses at your own risk; contacts hold substances against the eye and make it difficult to wash your eyes thoroughly. | |
| Treat living organisms with care and respect. | |
| Your instructor will tell you the locations of lab safety equipment, including fire extinguishers, fire blanket, eyewash stations, and emergency showers. Familiarize yourself with the location and operation of this equipment. | |
| If anything is splashed into your eyes, wash your eyes thoroughly and immediately. Tell your lab instructor what happened. | |
| Notify your instructor of any allergies to latex, chemicals, stings, or other substances. | |
| If you break any glassware, do not pick up the pieces of broken glass with your hands. Instead, use a broom and dustpan to gather the broken glass. Ask your instructor how to dispose of the glass. | |
| Unless told by your instructor to do otherwise, work only during regular, assigned hours when the instructor is present. Do not conduct any unauthorized experiments; for example, do not mix any chemicals without your instructor's approval. | |
| Do not leave any experiments unattended unless you are authorized by your instructor to do so. If you leave your work area, slide your chair under the lab table. Keep walkways and desktops clean and clear by putting books, backpacks, and so on along the edge of the room, in the hall, in a locker, or in an adjacent room. Keep your work area as clean and uncluttered as possible. | |
| Don't touch or put anything on the surface of hotplates unless told to do so. Many types of hotplates have no visible sign that they are hot. Assume they are hot. | |
| Know how to use the equipment in lab. Most of the equipment is expensive; you may be required to pay all or part of its replacement cost. Keep water and solutions away from equipment and electrical outlets. Report malfunctioning equipment to your instructor. Leave equipment in the same place and condition that you found it. If you have any questions about or problems with equipment, contact your instructor. | |
| Know what to do and whom to contact if there is an emergency. Know the fastest way to get out of the lab. Immediately report all injuries—no matter how minor—to your instructor. Seek medical attention immediately if needed. If any injury appears to be life-threatening, call 911 immediately. | |
| At the end of each lab, clean your work area, wash your hands thoroughly with soap, slide your chair under the lab table, and return all equipment and supplies to their original locations. Do not remove any chemicals or equipment from the lab. | |

Name _____

Lab Section _____

Your lab instructor may require that you submit this page at the end of today's lab.

1. In the space below, write an analysis of the data shown in figure 1.

After completing table 1, read and sign this statement:

2. I have read and I understand and agree to abide by the laboratory safety rules described in this exercise and discussed by my instructor. I know the locations of the safety equipment and materials. If I violate any of the laboratory safety rules, my instructor will lower my grade and/or remove me from the lab.

_____
Signature

_____
Name (printed)

_____
Date

# Scientific Method
## The Process of Science

Please visit **connect.mheducation.com** to review online resources tailored to this lab.

The word *science* brings to mind different things to different students. To some students, science is a textbook. To others, it's a microscope, a dissected frog, or a course that you take. In fact, science is none of those things. Some definitions are more useful than others, but for biological research a good definition of **science** is *the orderly process of posing and answering questions about the natural world through repeated and unbiased experiments and observations*. This definition emphasizes that science is a process rather than a book, course, or list of facts. Science is not a "thing." It's a way of thinking about and doing things—a powerful way of learning and knowing about the natural world (fig. 1.1). Science has helped us understand phenomena such as climate change and the coronavirus 2019 (COVID-19) pandemic, produce antibiotics and other drugs, and improve our lives in many, many ways.

Our definition also emphasizes that people do science by *asking questions* and then *doing experiments* to answer those questions. Questions and curiosity are part of human nature, and science is a human activity. Like any human task, it takes practice to do science effectively.

Finally, our definition emphasizes that science is a tool for learning about the *natural world*. It is ineffective for moral choices, ethical dilemmas, and untestable ideas. For example, the scientific method cannot tell us if pollution is good or bad. It can tell us the environmental *consequences* of pollution, but whether these consequences are "good" or "bad" is a judgment that we make based on our values or goals, not on science. Although this is an important limitation of the scientific method, science remains one of the most powerful ways of understanding our world.

### Question 1
What practices besides science are used among world cultures to learn about the natural world?

The questioning and testing inherent in science systematically sift through natural variation to find underlying patterns. The natural world includes much variation, and learning biology would be relatively easy if simple observations accurately revealed patterns of the natural world. But they usually don't—nature is too complicated to rely solely on simple observation. We can certainly learn much

Morsa Images/DigitalVision/Getty Images

**Figure 1.1** Science is a process of learning about the natural world. Doing experiments that involve gathering repeated and unbiased measurements (data) is at the heart of testing hypotheses and answering questions.

about our environment just by looking around us, but casual observations are often biased and misleading because nature varies from time to time and from organism to organism. Biologists need a structured and repeatable process for testing their ideas about the variation in nature. Science is that process.

**Question 2**
What factors might be responsible for variation in measurements of traits such as the heights of 10-year-old pine trees or the kidney filtration rates of 10 replicate lab-mice?

The process of science deals with variation primarily through an organized sequence of steps that maintains as much objectivity and repeatability as possible. Although these loosely organized steps, sometimes called the **scientific method,** vary from situation to situation, they are remarkably effective for research and problem solving. The typical steps in the process of science are:

- Make insightful observations
- Pose and clarify testable questions
- Formulate hypotheses
- Do experiments to gather data
- Quantify the data
- Test the hypotheses
- Refine hypotheses and retest
- Answer the questions and make conclusions

## DEVELOPMENT OF OBSERVATIONS, QUESTIONS, AND HYPOTHESES

### Make Insightful Observations

Good scientists make insightful observations. But that's not as easy as it seems. Consider these two observations:

Observation 1: There are fewer elk in Yellowstone National Park than there used to be.

Observation 2: The density of elk in Yellowstone National Park has declined during the consecutive dry years since the reintroduction of the native wolf population.

Which of these two observations is stronger and more useful? Both of them may be true, but the second one is much more insightful because it provides a context to the observation that the elk population is declining. It also suggests a relevant factor—that is, the reintroduction of the wolf population—as a productive topic for investigation. It also suggests a relationship between density of the elk population and the variation in the local environment.

### *Procedure 1.1* Make insightful observations

*1.* Consider the following two observations.

Observation 1: Fungi often grow on leftover food.

Observation 2: Fungi such as mold and yeast grow more on leftover bread than on leftover meat.

Which of the above observations is more useful for further investigation? Why? _____

_____

_____

**SAFETY FIRST** Before coming to lab, you were asked to read this exercise so you would know what to do and be aware of safety issues. Briefly list the safety issues associated with today's procedures. If you have questions about these issues, contact your laboratory assistant before starting work.

Record the more insightful of the two observations on Worksheet 1 on page 9.

*2.* Consider this observation: Pillbugs (sometimes called roly-poly bugs) often find food and shelter where fungi are decomposing leaf litter (fig. 1.2).

For this example we are interested in whether pillbugs are attracted to leaves or to fungi (including yeasts) growing on the leaves' surfaces.

Observation 1: Pillbugs often hide under things.

Propose a more productive observation.

Observation 2: _____

_____

Record Observation 2 on Worksheet 2 on page 10. You may revise this later.

### Pose and Clarify Testable Questions

Productive observations inspire questions. Humans think in terms of questions rather than abstract hypotheses or numbers. But phrasing a good question takes practice and experience, and the first questions that capture our attention are usually general. For example, "Which nutrients can yeast most readily metabolize?" is a general question that expands the observation posed in procedure 1.1. This question is broadly applicable and is the type of question that we ultimately want to understand. Enter this as the General Question in Worksheet 1.

Broad questions are important, but their generality often makes them somewhat vague. The best questions for the process of science are specific enough to answer clearly. Therefore, scientists usually refine and subdivide broad questions into more specific ones. For example, a more specific question is "What classes of biological molecules are most readily absorbed and metabolized by yeast?" Enter this as Specific Question 1 in Worksheet 1.

**Figure 1.2** Pillbugs are excellent experimental organisms to test hypotheses about microenvironments, such as those under logs and within leaf litter. Pillbugs are readily available and easily cultured in the lab (10×).

A further clarification might be "Does yeast absorb and metabolize carbohydrates better than it absorbs and metabolizes proteins?" This is a good, specific question because it clearly refers to organisms, processes, and variables that are likely involved. It also suggests a path for investigation—that is, it suggests an experiment. Enter this as Specific Question 2 in Worksheet 1.

### Question 3

Consider the questions "What color is your roommate's car?" and "How many legs do cats have?" To answer these questions, would you use the scientific method, or would you rely on observation? Why?

---

### *Procedure 1.2* Posing and refining questions

*1.* Examine the following two questions.

Question 1: Do songbird populations respond to the weather?

Question 2: Do songbirds sing more often in warm weather than in cold weather?

Which of those questions is more useful for further investigation? Why? _____

_____

_____

*2.* Examine the following general question, and record it in Worksheet 2.

General Question: What influences the distribution of pillbugs?

Propose a specific question that refers to the food of pillbugs as a variable, and record it here and in Worksheet 2. Know that you may revise this later.

---

Specific Question 1 _____

_____

Propose a more specific question that refers to pillbugs eating leaves, as opposed to pillbugs eating fungi growing on leaves. Record this question here and in Worksheet 2. Know that you may revise this later.

Specific Question 2 _____

_____

## Formulate Hypotheses

Well-organized experiments to answer questions require that questions be restated as testable hypotheses. A **hypothesis** is a statement that clearly states the relationship between biological variables. A good hypothesis identifies the organism or process being investigated, identifies the variables being recorded, and implies how the variables will be compared. A hypothesis is a statement rather than a question, and an analysis of your experimental data will ultimately determine whether you accept or reject your hypothesis. Remember that even though a hypothesis can be falsified, it can never be proved true.

Accepting or rejecting a hypothesis, with no middle ground, may seem like a rather coarse way to deal with questions about subtle and varying natural processes. But using controlled experiments to either accept or reject a hypothesis is effective. The heart of science is gathering and analyzing experimental data that lead to rejecting or accepting hypotheses relevant to the questions we want to answer.

In this exercise, you are going to do science as you investigate yeast nutrition and then experiment with food choice by pillbugs. As yeast ferments its food, $CO_2$ is produced as a by-product. Therefore, we can measure the growth of yeast by measuring the production of $CO_2$ (fig. 1.3).

A hypothesis related to our question about the growth of yeast might be:

$H_0$: $CO_2$ production by yeast fed sugar is not significantly different from the $CO_2$ production by yeast fed protein.

A related alternative hypothesis can be similarly stated:

$H_a$: Yeast produces more $CO_2$ when fed sugar than when fed protein.

**Figure 1.3** These tubes of yeast are fermenting nutrients provided in solution. The $CO_2$ produced by the yeast accumulates at the top of the test tubes and indicates that yeast's rate of metabolism. From left to right, the tubes include a control with no added nutrients, a tube with low nutrients, and a tube with high nutrients.

©BiologyImaging.com

The first hypothesis ($H_0$) is a **null hypothesis** because it states that there is *no difference*. This is the most common way to state a clear and testable hypothesis. (Your instructor may elaborate on why researchers state and test null hypotheses more effectively than alternative hypotheses.) Researchers usually find it more useful to associate statistical probabilities with null hypotheses rather than with alternative hypotheses. Enter the null hypothesis into Worksheet 1.

A well-written null hypothesis is useful because it is testable. In our experiment, the null hypothesis (1) specifies yeast as the organism, population, or group that we want to learn about; (2) identifies $CO_2$ production as the variable being measured; and (3) leads directly to an experiment to evaluate variables and compare means of replicated measurements.

### Procedure 1.3 Formulating hypotheses

1. Examine the following two hypotheses:

   Hypothesis 1: Songbirds sing more when the weather is warm.

   Hypothesis 2: The number of bird songs heard per hour during daylight temperatures above 80°F (27°C) is not significantly different from the number heard per hour at temperatures below 80°F (27°C).

   Which of these hypotheses is more useful for further investigation? Why? _____

   _____

   Which of these hypotheses is a null hypothesis? Why?

   _____

2. Examine the following hypothesis.

   Hypothesis 1: Pillbugs prefer leaves coated with a thin layer of yeast.

   Propose a more effective null hypothesis. Be sure that it is a null hypothesis, that it is testable, and that it includes the parameter you will control in an experiment.

   Hypothesis 2 ($H_0$): _____

   _____

   Enter your null hypothesis in Worksheet 2.

## EXPERIMENTATION AND DATA ANALYSIS: YEAST NUTRITION

### Gather Experimental Data

To test our hypothesis about yeast growth, we must design a controlled and repeatable experiment. The experiment suggested by our specific question and hypothesis involves offering sugar such as glucose to one population of yeast, offering protein to another population of yeast, and then measuring their respective growth rates. Fortunately, yeast grows readily in test tubes. As yeast grows in a closed, anaerobic container, it produces $CO_2$ in proportion to how readily it uses the available food. $CO_2$ production is easily measured by determining the volume of $CO_2$ that accumulates at the top of an inverted test tube (fig. 1.3).

Experiments provide data that determine if a hypothesis should be accepted or rejected. A well-designed experiment links a biological response to different levels of the variable being investigated. In this case, the biological response is $CO_2$ production, which indicates growth. The levels of the variable are sugar and protein. These levels are called **treatments,** and in our experiment they include glucose, protein, and a control. For this experiment the **treatment** (i.e., independent) **variable** being tested is the type of food molecule (i.e., protein, sugar), and the **response** (i.e., dependent) **variable** is the $CO_2$ production that indicates yeast growth.

An experiment that compensates for natural variation must be well designed. It should (1) include replications, (2) test only one treatment variable, and (3) include controls. **Replications** are repeated measures of each treatment under the same conditions. Replications effectively deal with naturally occurring variation. Usually the more replications, the better. Your first experiment today will include replicate test tubes of yeast, each being treated the same. Good design tests only one treatment variable at a time.

Good experimental design also requires **controls** to verify that the biological response we measure is a function of the variable being investigated and nothing else. Controls are standards for comparison. They are replicates with all of the conditions of an experimental treatment *except the treatment variable.* For example, if the treatment is glucose dissolved in water, then a control has only water (i.e., it lacks only glucose, the treatment variable). This verifies that the response is to glucose and not to the solvent. Controls validate that our results are due only to the treatment variable.

### Procedure 1.4 An experiment to determine the effects of food type on yeast growth

1. Label 12 test tubes as C1–C4, G1–G4, and P1–P4. See Worksheet 1.

2. To test tubes C1–C4 add 5 mL of water. These are control replicates.

3. To test tubes G1–G4 add 5 mL of 5% glucose solution. These are replicates of the glucose treatment.

4. To test tubes P1–P4 add 5 mL of 5% protein solution. These are replicates of the protein treatment.

5. Swirl the suspension of yeast until the yeast is distributed uniformly in the liquid. Then completely fill the remaining volume in each tube with the yeast suspension that is provided.

6. For each tube, slide an inverted, flat-bottomed test tube down over the yeast-filled tube. Hold the yeast-filled tube

firmly against the inside bottom of the cover tube and invert the assembly. Your instructor will demonstrate how to slip this slightly larger empty tube over the top of each yeast tube and invert the assembly. If done properly, no bubble of air will be trapped at the top of the tube of yeast after inversion.

7. Place the tubes in a rack and incubate them at 50°C.

8. Measure the height (mm) of the bubble of accumulated $CO_2$ after 10, 20, 40, and 60 minutes. Record your results in Worksheet 1 and graph them here:

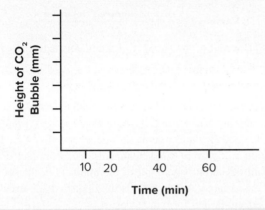

9. When you are finished, clean your work area and dispose of the contents of each tube as instructed by your lab instructor.

## Test Your Predictions by Analyzing the Experimental Data

Analysis begins with summarizing the raw data for biological responses to each treatment. The first calculation is the **mean** ($\bar{x}$), which is the average of a set of numbers (e.g., measurements) for replicates of each treatment and the control. That is, the mean is a single number that represents the central tendency of the response variable. Later the mean of each treatment will be compared to determine if the treatments had different effects.

The second step in data analysis is to calculate variation within each set of replicates. The simplest measure of variation is the **range,** which is the highest and lowest values in a set of replicates. A wide range indicates much variation in the data. The **standard deviation (SD),** another informative measure of variation, summarizes variation just as the range does, but the standard deviation is less affected by extreme values. Refer to the box "Variation in Replicate Measures" to learn how to calculate the standard deviation.

### Question 4

Even the seemingly simple question "How tall are mature males of the human species?" can be difficult to answer. How would you best express the answer?

## Procedure 1.5 Quantify and summarize the data

1. Examine your raw data in Worksheet 1.

2. Calculate the mean of the response variable ($CO_2$ production) for the four control replicates. To calculate the means for the four replicates, sum the four values and divide by four. Record the mean for the control replicates in Worksheet 1.

3. The $CO_2$ production for each glucose and protein replicate must be adjusted with the control mean. This ensures that the final data reflect the effects of only the treatment variable and not the solvent. Subtract the control mean from the $CO_2$ production of each glucose replicate and each protein replicate, and record the results in Worksheet 1.

4. Record in Worksheet 1 the range of adjusted $CO_2$ production for the four replicates of the glucose treatment and of the protein treatment.

5. Calculate the mean $CO_2$ production for the four adjusted glucose treatment replicates. Record the mean in Worksheet 1.

6. Calculate the mean $CO_2$ production for the four adjusted protein treatment replicates. Record the mean in Worksheet 1.

7. Refer to "Variation in Replicate Measures," and calculate the standard deviation for the four adjusted glucose treatment values and for the four adjusted protein treatment values. Record the two standard deviations in Worksheet 1.

## Test the Hypotheses

Our hypothesis about yeast growth is tested by comparing the mean $CO_2$ production by yeast fed glucose to the mean $CO_2$ production by yeast fed protein. However, only determining if one mean is higher than the other is not an adequate test because natural variation will always make the two means at least slightly different, even if the two treatments have the same effect on yeast growth. Therefore, the means and the variation about the means must be compared to determine if the means are not just different but **significantly different.** To be significantly different, the differences between means must be due to the treatment and not just due to natural variation. If the difference is significant, then the null hypothesis is rejected. If the difference is not significant, then the null hypothesis is accepted. Testing for significant differences is usually done with statistical methods.

Statistical methods calculate the probability that the means are significantly different. But these complex calculations are beyond the scope of this exercise. We will use a simpler method to test for a significant difference between the means of our two treatments. We will declare that two means are significantly different *if the means plus or minus 1/2 of the standard deviation do not overlap.*

Natural variation occurs in all processes of biology. This variation will inevitably produce different results in replicated treatments. One of the most useful measures of variation of values about the mean is **standard deviation.** It's easy to calculate: calculate the mean, calculate the deviation of each sample from the mean, square each deviation, and then sum the deviations. This summation is the sum of squared deviations. For example, data for $CO_2$ production by yeast in four replicate test tubes might be 22, 19, 18, and 21 mm. The mean is 20 mm.

| $CO_2$ Production (mm) | Mean | Deviation | Deviation$^2$ |
|---|---|---|---|
| 22 | 20 | 2 | 4 |
| 19 | 20 | −1 | 1 |
| 18 | 20 | −2 | 4 |
| 21 | 20 | 1 | 1 |

Sum of squared deviations = 10

The summary equation for the sum of squared deviations is

$$\text{Sum of squared deviations} = \sum_{i=1}^{N}(x_i - \bar{x})^2$$

where

$$N = \text{total number of samples}$$

$\bar{x} = $ **the sample mean**

$x_i = $ **measurement of an individual sample**

The summation sign ($\sum_{i=1}^{N}$) means to add up all the squared deviations from the first one ($i = 1$) to the last one ($i = N$).

The sum of squared deviations (10) divided by the number of samples minus one ($4 - 1 = 3$) produces a value of $10/3 = 3.3$ mm$^2$ (the units are millimeters squared). This is the variance:

$$\text{Variance} = \frac{\text{sum of squared deviations}}{N - 1}$$

The square root of the variance, 1.8 cm, equals the standard deviation

$$\text{SD} = \sqrt{\text{Variance}} = \sqrt{3.3} = 1.8$$

The standard deviation is often reported with the mean in statements such as, "The mean $CO_2$ production was $20 \pm 1.8$ mm." The standard deviation helps us understand the spread or variation among replicated treatments. For example, if the standard deviation is zero, all of the numbers in the set are the same. A larger standard deviation implies that individual numbers are farther from the mean.

For example, consider these two means and their standard deviations (SD):

Mean$_a$ = 10   SD = 5       Mean$_b$ = 20   SD = 10

Mean$_a$ − (½)SD = 7.5       Mean$_b$ − (½)SD = 15

Mean$_a$ + (½)SD = 12.5      Mean$_b$ + (½)SD = 25

Are Mean$_a$ and Mean$_b$ significantly different according to our test for significance? Yes they are, because 7.5 ↔ 12.5 does not overlap 15 ↔ 25.

## Answer the Questions

The results of testing the hypotheses are informative, but it still takes a biologist with good logic to translate these results into the answers of our specific and general questions. If your specific questions were well stated, then answering them based on the results of your experiment and hypothesis testing should be straightforward.

### Procedure 1.6   Testing hypotheses

1. Consider your null hypothesis and the data presented in Worksheet 1.

2. Calculate the glucose mean − (½)SD and the glucose mean + (½)SD. Record them in Worksheet 1.

3. Calculate the protein mean − (½)SD and the protein mean + (½)SD. Record them in Worksheet 1.

4. Do the half standard deviations surrounding the means of the two treatments overlap? Record your answer in Worksheet 1.

5. Are the means for the two treatments significantly different? Record your answer in Worksheet 1.

6. Is your null hypothesis accepted? Or rejected? Record your answer in Worksheet 1.

### Procedure 1.7   Answering the questions: yeast nutrition

1. Examine the results of hypothesis testing presented in Worksheet 1.

2. Specific Question 2 was "Does yeast absorb and metabolize carbohydrates better than it absorbs and metabolizes proteins?" Enter your answer in Worksheet 1.

3. Does your experiment adequately answer this question? Why or why not? _____

_____

_____

4. Specific Question 1 was "What classes of biological molecules are most readily absorbed and metabolized by yeast?" Enter your best response in Worksheet 1.

5. Does your experiment adequately answer Specific Question 1? Why or why not? _____
_____
_____

6. The General Question was "Which nutrients can yeast most readily metabolize?" After testing the hypotheses, are you now prepared to answer this general question? Why or why not? _____
_____
_____

## EXPERIMENTATION AND DATA ANALYSIS: FOOD PREFERENCE BY PILLBUGS

In the previous procedures you developed and recorded observations, questions, and hypotheses concerning food preference by pillbugs. Pillbugs may be attracted to dead leaves as food, or they may be attracted to fungi growing on the leaves as food. Leaves dipped in a yeast suspension can simulate fungi growing on leaves. Use the following procedures as a guide to the science of experimentation and data analysis to test the hypothesis you recorded in Worksheet 2.

### Procedure 1.8   Design an experiment to test food preference by pillbugs

1. Design an experiment to test your hypothesis in Worksheet 2 about food preference by pillbugs. To do this, specify:

   Experimental setup _____
   _____

   Treatment 1 to be tested _____
   _____

   Treatment 2 to be tested _____
   _____

   Control treatment _____
   _____

   Response variable _____
   _____

   Treatment variable _____
   _____

   Number of replicates _____
   _____

   Means to be compared _____
   _____

2. Conduct your experiment and record the data in Worksheet 2.

3. Analyze your data. Record the control means and adjusted treatment-means in Worksheet 2.

4. Calculate the range and standard deviation for your treatments, and record them in Worksheet 2.

5. Test your hypothesis. Determine if the null hypothesis should be accepted or rejected. Record the results in Worksheet 2.

6. Answer the Specific Question 2, Specific Question 1, and General Question posed in Worksheet 2.

### Procedure 1.9   Answering the questions: food preference by pillbugs

1. Examine the results of your hypothesis testing presented in Worksheet 2.

2. Enter your answer to Specific Question 2 in Worksheet 2. Does your experiment adequately answer this question? Why or why not? _____
_____
_____

3. Enter your best response to Specific Question 1 in Worksheet 2. Does your experiment adequately answer this question? Why or why not? _____
_____
_____

4. After testing the hypotheses, are you now prepared to answer your General Question "What influences the distribution of pillbugs?" Why or why not? _____
_____
_____

**Question 5**
What are some examples of biological theories?

Throughout this course you will make many predictions and observations about biology. When you account for a group of these observations with a generalized explanation, you have proposed a scientific theory.

In science, as opposed to common usage, a theory is a well-substantiated explanation of some aspect of the natural world that usually incorporates many confirmed observational and experimental facts. A scientific theory makes predictions consistent with what we see. It is not a guess; on the contrary, a scientific theory is widely accepted within the scientific community—for example, the germ theory claims that certain infectious diseases are caused by microorganisms. Scientific theories do not become facts; scientific theories *explain* facts.

## INQUIRY-BASED LEARNING

### *How do changes in temperature affect the production of $CO_2$ by yeast?*

Observation: Fermentation of nutrients by yeast produces $CO_2$, and the production rate of this $CO_2$ can be used to measure growth of the yeast. In this lab you've already investigated how the production of $CO_2$ is affected by different nutrients (i.e., sugar, protein). Here you'll investigate another variable: temperature.

Question: How is the production of $CO_2$ by yeast affected by temperature?

**a.** Establish a working lab group and obtain Inquiry-Based Learning Worksheet 1 from your instructor.

**b.** Discuss with your group well-defined questions relevant to the preceding observation and question. Choose and record your group's best question for investigation.

**c.** Translate your question into a testable hypothesis. Record this hypothesis.

**d.** Outline on Worksheet 1 your experimental design and supplies needed to test your hypothesis. Ask your instructor to review your proposed investigation.

**e.** Conduct your procedures, record your data, answer your question, and make relevant comments.

**f.** Discuss with your instructor any revisions to your questions, hypothesis, or procedures. Repeat your work as needed.

## Questions for Further Study and Inquiry

*1.* Consider the traits of science. Newspaper articles often refer to a discovery as "scientific" or claim that something has been proved "scientifically." What is meant by this description?

*2.* Experimental results in science are usually reviewed by other scientists before they are published. Why is this done?

*3.* Have all of our discoveries and understandings about the natural world been the result of testing hypotheses and applying the scientific method? How so?

*4.* Suppose that you hear that two means are *significantly* different. What does this mean? Can means be different but not significantly different? Explain your answer.

*5.* Why do scientists refrain from saying, "These results prove that…"?

*6.* How can science be used to address "big" issues such as climate change and COVID-19?

*7.* Some people dismiss evolution by natural selection as being "only a theory." Biologists often respond that yes, evolution *is* a scientific theory. What does this mean?

*8.* A hallmark of a scientific theory is that it is falsifiable. What does this mean, and why is it important?

*9.* Why is there no role for superstition in science?

**OBSERVATION** _____

**QUESTIONS**

General Question: _____

Specific Question 1: _____

Specific Question 2: _____

**HYPOTHESIS $H_0$:** _____

**EXPERIMENTAL DATA: Nutrient Use by Yeast**

| | Treatments | | | | | Treatments Minus Control $\bar{x}$ | |
| --- | --- | --- | --- | --- | --- | --- | --- |
| Replicate | Control $CO_2$ Production (mm) | Replicate | Glucose $CO_2$ Production (mm) | Replicate | Protein $CO_2$ Production (mm) | Glucose $CO_2$ Production Adjusted for the Control $\bar{x}$ | Protein $CO_2$ Production Adjusted for the Control $\bar{x}$ |
| C1 | _____ | G1 | _____ | P1 | _____ | _____ | _____ |
| C2 | _____ | G2 | _____ | P2 | _____ | _____ | _____ |
| C3 | _____ | G3 | _____ | P3 | _____ | _____ | _____ |
| C4 | _____ | G4 | _____ | P4 | _____ | _____ | _____ |

Control $\bar{x}$ = _____                     Protein $\bar{x}$ = _____
Glucose $\bar{x}$ = _____                     Protein range = _____ – _____
Glucose range = _____ – _____     Protein SD = _____
Glucose SD = _____

**TEST HYPOTHESIS**

Glucose $\bar{x}$ − (½)SD = _____            Protein $\bar{x}$ − (½)SD = _____

Glucose $\bar{x}$ + (½)SD = _____            Protein $\bar{x}$ + (½)SD = _____

Do the half standard deviations surrounding the means of the two treatments overlap? Yes _____ No _____

Are the means for the two treatments significantly different? Yes _____ No _____

Is the null hypothesis accepted? _____ or rejected? _____

**ANSWER QUESTIONS**

Answer to Specific Question 2 _____

Answer to Specific Question 1 _____

Answer to General Question _____

**OBSERVATION**_____

**QUESTIONS**

General Question: _____

Specific Question 1: _____

Specific Question 2: _____

**HYPOTHESIS H$_0$:** _____

**EXPERIMENTAL DATA: Food Preference by Pillbugs**

| | | Treatments | | | | Treatments Minus Control $\bar{x}$ | |
|---|---|---|---|---|---|---|---|
| Replicate | Control | Replicate | Treatment 1 | Replicate | Treatment 2 | Treatment 1 Adjusted for the Control $\bar{x}$ | Treatment 2 Adjusted for the Control $\bar{x}$ |
| 1 | _____ | 1 | _____ | 1 | _____ | _____ | _____ |
| 2 | _____ | 2 | _____ | 2 | _____ | _____ | _____ |
| 3 | _____ | 3 | _____ | 3 | _____ | _____ | _____ |
| 4 | _____ | 4 | _____ | 4 | _____ | _____ | _____ |

Control $\bar{x}$ = _____         Treatment 2 $\bar{x}$ = _____
Treatment 1 $\bar{x}$ = _____      Treatment 2 range = _____ – _____
Treatment 1 range = _____ – _____   Treatment 2 SD = _____
Treatment 1 SD = _____

**TEST HYPOTHESIS**

Treatment 1 $\bar{x} - (\frac{1}{2})SD$ = _____      Treatment 2 $\bar{x} - (\frac{1}{2})SD$ = _____

Treatment 1 $\bar{x} + (\frac{1}{2})SD$ = _____      Treatment 2 $\bar{x} + (\frac{1}{2})SD$ = _____

Do the half standard deviations surrounding the means of the two treatments overlap? Yes _____ No _____

Are the means for the two treatments significantly different? Yes _____ No _____

Is the null hypothesis accepted? _____ or rejected? _____

**ANSWER QUESTIONS**

Answer to Specific Question 2 _____

Answer to Specific Question 1 _____

Answer to General Question _____

# Measurements in Biology
## The Metric System and Data Analysis

## Learning Objectives

By the end of this exercise you should be able to:
1. Understand the difference between accuracy and precision in measurements.
2. Identify the metric units used to measure length, volume, mass, and temperature.
3. Measure length, volume, mass, and temperature in metric units.
4. Convert one metric unit to another (e.g., grams to kilograms).
5. Use measures of volume and mass to calculate density.
6. Practice the use of simple statistical calculations such as mean, median, range, and standard deviation.
7. Analyze sample data using statistical tools.

Please visit **connect.mheducation.com** to review online resources tailored to this lab.

Every day we're bombarded with numbers and measurements. They come at us from all directions, including while we're at the supermarket, gas station, golf course, and pharmacy, as well as while we're in our classrooms and kitchens. Virtually every package that we touch is described by a measurement.

Scientists use a standard method to collect data as well as use mathematics to analyze measurements. We must measure things before we can objectively describe what we are observing, before we can experiment with biological processes, and before we can predict how organisms respond, adjust to, and modify their world. Once we have made our measurements, we can analyze our data and look for variation and the sources of that variation. Then we can infer the causes and effects of the biological processes that interest us.

## ACCURACY AND PRECISION

Scientists strive to make accurate, precise measurements. The **accuracy** of a group of measurements refers to how closely the measured values agree with the true or correct value. In contrast, the **precision** of a group of measurements refers to how closely the measurements agree with each other. That is, precision is the degree to which the measurements produce the same results, regardless of their accuracy. Scientists strive to make measurements that are accurate and precise.

### Question 1
*a.* Can measurements be accurate but not precise? Explain.

*b.* Can measurements be precise but not accurate? Explain.

To help you check your answers, consider an analogy involving shooting arrows at a bull's-eye target (fig. 2.1). In this analogy, each arrow (represented by a dot on the target) would represent a measurement. Accuracy would be the

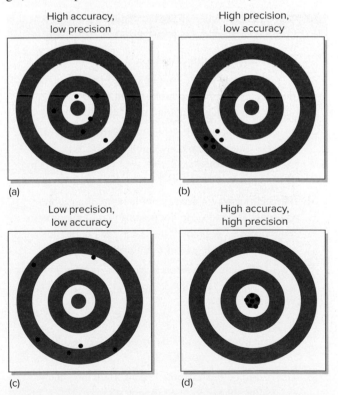

(a) High accuracy, low precision

(b) High precision, low accuracy

(c) Low precision, low accuracy

(d) High accuracy, high precision

**Figure 2.1** Precision and accuracy. Measurements can be (*a*) accurate but not precise, (*b*) precise but not accurate, (*c*) neither precise nor accurate, or (*d*) both precise and accurate.

closeness of the arrows to the center of the target; arrows closest to the bullseye would be most accurate. Precision would be the size of the cluster of arrows, regardless of how close they are to the center of the target.

## THE METRIC SYSTEM

Scientists throughout the world use the **metric system** to make measurements. The metric system is also used in everyday life virtually everywhere except the United States. With few exceptions (e.g., liter bottles of soda), most measurements in the United States use the antiquated English system of pounds, inches, feet, and so on. Check with your instructor about bringing to class common grocery store items with volumes and weights in metric units or examining those items on display.

The most modern form of the metric system is called the International System of Units (abbreviated SI). Metric measurement is used worldwide in science to improve communication in the scientific community. Scientists make all of their measurements in the metric system; they do not routinely convert from one system to another. When scientists have mixed metric units with English units, the results have often been confusing, and have sometimes been disastrous. For example, in 1999, the $125-million Mars Climate Orbiter was approaching Mars to study the planet's climate. Lockheed Martin Astronautics, which built the spacecraft, gave NASA critical flight information in English units, but software aboard the orbiter expected the data in metric units. As a result, the orbiter was sent into, rather than safely above, the Mars atmosphere, where it disintegrated.

### Hints for Using the Metric System

1. Use decimals, not fractions (e.g., 2.5 m, not 2½ m).

2. Express measurements in units requiring only a few decimal places. For example, 0.3 m is more easily manipulated and understood than 300000000 nm.

3. When measuring pure water, the metric system offers an easy and common conversion from volume measured in liters to volume measured in cubic meters to mass measured in grams: $1 \text{ mL} = 1 \text{ cm}^3 = 1 \text{ g}$.

4. The metric system uses symbols rather than abbreviations. Therefore, do not place a period after metric symbols (e.g., 1 g, not 1 g.). Use a period after a symbol only at the end of a sentence.

5. Do not mix units or symbols (e.g., 9.2 m, not 9 m 200 mm).

6. Metric symbols are always singular (e.g., 10 km, not 10 kms).

7. Except for degrees Celsius, always leave a space between a number and a metric symbol (e.g., 20 mm, not 20mm; 10°C, not 10° C).

8. Use a zero before a decimal point when the number is less than one (e.g., 0.42 m, not .42 m).

The following conversions will help give you a sense of how some common English units are related to their metric equivalents:

1 inch = 2.5 centimeters
1 foot = 30 centimeters
1 yard = 0.9 meter
1 mile = 1.6 kilometers
1 fluid ounce = 30 milliliters
1 pint = 0.47 liter
1 quart = 0.95 liter
1 gallon = 3.8 liters
1 cup = 0.24 liter

To learn more about these conversions, see Appendix II.

This exercise will introduce you to making metric measurements of length, mass, volume, and temperature. During this lab, you should spend your time making measurements, not reading background information. Therefore, *before lab, read this exercise carefully to familiarize yourself with the basic units of the metric system.*

Metric units commonly used in biology include:

meter (**m**)—the basic unit of length
liter (**L**)—the basic unit of volume
kilogram (**kg**)—the basic unit of mass
degrees Celsius (**°C**)—the basic unit of temperature

Unlike the English system with which you are already familiar, the metric system is based on units of ten. This simplifies conversions from one metric unit to another (e.g., from kilometers to meters). This base-ten system is similar to our monetary system, in which 10 cents equal a dime, 10 dimes equal a dollar, and so forth. Units of 10 in the metric system are indicated by Latin and Greek prefixes placed before the base units:

| Prefix (Latin) | | Division of Metric Unit | |
|---|---|---|---|
| deci | (d) | 0.1 | $10^{-1}$ = tenth |
| centi | (c) | 0.01 | $10^{-2}$ = hundredth |
| milli | (m) | 0.001 | $10^{-3}$ = thousandth |
| micro | (μ) | 0.000001 | $10^{-6}$ = millionth |
| nano | (n) | 0.000000001 | $10^{-9}$ = billionth |
| pico | (p) | 0.000000000001 | $10^{-12}$ = trillionth |

| Prefix (Greek) | | Multiple of Metric Unit | |
|---|---|---|---|
| deka | (da) | 10 | $10^1$ = ten |
| hecto | (h) | 100 | $10^2$ = hundred |
| kilo | (k) | 1000 | $10^3$ = thousand |
| mega | (M) | 1000000 | $10^6$ = million |
| giga | (G) | 1000000000 | $10^9$ = billion |
| tera | (T) | 1000000000000 | $10^{12}$ = trillion |

Thus, multiply by

0.01  to convert centimeters to meters
0.001  to convert millimeters to meters
1000  to convert kilometers to meters
0.1  to convert millimeters to centimeters

For example, there are 10 millimeters per centimeter. There-fore, to convert 62 centimeters to millimeters,

$$62 \text{ cm} \times \frac{10 \text{ mm}}{\text{cm}} = 620 \text{ mm}$$

In these conversion equations, the units being converted *from* (in this case, centimeters) cancel out, leaving you with the desired units (in this case, millimeters). Also note that when units are converted to *smaller* units, the number associated with the new units will *increase*, and vice versa. For example, 620 meters = 0.620 kilometer = 620,000 millimeters = 62,000 centimeters.

### Question 2

*a.* Make the following metric conversions:

1 meter = _____ centimeters = _____ millimeters
92.4 millimeters = _____ meters = _____ centimeters
82 centimeters = _____ meters = _____ millimeters
3.1 kilograms = _____ grams = _____ milligrams
281 milliliters = _____ liters = _____ deciliters

*b.* The spikes on a COVID-19 viral particle are about 11 nm long. How long, in micrometers, are these spikes?

## Length and Area

The **meter** (m) is the basic unit of length. Units of area are squared units (i.e., two-dimensional) of length.

1 m = 100 cm = 1000 mm = 0.001 km = $1 \times 10^{-3}$ km
1 km = 1000 m = $10^3$ m
1 cm = 0.01 m = $10^{-2}$ m = 10 mm
470 m = 0.470 km
1 cm² = 100 mm² (i.e., 10 mm × 10 mm = 100 mm²)

To help you appreciate the size of each of these units, here are the lengths and areas of some familiar objects:

### Length

| | |
|---|---|
| Housefly | 0.5 cm |
| Diameter of penny | 1.9 cm |
| Diameter of baseball | 7.4 cm |
| Soda can | 12.2 cm |
| Toyota Camry | 4.7 m |
| Mt. Everest | 8848 m |

### Area

| | |
|---|---|
| Credit card | 46 cm² |
| Total skin area of adult human male | 1.8 m² |
| Ping-pong table | 4.18 m² |
| Surface area of human lungs | 80 m² |
| Football field (goal line to goal line) | 4459 m² |
| Central Park (New York City) | 3.4 km² |

### Procedure 2.1  Make metric measurements of length and area

Most biologists measure lengths with metric rulers or metersticks.

*1.* Examine intervals marked on the metric rulers and metersticks available in the lab.
*2.* Make the following measurements. Be sure to include units for each measurement.

| | |
|---|---|
| Length of this page | _____ |
| Width of this page | _____ |
| Area of this page (Area = Length × Width) | _____ |
| Your height | _____ |
| Thickness of this manual | _____ |
| Height of a 200-mL beaker | _____ |
| Height of your chair | _____ |
| Length of your cell phone | _____ |

### Question 3

What are some potential sources of error in your measurements?

## Volume

**Volume** is the space occupied by an object. Units of volume are cubed (i.e., three-dimensional) units of length. The liter (L) is the basic unit of volume.

1 L = 1000 cm³ = 1000 mL
1 L = 0.1 m × 0.1 m × 0.1 m
1 cm³ = 0.000001 m³

To help you appreciate the size of each of these units, here are the volumes of some familiar objects:

| | |
|---|---|
| Chicken egg | 60 mL |
| Coke can | 355 mL |
| One breath of air | 500 cm³ |

Scientists often measure volumes with pipets and graduated cylinders. Pipets are used to measure small volumes, typically 25 mL or less. Liquid is drawn into a pipet using a bulb or pipet pump (fig. 2.2). Never pipet by mouth.

Graduated cylinders are used to measure larger volumes. To appreciate how to make a measurement accurately, pour 40–50 mL of water into a 100-mL graduated cylinder, and observe the interface between the water and air. This interface, called the **meniscus,** is curved because of surface tension and the adhesion of water to the sides of the cylinder. When measuring the liquid in a cylinder such as a graduated cylinder, always position your eyes level with the meniscus and read the volume at the lowest level (fig. 2.3).

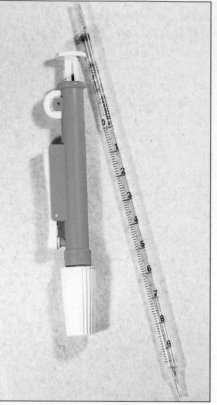

©BiologyImaging.com

**Figure 2.2** A pipet is used to extract and dispense volumes of liquid. A suction device (shown in green on the left) draws fluid into the pipet, and graduated markings on the pipet allow precise measurement of a fluid's volume. Never use your mouth to suck fluid into a pipet.

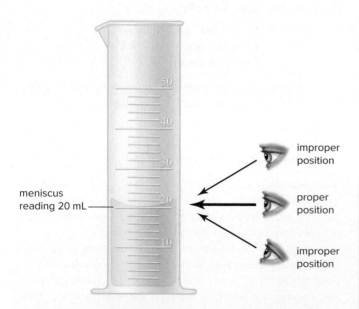

meniscus
reading 20 mL

improper position

proper position

improper position

**Figure 2.3** When measuring the volume of liquid in a graduated cylinder, always position your eye at the bottom of the meniscus. The correct volume is 20 mL.

### *Procedure 2.2* **Make metric measurements of volume**

1. Biologists often use graduated cylinders to measure volumes. Locate the graduated cylinders available in the lab to make the following measurements. Determine what measurements the markings on the graduated cylinder represent. Be sure to include units for each measurement.

2. Measure the volume (in mL) needed to fill a cup (provided in the lab). _____

3. Measure the volume (in L) needed to fill a gallon. _____

### *Procedure 2.3* **Measure the volume of a solid object by water displacement**

1. Obtain a 100-mL graduated cylinder, a thumb-sized rock, and a glass marble.

2. Fill the graduated cylinder with 70 mL of water.

3. Gently submerge the rock in the graduated cylinder. Notice that the volume of the contents rises.

4. Carefully observe the meniscus of the fluid and record its volume.

5. Calculate and record the volume of the rock by subtracting the original volume (70 mL) from the new volume.

   Rock volume _____

6. Repeat steps 2–5 to measure and record the volume of the marble.

   Marble volume _____

Biologists use pipets to measure and transfer small volumes of liquid from one container to another. The following procedure will help you appreciate the usefulness of pipets.

### *Procedure 2.4* **Learn to use a pipet**

1. Add approximately 100 mL of water to a 100-mL beaker.

2. Use a 5-mL pipet with a bulb or another filling device provided by your instructor to remove some water from the beaker.

3. Fill the pipet to the zero mark.

4. To read the liquid level correctly, your eye must be directly in line with the bottom of the meniscus.

5. Release the liquid into another container.

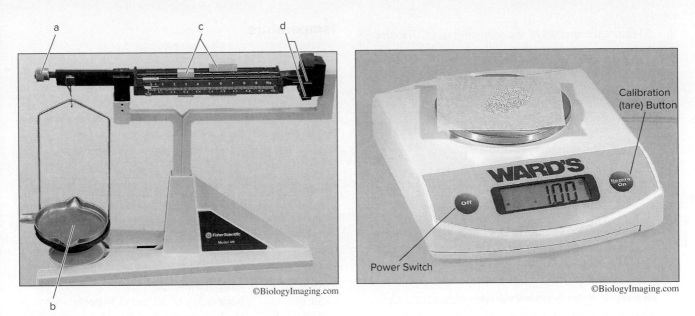

a     c     d

Calibration (tare) Button

Power Switch

©BiologyImaging.com      ©BiologyImaging.com

b

**Figure 2.4** Biologists use balances to measure mass. (A) The parts of a triple-beam balance include the (*a*) zero-adjustment knob, (*b*) measuring pan, (*c*) movable masses on horizontal beams, and (*d*) balance marks. (B) A top-loading balance has a measuring pan, a power switch, and a zero calibration ("tare") button.

**Question 4**
What volume of liquid did you measure?

## Mass

The **kilogram** (kg) is the basic unit of mass.[1] A kilogram is approximately equal to the mass of 1000 cubic centimeters (cm³) of water at 4°C. Similarly,

$$1 \text{ kg} = 1000 \text{ g} = 10^3 \text{ g} \quad 1 \text{ mg} = 0.001 \text{ g} = 10^{-3} \text{ g}$$

Note that the kilogram is the only basc-unit of SI that includes a prefix ("kilo," symbol "k") as part of its name (see Appendix II). Here are the approximate masses of some familiar objects:

| | | | |
|---|---|---|---|
| Housefly | 12 mg | Human heart | 300 g |
| Hummingbird | 1.6 g | Basketball | 0.62 kg |
| Ping-pong ball | 2.45 g | $1 bill | 1 g |
| Quarter | 6.25 g | Penny | 2.5 g |
| 9V battery | 40 g | | |

Biologists usually measure mass with a top-loading balance or a triple-beam balance (fig. 2.4). Locate the triple-beam balances or top-loading electronic balances in the lab. Triple-beam balances get their names from their three horizontal beams. Suspended from each of the three beams are movable masses. Each of the three beams of the balance is

marked with graduations: the closest beam has 0.1-g graduations, the middle beam has 100-g graduations, and the farthest beam has 10-g graduations.

**Procedure 2.5**    **Make metric measurements of mass**

1. *If you're using a triple-beam balance:* Before making any measurements, clean the weighing pan and move all of the suspended weights to the far left. The balance marks should line up to indicate zero grams; if they do not, turn the adjustment knob until they do. Measure the mass of an object by placing it in the center of the weighing pan and moving the suspended masses until the beams balance. The mass of the object is the sum of the masses indicated by the weights on the three beams.

2. *If you're using an electronic balance:* Turn on the balance and let it warm up for 5 minutes. Wait until the display reads 0.0 g; if the display does not read 0.0 g, press the "tare" button to reset the display to 0.0 g. If you are weighing an object such as a coin or pencil, place the object on the measuring pan. After the display has stabilized, read and record the object's mass.

3. If you are weighing a liquid, powder, or similar specimen, place an empty beaker (in which you will place the liquid) or a piece of weighing paper (on which you will place the powder) on the balance's measuring pan. After the display has stabilized, press the "tare" button to reset the display to 0.0 g. Place the liquid in the beaker (or the powder on the weighing paper). After the display has stabilized, read and record the mass.

---

1   *Remember that mass is not necessarily synonymous with weight. Mass measures an object's **potential** to interact with gravity, whereas weight is the force exerted by gravity on an object. Thus, a weightless object in outer space has the same mass as it has on earth.*

**4.** Measure the masses of the following items. Be sure to include units for each measurement.

Penny _____

Paper clip _____

Pencil _____

Rock (used in procedure 2.3) _____

100-mL beaker (empty) _____

100-mL beaker containing 50 mL of water _____

## Question 5

*a.* **Density** is mass per unit volume. Use data that you've gathered to determine the density of water at room temperature.

**Density of water = (mass/volume) = _____**

*b.* What is the density of the wooden pencil? Does it float? Why?

*c.* What is the density of the rock? Does it sink? Why?

## Temperature

Temperature is the measure of the kinetic energy of molecules—that is, the amount of heat in a system. Biologists measure temperature with a thermometer calibrated in degrees Celsius (°C). The Celsius scale is based on water freezing at 0°C and boiling at 100°C. You can interconvert °C and degrees Fahrenheit (°F) by using the formula 5(°F) = 9(°C) + 160. Here are some typical temperatures:

| | |
|---|---|
| −20°C | temperature in a freezer |
| −18°C | mixture of ice and salt |
| 0°C | water freezes |
| 4°C | temperature in a refrigerator |
| 22°C | room temperature |
| 30.6°C | butter melts |
| 37°C | human body temperature |
| 40°C | a hot summer day |
| 50°C | hottest day on record in Phoenix, AZ |
| 71°C | flash pasteurization of milk |
| 75°C | hot coffee |
| 100°C | water boils |
| 260°C | broiler temperature |

## *Procedure 2.6* Make metric measurements of temperature

*1.* Obtain a thermometer in the lab. Handle the thermometer with care. If it breaks, notify your instructor immediately.

## Significant Figures

Let's suppose that you're measuring the length of a bone, as shown in figure 2.5. How would you record this length—as 8 cm? 8.3 cm? 8.33 cm? 8.33333 cm? To answer this question, you need to know something about significant figures.

Significant figures are the number of figures required to record a measurement so that only the last digit in the number is in doubt. For example, if the ruler you're using is calibrated only in centimeters and you find that the object you're measuring is between 8 and 9 cm long (fig. 2.5), then you should estimate your measurement only to a tenth of a centimeter. That is, a measurement of 8.3 cm is acceptable, but 8.33 is not because it implies a precision that did not exist in the equipment you used to make the measurement. If, however, your ruler was calibrated in millimeters, then 8.33 cm would be acceptable. Remember this: *When recording measurements, include all of the digits you are sure of plus an estimate to the nearest one-tenth of the next smaller digit.*

Here are some other guidelines for using the correct number of significant figures in your measurements:

• When adding or subtracting measurements, the answer should have no more precision than the measurement having the least number of significant figures. For example, suppose the air temperature in an incubator drops from 8.663°C to 8.2°C. This is a difference of 8.663°C − 8.2°C = 0.5°C, not 0.463°C. If the second temperature reading had been 8.200°C, then the correct answer would have been 0.463°C.

• When converting measurements from one set of units to another, do not introduce precision that is not present in the first number. For example, 8.3 cm = 83 mm, not 83.0 mm.

• When manipulating two measurements simultaneously, the precision of the final measurement should not exceed that of the least number of significant figures. For example, the calculation for the mass of 17.2 mL of water is 17.2 mL × 0.997821 g mL$^{-1}$ = 17.2 g, not 17.162521 g.

**Figure 2.5** How long is this bone? 8 cm? 8.3 cm? 8.33 cm?

Do not change the value of the last significant digit if that digit is followed by a number that is less than 5. For example, if two significant figures are required, 6.449 rounds to 6.4, 66.449 rounds to 66, 66.641 rounds to 67, and 6.591 rounds to 6.6. Here is how an original measurement of 49.5149 rounds to various numbers of significant figures:

| | |
|---|---|
| Five significant figures: | 49.515 |
| Four significant figures: | 49.51 |
| Three significant figures: | 49.5 |
| Two significant figures: | 50 |
| One significant figure: | 50 |

Statisticians disagree on what to do when the number following the last significant figure is exactly 5, as in 89.5 (and, in this case, the precision is limited to two significant figures). Some round the measurement to the higher number, while others claim that doing so introduces bias into the data. You can decide which approach to take, but be consistent.

2. Determine the range of the temperatures that can be measured with your thermometer by examining the scale imprinted along the barrel of the thermometer.

3. Measure the following temperatures:

| | |
|---|---|
| Room temperature | _____ °C |
| Cold tap water | _____ °C |
| Hot tap water | _____ °C |
| Inside refrigerator | _____ °C |

## UNDERSTANDING NUMERICAL DATA

**Statistics** offer a way to organize, summarize, and describe data—the data are usually samples of information from a much larger population of values. Statistics and statistical tests allow us to analyze the sample and draw inferences about the entire population. Consequently, the use of statistics enables us to make decisions even though we have incomplete data about a population. Although this may seem unscientific, we do it all the time; for example, we diagnose diseases with a drop of blood. Decisions are based on statistics when it is impossible or unrealistic to analyze an entire population.

Let's say that you want to know the mass of a typical apple in your orchard. To obtain this information, you could analyze one apple, but how would you know that you'd picked a "typical" sample? After all, the batch from which you chose the apple may contain many others, each a little different. You'd get a better estimate of "typical" if you increased your sample size to a few hundred apples, or even to 10,000. Or, better yet, to 1,000,000.

The only way to be certain of your conclusions would be to accurately measure all the apples in your orchard. This is impossible, so you must choose apples that *represent* all of the other apples—that is, you must be working with a *representative sample*. A statistical analysis of those sample apples reduces the sample values to a few characteristic measurements (e.g., mean mass). As you increase the size of the sample, these characteristic measurements provide an ever-improving estimation of what is "typical."

There are a variety of software programs that perform statistical analyses of data; all you have to do is enter your data into a spreadsheet, select the data that you want to analyze, and perform the analysis. Although these software packages save time and can increase accuracy, you still need to understand a few of the basic variables that you'll use to understand your numerical data. We'll start with the mean and median:

The **mean** is the arithmetic average of a group of measurements. Chance errors in measurements tend to cancel themselves when means are calculated for relatively large samples; a value that is too high because of random error is often balanced by a value that is too low for the same reason.

The **median** is, after arranging the measurements from the smallest to the largest, the middle value that divides the set of measurements into two subsets of equal size. If there are an even number of measurements, the median is the mean of the two middle values. *In biology, the mean is usually preferred to the median when reporting descriptive statistics.*

The median is less sensitive to extreme values than is the mean. To appreciate this, consider a sample consisting of 14 leaves having the following lengths (all in mm):

80  69  62  74  69  51  45  40  9  64  65  64  61  67

The mean length is 58.6 mm. However, none of the leaves are that length, and most of the leaves are longer than 60 mm. In biology, the mean is usually preferred to the median when reporting descriptive statistics.

Consider these sets of data:

1 3 5 7 9 – the mean and median are both 5

1 3 5 7 14 – the median is 5, but the mean is 6

1 3 5 7 34 – the median is 5, but the mean is 10

**Question 6**

*a.* Does the mean always describe the "typical" measurement? Why or why not?

*b.* What information about a sample does a mean *not* provide?

Determine the median by arranging the measurements in numerical order:

9  40  45  51  61  63  64  64  65  67  69  69  73  80

The median is between the seventh and eighth measurements: 64 mm. In this sample, the mean differs from the median.

**Question 7**

*a.* What is responsible for this difference between the mean and median?

*b.* How would the median change if the 9-mm-long leaf was not in the sample?

*c.* How would the mean change if the 9-mm-long leaf was not in the sample?

*d.* Consider these samples:

Sample 1: 25   35   32   28

Sample 2: 15   75   10   20

What is the mean for Sample 1? _____

What is the mean for Sample 2? _____

In most of the exercises in this manual, you'll have time to make only one or two measurements of a biological structure or phenomenon. In these instances, a mean may be the only descriptor of the sample. However, if your class combines its data so that there are many measurements, you'll need to know how to do a couple of other calculations so that you understand the variation within your sample.

## Variability

As you can see, the samples in Question 7d are different, but their means are the same. Thus, the mean does not reveal all there is to know about these samples. To understand how these samples are different, you need other statistics: the range and standard deviation.

The **range** is the difference between the extreme measurements (i.e., smallest and largest) of the sample. In Sample 1, the range is $35 - 25 = 10$; in Sample 2 the range is $75 - 10 = 65$. The range provides a sense of the variation of the sample, but the range can be artificially inflated by one or two extreme values. Notice the extreme values in the sample of leaf measurements previously discussed. Moreover, ranges do not tell us anything about the measurements between the extremes.

**Question 8**

*a.* Could two samples have the same mean but different ranges? Explain.

*b.* Could two samples have the same range but different means? Explain.

The **standard deviation** indicates how measurements vary about the mean. The standard deviation is easy to calculate. Begin by calculating the mean, measuring the deviation of each sample from the mean, squaring each deviation, and then summing the deviations. This summation results in the **sum of squared deviations.** For example, consider a group of shrimp that are 22, 19, 18, and 21 cm long. The mean length of these shrimp is 20 cm.

| Sample Value | Mean | Deviation | (Deviation)² |
|---|---|---|---|
| 22 | 20 | 2 | 4 |
| 19 | 20 | −1 | 1 |
| 21 | 20 | 1 | 1 |
| 18 | 20 | −2 | 4 |

Sum of Squared Deviations = 10

The summary equation for the sum of squared deviations is:

$$\text{Sum of squared deviations} = \sum_{i=1}^{N}(x_i - \bar{x})^2$$

where

$N$ = total number of samples

$\bar{x}$ = the sample mean

$x_i$ = measurement of an individual sample

This formula is simple. The summation sign $(\sum_{i=1}^{N})$ means to add up all the squared deviations from the first one ($i = 1$) to the last one ($i = N$). The sum of squared deviations (10) divided by the number of samples minus one ($4 - 1 = 3$) produces a value of $10/3 = 3.3$ cm² (note that the units are centimeters squared). This is the **variance:**

$$\text{Variance} = \frac{\text{sum of squared deviations}}{N - 1}$$

The square root of the variance, 1.8 cm, equals the **standard deviation (SD):**

$$\text{SD} = \sqrt{\text{Variance}} = \sqrt{3.3} = 1.8$$

The standard deviation is usually reported with the mean in statements such as, "The mean length of the shrimp was $20 \pm 1.8$ cm."

The standard deviation helps us understand the spread or variation of a sample. For many distributions of measurements, the mean ± 1 SD includes 68% of the measurements, whereas the mean ± 2 SD includes 95% of the measurements.

## Procedure 2.7 Gather and analyze data statistically

1. Use a meterstick or tape measure to measure your height in centimeters. Record your height here:
   _____ cm

2. Record your height and gender (male or female) on the board in the lab.

3. After all of your classmates have reported their heights, calculate the following:

   Size of sample

   All classmates _____

   Male classmates _____

   Female classmates _____

   Mean height

   All classmates _____

   Male classmates _____

   Female classmates _____

   Median height

   All classmates _____

   Male classmates _____

   Female classmates _____

Range

All classmates _____ to _____

Male classmates _____ to _____

Female classmates _____ to _____

Standard deviation

All classmates _____ ± _____

Male classmates _____ ± _____

Female classmates _____ ± _____

If there is sufficient time, obtain a newspaper that advertises cars, groceries, or other common commodities. Choose one example (e.g., new cars) and determine its average price (e.g., determine the average price of a new car).

### Question 9
*a.* What does your calculation tell you?

*b.* What are the limitations of your sample?

Your instructor may ask you to do other statistical tests, such as Student's *t*, chi-square, and analysis of variance (ANOVA). The type of test you'll do will depend on the amount and type of data you analyze, as well as the hypotheses you are trying to test.

## INQUIRY-BASED LEARNING

### How much do the areas and shapes of leaves vary?

Observation: Leaves, which are the primary photosynthetic organ of most plants, are adapted for absorbing light. This involves exposing large surface areas to the environment.

Question: How do the surface area and shape of leaves vary on different parts of plants?

a. Establish a working lab group and obtain Inquiry-Based Learning Worksheet 2 from your instructor.

b. Discuss with your group well-defined questions relevant to the preceding observation and question. If leaves are not available from outdoor plants (e.g., during winter), use the plants provided by your instructor that were grown indoors. Choose and record your group's best question for investigation.

c. Translate your question into a testable hypothesis and record it.

d. Outline on Worksheet 2 your experimental design and supplies needed to test your hypothesis. Ask your instructor to review your proposed investigation.

e. Conduct your procedures, record your data, answer your question, and make relevant comments.

f. Discuss with your instructor any revisions to your questions, hypothesis, or procedures. Repeat your work as needed.

1.  What are the advantages and disadvantages of using the metric system of measurements?

2.  Why is it important for all scientists to use a standard system of measures rather than the system that may be most popular in their home country or region?

3.  Do you lose or gain information when you use statistics to reduce a population to a few characteristic numbers? Explain your answer.

4.  Suppose that you made repeated measurements of your height. If you used good technique, would you expect the range to be large or small? Explain your answer.

5.  Suppose that a biologist states that the average height of undergraduate students at your university is 205 cm plus or minus a standard deviation of 17 cm. What does this mean?

6.  What does a small standard deviation signify? What does a large standard deviation signify?

7.  Is it possible to make a perfectly precise measurement? Explain.

8.  When in our everyday lives do we *not* want precise measurements?

9.  Consider these measurements of the diameter (in nm) of viral particles ("virons") of COVID-19: 0.12, 0.14, 0.10, 0.12, 0.11, 0.13, 0.14, 0.13, 0.08, 0.13, 0.13. What is the mean? What is the range? What is the median?

# The Microscope
## Basic Skills of Light Microscopy

## Learning Objectives

By the end of this exercise you should be able to:
1. Identify and explain the functions of the primary parts of a compound microscope and dissecting (stereoscopic) microscope.
2. Carry and focus a microscope properly.
3. Use a compound microscope and dissecting microscope to examine biological specimens.
4. Prepare a wet mount, determine the magnification and size of the field of view, and determine the depth of field.

Please visit **connect.mheducation.com** to review online resources tailored to this lab.

Many organisms and biological structures are too small to be seen with the unaided eye (fig. 3.1). Biologists often use a light microscope to observe such specimens. A **light microscope** is a coordinated system of lenses arranged to produce an enlarged, focusable image of a specimen. A light microscope **magnifies** a specimen, meaning that it increases its apparent size. Magnification with a light microscope is usually accompanied by improved **resolution,** which is the ability to distinguish two points as separate points. Thus, the better the resolution, the sharper or crisper the image appears. The resolving power of the unaided eye is approximately 0.1 mm (1 in = 25.4 mm), meaning that our eyes can distinguish two points that are 0.1 mm apart. A light microscope, used properly, can improve resolution as much as 1000-fold (i.e., to 0.1 μm).

The ability to discern detail also depends on **contrast,** which is the difference between the lightest and darkest parts of an image. Therefore, many specimens examined with a light microscope are stained with artificial dyes that increase contrast and make the specimen more visible.

The invention of the light microscope was profoundly important to biology because it was used to formulate the cell theory and study biological structure at the cellular level. Light microscopy has revealed a vast new world to the human eye and mind (fig. 3.2). Today, the light microscope is the most fundamental tool of many biologists.

## THE COMPOUND LIGHT MICROSCOPE

Study and learn the parts of the typical compound light microscope shown in figure 3.3. A light microscope has two, sometimes three, systems: an illuminating system, an imaging system, and possibly a viewing and recording system.

## Illuminating System

The illuminating system, which concentrates light on the specimen, usually consists of a light source, condenser lens, and iris diaphragm. The **light source** is a lightbulb located

### Caring for Your Microscope

Microscopes are powerful tools for understanding biology. However, they're also expensive and fragile and require special care. When you use your microscope, always do the following to ensure optimal performance and care:

- Always carry your microscope upright with both hands—one hand under the base and the other around the microscope's arm (fig. 3.3).
- Always begin by cleaning the ocular and objective lenses with lens paper.
- Always start your examinations with the low-power objective in place.
- If you shift to the high-power objective, rotate the objective into place carefully. Never force the objective lens into place. If the objective lens contacts the slide, stop and restart your examination with the low-power objective lens.
- After shifting to the high-power objective, always use only the fine adjustment to focus the image.
- When you've completed your work with the microscope, clean the lenses with lens paper, wrap the electrical cord securely around the microscope's arm, and return your microscope to its storage area.

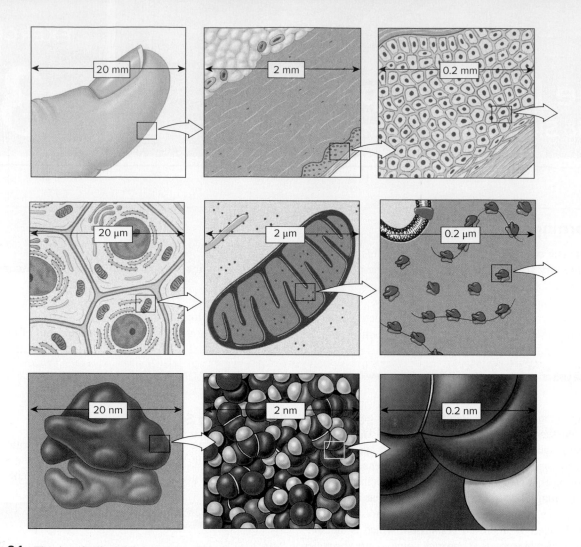

**Figure 3.1** The size of cells and their contents. This diagram shows the size of human skin cells, organelles, and molecules. In general, the diameter of a human skin cell is about 20 micrometers (μm), of a mitochondrion is 2 μm, of a ribosome is 20 nanometers (nm), of a protein molecule is 2 nm, and of an atom is 0.2 nm.

at the base of the microscope. The light source illuminates the specimen by passing light through a thin, almost transparent part of the specimen. The **condenser lens,** located immediately below the specimen, focuses light from the light source onto the specimen. Just below the condenser is the **condenser iris diaphragm,** a knurled ring or lever that can be opened and closed to regulate the amount of light reaching the specimen. When the condenser iris diaphragm is open, the image will be bright; when closed, the image will be dim.

## Imaging System

The imaging system improves resolution and magnifies the image. It consists of the objective and ocular (eyepiece) lenses and a body tube. The objectives are three or four lenses mounted on a revolving nosepiece. Each objective is a series of several lenses that magnify the image, improve

resolution, and correct aberrations in the image. The most common configuration for student microscopes includes four objectives: low magnification (4×), medium magnification (10×), high magnification (40×), and oil immersion (100×). Using the oil immersion objective requires special instructions, as explained in Exercise 24 to study bacteria. To avoid damaging your microscope, do not use the oil immersion objective during this exercise.

The magnifying power of each objective is etched on the side of the lens (e.g., 4×). The **ocular** is the lens that you look through. Microscopes with one ocular are **monocular** microscopes, and those with two are **binocular** microscopes. Oculars usually magnify the image 10 times. The **body tube** is a metal casing through which light passes to the oculars. In microscopes with bent body-tubes and inclined oculars, the body tube contains mirrors and a prism that redirect light to the oculars. The **stage** secures the glass slide on which the specimen is mounted.

**Figure 3.2** "Egad, I thought it was tea, but I see I've been drinking a blooming micro-zoo!" says this horrified, proper 19th-century London woman when she used a microscope to examine her tea. People were shocked to learn that there is an active, living world too small for us to see.

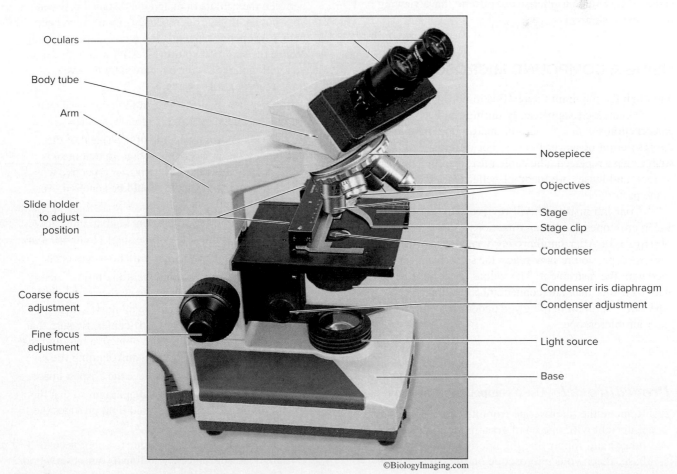

**Figure 3.3** Major parts of a compound light microscope.

## Viewing and Recording System

The viewing and recording system, if present, converts radiation to a viewable and/or permanent image. The viewing and recording system usually consists of a camera or video screen. Most student microscopes do not have viewing and recording systems.

## USING A COMPOUND MICROSCOPE

Although the maximum magnification of light microscopes has not increased significantly during the last century, the construction and design of light microscopes have improved the resolution of newer models. For example, built-in light sources have replaced adjustable mirrors in the illuminating system, and lenses are made of better glass than they were in the past.

Your lab instructor will review with you the parts of the microscopes (and their functions) you will use in the lab (fig. 3.3). After familiarizing yourself with the parts of a microscope, you're now ready for some hands-on experience with the instrument. The videos at the website associated with this manual (connect.mheducation.com) will be especially useful for helping you understand how to properly use your microscope.

### *Procedure 3.1*   Use a compound microscope

1. Remove the microscope from its cabinet and carry it upright with one hand grasping the arm and your other hand supporting the microscope below its base. Place your microscope on the table in front of you.

 Do not use paper towels or Kimwipes to clean the lenses of your microscope; they can scratch the lenses. Clean the lenses only with lens paper.

2. Plug in the microscope and turn on the light source.
3. If it isn't already in position, rotate the nosepiece until the lowest-power (4×) objective is in line with the light source. (The 4× objective is often called the "scanning objective" because it enables users to scan large areas of a specimen.) You'll feel the objective click into place when it is positioned properly. *Always begin examining slides with the lowest-power objective.*
4. Locate the coarse adjustment knob on the side of the microscope. Depending on the type of microscope that you're using, the coarse adjustment knob moves either the nosepiece (with its objectives) or the stage to focus the lenses on the specimen. Only a partial turn of the coarse adjustment knob moves the stage or nosepiece a relatively large distance. *The coarse adjustment should only be used when you're viewing a specimen with the 4× or 10× objective lens.*
5. If your microscope is binocular, adjust the distance between the oculars to match the distance between your pupils. If your microscope is monocular, keep both eyes open when using the microscope; this will reduce the strain on your eyes. After a little practice you will ignore the image received by the eye not looking through the ocular.
6. Adjust the light intensity for comfort and visual quality. Focus a specimen by using the following steps:
   a. Place a microscope slide of newsprint of the letter *e* on the horizontal stage so that the *e* is directly below the lowest-power objective lens and is right side up. It should be centered over the hole in the stage.
   b. Rotate the coarse adjustment knob to move the objective within 1 cm of the stage (1 cm = 0.4 in).
   c. Look through the oculars with both eyes open.
   d. Rotate the coarse adjustment knob (i.e., raising the objective lens or lowering the stage) until the *e* comes into focus. If you don't see an image, the *e* is probably off center. Be sure that the *e* is directly below the objective lens and that you can see a spot of light surrounding the *e*.
   e. Focus up and down to achieve the crispest image.
   f. Adjust the condenser iris diaphragm so that the brightness of the transmitted light provides the best view.
   g. Observe the letter, then rotate the nosepiece to align the 10× objective to finish your observation. Do not use the oil immersion objective.

## Question 1

*a.* As you view the letter *e,* how is it oriented? Upside down or right side up?

*b.* How does the image move when the slide is moved to the right or left? Toward you or away from you?

*c.* What happens to the brightness of the view when you go from 4× to 10×?

## Magnification

### Procedure 3.2 Determine magnification

1. Estimate the magnification of the *e* by looking at the magnified image on lowest magnification (4×), and then at the *e* without using the microscope.

2. Examine each objective and record the magnifications of the objectives and oculars of your microscope in table 3.1.

3. Calculate and record in table 3.1 the total magnification for each objective following this formula:

$$\text{Mag}_{\text{Tot}} = \text{Mag}_{\text{Obj}} \times \text{Mag}_{\text{Ocu}}$$

where

$\text{Mag}_{\text{Tot}}$ = **total magnification of the image**

$\text{Mag}_{\text{Obj}}$ = **magnification of the objective lens**

$\text{Mag}_{\text{Ocu}}$ = **magnification of the ocular lens**

For example, if you're viewing the specimen with a 4× objective lens and a 10× ocular, the total magnification of the image is 4 × 10 = 40×. That is, the specimen appears 40 times larger than it is.

4. Slowly rotate the high-power (i.e., 40×) objective into place. *Be sure that the objective does not touch the slide!* If the objective does not rotate into place without touching the slide, do not force it; ask your lab instructor to help you. After the 40× objective is in place, you should notice that the image remains near focus and that the field-of-view has gotten smaller. Most light microscopes are **parfocal,** meaning that the image will remain nearly focused after the 40× objective lens is moved into place. Most light microscopes are also **parcentered,** meaning that the image will remain centered in the field of view after the 40× objective lens is in place.

5. You may need to readjust the iris diaphragm because the high-magnification objective allows less light to pass through to the ocular.

6. To fine-focus the image, locate the **fine adjustment knob** on the side of the microscope. Turning this knob changes the specimen-to-objective distance slightly and therefore makes it easy to fine-focus the image. **Use only the fine adjustment when using the 40× (or higher) objective.**

 Never use the coarse adjustment knob to focus an image on high power.

## Question 2

*a.* How many times is the image of the *e* magnified when viewed through the high-power objective?

*b.* If you didn't already know what you were looking at, could you determine at this magnification that you were looking at a letter *e*? How?

## Determine the Size of the Field of View

The **field of view** is the area that you can see through the ocular and objective (fig. 3.4). Knowing the size of the field of view is important because you can use it to estimate the size of an object you are examining. The field of view can be measured with ruled **micrometers** (fig. 3.5). An **ocular micrometer** is a small glass disk with thin lines numbered and etched in a row. It was put into an ocular on your microscope so that the lines superimpose on the image and

---

**Table 3.1**

**Total Magnifications and Areas of Field of View (FOV) for Three Objective Lenses**

| Objective Power | Objective Magnification | × | Ocular Magnification | = | Total Magnification | FOV Diameter (mm) | FOV Area (mm²) | Measurement (mm) for 1 Ocular Space |
|---|---|---|---|---|---|---|---|---|
| 4× | _____ | × | _____ | = | _____ | _____ | _____ | _____ |
| 10× | _____ | × | _____ | = | _____ | _____ | _____ | _____ |
| 40× | _____ | × | _____ | = | _____ | _____ | _____ | _____ |

©BiologyImaging.com

**Figure 3.4** The circular, illuminated field of view of a compound light microscope. Shown here is the letter *e* from newsprint that is magnified 40 times (i.e., 40×).

allow you to measure the specimen. Before you can use the micrometer you must determine for each magnification the apparent distance between the lines on the ocular micrometer. This means that you must calibrate the ocular micrometer by comparing its lines to those lines on a standard ruler called a **stage micrometer.** A stage micrometer is a glass slide having precisely spaced lines etched at known intervals.

**Procedure 3.3** Use a stage micrometer to calibrate the ocular micrometer, and determine the size of the field of view

*1.* Rotate the ocular until the lines of the ocular micrometer parallel those of the stage micrometer (fig. 3.5).

*2.* Align lines at the left edges (0 lines) of the two micrometers by moving the stage micrometer (fig. 3.5).

*3.* Count how many spaces on the stage micrometer fit precisely in a given number of spaces on the ocular micrometer. Record the values here:

**y ocular spaces = x stage spaces**

$$y = \text{_____}$$

$$x = \text{_____}$$

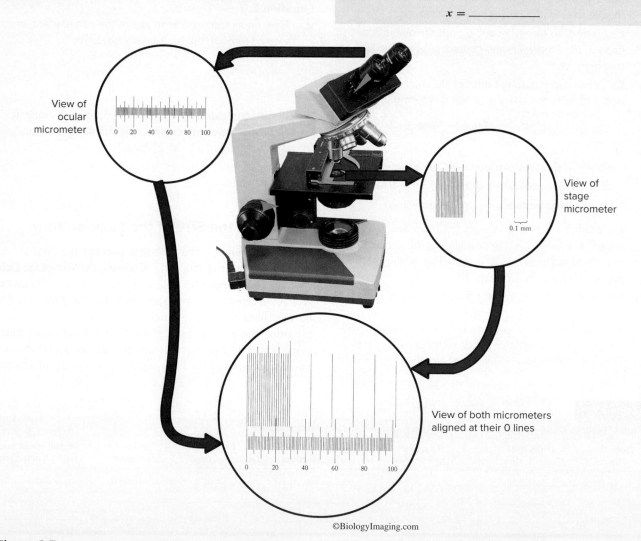

©BiologyImaging.com

**Figure 3.5** Stage and ocular micrometers. A stage micrometer is used to calibrate a microscope with its ocular micrometer to measure the size of specimens.

The smallest space on a stage micrometer = 0.01 mm, so

$$y \text{ ocular spaces (mm)} = x \text{ stage spaces} \times 0.01$$

$$1 \text{ ocular space (mm)} = (x/y) \times 0.01$$

4. Calculate the distance in millimeters between lines of the ocular micrometer. For example, if the length of 10 spaces on the ocular micrometer equals the length of seven spaces on the stage micrometer, then

$$y = 10$$

$$x = 7$$

$$10 \text{ ocular spaces (mm)} = 7 \text{ stage spaces} \times 0.01 \text{ mm}$$

$$1 \text{ ocular space (mm)} = (7 \times 0.01 \text{ mm})/10$$

$$1 \text{ ocular space (mm)} = 0.007 \text{ mm}$$

$$1 \text{ ocular space} = 7 \text{ μm}$$

Therefore, if a specimen spans eight spaces on your ocular micrometer with that objective in place, that specimen is 56 μm long.

5. Calibrate the ocular micrometer for each objective on your microscope. Record in table 3.1 the diameter of the field of view (FOV) for each objective. Also record for each objective lens in table 3.1 the measurement (mm) for 1 ocular space. You can use this information in future labs as you measure the sizes of organisms and their parts.

6. Calculate the radius, which is half the diameter.

7. Use this information to determine the area of the circular field of view with the following formula:

$$\text{Area of circle} = \pi \times \text{radius}^2$$
$$(\pi = 3.14)$$

8. Record your calculated field of view areas in table 3.1.

## Alternate Procedure 3.3   Use a transparent ruler to determine the size of the field of view

1. Obtain a clear plastic ruler with a metric scale.

2. Place the ruler on the stage and under the stage clips of your microscope. If your microscope has a mechanical stage, ask your instructor how to place the ruler to avoid damage. *Carefully rotate the nosepiece to the objective of lowest magnification.*

3. Slowly focus with the coarse adjustment, and then with the fine adjustment, until the metric markings on the ruler are clear.

4. Align the ruler to measure the diameter of the circular field of view. The space between each line on the ruler should represent a 1-mm interval.

5. Record in table 3.1 the diameter of this low-magnification field of view. Also calculate the radius, which is half the diameter.

6. The ruler cannot be used to measure the diameters of the field of view at medium and high magnifications because the markings are too far apart. Therefore, these diameters must be calculated using the following formula:

$$\text{FOV}_{low} \times \text{Mag}_{low} = \text{FOV}_{high} \times \text{Mag}_{high}$$

where

$\text{FOV}_{low}$ = diameter of the field of view of the low-power objective

$\text{Mag}_{low}$ = magnification of the low-power objective (Be consistent and use the magnification of the objective, not total magnification.)

$\text{FOV}_{high}$ = diameter of the field of view of the high-power objective

$\text{Mag}_{high}$ = magnification of the high-power objective

For example, if 3.0 mm is the diameter of the field of view for a 4× low-power objective, then what is the diameter of the field of view of the 40× high-power objective?

$$3.0 \text{ mm} \times 4 = \text{FOV}_{high} \times 40$$
$$0.30 \text{ mm} = \text{FOV}_{high}$$

7. Calculate and record in table 3.1 the diameters of the field of view for the 10× and 40× magnifications.

8. Calculate and record in table 3.1 the circular area of the field of view for the three magnifications by using the following formula.

$$\text{Area of circle} = \pi \times \text{radius}^2$$
$$(\pi = 3.14)$$

### Question 3

a. Which provides the largest field of view, the 10× or 40× objective?

b. How much more area can you see with the 4× objective than with the 40× objective?

c. Why is it more difficult to locate an object starting with the high-power objective than with the low-power objective?

d. Which objective should you use to initially locate the specimen? Why?

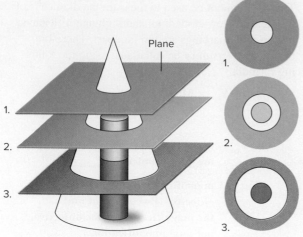

Plane

1.

2.

3.

1.

2.

3.

Specimen
on slide

Microscope image
at different levels
of focus

(a)

(b)                                            ©BiologyImaging.com

**Figure 3.6** How focusing at different planes of a specimen would produce three different images. (*a*) Focusing up and down when you view a specimen can help you to understand its three-dimensional structure. (*b*) A thin depth of field is apparent in this 100× image of cells of *Closterium*, a green alga. The upper and lower layers of cells are out of focus, while the midlayer of cells is within the thin depth of field and is clearly focused.

## Determine the Depth of Field

**Depth of field** is the thickness of the object in sharp focus (fig. 3.6). Depth of field varies with different objectives and magnifications.

*Procedure 3.4* **Determine the depth of the field of view**

*1.* Using the low-power objective, examine a prepared slide of three colored threads mounted on top of each other.

*2.* Focus up and down and try to determine the order of the threads from top to bottom. The order of the threads will not be the same on all slides.

*3.* Re-examine the threads using the high-power objective lens.

### Question 4

*a.* Are all three colored threads in focus at low power?

*b.* Can all three threads be in focus at the same time using the high-power objective?

*c.* Which objective, high or low power, provides the greatest depth of field?

## Preparing a Wet Mount of a Biological Specimen

*Procedure 3.5* **Prepare a wet mount of a biological specimen**

*1.* Place a drop of water containing algal cells from a culture labeled "Algae" on a clean microscope slide.

*2.* Place the edge of a clean coverslip at an edge of the drop at a 45° angle; then slowly lower the coverslip onto the drop so that no air bubbles are trapped (fig. 3.7). (Your instructor will demonstrate this technique.) The coverslip holds the specimen in place and prevents the lens of an objective from contacting the water and the specimen. This fresh preparation is called a **wet mount** and can be viewed with your microscope.

*3.* Experiment with various intensities of illumination. To do this, rotate the 4× objective into place and adjust the condenser iris diaphragm to produce the least illumination. Observe the image; note its clarity, contrast, and color. Repeat these observations with at least four different levels of illumination. The fourth level should have the diaphragm completely open.

*4.* Repeat step 3 for the 10× and 40× objectives.

### Question 5

*a.* Is the image always best with the highest illumination?

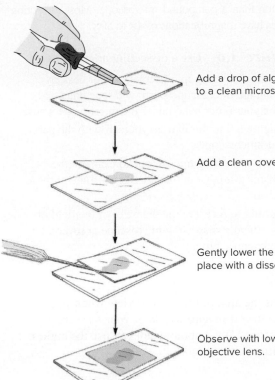

Add a drop of algal culture to a clean microscope slide.

Add a clean coverslip.

Gently lower the coverslip into place with a dissecting needle.

Observe with low-power objective lens.

(a)

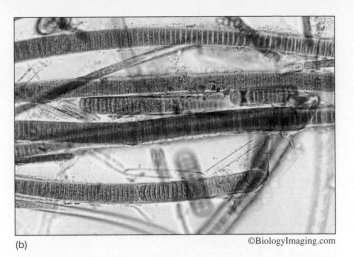

(b)                                                    ©BiologyImaging.com

**Figure 3.7** (*a*) Preparing a wet mount of a biological specimen. (*b*) A wet mount of pond water will often include the common cyanobacterium *Oscillatoria* (200×). See also figures 3.6 and 25.1–25.4.

***b.*** Is the same level of illumination best for all magnifications?

***c.*** Which magnifications require the most illumination for the best clarity and contrast?

**5.** Examine your preparation of algae, and sketch in the following field of view the organisms that you see. Don't mistake air bubbles for organisms! Air bubbles appear as uniformly round structures with dark, thick borders.

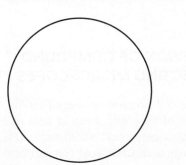

**6.** Prepare a wet mount of some newly hatched brine shrimp (*Artemia,* which are popularly referred to as "sea monkeys") and their eggs. Sketch in the

following field of view what you see. Use your calculations for the diameter of the field of view to estimate the length of the shrimp.

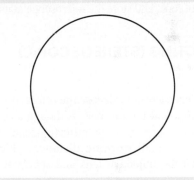

Approximate length of the shrimp: _____

**Question 6**

***a.*** Why is it important to put a coverslip over the drop of water when you prepare a wet mount? That is, what are the functions of a coverslip?

***b.*** Approximately how long and wide is a brine shrimp?

## Practice

For practice using your microscope, prepare some wet mounts of pond water or a hay infusion to view the diversity of protozoa and algae (fig. 3.8). If the protozoa are moving too fast for you to examine carefully, add a drop of methylcellulose (often sold commercially as Proto-Slo) to your sample. (The

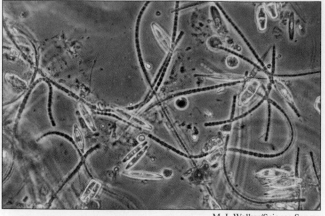

M. I. Walker/Science Source

**Figure 3.8** The diversity of organisms in pond water (200×).

methylcellulose will slow the movement of the protozoa.) Also examine the prepared slides available in the lab. You'll examine these slides in more detail in the coming weeks, so don't worry about their contents. Rather, use this exercise to familiarize yourself with the microscope. Also prepare wet mounts of the cultures available in the lab and sketch the organisms that you see. When you've finished, turn off the light source, cover your microscope, and store the microscope in its cabinet.

## THE DISSECTING (STEREOSCOPIC) MICROSCOPE

A **dissecting (stereoscopic) microscope** offers some advantages over a compound microscope. Although a compound microscope can produce high magnifications and excellent resolution, it has a small **working distance,** which is the distance between the objective lens and specimen. Therefore, it is difficult to manipulate a specimen while observing it with a compound microscope. Specimens that can be observed with a compound microscope are limited to those thin enough for light to pass through them. In contrast, a dissecting microscope is used to view objects that are opaque or too large to see with a compound microscope.

A dissecting microscope provides a much larger working distance than does a compound microscope. This distance is usually several centimeters (compared to a centimeter or less for a compound microscope), making it possible to dissect and manipulate most specimens. Also, most specimens for dissection are too thick to observe with transmitted light from a light source below the specimen. Therefore, many dissecting microscopes use a light source above the specimen; the image you see is formed from reflected light.

Dissecting microscopes are always binocular (fig. 3.9). Each ocular views the specimen at different angles through one or more objective lenses. This arrangement provides a three-dimensional image with a large depth of field. This is in contrast to the image in a compound microscope, which is basically two-dimensional. However, the advantages of a stereoscopic microscope are often offset by lower resolution and

magnification than a compound microscope. Most dissecting microscopes have magnifications of 4× to 50×.

### *Procedure 3.6* Use a dissecting microscope

1. Carry the dissecting microscope to your desk by grasping the microscope's arm with one hand and placing your other hand under the microscope's base.

2. Use figure 3.9 to familiarize yourself with the parts of your microscope.

3. Use your dissecting microscope to examine the organisms available in the lab. Sketch some of these organisms.

4. Use a ruler to measure the diameter of the field of view with your dissecting microscope at several levels of magnification.

### Question 7

*a.* What is the area of the field of view when you use the lowest magnification of your dissecting microscope? What about when you use the highest magnification?

*b.* Place a microscope slide of the letter *e* on the stage. As you view the letter *e,* how is it oriented?

*c.* How does the image through a dissecting microscope move when the specimen is moved to the right or left? Toward you or away from you?

*d.* How does the direction of illumination differ in dissecting as opposed to compound microscopes?

*e.* What are some examples of biological specimens that would be best examined with a dissecting microscope instead of a compound microscope?

### A COMPARISON OF COMPOUND AND DISSECTING MICROSCOPES

Complete table 3.2 comparing magnification, resolution, size of the field of view, and depth of field of a dissecting microscope and a compound microscope. Use the terms *high, low,* or *same* to describe your comparisons.

### Question 8

What other differences are there between compound and dissecting microscopes?

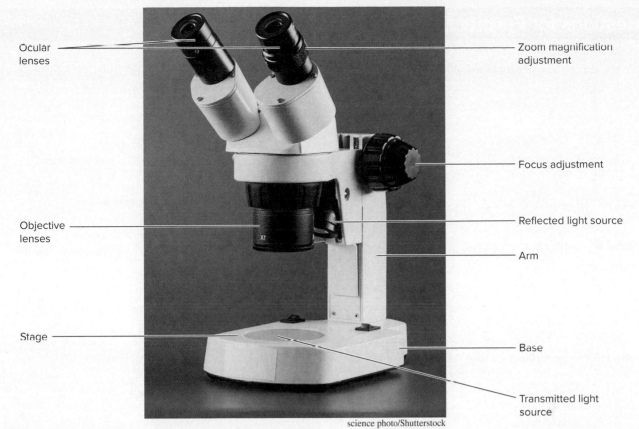

Ocular
lenses

Zoom magnification
adjustment

Focus adjustment

Objective
lenses

Reflected light source

Arm

Stage

Base

Transmitted light
source

science photo/Shutterstock

**Figure 3.9** Major parts of a dissecting (stereoscopic) microscope.

<table>
<tr><th colspan="3">Table 3.2</th></tr>
<tr><th colspan="3">A Comparison of Dissecting and Compound Microscope</th></tr>
<tr><th>Characteristic</th><th>Dissecting Microscope</th><th>Compound Microscope</th></tr>
<tr><td>Magnification</td><td></td><td></td></tr>
<tr><td>Resolution</td><td></td><td></td></tr>
<tr><td>Size of field of view</td><td></td><td></td></tr>
<tr><td>Depth of field</td><td></td><td></td></tr>
</table>

# INQUIRY-BASED LEARNING

## *What are the shapes, surface areas, and volumes of red blood cells?*

Observation: Red blood cells, which are the most common type of blood cell, are used by vertebrates to deliver oxygen to body tissues. Red blood cells are filled with hemoglobin, which gives them their characteristic color.

Question: What are the shapes, surface areas, and volumes of red blood cells?

a. Establish a working lab group and obtain Inquiry-Based Learning Worksheet 3 from your instructor.

b. Discuss with your group well-defined questions relevant to the preceding observation and question. Choose and record your group's best question for investigation.

c. Translate your question into a testable hypothesis and record it.

d. Outline on Worksheet 3 your experimental design and supplies needed to test your hypothesis. Ask your instructor to review your proposed investigation.

e. Conduct your procedures, record your data, answer your question, and make relevant comments.

f. Discuss with your instructor any revisions to your questions, hypothesis, or procedures. Repeat your work as needed.

1. What are the advantages of knowing the diameter of the field of view at a given magnification?

2. Why must specimens viewed with a compound microscope be thin? Why are they sometimes stained with dyes?

3. Why is depth of field important in studying biological structures? How can it affect your ability to find and examine a specimen?

4. What is the importance of adjusting the light intensity when viewing specimens with a compound microscope?

5. What is the function of each of the following parts of a compound and dissecting microscope?
   Oculars
   Objectives
   Condenser
   Iris diaphragm
   Stage
   Coarse adjustment
   Fine adjustment

6. Examine the micrograph of the letter *e* shown in figure 3.4. This letter is magnified 40×. What is the actual height of the letter?

## WRITING TO LEARN BIOLOGY
The smallest structures of cells are best seen with a transmission electron microscope. What are some examples of cellular structures that were discovered with a transmission electron microscope? Refer to your textbook or other book and describe how an electron microscope can resolve such small structures. Write a short essay about the advantages and limitations of a **transmission electron microscope.**

# The Cell
## Structure and Function

## Learning Objectives

By the end of this exercise you should be able to:
1. Understand the differences between prokaryotes and eukaryotes and identify structures characteristic of each.
2. Prepare a wet mount to view cells with a compound microscope.
3. Explain the function of organelles within eukaryotic cells visible with a light microscope.
4. Examine a cell's structure and determine whether it is from a plant or animal.
5. Observe representatives of the protists, a large group of eukaryotic organisms that are heterotrophic or autotrophic.
6. Understand the importance of a cell's surface-area-to-volume ratio.

Please visit **connect.mheducation.com** to review online resources tailored to this lab.

**C**ells are the basic unit of living organisms because they perform all of the processes we collectively call "life." All organisms are made of cells. Although most individual cells are visible only with the aid of a microscope, some may be a meter long (e.g., nerve cells) or as large as a small orange (e.g., the yolk of an ostrich egg). Despite these differences, all cells are designed similarly and share fundamental features.

**Cytology** is the study of cellular structure and function. The major tools of cytologists are light microscopy, electron microscopy, and cell chemistry. By studying the anatomy of a cell, we can find clues to how the cell works.

In today's lab you will study some of the features and variations among living cells. Prior to this exercise, review in your textbook the general features of cellular structure and function.

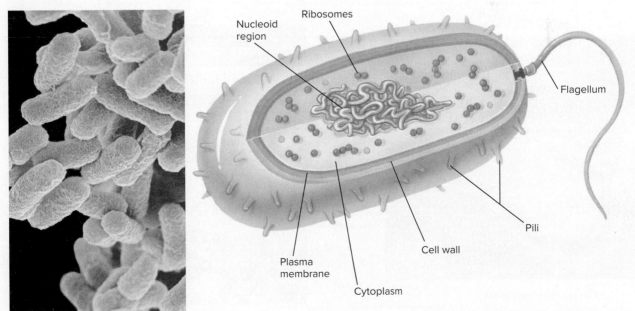

Steve Gschmeissner/Science Photo Library/Getty Images

**Figure 4.1** The structure of a bacterial cell. Bacteria lack a nuclear membrane. All prokaryotic (bacterial) cells have a nucleoid region, ribosomes, plasma membrane, cytoplasm, and cell wall, but not all have flagella (1500×). Many bacterial cells are surrounded by a gelatinous capsule and have pili as well as flagella.

# PROKARYOTIC CELLS

Bacteria and cyanobacteria are **prokaryotes** (fig. 4.1), and their diversity is considerable (>5000 species). Prokaryotes do not contain a membrane-bound nucleus or any other membrane-bound **organelles.** Organelles are organized structures of macromolecules having specialized functions and are suspended in the **cytoplasm.** The cytoplasm of prokaryotes is enclosed in a **plasma membrane** (cellular membrane) and is surrounded by a supporting **cell wall** covered by a gelatinous **capsule. Flagella** and hairlike outgrowths called **pili** are common in prokaryotes; flagella are used for movement, and pili are used to attach some types of bacteria to surfaces or to exchange genetic material with other bacteria. Within the cytoplasm are **ribosomes** (small organelles involved in protein synthesis) and **nucleoid regions** (concentrations of DNA). Prokaryotes do not reproduce sexually, but they have mechanisms for genetic recombination (see Exercise 16).

## Cyanobacteria

The largest prokaryotes are **cyanobacteria,** formerly called blue-green algae. They contain chlorophyll *a* and accessory pigments for photosynthesis, but these pigments are not contained in membrane-bound chloroplasts. Instead, the pigments are held in photosynthetic membranes called **thylakoids** (fig. 4.2). Cyanobacteria are often surrounded by a **mucilaginous sheath.** Their ability to photosynthesize made them the primary contributors to the early oxygenation of the ancient earth's atmosphere.

 **SAFETY FIRST** Before coming to lab, you were asked to read this exercise so you would know what to do and be aware of safety issues. In the space below, briefly list the safety issues associated with today's procedures. If you have questions about these issues, contact your laboratory assistant before starting work. Cheek scraping is a potential biohazard. Check with your instructor before you use this technique.

---

## *Procedure 4.1* Examine cyanobacteria

1. Examine a prepared slide of *Oscillatoria,* a filament of cells, and one of *Gloeocapsa,* a loosely arranged colony (fig 4.3). Review Exercise 3 and the associated videos for the proper way to use the microscope.
2. Focus with the low-power objective.
3. Rotate the high-power objective into place to see filaments and masses of cells.

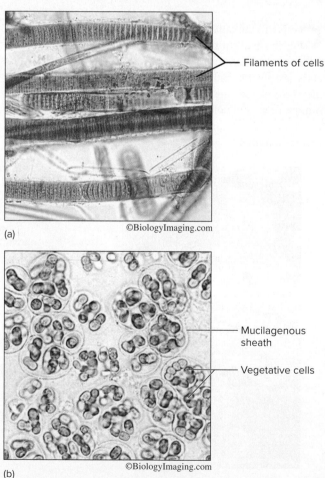

(a) ©BiologyImaging.com

— Filaments of cells

— Mucilagenous sheath

— Vegetative cells

(b) ©BiologyImaging.com

**Figure 4.3** Common cyanobacteria. (*a*) *Oscillatoria* (100×). (*b*) *Gloeocapsa* (400×).

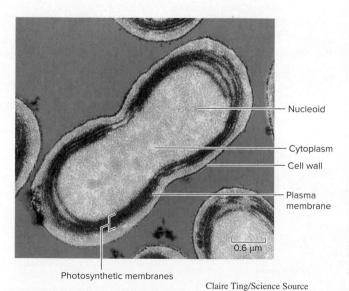

— Nucleoid

— Cytoplasm

— Cell wall

— Plasma membrane

0.6 µm

Photosynthetic membranes

Claire Ting/Science Source

**Figure 4.2** Electron micrograph of a photosynthetic bacterial cell. Extensive folded photosynthetic membranes are shown in green in this false-colored electron micrograph of a *Prochloron* cell.

4. Prepare a wet mount of *Oscillatoria* and one of *Gloeocapsa*. Review procedure 3.5 in Exercise 3 for preparing a wet mount.

5. Observe the cellular structures and draw the cellular shapes and relative sizes of *Oscillatoria* and *Gloeocapsa* in the following space. Use an ocular micrometer to measure their dimensions.

    *Oscillatoria*          *Gloeocapsa*

3. Focus with the low-power objective.

4. Rotate the high-power objective (40×) into place to see masses of rod-shaped cells.

5. Observe the simple, external structure of the bacteria and draw their cellular shapes in the following space:

## Question 1
*a.* Where are the pigments located in these cyanobacteria?

*b.* Are nuclei visible in cyanobacterial cells?

*c.* Which of these two genera has the most prominent mucilaginous sheath?

*d.* How many cells are held within one sheath of *Gloeocapsa*?

## Bacteria

Most bacteria are much smaller than cyanobacteria and do not contain chlorophyll. Yogurt is a nutrient-rich culture of bacteria. The bacterial cells composing most of the yogurt are *Lactobacillus,* a bacterium adapted to live on milk sugar (lactose). *Lactobacillus* converts milk to yogurt. Yogurt is acidic and keeps longer than milk. Historically, *Lactobacillus* has been used in many parts of the world by peoples deficient in lactase, an enzyme that breaks down lactose. Many Middle Eastern and African cultures use the more digestible yogurt in their diets instead of milk.

## Procedure 4.2  Examine bacteria
1. Place a tiny dab of yogurt on a microscope slide.
2. Mix this small amount of yogurt in a drop of water, add a coverslip, and examine the yogurt with a compound microscope. Review Exercise 3.

## Question 2
How does the size of *Lactobacillus* compare with that of *Oscillatoria* and *Gloeocapsa?*

## EUKARYOTIC CELLS

Eukaryotic cells are structurally more complex than prokaryotic cells. Although some features of prokaryotic cells are in eukaryotic cells (e.g., ribosomes, cell membrane), eukaryotic cells also contain several organelles not found in prokaryotic cells (table 4.1).

Eukaryotic cells contain membrane-bound **nuclei** and other organelles (figs. 4.4, 4.5). Nuclei contain genetic material of a cell and control metabolism. **Cytoplasm** forms the matrix of the cell and is contained by the plasma membrane. Within the cytoplasm are a variety of organelles. **Chloroplasts** are elliptical green organelles in plant cells. Chloroplasts are the site of photosynthesis in plant cells and are green because they contain chlorophyll, a photosynthetic pigment capable of capturing light energy. **Mitochondria** are organelles found in plant and animal cells. These organelles are where aerobic respiration occurs. When viewed with a conventional light microscope, mitochondria are small, dark, and often difficult to see. All of the material and organelles contained by the plasma membrane are collectively called the **protoplast.**

## PLANT CELLS

### Procedure 4.3  Examine living *Elodea* cells and chloroplasts
1. Remove a young leaf from the tip of a sprig of *Elodea*. *Elodea* is a common pond-weed used frequently in studies of photosynthesis, cellular structure, and cytoplasmic streaming.
2. Place this leaf, with the top surface facing up, in a drop of water on a microscope slide. The cells on the

Table 4.1

**Some of the Major Differences between Prokaryotic and Eukaryotic Cells and between Plant and Animal Cells**

| | Prokaryote | Eukaryote | |
| --- | --- | --- | --- |
| | | Animal | Plant |
| **Exterior Structures** | | | |
| Cell wall | Present (protein-polysaccharide) | Absent | Present (cellulose) |
| Cell membrane | Present | Present | Present |
| Flagella | May be present (single strand) | May be present | Absent except in sperm of a few species |
| **Interior Structures** | | | |
| ER | Absent | Usually present | Usually present |
| Ribosomes | Present | Present | Present |
| Microtubules | Absent | Present | Present |
| Centrioles | Absent | Present | Absent |
| Golgi complex | Absent | Present | Present |
| **Other Organelles** | | | |
| Nucleus | Absent | Present | Present |
| Mitochondria | Absent | Present | Present |
| Chloroplasts | Absent | Absent | Present |
| Chromosomes | A single circle of naked DNA | Multiple; DNA-protein complex | Multiple; DNA-protein complex |
| Vacuoles | Absent | Absent or small | Usually a large single vacuole |

upper surface are larger and more easily examined. Add a coverslip, but do not let the leaf dry. Add another drop of water if necessary.

3. Examine the leaf with your microscope. Review Exercise 3 and the associated videos. First use low, then high, magnification to bring the upper layer of cells into focus (fig. 4.6). Each of the small, regularly shaped units you see are cells surrounded by cell walls made primarily of **cellulose** (fig. 4.7). Cellulose is a complex carbohydrate made of glucose molecules attached end-to-end. The plasma membrane lies just inside the cell wall. Sketch what you see.

*b.* Examine various layers of cells by focusing up and down through the layers. About how many cells thick is the leaf that you are observing?

*c.* What are the functions of the cell wall?

*d.* Use an ocular micrometer or refer to the dimensions of the field of view calculated in Exercise 3 to measure the dimensions of an *Elodea* cell. What are the cell's approximate dimensions?

**Question 3**

*a.* What three-dimensional shape are *Elodea* cells?

4. Chloroplasts appear as moderately sized green spheres within the cells (figs. 4.6, 4.8). Locate and

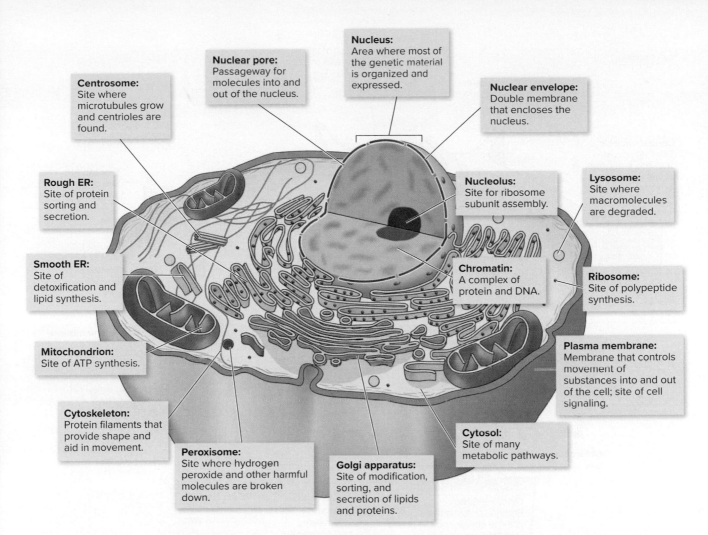

**Centrosome:** Site where microtubules grow and centrioles are found.

**Nuclear pore:** Passageway for molecules into and out of the nucleus.

**Nucleus:** Area where most of the genetic material is organized and expressed.

**Nuclear envelope:** Double membrane that encloses the nucleus.

**Rough ER:** Site of protein sorting and secretion.

**Nucleolus:** Site for ribosome subunit assembly.

**Lysosome:** Site where macromolecules are degraded.

**Smooth ER:** Site of detoxification and lipid synthesis.

**Chromatin:** A complex of protein and DNA.

**Ribosome:** Site of polypeptide synthesis.

**Mitochondrion:** Site of ATP synthesis.

**Plasma membrane:** Membrane that controls movement of substances into and out of the cell; site of cell signaling.

**Cytoskeleton:** Protein filaments that provide shape and aid in movement.

**Cytosol:** Site of many metabolic pathways.

**Peroxisome:** Site where hydrogen peroxide and other harmful molecules are broken down.

**Golgi apparatus:** Site of modification, sorting, and secretion of lipids and proteins.

**Figure 4.4** Structure of animal cells. Cells are surrounded by a bilayered plasma membrane containing phospholipids and proteins. The nucleus houses chromosomal DNA and is surrounded by a double-membraned nuclear envelope. Centrioles organize spindle fibers during cell division. Endoplasmic reticulum (ER) is a system of membranes inside the cell. Rough ER has many ribosomes, and smooth ER has fewer ribosomes. Mitochondria are sites of oxidative respiration and ATP synthesis. Microvilli are cytoplasmic projections that increase the surface area of some specialized animal cells. Golgi complexes are flat sacs and vesicles that collect and package substances made in the cell. Ribosomes are aggregations of rRNA and proteins that make protein. Lysosomes contain enzymes important in recycling cellular debris.

sketch cells having many chloroplasts; estimate the number of chloroplasts in a healthy cell. Remember that a cell is three-dimensional, and some chloroplasts may obscure others.

**Question 4**

*a.* What shape are the chloroplasts? What is their function?

*b.* Where are the chloroplasts located within the *Elodea* cell—toward the perimeter or center of the cell?

5. Determine the spatial distribution of chloroplasts within a cell. They may be pushed against the margins of the cell by the large **central vacuole** containing mostly water and bounded by a **vacuolar membrane.** The vacuole occupies about 90% of the volume of a mature cell. Its many functions include storage of organic and inorganic molecules, ions, water, enzymes, and waste products.

6. Search for a **nucleus;** it may or may not be readily visible. The nucleus contains most of the cell's genetic material; in humans, this amounts to more than 20,000 genes that provide the blueprint of life. Nuclei usually

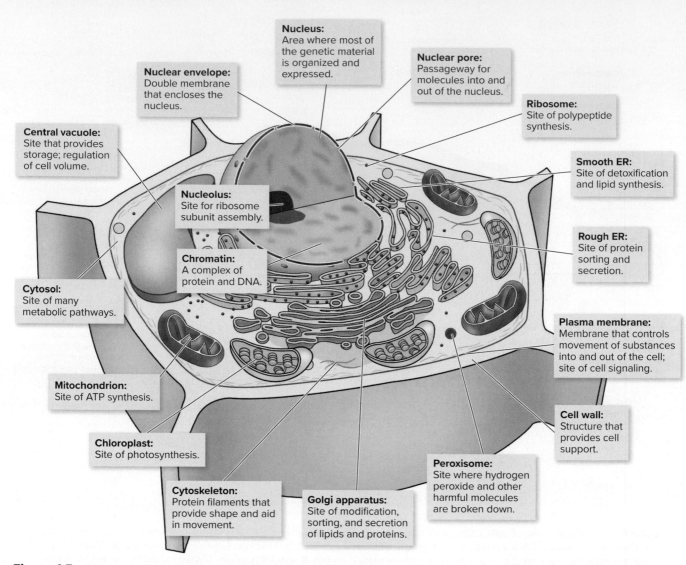

**Nucleus:**
Area where most of the genetic material is organized and expressed.

**Nuclear pore:**
Passageway for molecules into and out of the nucleus.

**Ribosome:**
Site of polypeptide synthesis.

**Nuclear envelope:**
Double membrane that encloses the nucleus.

**Smooth ER:**
Site of detoxification and lipid synthesis.

**Central vacuole:**
Site that provides storage; regulation of cell volume.

**Nucleolus:**
Site for ribosome subunit assembly.

**Rough ER:**
Site of protein sorting and secretion.

**Chromatin:**
A complex of protein and DNA.

**Cytosol:**
Site of many metabolic pathways.

**Plasma membrane:**
Membrane that controls movement of substances into and out of the cell; site of cell signaling.

**Mitochondrion:**
Site of ATP synthesis.

**Cell wall:**
Structure that provides cell support.

**Chloroplast:**
Site of photosynthesis.

**Peroxisome:**
Site where hydrogen peroxide and other harmful molecules are broken down.

**Cytoskeleton:**
Protein filaments that provide shape and aid in movement.

**Golgi apparatus:**
Site of modification, sorting, and secretion of lipids and proteins.

**Figure 4.5**  Structure of plant cells. Most mature plant cells contain large central vacuoles, which occupy most of the volume of the cell. Cytoplasm is often a thin layer between the vacuole and the plasma membrane. Cytoplasm contains the cell's organelles.

are appressed to the cell wall as a faint gray sphere the size of a chloroplast or larger. Staining the cells with a drop of iodine may enhance the nucleus. If your preparation is particularly good, a **nucleolus** may be visible as a dense spot in the nucleus.

7. Search for some cells that may appear pink due to water-soluble pigments called anthocyanins. These pigments give many flowers and fruits their bright reddish color.

8. Warm the slide with intense light for about 10 min and search for movement of the chloroplasts. You may need to search many cells or make a new preparation. This movement is called **cytoplasmic streaming,** or **cyclosis.** Chloroplasts are not motile; instead, they are being moved by the activity of the cytoplasm. Add water if the cells appear to be drying out.

9. In the following space sketch a few cells of *Elodea*; compare the cells with those shown in fig. 4.6.

10. When you are finished examining *Elodea*, dispose of the *Elodea* as specified by your instructor.

**Question 5**

*a.* Can you see nuclei in *Elodea* cells?

*b.* What are the functions of nuclei?

*c.* Which are larger, chloroplasts or nuclei?

**Figure 4.6** (*a*) *Elodea* cells containing abundant chloroplasts (150×). (*b*) The cellular structure of *Elodea* (400×).

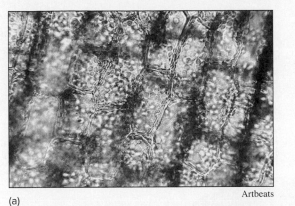

Artbeats

(a)

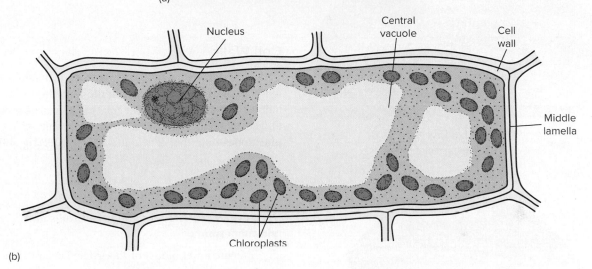

(b)

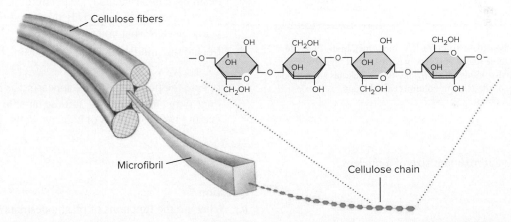

**Figure 4.7** Cellulose is the most abundant organic compound on earth and is a polymer of glucose molecules. Free hydroxyl ($OH^-$) groups of the glucose molecules form hydrogen bonds between adjacent cellulose molecules to form cohesive microfibrils. Microfibrils align to form strong cellulose fibers that resist metabolic breakdown. Because humans cannot hydrolyze the bonds between glucose molecules of cellulose, cellulose is indigestible and its energy is unavailable. Cellulose passes through the human digestive tract as bulk fiber.

## Question 6

*a.* Are all cellular components moving in the same direction and rate during cytoplasmic streaming?

*b.* What do you conclude about the uniformity of cytoplasmic streaming?

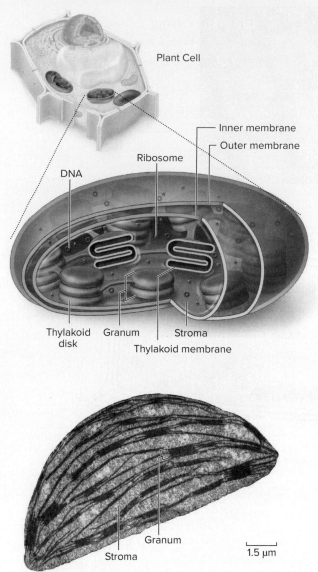

**Figure 4.8** Chloroplast structure. The inner membrane of a chloroplast is fused to form stacks of closed vesicles called thylakoids. Photosynthesis occurs within these thylakoids. Thylakoids are typically stacked one on top of the other in columns called grana.

Dr Jeremy Burgess/Science Source

## Cell Walls

Cell walls include an outer **primary cell wall** deposited during growth of the cell and a **middle lamella,** the substance holding walls of two adjacent cells together. The protoplasm of adjacent cells is connected by cytoplasmic strands called **plasmodesmata** that penetrate the cell walls (fig. 4.9).

> ### *Procedure 4.4* Examine cell walls and plasmodesmata
>
> 1. Prepare a wet mount of *Elodea* and examine the cell walls. Always begin your examination at the lowest magnification and cautiously move to higher magnifications. The middle lamella may be visible as a faint line between cells.
> 2. Obtain a prepared slide of tissue showing plasmodesmata. This tissue may be persimmon (*Diospyros*) endosperm, which has highly thickened primary walls. Sketch what you see.
> 3. Locate the middle lamella as a faint line between cell walls.
> 4. Locate the plasmodesmata, appearing as darkened lines perpendicular to the middle lamella and connecting the protoplasts of adjacent cells (fig. 4.9).

*d.* What is the approximate size of a nucleus?

*e.* Why is the granular-appearing cytoplasm more apparent at the sides of a cell rather than in the middle?

## Question 7

*a.* What are the functions of plasmodesmata?

*b.* Why do you suspect that there are so many plasmodesmata connecting the cells in this fruit?

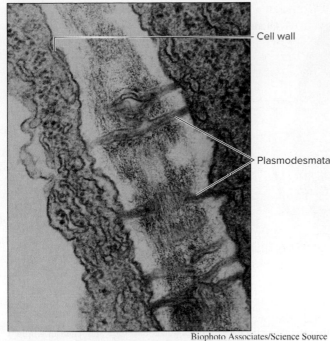

Figure 4.9  This electron micrograph of the thickened primary cell walls of persimmon endosperm shows plasmodesmata connecting adjacent cells (130,000×).

Biophoto Associates/Science Source

## Onion Cells

Staining often reveals the structure of cells and cell organelles more clearly. A specimen is **stained** by adding a dye that preferentially colors some parts of the specimen but not others. Neutral red is a common stain that accumulates in the cytoplasm of the cell, leaving the cell walls clear. Nuclei appear as dense bodies in the translucent cytoplasm of the cells.

### Procedure 4.5  Examine stained onion cells

1. Cut a red onion into eighths and remove a fleshy leaf.
2. Snap the leaf backward and remove the thin piece of the inner epidermis formed at the break point (fig. 4.10), as demonstrated by your lab instructor.
3. Place this epidermal tissue in a drop of water on a microscope slide, add a coverslip, and examine the tissue. This preparation should be one cell thick. Always begin your examination with the lowest magnification.
4. Stain the onion cells by placing a small drop of 0.1% neutral red at the edge of the coverslip. Draw the neutral red across the specimen by wicking. To wick the solution, hold the edge of a small piece of paper towel at the opposite edge of the coverslip and it will withdraw some fluid. This will cause the neutral red to flow over the onion and will not disturb the tissue under the coverslip.
5. Stain the tissue for 5–10 min.
6. Carefully focus to distinguish the vacuole surrounded by the stained cytoplasm.
7. Search for the nucleus of a cell (fig. 4.11). The nucleus may appear circular in the central part of the cell. In other cells it may appear flattened.

### Question 8
How do you explain the differences in the apparent shapes and positions of the nuclei in different cells?

8. Repeat steps 1–7 and stain a new preparation of onion cells with other available stains, such as methylene blue.
9. In the following space sketch a few of the stained onion cells.

Figure 4.10  Preparing a wet mount of an onion epidermis.

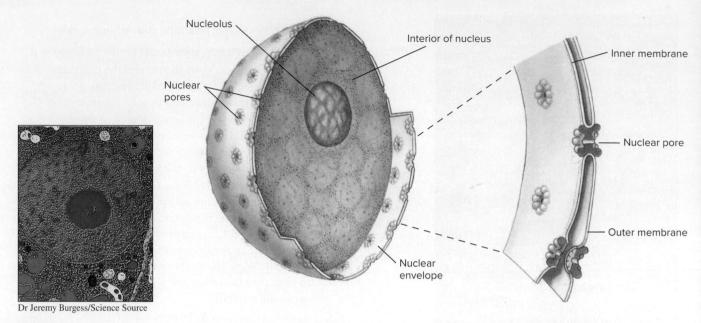

Nucleolus

Interior of nucleus

Inner membrane

Nuclear pores

Nuclear pore

Nuclear envelope

Outer membrane

Dr Jeremy Burgess/Science Source

**Figure 4.11** The nucleus. The nucleus consists of a double membrane, called a nuclear envelope, enclosing a fluid-filled interior containing the DNA. In the cross section, the individual nuclear pores extend through the two membrane layers of the nuclear envelope; the material within the pore is protein, which controls access through the pore (1765×).

## Question 9

**a.** What cellular structures of onion are more easily seen in stained as compared to unstained preparations?

**b.** Which of the available stains enhanced your observations the most?

**c.** Do onion cells have chloroplasts? Explain.

**d.** Use an ocular micrometer or the dimensions of the field of view (FOV) calculated in Exercise 3 to measure the dimensions of an onion epidermal cell. Are these cells larger or smaller than the *Elodea* cells you examined in procedure 4.3?

## Mitochondria

Mitochondria are surrounded by two membranes (fig 4.12). The inner membrane folds inward to form **cristae,** which hold respiratory enzymes and other large respiratory molecules in place. Some DNA also occurs in mitochondria. Chloroplasts also are double-membraned and contain DNA.

> ### *Procedure 4.6* Examine mitochondria in onion cells
>
> *1.* On a clean glass slide mix two or three drops of the stain Janus Green B with one drop of 7% sucrose.
>
> *2.* Prepare a thin piece of onion epidermis (as instructed in procedure 4.5) and mount it in the staining solution. The preparation should be one cell thick. For mitochondria to stain well, the onion cells must be healthy and metabolically active. Add a coverslip.
>
> *3.* Search the periphery of cells to locate stained mitochondria. They are small blue spheres about 1 μm in diameter. The color will fade in 5–10 min, so examine your sample quickly and make a new preparation if needed.
>
> *4.* Label the following structures in the micrograph of the mitochondrion in fig. 4.12: inner membrane, outer membrane, intermembrane space, matrix, and cristae.

## Plastids

**Plastids** are organelles where food, especially sugars and starch, is made and stored. You have already examined chloroplasts, a type of plastid in which photosynthesis occurs.

Other plastids have different functions. We will examine **amyloplasts,** plastids that store starch and therefore will stain darkly with iodine.

Procedure 4.7 **Examine amyloplasts**

1. Use a razor blade to make a thin section of a potato tuber. Make the section as thin as you can.

2. Place the section in a drop of water on a microscope slide and add a coverslip. Add another drop of water to the edge if needed.

3. Locate the small, clam-shaped amyloplasts within the cells. High magnification may reveal the eccentric lines distinguishing layers of deposited starch on the grains.

4. Stain the section by adding a drop of iodine to the edge of the coverslip. Iodine is a stain specific for starch (see Exercise 6, "Biologically Important Molecules"). If necessary, pull the stain under the coverslip by touching a paper towel to the water at the opposite edge of the coverslip.

**Question 10**

*a.* Are any cellular structures other than amyloplasts stained intensely by iodine?

*b.* What can you conclude about the location of starch in storage cells of potato?

*c.* What are the functions of amyloplasts in potatoes?

*d.* Why are potatoes a good source of carbohydrates?

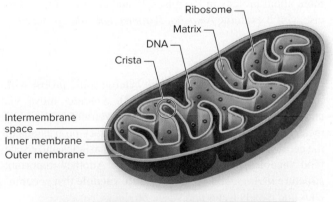

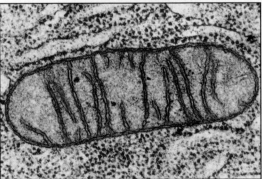

CNRI/Science Photo Library/Getty Images

**Figure 4.12** Mitochondrion in cross section. Mitochondria evolved from bacteria that long ago took up residence within the ancestors of present-day eukaryotes (80,000×).

## ANIMAL CELLS

Animals, like plants, are eukaryotes. They share many similarities, and also have several differences (see table 4.1).

### Human Epithelial Cells

Human epithelial cells are sloughed from the inner surface of your mouth. They are flat cells with a readily visible nucleus.

Procedure 4.8 **Examine human epithelial cells**

1. Check with your instructor about proper technique and disposal of materials for this procedure. Your instructor may substitute examination of a prepared slide for preparation of a wet mount.

2. Gently scrape the inside of your cheek with the broad end of a clean toothpick.

3. Stir the scrapings into a drop of water on a microscope slide, add a coverslip, and examine with your compound microscope. Dispose of used toothpicks in a container designated by your instructor.

4. Stain the cells by placing a small drop of methylene blue at one edge of the coverslip and drawing it under the coverslip with a piece of absorbent paper towel placed at the opposite side of the coverslip.

# Cellular Structures and Human Disease

A variety of human diseases are linked with defective cellular organelles and membranes. For example, examine the animal cell shown in fig. 4.4. Note the small lysosomes, which are where macromolecules—many of which are toxic—are digested. In Tay-Sachs, which is a rare, inherited disease, a child's lysosomes lack the enzyme that breaks down a fatty substance called GM2 ganglioside. As a result, GM2 ganglioside accumulates to toxic levels, especially in the neurons of the brain and spinal cord. When they are 3–6 months old, children having Tay-Sachs become listless, and then progressively lose motor skills (e.g., they can't crawl or turn themselves over), experience seizures, and become blind, deaf, and paralyzed. Children with Tay-Sachs usually die by the time they are 6 years old. This tragedy results from a tiny genetic difference that results in their lysosomes missing one enzyme.

Defective organelles and membranes are also associated with diseases such as diabetes (problems controlling blood glucose, or blood sugar), progeria (an inherited condition characterized by rapid, premature aging), cystic fibrosis (an accumulation of thick mucus in several organs), and familial hypercholesteremia (high inherited levels of cholesterol). Your instructor may ask you to write a short report about one of these diseases and its cellular basis.

---

5. Prepare another slide and stain the cells with Janus Green B. Observe the mitochondria.

6. Use an ocular micrometer or the dimensions of the FOV calculated in Exercise 3 to measure the dimensions of a human epithelial cell.

7. After viewing the preparation, put the slides and coverslips in a container of 10% bleach.

## Question 11

*a.* What structures visible in the stained preparation were invisible in the unstained preparation?

*b.* Were mitochondria as abundant in human epithelial cells as in onion epidermal cells (procedure 4.6)? Explain.

*c.* What similarities and differences are there between plant and animal cells?

*d.* How do the size and shape of a human epithelial cell differ from those of the *Elodea* and onion cells that you examined earlier?

*e.* Why do *Elodea* and onion cells have more consistent shapes than human epithelial cells?

## PROTISTS

*Amoeba, Paramecium,* and *Spirogyra* are members of a large group of eukaryotic organisms called protists. You will learn more about protists in Exercises 25 and 26. In today's exercise, you'll examine *Amoeba, Paramecium,* and *Spirogyra.*

### Amoeba

*Amoeba* is an irregularly shaped, heterotrophic protist with many internal organelles (fig. 4.13). *Amoeba* move via amoeboid movement. **Amoeboid movement** occurs by means of **pseudopodia,** which are temporary protrusions of the cell. Pseudopodia also surround food particles and create food vacuoles, where food is digested. Another important structure in *Amoeba* is the **contractile vacuole** that accumulates and expels water and waste products.

### *Procedure 4.9* Examine *Amoeba*

1. Use an eyedropper to obtain a few drops from the bottom of an *Amoeba* culture. Examining the culture with a dissecting microscope may help you locate some organisms.

2. Place the organisms on a microscope slide.

3. Add a coverslip and use a compound microscope to locate a living *Amoeba.* Your instructor may allow you to view the *Amoeba* without using a coverslip, but view them *only* on 4× or 10× magnification.

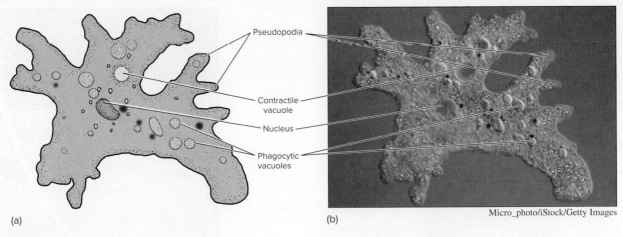

(a)                                                                              (b)    Micro_photo/iStock/Getty Images

**Figure 4.13**   *Amoeba.* (*a*) Diagram of *Amoeba.* (*b*) Light micrograph of a living *Amoeba* (160×).

4. Decrease the light intensity and observe an *Amoeba* for a few minutes.

5. Locate the structures shown in fig. 4.13.

6. Examine a prepared slide of stained *Amoeba;* then observe a demonstration of *Amoeba* on a dark-field microscope if one is available.

7. Sketch an *Amoeba* in the following space.

**Question 12**

*a.* List the organelles found in plant cells, in *Amoeba,* and common to both.

*b.* Does *Amoeba* have a cell wall? How can you tell?

*c.* How do the appearances of *Amoeba* differ in live cells and preserved cells?

## *Paramecium*

Like *Amoeba, Paramecium* is also a single-celled, heterotrophic organism (fig. 4.14).

## *Procedure 4.10*   Examine *Paramecium*

1. Place a small ring of methylcellulose on a microscope slide to slow the *Paramecium.*

2. Place a drop from a culture containing *Paramecium* inside the methylcellulose ring.

3. Use a toothpick to mix the methylcellulose with the drop of water from the culture of *Paramecium.*

4. Add a coverslip and examine *Paramecium* with your compound microscope. On the surface of *Paramecium* are cilia, which are short hairlike structures used for locomotion.

5. Examine a prepared slide of stained *Paramecium.*

6. In the following space, sketch a *Paramecium.*

**Question 13**

*a.* How does movement of *Paramecium* compare to that of *Amoeba?*

*b.* How do shape and body consistency differ between *Amoeba* and *Paramecium?*

*c.* What structures in *Amoeba* and *Paramecium* also occur in plant cells? What structures in *Amoeba* and *Paramecium* do not occur in plant cells?

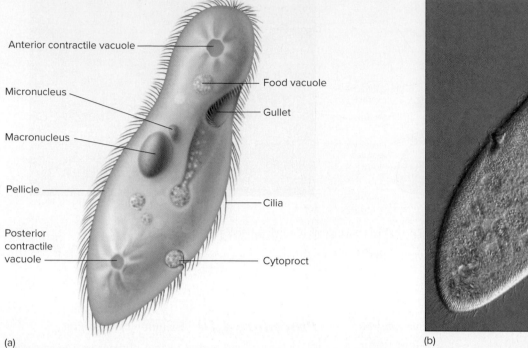

Anterior contractile vacuole

Micronucleus

Macronucleus

Pellicle

Posterior contractile vacuole

Food vacuole

Gullet

Cilia

Cytoproct

(a)

(b)

**Figure 4.14** *Paramecium.* (*a*) Diagram of *Paramecium* (150×). (*b*) Light micrograph of a living *Paramecium*. Note the abundant cilia (150×).

## *Spirogyra*

*Spirogyra* (fig. 4.15) is a filamentous, autotrophic green alga that is named for the spiral arrangement of its chloroplasts. *Spirogyra* is common in freshwater ponds and streams, where it is a major part of "pond scum."

(a)

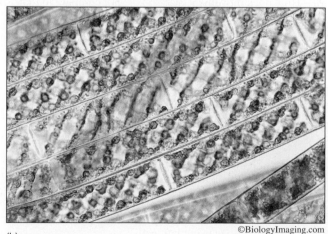

(b)

**Figure 4.15** *Spirogyra.* (*a*) Diagram of a *Spirogyra* cell (250×). (*b*) Light micrograph of a living *Spirogyra*. Note the spiral-shaped chloroplast for which the alga is named (200×).

### Procedure 4.11   Examine *Spirogyra*

1. Place a drop from a culture containing *Spirogyra* on a microscope slide.
2. Add a coverslip and examine *Spirogyra* with your compound microscope.
3. Sketch *Spirogyra* in the space below.

### Question 14

*a.* Filaments of some algal species have the ends of each cell linked to one other cell. In contrast, the cells of other species sometimes give rise to two cells and form forked or branched filaments. Is *Spirogyra* branched or unbranched?

*b.* In what shapes are the cells?

*c.* Do the cells have a cell wall? If so, how can you tell?

*d.* What organelles visible in *Spirogyra* are not visible in *Amoeba* and *Paramecium*?

**Procedure 4.12** Identifying an unknown organism

You will be given a slide of an unknown organism. Use what you've learned in today's lab to identify the cells as prokaryotic or eukaryotic; if eukaryotic, identify the cells as plant, animal, or protist. Complete table 4.2 before leaving the lab. If instructed to do so, turn in table 4.2 before leaving the lab.

## Why Are Cells So Small?

As you have seen throughout this laboratory exercise, all cells are small. The 37 trillion cells in our bodies are no exception; for example, a red blood cell has a volume of about 100 μm³, a neutrophil (a type of white blood cell) has a volume of about 300 μm³, an osteoblast (a cell that secretes bone) has volume of about 4000 μm³, and an adipocyte (a cell that stores fat) has a volume of about 600,000 μm³. A human egg cell is one of the largest cells in the human body; it is about 0.1 mm in diameter, which is about the same as a strand of hair. All cells exchange materials such as oxygen, wastes (e.g., the $CO_2$ that we exhale), water, and sugars with their surroundings. Cells also sense stimuli, release hormones, carry electrical impulses, and protect us from invaders (e.g., bacteria, viruses). If they do all of these things, why are they so small? The answer lies in the ratio of their volume to their surface area.

All cells have a volume and surface area. A cell's volume is the space inside the cell, and a cell's surface area is the area of the outside of the cell (i.e., the area of the cell's plasma membrane). To make an analogy with a balloon, the volume is the amount of air in the balloon, and the surface area is the area of the latex that forms the balloon.

As a cell enlarges, its volume and surface area increase. However, its surface area (i.e., units²) increases slower than does its volume (i.e., units³). This means that, as a cell enlarges, the ratio of its surface area to its volume decreases. To appreciate this, consider the volumes and surface areas of these cubes. The surface area of a cube = 6 × (length of a side)², and the volume of a cube = (length of a side)³.

As a cuboidal cell gets larger, the ratio of its surface area to its volume gets smaller.

The surface-area-to-volume ratio is important because it defines how much surface area is available to move materials into and out of the cell's volume. If a cell grows larger past a certain point, not enough materials can cross the plasma membrane to meet the needs of the cell's growing volume. This is why cells remain small.

Now consider this arrangement of cells:

### Surface area increases, volume stays constant

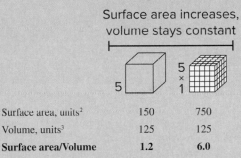

| | | |
|---|---|---|
| Surface area, units² | 150 | 750 |
| Volume, units³ | 125 | 125 |
| **Surface area/Volume** | **1.2** | **6.0** |

The presence of many small cells instead of one large cell maintains a high surface-area-to-volume ratio, thereby enabling the cells to continue to function.

Some cells, such as those in our lungs and intestine, are specialized to exchange large amounts of materials with their surroundings. These cells have extensions that are long and thin, thereby increasing their surface-area-to-volume ratio and providing a large area across which materials such as nutrients, gases, and water can move.

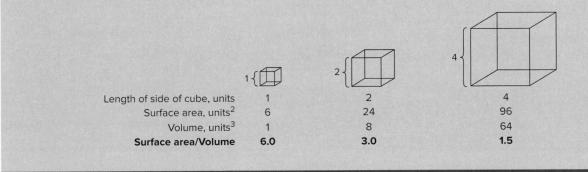

| | | | |
|---|---|---|---|
| Length of side of cube, units | 1 | 2 | 4 |
| Surface area, units² | 6 | 24 | 96 |
| Volume, units³ | 1 | 8 | 64 |
| **Surface area/Volume** | **6.0** | **3.0** | **1.5** |

1.  What is a cell?

2.  Describe the structure and function of each of these parts of a cell:

    Cell wall
    Plasma membrane
    Mitochondria
    Nucleus

    Chloroplasts
    Vacuole
    Cell wall
    Flagella

3.  What structures do plant and animal cells have in common?

4.  Would you expect a cell of a multicellular organism to be more complex than the cell of a unicellular organism? Less complex? Why?

5.  What is the purpose of using a biological stain when microscopically examining cellular components?

6.  How are eukaryotic cells different from prokaryotic cells? How are they similar?

7.  *Amoeba, Paramecuim*, and *Spirogyra* are diverse. Why, then, are they all classified as protists?

**Table 4.2**

**Using Distinguishing Features to Identify an Unknown Organism**

**Overall Description of Specimen:**

Name _____

Unknown No: _____

Lab Section: _____

**Based on the above, my unknown organism is a:**          Prokaryote          Eukaryote
      (Circle One)

**If the specimen is a eukaryote, it is a(n):**          Plant          Animal          Protist
      (Circle One)

## INQUIRY-BASED LEARNING

*How do single-celled organisms respond to environmental stimuli?*

Observation: Single-celled protists such as *Paramecium* and *Amoeba* live in water and are sensitive to environmental stimuli.

Question: How are the movements of single-celled protists affected by temperature?

a. Establish a working lab group and obtain Inquiry-Based Learning Worksheet 4 from your instructor.

b. Discuss with your group well-defined questions relevant to the preceding observation and question. Choose and record your group's best question for investigation.

c. Translate your question into a testable hypothesis and record it.

d. Outline on Worksheet 4 your experimental design and supplies needed to test your hypothesis. Ask your instructor to review your proposed investigation.

e. Conduct your procedures, record your data, answer your question, and make relevant comments.

f. Discuss with your instructor any revisions to your questions, hypothesis, or procedures. Repeat your work as needed.

## DOING BIOLOGY YOURSELF

Determine the total surface areas and volumes of the chloroplasts in a typical *Elodea* cell. Assume that each chloroplast is a sphere of 5 μm diameter. (The surface area of a sphere = $\pi d^2$; the volume of a sphere = $(\frac{4}{3})\pi r^3$.) What is the significance of these surface areas and volumes? Would it be advantageous for a cell to be filled with chloroplasts? Why or why not?

## WRITING TO LEARN BIOLOGY

What criteria might you use to distinguish colonial organisms, such as many cyanobacteria, from truly multicellular organisms?

# Solutions, Acids, and Bases
## The pH Scale

### Learning Objectives

By the end of this exercise you should be able to:
1. Apply the concepts of mole and molarity to prepare solutions.
2. Measure the pH of various liquids.
3. Demonstrate that buffers stabilize the pH of a liquid.
4. Measure the ability of commercial antacids to buffer the pH of a liquid.

Please visit **connect.mheducation.com** to review online resources tailored to this lab.

Chemicals in living systems are in solution. Biologists experiment with solutions because dissolved chemicals react more readily than solid, crystalline chemicals. A **solution** consists of a **solute**(s) dissolved in a **solvent.** For example, salt water is a solution in which salt (i.e., the solute) is dissolved in water (i.e., the solvent).

The concentration of a solute is often expressed as a percentage of the total solution (e.g., weight/volume or grams solute/100 mL solution). For example, a 3% (weight/volume) solution of sucrose is prepared by dissolving 30 g of sucrose in water for a total solution volume of 1 L (or 3 g of sucrose in water for a total volume of 100 mL).

### Question 1

*a.* How many grams of sucrose would you dissolve in water for a total volume of 500 mL to make a 5% (weight/volume) solution?

*b.* How many grams of calcium chloride would you add to water for a total volume of 500 mL to make a 5% (weight/volume) solution?

*c.* How many grams of calcium chloride would you add to water for a total volume of 100 mL to make a 5% (weight/volume) solution?

Molarity is the most common measure of concentration. To understand how to prepare a molar solution you must first understand what is meant by a **mole** of a chemical. A mole is a standard measure of the amount of a chemical—one mole of any substance has $6.02 \times 10^{23}$ molecules (Avogadro's number). One mole of NaCl and one mole of sucrose contain the same number of molecules. However, a mole of NaCl and a mole of sucrose weigh different amounts. This is because each chemical has a different **molecular weight.** The weight of 1 mole of a chemical equals that chemical's molecular weight in grams. For example, the molecular weight of water ($H_2O$) is 18 g ($2H = 2 \times 1 = 2$; $O = 16$; $16 + 2 = 18$). A mole of water weighs 18 g. A mole of NaCl weighs 58.5 g (fig. 5.1). A chemical's molecular weight is the sum of the atomic weights of its component elements.

To further understand why biologists usually prepare solutions in molar concentrations rather than as percentages you must remember that chemicals react on a molecule by molecule basis—that is, the number of molecules is more critical than the weight. It follows that expressing a solution's concentration in moles is a better measure of how much chemical is available to react. A solution that contains one mole of a chemical in 1 liter of solution has $6.02 \times 10^{23}$ molecules available and is a 1-molar (1 M) solution. For example, a liter of solution containing 58.5 g of NaCl is a 1 M solution of NaCl (fig. 5.2).

Use this formula to answer Questions 2 and 3 before coming to lab:

$$\text{Molarity (M)} = \frac{\text{moles of solute}}{\text{liters of solution}}$$

Your instructor may want to check your answers during the lab period.

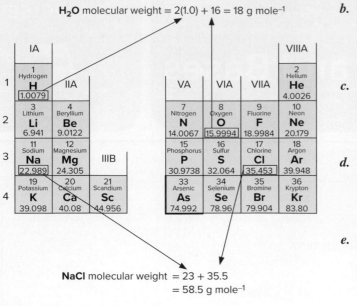

$H_2O$ molecular weight = 2(1.0) + 16 = 18 g mole$^{-1}$

**NaCl** molecular weight = 23 + 35.5
= 58.5 g mole$^{-1}$

**Figure 5.1** The atomic weights of elements are listed in the periodic table. Shown here are the portions of the periodic table that would be used to calculate the molecular weights of water ($H_2O$) and table salt (sodium chloride, NaCl). Note that g mole$^{-1}$ = grams per mole.

## Question 2

**a.** How many grams of NaCl (molecular weight = 58.5 g mole$^{-1}$) would you dissolve in water to make a 0.5 M NaCl solution with 500 mL final volume?

_____ g

**b.** How many grams of NaCl (molecular weight = 58.5 g mole$^{-1}$) would you dissolve in water to make a 50 mM NaCl solution with 500 mL final volume?

_____ g

**c.** How many grams of sucrose (molecular weight = 342 g mole$^{-1}$) would you dissolve in water to make a 0.22 M sucrose solution with 1 L final volume?

_____ g

**d.** How many grams of sucrose (molecular weight = 342 g mole$^{-1}$) would you dissolve in water to make a 0.22 mM sucrose solution with 100 mL final volume?

_____ g

**e.** How many grams of calcium chloride ($CaCl_2$; molecular weight = 111 g mole$^{-1}$) would you dissolve in water to make a 0.111 M $CaCl_2$ solution with 1 L final volume?

_____ g

**f.** How many grams of calcium chloride ($CaCl_2$; molecular weight = 111 g mole$^{-1}$) would you dissolve in water to make a 0.2 M $CaCl_2$ solution with 200 mL final volume?

_____ g

**g.** If you were presented with 2 L of a 2 M sucrose stock solution, how many grams of sugar would be in a 100 mL aliquot?

_____ g

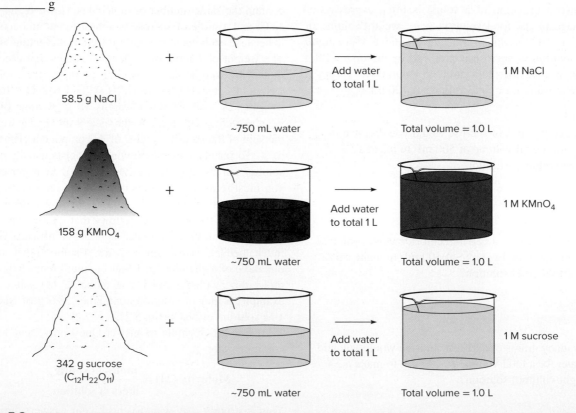

**Figure 5.2** Preparing 1.0 M solutions of sodium chloride (NaCl; molecular weight = 58.5 g mole$^{-1}$), potassium permanganate ($KMnO_4$; molecular weight = 158 g mole$^{-1}$), and sucrose ($C_{12}H_{22}O_{11}$; molecular weight = 342 g mole$^{-1}$). Each of these solutions contains the same number of units of solutes (i.e., $6.02 \times 10^{23}$ molecules).

*h.* To prepare the 5% sucrose solution called for in Question 1a, how many moles of sugar did you add? What was the molarity of that solution?

*i.* To prepare the 5% calcium chloride solution called for in Question 1b, how many moles of calcium chloride did you add? What was the molarity of that solution?

*j.* How many milliliters of a 2 M sucrose solution would contain 1 mole of sucrose?

## Dilutions

To save time and space, biologists often prepare commonly used solutions in concentrated forms called **stock solutions.** These stock solutions are then diluted with water to make new solutions having a desired molarity. This process is called **dilution.**

Dilution involves spreading a given amount of solute throughout a larger solution. The number of moles of solute doesn't change when a solution is diluted but the volume of solution containing those moles increases. This means that the product of the initial volume ($V_i$) and initial molarity ($M_i$) must equal the product of the final volume ($V_f$) and final molarity ($M_f$):

$$V_i M_i = V_f M_f$$

where

$$V_i = \textbf{initial volume}$$
$$M_i = \textbf{initial molarity}$$
$$V_f = \textbf{final volume}$$
$$M_f = \textbf{final molarity}$$

Let's now use this simple equation to solve a dilution problem. Suppose we want to know how much water to add to 25 mL of a 0.50 M KOH solution to produce a solution having a KOH concentration of 0.35 M. In this case,

$$M_i = \textbf{0.5 M}$$
$$V_i = \textbf{25 mL}$$
$$M_f = \textbf{0.35 M}$$
$$V_f = \textbf{?}$$

We can now solve the problem:

$$V_i M_i = V_f M_f$$
$$\textbf{(25 mL)(0.5 M)} = \textbf{(}V_f\textbf{)(0.35 M)}$$
$$V_f = \textbf{35.7 mL}$$

The initial volume ($V_i$) was 25 mL, so we must subtract 25 mL from 35.7 mL to get our answer: 35.7 mL − 25 mL = 10.7 mL of water to produce a KOH solution having a concentration of 0.35 M.

### Question 3

*a.* How many milliliters of concentrated (18 M) sulfuric acid ($H_2SO_4$) are required to prepare 750 mL of 3 M sulfuric acid?

*b.* How would you prepare 100 mL of 0.4 M $MgSO_4$ from a stock solution of 2 M $MgSO_4$?

*c.* How many milliliters of water would you add to 100 mL of 1.0 M HCl to prepare a final solution of 0.25 M HCl?

## ACIDS AND BASES

One of the most important applications of molarity involves the concentration of hydrogen ions ($H^+$) in a solution. Pure water is the standard by which all other solutions are compared because pure water is an ionically neutral solution. This neutrality is not due to the absence of ions, but rather to the equal concentrations of positive and negative ions. When the oxygen of water pulls hard enough on an electron from one of its hydrogens, two ions form:

$$H_2O \leftrightarrow H^+ + OH^-$$

This dissociation of water is rare and reversible, but it happens often enough for the concentration of $H^+$ in pure water to be $10^{-7}$ M. The solution is neutral because the concentration of $OH^-$ is also $10^{-7}$ M. The sum of $H^+$ and $OH^-$ ions will always equal $10^{-14}$.

**Acids** are molecules that release hydrogen ions ($H^+$) when dissolved in water. Acids increase the concentration of $H^+$ in a solution. **Bases** are molecules that remove $H^+$ from solution. Bases decrease the concentration of $H^+$ in a solution. Acids neutralize bases, and bases neutralize acids.

When the concentration of $H^+$ increases, the concentration of $OH^-$ becomes proportionately less. For example, hydrochloric acid (HCl) quickly ionizes in water and increases the concentration of $H^+$; therefore, HCl is an acid. In contrast, sodium hydroxide (NaOH) is a base because it ionizes and increases the concentration of $OH^-$, thereby lowering the relative proportion of $H^+$. Thus, if enough acid is added to water to raise the $H^+$ concentration to $10^{-6}$ M, the $OH^-$ concentration would decrease to $10^{-8}$ M.

By general agreement, the scale we use to measure acidity is the **pH scale** (*pH* stands for the potential of hydrogen ions). The pH is the negative logarithm of the concentration of $H^+$; that is,

$$pH = -\log [H^+]$$

As pH goes up, the concentration of $H^+$ goes down. (The brackets indicate concentration of hydrogen ions.) The pH scale ranges from 0 ($-\log 10^\infty$; most acidic) to 14 ($-\log 10^{-14}$; most basic). On this scale, pure water has a pH of 7 ($-\log 10^{-7}$); pH values less than 7 are acidic, whereas those above 7 are basic (fig. 5.3).

Figure 5.3 shows the pHs of some common (and a few not-so-common) substances. The pH scale is a logarithmic scale; each unit represents a change of tenfold. Thus, a lime with a pH of 2 is ten times more acidic than an apple with a pH of 3 and 100 times more acidic than a tomato having a pH of 4. Each decrease of 1.0 pH unit represents a tenfold increase in acidity. Each increase of 1.0 pH unit represents a tenfold decrease in acidity.

**Question 4**

*a.* Vinegar has a pH of 3, and household ammonia has a pH of 11. Is the concentration of $H^+$ greater in the vinegar or ammonia?

*b.* How many times different is the concentration?

## MEASURING pH

A convenient way of measuring the pH of a solution is with **pH paper.** pH paper is treated with a chemical indicator that changes colors depending on the concentration of $H^+$ in

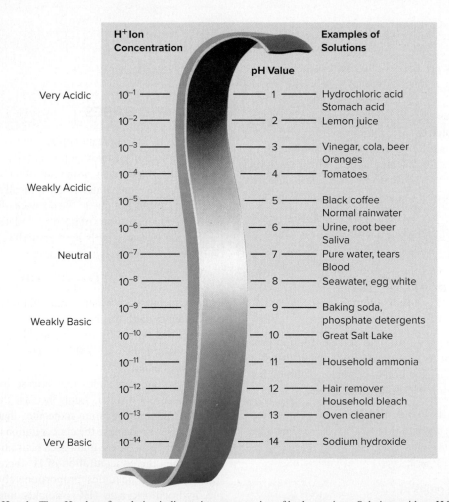

**Figure 5.3** The pH scale. The pH value of a solution indicates its concentration of hydrogen ions. Solutions with a pH less than 7 are acidic, whereas those with a pH greater than 7 are basic. The pH scale is logarithmic: A pH change of 1 means a tenfold change in the concentration of hydrogen ions. Thus, lemon juice is 100 times more acidic than tomato juice, and seawater is 10 times more basic than pure water, which has pH of 7.

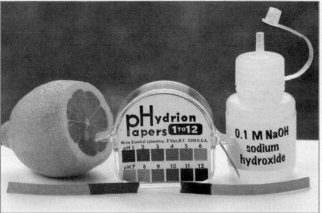

©BiologyImaging.com

**Figure 5.4** Indicator pH paper is embedded with chemicals that change color according to the pH of a solution. According to the color chart provided on the container of pH paper, the lemon juice sampled with the paper strip on the left has a pH of 2. The pH test strip on the right indicates that the sodium hydroxide solution has a pH of 12.

the solution that it has contacted (fig. 5.4). The color chart on the container of pH paper relates the color of the pH paper to the pH of the solution. Here are some examples of pH indicators:

| Indicator | Range | Color Change |
|---|---|---|
| Methyl violet | 0.2–3.0 | yellow to blue-violet |
| Bromophenol blue | 3.0–4.6 | yellow to blue |
| Methyl red | 4.4–6.2 | red to yellow |
| Litmus | 4.5–8.3 | red to blue |
| Bromcresol purple | 5.2–6.8 | yellow to purple |
| Phenol red | 6.8–8.0 | yellow to red |
| Thymol blue | 8.0–9.6 | yellow to blue |
| Phenolphthalein | 8.3–10.0 | colorless to red |

**SAFETY FIRST** Before coming to lab you were asked to read this exercise so you would know what to do and be aware of safety issues. In the space below, briefly list the safety issues associated with today's procedures. If you have questions about these issues, contact your laboratory assistant before starting work.

Handle all of the solutions carefully. Although some are harmless (e.g., water, milk), others are caustic and can stain clothes and burn your skin.

## Procedure 5.1   Measure the pH of liquids

Use pH papers to measure the pH of the following liquids. Be as accurate as possible and use a fresh piece of pH paper or pH dipstick for each test.

Vinegar _____

Skim milk _____

Apple juice _____

Grapefruit juice _____

Buttermilk _____

Black coffee _____

Sprite _____

Household bleach _____

Mixture of Sprite and baking soda _____

10 mM hydrochloric acid _____

1.0 mM hydrochloric acid _____

0.01 mM hydrochloric acid _____

Distilled water _____

Tap water _____

Dissolved aspirin _____

Soap solution _____

Shampoo _____

Mouthwash _____

Deodorant _____

Check your measurements of the hydrochloric acid solutions by comparing them with calculations using the following formula. For example,

$$pH = -\log[H^+]$$
$$10 \text{ mM HCl} = 10^{-2} \text{ M HCl}$$
$$pH = -\log[10^{-2}]$$
$$pH = 2$$

**Question 5**
Are your measured pH values similar to the calculated pH values? What are possible sources of error?

## Buffers

In most organisms, the pH is kept relatively constant by **buffers,** which are mixtures of a weak acid and a weak base that can combine with a strong acid or base to limit changes in pH. That is, buffers absorb excess $H^+$ as the pH decreases or release $H^+$ as the pH increases. Buffers minimize changes in pH (fig. 5.5). The addition of a small amount of acid to a buffered solution produces a small change in pH, whereas adding the same amount of acid to an unbuffered solution changes the pH drastically. Most biological fluids

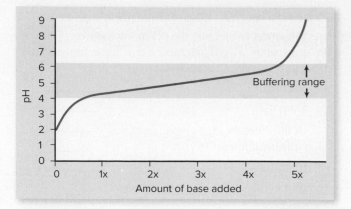

**Figure 5.5** Buffers minimize changes in pH. Adding a base to a solution will raise the pH (neutralize some of the acid present). Thus, as more and more base is added, the pH continues to rise. However, a buffer makes the curve rise or fall very slowly over a portion of the pH scale, called the "buffering range" of that buffer.

(e.g., milk, blood) contain buffers, the most important of which is bicarbonate:

$$CO_2 + H_2O \rightleftharpoons H_2CO_3 \rightleftharpoons H^+ + HCO_3^-$$

For example, human blood contains buffers that maintain a pH of 7.3–7.5; blood pH above 7.5 produces alkalosis, and blood pH below 7.3 produces acidosis. Both of these conditions can cause illness and even death.

## Procedure 5.2 Test the ability of buffers to stabilize pH

1. Obtain and label four test tubes to receive the four solutions listed in table 5.1.

2. Place 5 mL of each solution into its appropriately labeled tube.

3. Measure the pH of each of the solutions in the tubes and record these initial values in table 5.1.

4. Add 5 drops of acid (0.1 M HCl) to the first tube. Cover the tube with Parafilm and swirl the tube gently to mix the contents.

5. Measure the pH of the acidified solution and record it in table 5.1.

6. Repeat steps 4 and 5 for each of the remaining tubes. Record your results in table 5.1.

### Question 6

**a.** Compare the initial pH and the pH after adding acid to each sample. Which is the most effective buffer? Which is least effective?

**b.** What accounts for the different buffering capacities of these fluids?

**c.** What is the biological importance of what you observed?

## Procedure 5.3 Test the effectiveness of commercial antacids and other products

Commercial antacids such as *Alka-Seltzer, Rolaids,* and *Tums* claim to "neutralize stomach acid" by absorbing excess $H^+$ (produced as hydrochloric acid by the stomach; fig 5.6). To test the abilities of these products to absorb acids, do the following:

1. Use a mortar and pestle to pulverize the amount of antacid that is listed as one dose. Dissolve the crushed antacid in 100 mL of distilled water. Some of the products may require extensive stirring to get most or all of the powder to dissolve.

2. Use a pipet or 10-mL graduated cylinder to add 5 mL of the antacid solution into a test tube. Add 4 drops of the indicator bromcresol purple to the tube. Cover the tube with Parafilm and invert the tube to mix the contents.

3. Add 0.1 M hydrochloric acid (HCl) dropwise to the tube; mix after each drop. Continue this process until the solution turns yellow, indicating an acidic solution.

### Table 5.1

**Testing the Buffering Capacity of Various Solutions**

| Procedure 5.2 Solution | Initial pH | pH after Adding Acid | Procedure 5.3 Solution | Drops of Acid |
|---|---|---|---|---|
| Water | _____ | _____ | *Alka-Seltzer* | _____ |
| 0.1 M NaCl | _____ | _____ | *Rolaids* | _____ |
| Skim milk | _____ | _____ | *Tums* | _____ |
| 0.1 M phosphate buffer | _____ | _____ | | |

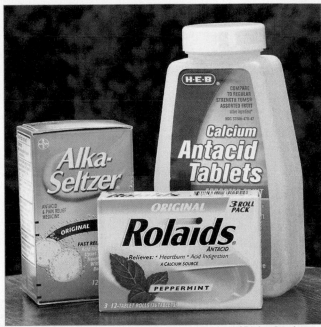

**Figure 5.6** These commercially available products soothe or prevent acid indigestion, upset stomach, and related problems. How do the products shown here do this?

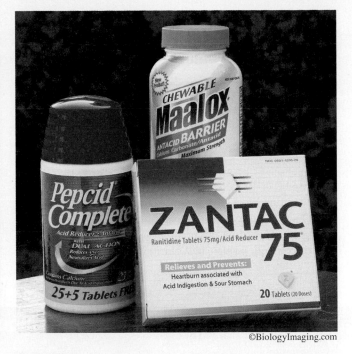

**Figure 5.7** Several over-the-counter products claim to reduce heartburn and acid indigestion. How do the products shown here do this?

4. Record in table 5.1 the number of drops of acid needed to generate the change of color. This number of drops is an index to the amount of acid ($H^+$) that the solution neutralizes before the pH drops below the yellow endpoint of bromcresol purple (pH 5.2).

**Question 7**

*a.* Which antacid neutralizes the most acid? Which neutralizes the least acid?

*b.* What is the effect of dose (for example, the size of tablets or the amount of antacid per tablet) on your results and conclusions?

*c.* Examine the packages of the products you tested. What are the active ingredients of each product? What does this tell you about how these products work?

Many people also use products such as *Zantac, Pepcid AC, Gaviscon, Prilosec, Tagamet, Pepcid AC Complete, Maalox,* and *Zantac 75* to soothe upset stomachs (fig. 5.7). Examine these products in lab, noting their claims and active ingredients. Based on your observations, write a hypothesis predicting each product's ability to absorb acid.

Now use procedure 5.3 to test each product's ability to absorb acid. List your results here.

**Question 8**

*a.* How accurate were your hypotheses?

*b.* How does each product work?

## INQUIRY-BASED LEARNING

### *How effective is* **Phillips' Milk of Magnesia** *at neutralizing acid?*

Observation: *Phillips' Milk of Magnesia* is a milky-white liquid that is a popular over-the-counter laxative and antacid. *Phillips' Milk of Magnesia* is often taken by people suffering from "acid indigestion."

Question: How effective is *Phillips' Milk of Magnesia* at neutralizing acid?

a. Establish a working lab group and obtain Inquiry-Based Learning Worksheet 5 from your instructor.

b. Discuss with your group well-defined questions relevant to the preceding observation and question. Choose and record your group's best question for investigation.

c. Translate your question into a testable hypothesis and record it.

d. Outline on Worksheet 5 your experimental design and supplies needed to test your hypothesis. Ask your instructor to review your proposed investigation.

e. Conduct your procedures, record your data, answer your question, and make relevant comments.

f. Discuss with your instructor any revisions to your question, hypothesis, or procedures. Repeat your work as needed.

## Questions for Further Study and Inquiry

*1.* What do buffers do and why are they important in biological systems?

*2.* Our stomachs secrete hydrochloric acid. Knowing the function of antacids, what do you think causes most "upset stomachs"?

*3.* The soft drink *Mr. Pibb* contains (among other things) 39 g of sucrose in 355 mL of solution. What is the molarity of this sucrose solution? What is the percentage (weight/volume) of sucrose in the solution?

*4.* Our stomachs secrete hydrochloric acid. What functions does this hydrochloric acid serve?

*5.* Suppose that the concentration of $H^+$ in Solution #1 is 10,000 times greater than in Solution #2. What can you conclude about the difference in pH of these two solutions?

*6.* What is the active ingredient in *Phillips' Milk of Magnesia?* How is this different from that of products such as *Tums?*

*7.* Normally, rainwater has a pH of approximately 5.6 because $CO_2$ in the air dissolves in rain to produce a weak solution of carbonic acid ($H_2CO_3$). What could produce "acid rain" having a pH of 4.5?

# Biologically Important Molecules

## Carbohydrates, Proteins, Lipids, and Nucleic Acids

## Learning Objectives

By the end of this exercise you should be able to:
1. Perform qualitative tests to detect the presence of biologically important carbohydrates, proteins, lipids, and nucleic acids.
2. Explain the importance of a positive and a negative control in biochemical tests.
3. Use biochemical tests to identify an unknown compound.

Please visit **connect.mheducation.com** to review online resources tailored to this lab.

**M**ost organic compounds in living organisms are **carbohydrates, proteins, lipids,** or **nucleic acids.** Each of these macromolecules is made of smaller subunits. These subunits are linked by **dehydration synthesis,** which is an energy-requiring process in which a molecule of water is removed and the two subunits are bonded covalently (fig. 6.1). Similarly, breaking the bond between the subunits requires the addition of a water molecule and releases energy. This energy-releasing process is called **hydrolysis.**

The subunits of macromolecules are held together by covalent bonds and have different structures and properties. For example, lipids (made of fatty acids) have many C—H bonds and relatively little oxygen, while proteins (made of amino acids) have amino groups (—NH$_2$) and carboxyl (—COOH) groups. These characteristic subunits and groups impart different chemical properties to macromolecules—for example, monosaccharides such as glucose are polar and soluble in water, whereas lipids are nonpolar and insoluble in water.

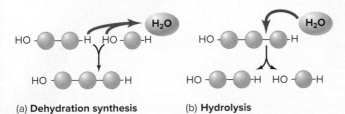

(a) **Dehydration synthesis**  (b) **Hydrolysis**

**Figure 6.1** Making and breaking macromolecules. (*a*) **Dehydration synthesis**. Many biological macromolecules are polymers formed by linking subunits together. The covalent bond between the subunits is formed by dehydration synthesis, an energy-requiring process that creates a water molecule for every bond formed. (*b*) **Hydrolysis**. Breaking the bond between subunits requires the returning of a water molecule with a subsequent release of energy, a process called hydrolysis.

## CONTROLLED EXPERIMENTS TO IDENTIFY ORGANIC COMPOUNDS

Scientists have devised several biochemical tests to identify the major types of organic compounds in living organisms. Each of these tests involves two or more treatments: (1) an **unknown solution** to be identified and (2) **controls** to provide standards for comparison. As its name implies, an unknown solution may or may not contain the substance that the investigator is trying to detect. Only a carefully conducted experiment will reveal its contents. In contrast, controls are known solutions. We use controls to validate that our procedure is detecting what we expect it to detect and nothing more. During the experiment we compare the unknown solution's response to the experimental procedure with the control's response to that same procedure.

A **positive control** contains the variable for which you are testing; it reacts positively and demonstrates the test's ability to detect what you expect. For example, if you are testing for protein in unknown solutions, then an appropriate positive control is a solution known to contain protein. A positive reaction shows that your test reacts as expected; it also shows you what a positive test looks like.

A **negative control** does not contain the variable for which you are searching. It contains only the solvent (often distilled water with no solute) and does not react in the test. A negative control shows you what a negative result looks like.

The tests you will do in this lab are **qualitative tests.** This means that the tests will show whether a particular substance is present in a sample but will not indicate how much of the substance is present.

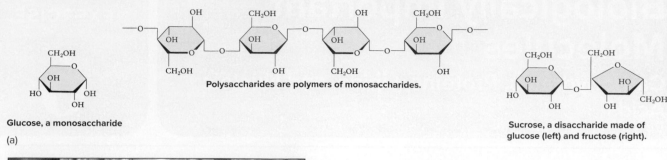

Glucose, a monosaccharide

(a)

Polysaccharides are polymers of monosaccharides.

Sucrose, a disaccharide made of glucose (left) and fructose (right).

Benedict's test for sugar

Dave Moyer

(b)

**Figure 6.2** Carbohydrates and Benedict's test. (*a*) Carbohydrates consist of subunits of mono- or disaccharides. These subunits can be combined by dehydration synthesis (see fig. 6.4) to form polysaccharides. (*b*) Benedict's test is a test for reducing sugars such as glucose. No reducing sugars produce a blue color, low levels produce a greenish color, high levels produce an orange color, and very high levels produce a red color.

## CARBOHYDRATES

### Benedict's Test for Reducing Sugars

Carbohydrates are molecules made of C, H, and O in a ratio of 1:2:1 (e.g., the chemical formula for glucose is $C_6H_{12}O_6$). Carbohydrates are made of **monosaccharides,** or simple sugars (fig. 6.2*a*). Paired monosaccharides form **disaccharides**—for example, sucrose (table sugar) is a disaccharide of glucose linked to fructose. Similarly, linking three or more monosaccharides forms a **polysaccharide** such as starch, glycogen, or cellulose (fig. 6.3).

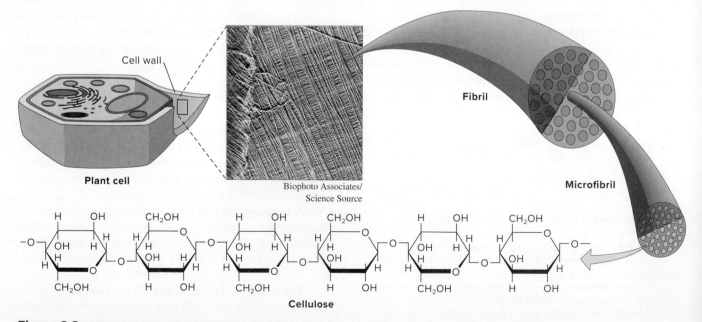

**Figure 6.3** Plant cell walls made of cellulose arranged in fibrils and microfibrils. The scanning electron micrograph shows the fibrils in a cell wall of the green alga *Chaetomorpha* (30,000×).

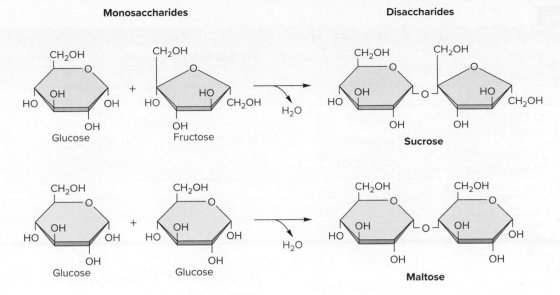

**Monosaccharides**   **Disaccharides**

Glucose + Fructose → Sucrose + $H_2O$

Glucose + Glucose → Maltose + $H_2O$

**Figure 6.4**   Dehydration synthesis is used to link monosaccharides (such as glucose and fructose) into disaccharides. The disaccharides shown here are maltose (malt sugar) and sucrose (table sugar).

## Question 1

Examine figure 6.2*a*. Which groups of a glucose molecule are involved in forming a polysaccharide? Shade the groups with a pencil.

As already mentioned, the linkage of subunits in carbohydrates, as well as other macromolecules, involves the removal of a water molecule (dehydration). Figure 6.4 depicts how dehydration synthesis is used to make maltose and sucrose, two common disaccharides.

Many monosaccharides such as glucose and fructose are **reducing sugars,** meaning that they possess free aldehyde (−CHO) or ketone (−C=O) groups that reduce weak oxidizing agents such as the copper in Benedict's reagent. **Benedict's reagent** contains cupric (copper) ion complexed with citrate in alkaline solution. Benedict's test identifies reducing sugars based on their ability to reduce the cupric ($Cu^{2+}$) ions to cuprous ($Cu^+$) oxide at basic (high) pH. Cuprous oxide is green to reddish orange (fig. 6.2*b*).

Oxidized Benedict's reagent ($Cu^{2+}$) + Reducing sugar (R-COH)

(blue)

Heat
High pH ↓

Reduced Benedict's reagent ($Cu^+$) + Oxidized sugar (R-COOH)

(green to reddish orange)

A green solution indicates a small amount of reducing sugars, and reddish orange indicates an abundance of reducing sugars. Nonreducing sugars such as sucrose produce no change in color (i.e., the solution remains blue).

 **SAFETY FIRST**  Before coming to lab, you were asked to read this exercise so you would know what to do and be aware of safety issues. In the space below, briefly list the safety issues associated with today's procedures. If you have questions about these issues, contact your laboratory assistant before starting work.

## *Procedure 6.1*   Perform the Benedict's test for reducing sugars

1.  Obtain seven test tubes and number them 1–7.

2.  Add to each tube the materials to be tested (the quantities of these materials are listed in table 6.1). Your instructor may ask you to test some additional materials. If so, include additional numbered test tubes. Add 3 mL of Benedict's solution to each tube. Swirl the contents of each tube.

3.  Place all of the tubes in a gently boiling water-bath for 3 min and observe color changes during this time.

4.  After 3 min, use a test-tube holder to remove the tubes from the water-bath. After giving the tubes ample time to cool to room temperature, record the color of their contents in table 6.1.

5.  When you are finished, dispose of the contents of each tube as instructed by your instructor.

## Table 6.1

**Solutions and Color Reactions for (1) Benedict's Test for Reducing Sugars, and (2) Iodine Test for Starch**

| Tube | Solution | Benedict's Solution | Benedict's Color Reaction | Iodine Color Reaction |
|------|----------|---------------------|---------------------------|-----------------------|
| 1 | 10 drops potato juice | 3 mL | | |
| 2 | 10 drops onion juice | 3 mL | | |
| 3 | 10 drops sucrose solution | 3 mL | | |
| 4 | 10 drops glucose solution | 3 mL | | |
| 5 | 10 drops distilled water | 3 mL | | |
| 6 | 10 drops reducing-sugar solution | 3 mL | | |
| 7 | 10 drops starch solution | 3 mL | | |
| 8 | | 3 mL | | |
| 9 | | 3 mL | | |

### Question 2

*a.* Which of the solutions is a positive control? Negative control?

*b.* Which is a reducing sugar, sucrose or glucose? How do you know?

*c.* Which contains more reducing sugars, potato juice or onion juice? How do you know?

*d.* What does this tell you about how sugars are stored in onions and potatoes?

### Iodine Test for Starch

Staining by iodine (iodine-potassium iodide, $I_2KI$, also called Lugol's iodine) distinguishes starch from monosaccharides, disaccharides, and other polysaccharides. The basis for this test is that starch is a coiled polymer of glucose; iodine interacts with these coiled molecules and becomes bluish black. Iodine does not react with carbohydrates that are not coiled and remains yellowish brown. Therefore, a bluish-black color is a positive test for starch, and a yellowish-brown color (i.e., no color change) is a negative test for starch. Glycogen, a common polysaccharide in animals, has a slightly different structure than does starch and produces only an intermediate color reaction.

### *Procedure 6.2* **Perform the iodine test for starch**

1. Obtain seven test tubes and number them 1–7.
2. Add to each tube the materials to be tested (table 6.1). Your instructor may ask you to test some additional materials. If so, include additional numbered test tubes.
3. Add three to six drops of iodine to each tube.
4. Record the color of the tubes' contents in table 6.1.

### Question 3

*a.* Which of the solutions is a positive control? Which is a negative control?

*b.* Which colors more intensely, onion juice or potato juice? Why?

*c.* In what parts of a plant is the most starch typically stored?

*d.* What are the functions of carbohydrates in living organisms?

## PROTEINS

Proteins are remarkably versatile structural molecules found in all life forms (fig. 6.5). Proteins are made of amino acids (fig. 6.6), each of which has an amino group (—$NH_2$), a

# Detecting Counterfeit Money

Merchants are always on the lookout for counterfeit money. According to the U.S. Department of the Treasury, there is about $70 million in counterfeit bills in circulation. About 1 of every 10,000 bills in circulation is counterfeit.

A quick, cost-effective way for merchants to check a bill's authenticity is to use a counterfeit detector pen, which is based on the iodine test you learned to do in this lab. Genuine bills are printed on paper made of cotton fibers, which includes no starch that reacts with iodine. Thus, when a counterfeit detector pen applies iodine-based ink on a genuine bill, the result is a yellow or colorless line. Counterfeit bills printed on copier paper, which includes much starch, turn black when marked by the iodine in the pen's ink.

To avoid having their counterfeit bills detected, sophisticated counterfeiters bleach small denomination bills (e.g., $1) and print more valuable bills on the blank paper. This is why many currencies use different-sized notes for different denominations.

andreynekrasov/123RF

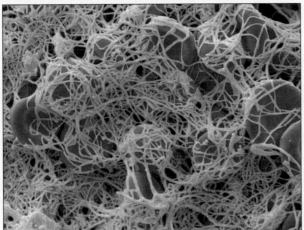

(a) Fibrin

Steve Gschmeissner/Science Photo Library/Getty Images

(b) Collagen

Craig Veltri/iStock/Getty Images

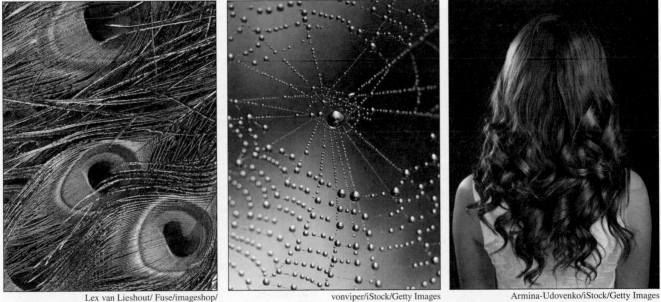

(c) Keratin

Lex van Lieshout/ Fuse/imageshop/ PunchStock/Getty Images

(d) Spider silk

vonviper/iStock/Getty Images

(e) Hair

Armina-Udovenko/iStock/Getty Images

**Figure 6.5** Common structural proteins. (*a*) Fibrin. This electron micrograph shows a red blood cell caught in threads of fibrin (800×). Fibrin is important in the formation of blood clots. (*b*) Collagen. The so-called cat-gut strings of a tennis racket are made of collagen. (*c*) Keratin. This type of protein makes up bird feathers, such as this peacock feather. (*d*) Spider silk. The web spun by this agile spider is made of protein. (*e*) Keratin. Hair is also made of protein.

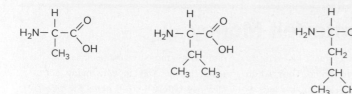

Alanine      Valine      Leucine

**Figure 6.6** Structures of three amino acids common in proteins. Each amino acid has one carbon bonded to both an amine group (—NH$_2$) and a carboxyl group (—COOH). The side chains that make each amino acid unique are shown in red.

| Table 6.2 | | |
|---|---|---|
| **Solutions and Color Reactions for the Biuret Test for Proteins** | | |
| **Tube** | **Solution** | **Color** |
| 1 | 2 mL egg albumen | |
| 2 | 2 mL honey | |
| 3 | 2 mL amino acid solution | |
| 4 | 2 mL distilled water | |
| 5 | 2 mL protein solution | |
| 6 | | |
| 7 | | |

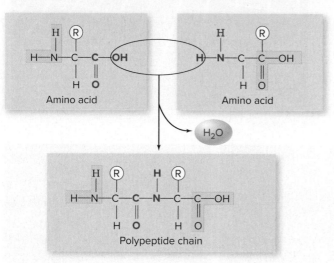

Amino acid      Amino acid

H$_2$O

Polypeptide chain

**Figure 6.7** A peptide bond joins two amino acids, and peptide bonds link many amino acids to form polypeptides, or proteins. The formation of a peptide bond (i.e., between the carbon of one amino acid's carboxyl group and the nitrogen of another amino acid's amino group) liberates a water molecule. The R in these amino acids represents a variable side chain that characterizes each type of amino acid.

carboxyl (acid) group (—COOH), and a variable side chain (—R). A **peptide bond** (fig. 6.7) forms between the amino group of one amino acid and the carboxyl group of an adjacent amino acid and is identified by a **Biuret test.** Specifically, peptide bonds (C—N bonds) in proteins complex with Cu$^{2+}$ in Biuret reagent and produce a violet color. A Cu$^{2+}$ must complex with four to six peptide bonds to produce a color; therefore, individual amino acids do not react

positively. Long-chain polypeptides (proteins) have many peptide bonds and produce a positive reaction.

Biuret reagent is a 1% solution of CuSO$_4$ (copper sulfate). A violet color is a positive test for the presence of protein; the intensity of color relates to the number of peptide bonds that react.

### Question 4
Examine figure 6.6. Shade with a pencil the reactive amino and carboxyl groups on the three common amino acids shown.

### *Procedure 6.3* Perform the Biuret test for protein

1. Obtain five test tubes and number them 1–5. Your instructor may ask you to test some additional materials. If so, include additional numbered test tubes.

2. Add the materials listed in table 6.2.

3. Add 2 mL of 2.5% sodium hydroxide (NaOH) to each tube. This raises the pH to levels required for the color reaction to work.

 Do not spill the NaOH—it is extremely caustic. Rinse your skin if it comes in contact with NaOH.

4. Add three drops of Biuret reagent to each tube and mix.

5. Record the color of the tubes' contents in table 6.2.

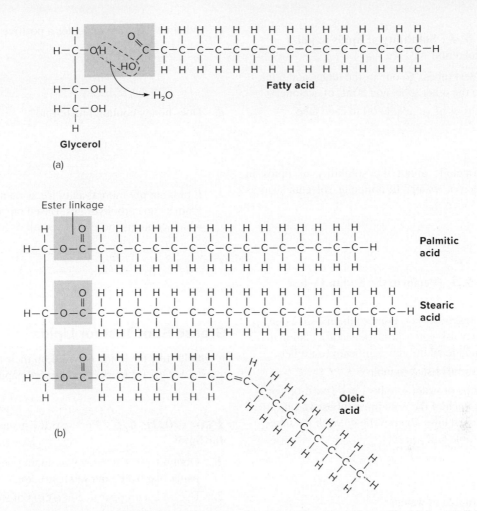

**Figure 6.8** Triglycerides, which are also called triacylglycerides, consist of glycerol and three fatty acids. (*a*) An ester linkage forms when the carboxyl group of a fatty acid links to the hydroxyl group of glycerol, with the removal of a water molecule. (*b*) Triacylglycerides are fats whose fatty acids vary in length and vary in the presence and location of carbon–carbon double bonds.

## Question 5

*a.* Which of the solutions is a positive control? Which is a negative control?

*b.* Which contains more protein (C—N bonds), egg albumen or honey? How can you tell?

*c.* Do free amino acids have peptide bonds?

*d.* What are the functions of proteins in living organisms?

## LIPIDS

Lipids include a variety of molecules that dissolve in nonpolar solvents such as ether, acetone, methanol, or ethanol, but not as well in polar solvents such as water. Triglycerides (fats) are abundant lipids made of glycerol and three fatty acids (fig. 6.8). Tests for lipids are based on a lipid's ability to selectively absorb pigments in fat-soluble dyes such as Sudan IV.

### Question 6

Examine figure 6.8. What are the reactive groups of the fatty acids?

Handle acetone carefully; it is toxic.

### Procedure 6.4 Solubility of lipids in polar and nonpolar solvents

1. Obtain two test tubes. To one of the tubes, add 5 mL of water. To the other tube, add 5 mL of acetone.
2. Add a few drops of vegetable oil to each tube.

### Question 7

What do you conclude about the solubility of lipids in polar solvents such as water? In nonpolar solvents such as acetone?

### Procedure 6.5 Perform the Sudan IV test for lipids

1. Obtain five test tubes and number them 1–5. Your instructor may ask you to test some additional materials. If so, include additional numbered test tubes.
2. Add the materials listed in table 6.3.
3. Add five drops of water to tube 1 and five drops of Sudan IV to each of the remaining tubes. Mix the contents of each tube. Record the color of the tubes' contents in table 6.3.

### Question 8

a. Is salad oil soluble in water?

b. Compare tubes 1 and 2. What is the distribution of the dye with respect to the separated water and oil?

c. What observation indicates a positive test for lipid?

d. Does honey contain much lipid?

e. Lipids supply more than twice as many calories per gram as do carbohydrates. Based on your results, which contains more calories, oil or honey?

## Grease-Spot Test for Lipids

A simpler test for lipids is based on their ability to produce translucent grease marks on unglazed paper.

### Procedure 6.6 Perform the grease-spot test for lipids

1. Obtain a piece of brown wrapping paper or brown paper bag from your lab instructor.
2. Use an eyedropper to add a drop of salad oil near a corner of the piece of paper.
3. Add a drop of water near the opposite corner of the paper.
4. Let the fluids evaporate.
5. Look at the paper as you hold it up to a light.
6. Test other food products and solutions available in the lab in a similar way and record your results in table 6.4.

### Table 6.3

**Solutions and Color Reactions for the Sudan IV Test for Lipids**

| Tube | Solution | Description of Reaction |
|------|----------|-------------------------|
| 1 | 1 mL salad oil + water | |
| 2 | 1 mL salad oil + Sudan IV | |
| 3 | 1 mL honey + Sudan IV | |
| 4 | 1 mL distilled water + Sudan IV | |
| 5 | 1 mL known lipid solution + Sudan IV | |
| 6 | | |
| 7 | | |

# Dietary Fats

We need fats in our diet. These molecules provide us with energy, help us absorb some nutrients, help keep us warm, and help us produce several hormones. Although all fats contain roughly the same amount of energy (about 9 calories per gram), they have different effects on our health. To appreciate this, examine these saturated and unsaturated fats:

In saturated fats (fig. 6.9), there are no double-bonds between carbon atoms in the fatty acids. That is, the carbon chains of the fatty acids are saturated with hydrogen. As a result, the molecules have a rigid shape and pack together tightly, thereby explaining why saturated fats are solid at room temperature. Most saturated fats come from animals; examples of saturated fats are butter and the whitish-colored fat in red meats such as bacon and steak.

In unsaturated fats (fig. 6.9), there are double-bonds between carbon atoms in the fatty acids. That is, the carbon chains of fatty acids are *not* saturated with hydrogen. If there is one such double-bond, the fat is a monounsaturated fat, and if there are two or more double-bonds, the fat is a polyunsaturated fat. The double-bonds produce a "kink" in the fatty acid that prevents the molecules from packing tightly together; this is why unsaturated fats are liquids at room temperature. Unsaturated fats come from plants; examples of unsaturated fats are corn oil, sunflower oil, and peanut oil.

Diets high in saturated fats are associated with heart disease, whereas diets high in some unsaturated fats are associated with lower levels of heart disease. Processed foods often contain large amounts of saturated fats. Why, then, do so many processed foods include saturated fats? If unsaturated fats are healthier for us, why don't manufacturers use unsaturated fats in their products?

The answer is a practical one: Unsaturated fats (i.e., plant oils) in processed food have a relatively short shelf-life and, after a while, they become rancid, thereby ruining the product. To avoid this problem, manufacturers start production with unsaturated fats, but then heat these oils and expose them to hydrogen gas. This saturates the oil; this is why so many labels on processed foods list "hydrogenated vegetable oil" as an ingredient. The "hydrogenation" process converts the unsaturated fat into a saturated fat and, in the process, gives the product a longer shelf-life.

An unintended consequence of hydrogenation is the creation of trans fats, which are also associated with elevated levels of heart disease.

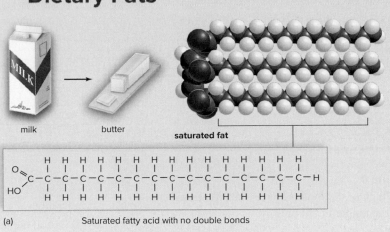

milk     butter     **saturated fat**

(a)     Saturated fatty acid with no double bonds

corn     corn oil

(b) Unsaturated fatty acid with double bonds (yellow) and a trans bond (red)

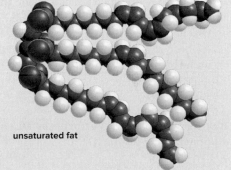

**unsaturated fat**

(a, b) McGraw Hill

**Figure 6.9** Saturated and unsaturated fatty acids. A fatty acid has a carboxyl group attached to a long hydrocarbon chain. (*a*) If the chain of carbons has no double bonds, then the fatty acid is saturated. (*b*) If there are double bonds between some of the carbons (yellow highlights), the fatty acid is unsaturated and a kink occurs in the chain. Trans fats (red highlight) possess a double bond with hydrogen atoms on either side of the carbon chain.

## Table 6.4

**Materials and Grease-Spot Reaction as a Test for Lipid Content**

| | Food Product | Description of Grease-Spot Reaction |
|---|---|---|
| 1 | | |
| 2 | | |
| 3 | | |
| 4 | | |
| 5 | | |
| 6 | | |

### Question 9

*a.* Which of the food products that you tested contain large amounts of lipid?

*b.* What are the functions of lipids in living organisms?

## NUCLEIC ACIDS

DNA and RNA are nucleic acids made of nucleotide sub-units (fig. 6.10). DNA can be identified chemically with the **Dische diphenylamine test.** Acidic conditions convert deoxyribose to a molecule that binds with diphenylamine to form a blue complex. The intensity of the blue color is proportional to the concentration of DNA.

### *Procedure 6.7* **Perform the Dische diphenylamine test for DNA**

*1.* Obtain four test tubes and number them 1–4. Your instructor may ask you to test some additional materials. If so, include additional numbered test tubes.

*2.* Add the materials listed in table 6.5.

*3.* Add 2 mL of the Dische diphenylamine reagent to each tube and mix thoroughly.

 Handle the Dische diphenylamine reagent carefully; it is toxic. Wash your hands after the procedure.

*4.* Place the tubes in a gently boiling water-bath to speed the reaction.

*5.* After 10 min, transfer the tubes to an ice bath. Gently mix and observe the color of their contents as the tubes cool. Record your observations in table 6.5.

### Question 10

*a.* How does the color compare between tubes 1 and 2? Why?

## Table 6.5

**Solutions and Color Reactions for Dische Diphenylamine Test for DNA**

| Tube | Solution | Color |
|---|---|---|
| 1 | 2 mL DNA solution | |
| 2 | 1 mL DNA solution, 1 mL water | |
| 3 | 2 mL RNA solution | |
| 4 | 2 mL distilled water | |
| 5 | | |
| 6 | | |

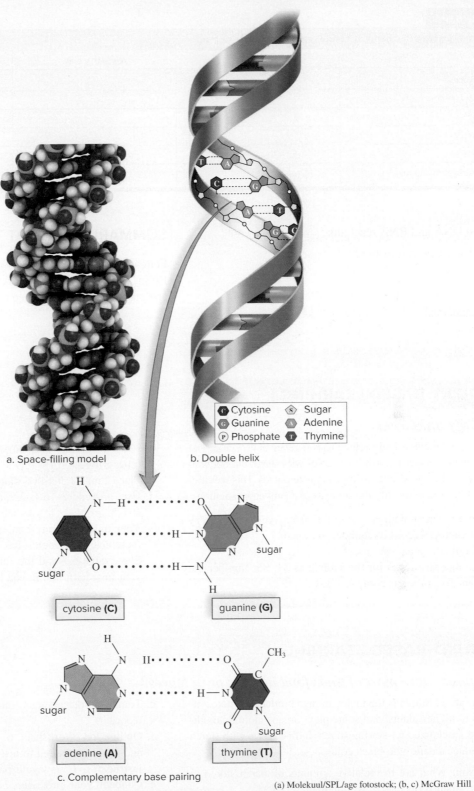

a. Space-filling model

b. Double helix

| C Cytosine | S Sugar |
| G Guanine | A Adenine |
| P Phosphate | T Thymine |

cytosine (C)          guanine (G)

adenine (A)          thymine (T)

c. Complementary base pairing

(a) Molekuul/SPL/age fotostock; (b, c) McGraw Hill

**Figure 6.10** The structure of DNA. (*a*) This image shows a three-dimensional, space-filling model of DNA. (*b*) DNA is a double-helix in which two strands of nucleotides twist about each other. (*c*) The two strands are held together by hydrogen bonds (dotted lines) between the paired bases. Adenine (A) always bonds with thymine (T), and guanine (G) always bonds with cytosine (C).

**Table 6.6**

**Summary of the Biochemical Tests Performed in this Exercise**

| Test | Tests for | Negative test | Positive test |
|---|---|---|---|
| Benedict's | | | |
| Iodine | | | |
| Biuret | | | |
| Sudan IV | | | |
| Grease | | | |
| Dische diphenylamine | | | |

*b.* Do DNA and RNA react alike? Why or why not?

*c.* What are the functions of nucleic acids in living organisms?

## SUMMARIZING WHAT YOU'VE LEARNED

For each of the following tests you performed in today's lab, list in Table 6.6 what the test is for, the appearance of a negative test, and the appearance of a positive test.

## INQUIRY-BASED LEARNING I

### Identify unknowns

Each of the previously described tests is relatively specific; that is, iodine produces a bluish-black color with starch but not with other carbohydrates, protein, lipid, or nucleic acids. This specificity can be used to identify the contents of an unknown solution.

a. Obtain an unknown solution from your laboratory instructor. Record its number in table 6.7.

b. Obtain 10 clean test tubes.

c. Number five tubes for the sample as S1–S5. Number the other five tubes as controls C1–C5.

d. Place 2 mL of your unknown solution into each of tubes S1–S5.

e. Place 2 mL of distilled water into each of tubes C1–C5.

f. Use procedures 6.1–6.5 to detect reducing sugars, starch, protein, DNA, and lipids in your unknown. Your unknown may contain one, none, or several of these macromolecules. Record your results in table 6.7. Show table 6.7 and the following report (page 71) to your instructor before you leave the lab.

## INQUIRY-BASED LEARNING II

### Do plants store most of their food in roots or in leaves?

Observation: Starch is the major storage product of photosynthesis in higher plants, and some plant organs more than others are specialized for storing starch. Iodine reacts with starch to produce a dark blue-black color.

Question: What are the relative amounts of starch stored in leaves versus roots of a flowering plant?

a. Establish a working lab group and obtain Inquiry-Based Learning Worksheet 6 from your instructor.

b. Discuss with your group well-defined questions relevant to the preceding observation and question. Choose and record your group's best question for investigation.

c. Translate your question into a testable hypothesis and record it.

d. Outline on Worksheet 6 your experimental design and supplies needed to test your hypothesis. Ask your instructor to review your proposed investigation.

e. Conduct your procedures, record your data, answer your question, and make relevant comments.

f. Discuss with your instructor any revisions to your questions, hypothesis, or procedures. Repeat your work as needed.

Name ______________________

Lab Section ______________

Unknown No. ______________

**Table 6.7**

**Chemical Testing to Identify an Unknown**

| Biochemical Test | Color | | Unknown Result |
|---|---|---|---|
| | Sample | Control | (+/−) |
| Benedict's test (reducing sugars) | | | |
| Iodine (starch) | | | |
| Biuret test (protein) | | | |
| Dische diphenylamine test (DNA) | | | |
| Sudan IV (lipid) | | | |

## Report: Identity of Unknown

Indicate which of the following are in your unknown:

Reducing sugars

Starch

Protein

DNA

Lipid

## Comments:

1. What is the importance of a positive control? What is the importance of a negative control?

2. What controls were used in each procedure that you performed in today's lab?

3. Why did you include controls in all of your tests?

4. Are controls always necessary? Why or why not?

5. What is a phospholipid? What functions do phospholipids have in cells?

6. What does a "dehydration synthesis" do?

7. Food labels list the amounts of (and calories from) carbohydrates, fats, and proteins, but not nucleic acids. Why not?

### DOING BIOLOGY YOURSELF

Design a procedure to indicate the amount of starch present in various plant tissue samples. How would you weigh your samples? How would you treat your samples? How would you quantify the iodine test?

### WRITING TO LEARN BIOLOGY

What are the limitations of these common techniques in detecting the presence of a class of macromolecules? Do biologists who study plant cells commonly use the iodine test for starch? Why or why not?

# Separating Organic Compounds
## Column Chromatography, Paper Chromatography, and Gel Electrophoresis

## Learning Objectives

By the end of this exercise you should be able to:

1. Explain how column chromatography, paper chromatography, and gel electrophoresis are used to separate compounds from mixtures.
2. Use column chromatography, paper chromatography, and gel electrophoresis to separate organic compounds from mixtures.

Please visit **connect.mheducation.com** to review online resources tailored to this lab.

Cells are a mixture of the types of organic compounds that you studied in Exercise 6 ("Biologically Important Molecules"), including carbohydrates, proteins, lipids, and nucleic acids. Biologists characterize and study these compounds to understand how organisms function. This requires that biologists separate the compounds, such as amino acids and nucleotides, from mixtures.

Biologists often use **chromatography** to separate mixtures. In this procedure, the mixture is dissolved in a fluid that moves through a matrix made of materials such as beads, paper, or a gel. During the process, the different parts of the mixture move at different speeds, causing them to separate. In today's exercise you will use column chromatography, paper chromatography, and gel electrophoresis to separate compounds from mixtures. The procedures are simple and model how these techniques are used by biologists in their research.

 **SAFETY FIRST** Before coming to lab, you were asked to read this exercise so you would know what to do and be aware of safety issues. In the space below, briefly list the safety issues associated with today's procedures. If you have questions about these issues, contact your laboratory assistant before starting work.

## PAPER CHROMATOGRAPHY

Biologists often analyze the amino acid content of samples to determine protein sequences and enzyme structures. Amino acids can be separated by partitioning them between the stationary and mobile phases of paper chromatography. The **stationary phase** is the paper fibers, and the **mobile phase** is an organic solvent that moves along the paper.

Separation by paper chromatography begins by applying a liquid sample to a small spot on an origin line at one end of a piece of chromatography paper. The edge of the paper is then placed in a solvent. As the solvent moves up the paper, any sample molecules that are soluble in the solvent will move with the solvent. However, some molecules move faster than others based on their solubility in the mobile phase and their attraction to the stationary phase. These competing factors are different for different molecular structures, so each type of molecule moves at a different speed and occurs at a different position on the finished chromatogram.

Amino acids in solution have no color but react readily with molecules of ninhydrin to form a colored product. A completed chromatogram is sprayed with a ninhydrin solution and heated to detect the amino acids. The distance of these spots from the origin is measured and used to quantify the movement of a sample. The resulting $R_f$ value (retardation factor) characterizes a known molecule in a known solvent under known conditions and is calculated as follows:

$$R_f = \frac{\textbf{Distance moved by sample}}{\textbf{Distance from origin to solvent front}}$$

### Procedure 7.1 Separate amino acids and identify unknowns by paper chromatography

*NOTE: This procedure will take about two hours, so start this procedure as soon as you can after arriving at lab and getting instructions from your lab instructor.*

1. Obtain a piece of chromatography paper 15 cm square. Avoid touching the paper with your fingers. Use gloves, tissue, or some other means to handle the paper because oils from your skin will alter the migration of the molecules on the paper.

## Table 7.1

**Chromatography Data for Determining Amino Acid Unknowns**

| Tick Mark Number | Amino Acid or Sample Number | Distance to Solvent Front | Distance Traveled by Sample | $R_f$ | Identity of Unknown |
|---|---|---|---|---|---|
| 1 | | | | | |
| 2 | | | | | |
| 3 | | | | | |
| 4 | | | | | |
| 5 | | | | | |

2. Lay the paper on a clean paper towel. Then use a pencil to draw a light line 2 cm from the bottom edge of the paper.

3. Draw five tick marks at 2.5 cm intervals from the left end of the line. Lightly label the marks 1–5 below the line.

4. Locate the five solutions available for the chromatography procedure. Three of the solutions are known amino acids. One solution is an unknown. The last solution is a plant extract or another unknown.

5. Use a wooden or glass applicator stick to "spot" one of the solutions on mark #1. To do this, dip the stick in the solution and touch it to the paper to apply a small drop (2–3 mm in diameter). Let the spot dry; then make three to five more applications on the same spot. Dry between each application. Record in table 7.1 the name of the solution next to the appropriate mark number.

6. Repeat step 5 for each of the other solutions.

7. Staple or paper clip the edges of the paper to form a cylinder with the spots on the outside and at the bottom.

8. Obtain a quart jar containing the chromatography solvent. The solvent should be 1 cm or less deep. The solvent consists of butanol, acetic acid, and water (2:1:1).

9. Place the cylinder upright in the jar (fig. 7.1). *The solvent must be below the pencil line and marks.* Close the lid to seal the jar.

10. Keep the jar out of direct light and heat. Allow the solvent to move up the paper for 2 hours (h) but not all the way to the top.

11. Open the jar and remove the chromatogram. Unclip and flatten the paper. Dry it with a fan or hair dryer. Work under a fume hood if possible to avoid breathing the solvent vapors.

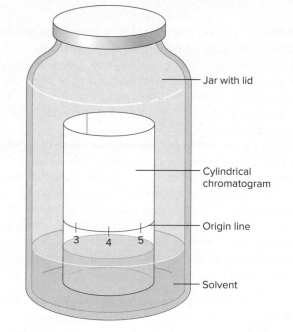

**Figure 7.1** Apparatus for paper chromatography. Numbers on the chromatogram indicate the positions of multiple samples applied to the chromatogram. The samples will move up the chromatogram along with the solvent.

12. Spray the chromatogram with ninhydrin. Carefully dry the chromatogram with warm air.

13. Circle with a pencil each of the spots. Measure the distance each of the spots has traveled and calculate the $R_f$ for each spot. Record the values in table 7.1.

14. Determine the contents of the unknown solutions by comparing $R_f$ values. Record the results in table 7.1.

# COLUMN CHROMATOGRAPHY

Column chromatography often separates molecules according to their size and shape. The procedure is simple and involves placing a sample onto a matrix that is a column of beads having tiny pores. Molecules can move through the column of beads in two ways: a fast route between the beads or a slower route through the tiny pores of the beads. Molecules too big to fit into the beads' pores move through the column quickly, whereas smaller molecules enter the beads' pores and move through the column more slowly (fig. 7.2). Movement of the molecules is analogous to going through or walking around a maze: It takes more time to walk through a maze than to walk around it.

The apparatus used for column chromatography is shown in figure 7.3 and consists of a chromatography column, a matrix, and a buffer.

- The **chromatography column** is a tube having a frit and a spout at its bottom. The frit is a membrane or porous disk that supports and keeps the matrix in the column but allows water and solutes to pass.

- The **matrix** is the material in the column that fractionates, or separates, the chemicals mixed in the sample. The matrix consists of beads having tiny pores and internal channels. The size of the beads' pores determines the matrix's **fractionation range,** which is the range of molecular weights the matrix can separate. These molecular weights are measured in units called daltons; 1 dalton $\approx$ 1 g mole$^{-1}$. Different kinds of matrices have different fractionation ranges. In today's exercise you'll use a matrix having a fractionation range of 1000 to 5000 daltons. As they move through

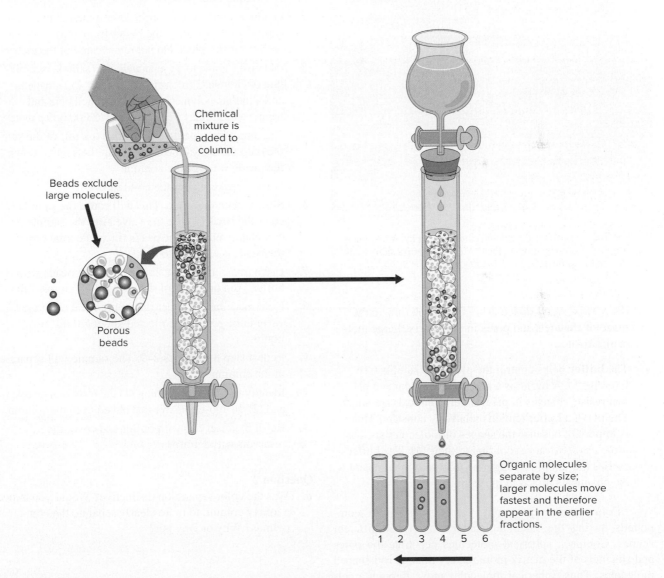

Chemical mixture is added to column.

Beads exclude large molecules.

Porous beads

Organic molecules separate by size; larger molecules move fastest and therefore appear in the earlier fractions.

1  2  3  4  5  6

**Figure 7.2**   Separation of organic molecules by column chromatography. As the solution flows through the column, the smaller molecules are slowed down as they pass through the pores of the beads. Medium-sized molecules will pass through a bead with pores less frequently, and the largest molecules will quickly flow around all the beads. The exiting fluid is collected in fractions. The first fractions collected will contain the largest molecules.

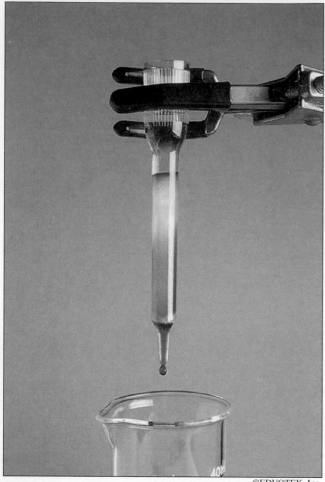

**Figure 7.3** Apparatus for column chromatography. A fraction is being collected, drop by drop, in the beaker. Smaller fractions would be collected in test tubes.

the matrix, small molecules spend much time in the maze of channels and pores in the matrix. Large molecules do not.

- The **buffer** helps control the pH of the sample (see Exercise 5). A buffer is a solution with a known pH that resists changes in pH if other chemicals are added. The pH of a buffer remains relatively constant. This is important because the shapes of molecules such as proteins often vary according to their pH. The buffer carries the sample through the matrix, which separates the chemicals mixed in the sample.

Column chromatography can also separate compounds having the same molecular weight but different shapes. Compact, spherical molecules penetrate the pores and channels of the matrix more readily than do rod-shaped molecules. Thus, spherical molecules move through a column more slowly than do rod-shaped molecules.

During column chromatography, the buffer containing the sample mixture of chemicals moves through the column and is collected sequentially in test tubes from the bottom of the column. Biologists then assay the content of the tubes to determine which tubes contain the compounds in which they are interested.

**Question 1**

In today's exercise you will isolate colored compounds from mixtures. However, it is important to note that most biological samples are colorless. How would you determine the contents of the test tubes if all of the samples were transparent?

---

*Procedure 7.2* **Separate compounds by column chromatography**

1. Label nine microtubes 1–9.

2. Obtain an apparatus for column chromatography and carefully remove all of the buffer from above the beads with a transfer pipet. Do not remove any of the matrix.

3. Obtain a sample to be separated. The sample is a mixture of Orange G (molecular weight = 452 g mole$^{-1}$) and a rodlike polymer of glucose stained blue and having a molecular weight of about 2,000,000 g mole$^{-1}$.

4. Use a transfer pipet to slowly load 0.2 mL of the sample onto the top of the beads. Drip the sample down the inside walls of the column.

5. Place a beaker under the column.

6. Slowly open the valve. This will cause the sample to enter the beads. Close the valve after the sample has completely entered the beads (i.e., when the top of the beads is exposed to air).

7. Use a transfer pipet to slowly cover the beads with buffer. Add buffer until the reservoir is almost full.

8. Hold microtube 1 under the column and open the valve until you have collected about 1.0 mL of liquid.

9. Repeat step 8 for tubes 2–9. The sample will separate in the column.

10. Identify the tubes containing (1) the most orange dye and (2) the most blue dye that eluted from the column.

11. Refill the reservoir with buffer and cover the reservoir with Parafilm.

**Question 2**

*a.* Was the color separation distinctive? Would you expect a longer column to more clearly separate the compounds? Why or why not?

*b.* Suppose your sample had consisted of a mixture of compounds having molecular weights of 50,000, 100,000, and 1,000,000 g mole$^{-1}$. What type of results would you predict? Explain your answer.

## GEL ELECTROPHORESIS

Gel electrophoresis separates molecules according to their charge, shape, and size (fig. 7.4). Buffered samples (mixtures of organic chemicals) are loaded into a Jello-like gel, after which an electrical current is placed across the gel. This current moves the charged molecules toward either the cathode or anode of the electrophoresis apparatus. The speed, direction, and distance that each molecule moves are related to its charge, shape, and size.

The apparatus for gel electrophoresis is shown in figure 7.5 and consists of an electrophoresis chamber, gel, buffer, samples, and a power supply.

- The gel is made by dissolving agarose powder (a derivative of agar) in hot buffer. When the solution cools, it solidifies into a gel having many pores that function as a molecular sieve. The gel is submerged in a buffer-filled chamber containing electrodes.

- The buffer conducts electricity and helps control the pH. The pH affects the stability and charge of the samples.

- The samples are mixtures of chemicals loaded into wells in the gel. These samples move in the gel during electrophoresis. Samples are often mixed with glycerol or sucrose to make them denser than the buffer so that they will not mix with the buffer.

- The power supply provides a direct current across the gel. Charged molecules respond to the current by moving from the sample wells into the gel. Negatively charged molecules move through the gel toward the positive electrode (anode), whereas positively charged molecules move through the gel toward the negative electrode (cathode). The greater the voltage, the faster the molecules move.

The sieve properties of the gel affect the rate of movement of a sample through the gel. Small molecules move more easily through the pores than do larger molecules. Consequently, small, compact (e.g., spherical) molecules move faster than do large, rodlike molecules. If molecules have similar shapes and molecular weights, the particles having the greatest charge move fastest and, therefore, the farthest.

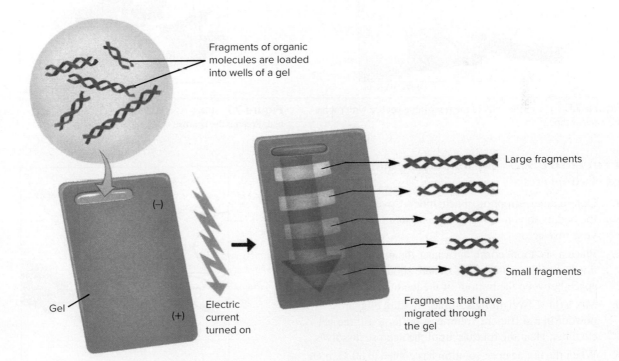

**Figure 7.4** Gel electrophoresis. This process separates DNA fragments, protein fragments, and other organic compounds by causing them to move through an electrically charged gel. Because DNA molecules are negatively charged, the electrical field will push the molecules toward the positive electrode. The fragments also move according to their size and shape, and some fragments move slowly and some move quickly. When their migration is complete, the fragments can be stained and visualized easily. In the example shown here, the DNA fragments were separated by size.

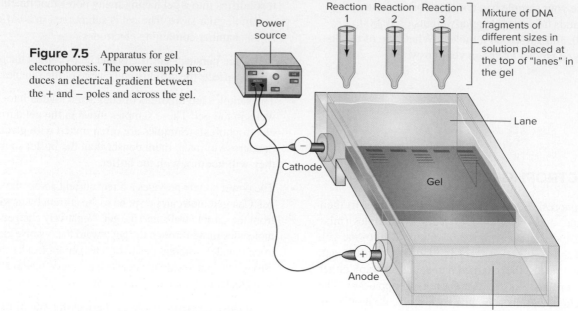

**Figure 7.5** Apparatus for gel electrophoresis. The power supply produces an electrical gradient between the + and − poles and across the gel.

Power source

Reaction 1    Reaction 2    Reaction 3

Mixture of DNA fragments of different sizes in solution placed at the top of "lanes" in the gel

Lane

Cathode

−

Gel

Anode

+

Buffer

©EDVOTEK, Inc.

**Figure 7.6** Cover the ends of the removable gel bed with rubber end-caps or tape.

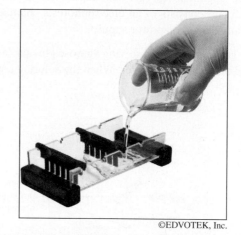

©EDVOTEK, Inc.

**Figure 7.7** Place comb near the center set of notches of the gel bed. Prepare the agarose solution and pour the gel.

## *Procedure 7.3* Separate organic molecules by gel electrophoresis

1. Obtain an electrophoresis chamber. Cover the ends of the bed as shown in figure 7.6 and demonstrated by your instructor.

2. Place a six-tooth comb in or near the middle set of notches of the gel-cast bed. There should be a small space between the bottom of the teeth and the bed.

3. Mix a 0.8% (weight by volume) mixture of agarose powder in a sufficient volume of buffer to fill the gel chamber. Heat the mixture until the agarose dissolves.

4. When the hot agarose solution has cooled to 50°C, pour the agarose solution into the gel-cast bed (fig. 7.7).

5. After the gel has solidified, gently remove the comb by pulling it straight up (fig. 7.8). Use of a plastic

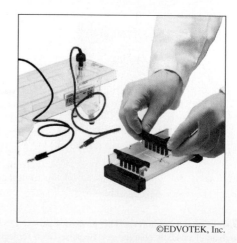

©EDVOTEK, Inc.

**Figure 7.8** After the gel solidifies, gently remove the rubber end-caps (or tape) and pull the combs straight up from the gel.

(−)

1   2   3   4   5   6

1   2   3   4   5   6

(+)

**Figure 7.9** Sketch of the wells formed in the gel by the comb as viewed from above.

spatula may help prevent tearing the gel. Use the sketch in figure 7.9 to label the wells formed in the gel by the comb.

6. Submerge the gel under the buffer in the electrophoresis chamber.

7. You will study six samples:

Sample 1: Bromophenol blue (molecular weight = 670 g mole$^{-1}$)

Sample 2: Methylene blue (molecular weight = 320 g mole$^{-1}$)

Sample 3: Orange G (molecular weight = 452 g mole$^{-1}$)

Sample 4: Xylene cyanol (molecular weight = 555 g mole$^{-1}$)

Samples 5 and 6: Unknowns

Use a micropipettor or a simple pipet and bulb to load the samples into the wells of the gel. If you use a micropipettor, your instructor will demonstrate its use. If you use a simple pipet and bulb, gently squeeze the pipet bulb to draw Sample 1 into the pipet. Be sure that the sample is in the lower part of the pipet. If the sample becomes lodged in the bulb, tap the pipet until the sample moves into the lower part.

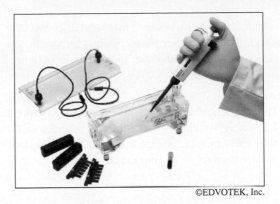

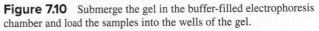
©EDVOTEK, Inc.

**Figure 7.10** Submerge the gel in the buffer-filled electrophoresis chamber and load the samples into the wells of the gel.

8. To eliminate excess air, hold the pipet above the sample tube and slowly squeeze the bulb until the sample is near the pipet's opening.

9. Place the pipet tip into the electrophoresis buffer so it is barely inside sample well 1 (fig. 7.10). Do not touch the bottom of the sample well. Maintain pressure on the pipet bulb to avoid pulling buffer into the pipet.

10. Slowly inject the sample into the sample well. Stop squeezing the pipet when the well is full. Do not release the pressure on the bulb. Remove the pipet from the well.

11. Thoroughly rinse the pipet with distilled water.

12. Load the remaining five samples into the gel by repeating steps 6–10 (fig. 7.10). Load Sample 2 into the second well, Sample 3 into the third well, etc.

13. Carefully snap on the cover of the electrophoresis chamber (fig. 7.11). The red plug in the cover should be placed on the terminal indicated by the red dot. The black plug in the cover should be placed on the terminal indicated by the black dot.

14. Insert the plug of the black wire into the black (negative) input of the power supply. Insert the plug of the red wire into the red (positive) input of the power supply.

15. Turn on the power and set the voltage at 90 V. You'll soon see bubbles forming on the electrodes. Examine the gel every 10 min.

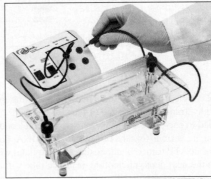
©EDVOTEK, Inc.

**Figure 7.11** Attach the safety cover, connect the power source, and run the electrophoresis.

*Separating Organic Compounds* **79**

**16.** After 30 min, turn off the power and disconnect the leads from the power source. Gently remove the cover from the chamber and sketch your results in figure 7.9.

### Question 3

**a.** Bromophenol blue, Orange G, and xylene cyanol each has a negative charge at neutral pH, whereas methylene blue has a positive charge at neutral pH. How does this information relate to your results?

**b.** Did Orange G, bromophenol blue, and xylene cyanol move the same distance in the gel? Why or why not?

**c.** What compounds do you suspect are in Samples 5 and 6? Explain your answer.

## INTERPRETING A DNA-SEQUENCING GEL

Examine figure 7.12, which includes a photograph of a gel used to determine the order, or sequence, of nucleotides in a strand of DNA. To prepare the sample for

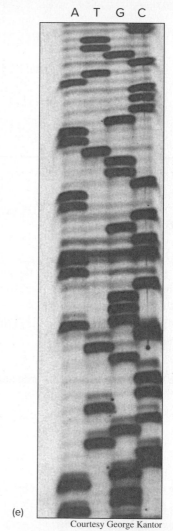

Courtesy George Kantor

**Figure 7.12** Determining the sequence of nucleotides in DNA. (*a*) Treating DNA with sodium hydroxide (NaOH) denatures double-stranded DNA into single-stranded DNA. One of the single strands of DNA to be sequenced is placed in each of four tubes. (*b*) The enzyme DNA polymerase is added to each tube along with a specific nucleotide-terminator. As polymerase replicates the DNA, the terminators are incorporated and will terminate various lengths of fragments of DNA. For example, the terminator ddATP will halt the reaction wherever adenosine occurs. The terminator ddATP (dideoxy adenosine triphosphate) will terminate a growing strand because it lacks a 3′ hydroxyl group and therefore cannot bond with the next deoxynucleotide. (*c*) Each tube will contain a sample of all possible replicated fragment lengths corresponding to the positions of that specific nucleotide. The sequences in red are the complement strands. (*d*) During electrophoresis, the fragments migrate at different rates according to their length. (*e*) The lanes of the resulting gel are labeled according to their base: A, adenine; T, thymine; G, guanine; and C, cytosine. This technique is usually referred to as "Sanger" sequencing in honor of Fred Sanger, a Nobel laureate who, in 1977, first sequenced a piece of DNA.

electrophoresis, samples of the DNA being investigated were put into each of four tubes and induced to replicate. Also, into the first tube, an adenine-terminator was added in addition to all the other nucleotides. As the complementary strand was being constructed, the terminators were occasionally incorporated wherever an adenine nucleotide was used. This random incorporation resulted in all possible lengths of DNA pieces that had an adenine on the end. The same process was conducted in the other tubes with thymine-, guanine-, and cytosine-terminators; one treatment for each of the four lanes in the gel. Electrophoresis separated the replicated pieces of DNA by size. Staining the gel revealed which lengths of the complementary DNA were terminated by which nucleotide-terminators. Examine figure 7.12d.

The gel consists of four "lanes," labeled A, T, G, and C, indicating either adenine-, thymine-, guanine-, or cytosine-terminated pieces of DNA. By "reading" down the gel, you can determine the sequence of nucleotides in the DNA. For example, the uppermost band of the gel is in the T (thymine) lane. Therefore, the first base of the piece of DNA is thymine. Similarly, the next bands are in the A, C, G, and A lanes. Thus, the first five bases of the complementary strand DNA are T-A-C-G-A. List the next seven nucleotides of the DNA as indicated by the gel. Also list the sequence of the first 12 nucleotides in the original DNA being investigated.

**Question 4**

*a.* How did the sequence of nucleotides revealed on the gel differ from the sequence of the original strand of DNA?

*b.* Assume that the gel shown in Figure 7.12d is from blood collected at a murder scene. This blood does not match that of the victim. You have collected DNA from five people suspected of murder. Gels comparable to the one shown in Figure 7.12d read as follows for each of the suspects:

Suspect #1: T-A-C-G-A-T-A-C-G-A-C
Suspect #2: T-A-C-G-A-T-A-C-G-A-C
Suspect #3: T-A-C-G-A-C-A-C-G-C-G
Suspect #4: T-A-C-G-A-T-G-C-G-A-C
Suspect #5: T-A-C-G-A-T-C-C-G-T-C

What do you conclude from this evidence?

# INQUIRY-BASED LEARNING I

## *Is there always room for improvement in laboratory techniques?*

Carefully planned and refined procedures are critical for laboratory techniques such as paper chromatography. The sensitivity of these techniques depends on a variety of factors, including the many parameters associated with timing, chemicals, measurements, and temperatures. In procedure 7.1 you were given a rather standardized protocol, but it can always be improved for specific experiments. For example, how would you modify the paper chromatography procedure to better resolve two amino acids having approximately the same $R_f$ values? What parameter(s) of the experimental design might be tweaked to increase the technique's resolving power? We suggest that you begin your investigation in the following way:

a. List the parameters involved in paper chromatography. Think carefully; many factors are involved.

b. Choose one or two parameters that you can test for their impact on the chromatography results. Why did you choose these?

c. Choose two amino acids for experimentation. Why did you choose these two?

d. Choose your treatment levels for each parameter, and then do your experiment.

e. What did you conclude?

## INQUIRY-BASED LEARNING II

### *What's the best column length for effective column chromatography?*

Observation: Column chromatography is a common means of separating molecules according to their size and shape. The movement of molecules through a column is affected by several factors, including the column's matrix and the column's length.

Question: How does the length of a column affect the separation of molecules via column chromatography?

a. Establish a working lab group and obtain Inquiry-Based Learning Worksheet 7 from your instructor.

b. Discuss with your group well-defined questions relevant to the preceding observation and question. Choose and record your group's best question for investigation.

c. Translate your question into a testable hypothesis and record it.

d. Outline on Worksheet 7 your experimental design and supplies needed to test your hypothesis. Ask your instructor to review your proposed investigation.

e. Conduct your procedures, record your data, answer your question, and make relevant comments.

f. Discuss with your instructor any revisions to your questions, hypothesis, or procedures. Repeat your work as needed.

## Questions for Further Study and Inquiry

*1.* How are column chromatography, paper chromatography, and gel electrophoresis different? How are they similar?

*2.* How would the results of electrophoresis vary if the voltage was increased? If the agarose was made more dense? Or if the migration was allowed to run twice as long?

*3.* How could knowing the nucleotide base sequence of a piece of DNA be important to a biologist?

*4.* How could knowing the nucleotide base sequence of a piece of DNA be important to someone trying to solve a crime?

*5.* How could knowing the nucleotide base sequence of a piece of DNA be important for someone studying a hereditary disease?

*6.* How could knowing the nucleotide base sequence of a piece of DNA be important for someone wanting to improve the yield of a crop such as corn?

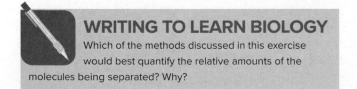

### WRITING TO LEARN BIOLOGY
Which of the methods discussed in this exercise would best quantify the relative amounts of the molecules being separated? Why?

# Spectrophotometry
## Identifying Solutes and Determining Their Concentration

## Learning Objectives

By the end of this exercise you should be able to:
1. Operate a spectrophotometer.
2. Describe the parts of a spectrophotometer and the function of each.
3. Construct absorption spectra for cobalt chloride and chlorophyll.
4. Construct and use a standard curve to determine the unknown concentration of a dissolved chemical.

Please visit **connect.mheducation.com** to review online resources tailored to this lab.

**A**bsorption and reflection of different wavelengths of light are part of your everyday experience. Different materials absorb and reflect different wavelengths of light; therefore they have different colors. When you recognize things with color, you are observing that they absorb, transmit, and reflect different wavelengths of light. For example, a red object reflects red wavelengths to your eyes and absorbs the other wavelengths.

Light includes the visible wavelengths of the electromagnetic spectrum and is only a small part of the total spectrum. The entire electromagnetic spectrum includes radiation with wavelengths from less than 1 to more than 1 million nanometers. Visible light represents wavelengths

between 380 and 700 nm (fig. 8.1). In this exercise you will work within the visible portion of the electromagnetic spectrum.

## Question 1

*a.* Chlorophyll reflects and transmits green light (540–560 nm) and absorbs other wavelengths. What biologically important molecules other than chlorophyll absorb and reflect certain colors?

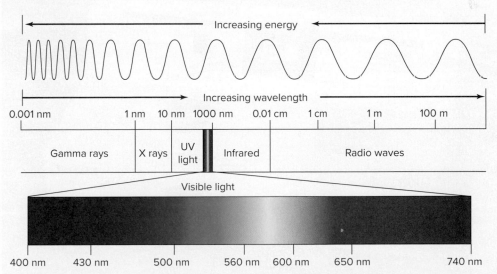

**Figure 8.1** The electromagnetic spectrum. Light is a form of electromagnetic energy and is conveniently thought of as a wave. The shorter the wavelength of light, the greater the energy. Visible light, which represents a small part of the electromagnetic spectrum, can be separated by a prism. The wavelengths of visible light range between 380 and 700 nanometers (nm). (UV stands for ultraviolet light.) Also see figure 8.5.

***b.*** Is the absorbance of light critical to these molecules' functions or just a consequence of their molecular structure? Explain your answer.

**Spectrophotometry** is one of our most versatile and precise techniques for assays ranging from blood chemistry to pollutants in lake water. Spectrophotometry is based on the principle that every different atom, molecule, or chemical bond absorbs unique wavelengths of light. For example, the nitrogenous base cytosine absorbs a different pattern of light than does adenine, uracil, or any other molecule with a different structure. As part of this pattern, some wavelengths are absorbed and some are not absorbed. Conversely, a unique pattern of light is reflected as well as absorbed by each chemical. Each chemical has a unique pattern or "fingerprint" of various wavelengths that it absorbs and/or reflects. In this exercise you will (a) determine the unique pattern of absorption for two common molecules and then (b) use spectrophotometry to measure the concentrations of one of these molecules.

## SPECTROPHOTOMETER

An instrument called a **spectrophotometer** measures the amount of light absorbed and transmitted by a dissolved chemical. For solutions we usually refer to the nonabsorbed light as transmitted light rather than reflected light. By measuring the pattern of absorbance or transmittance we can identify a chemical and its concentration.

A spectrophotometer separates white light into a spectrum of colors (wavelengths). It then directs a specific wavelength of light at a tube containing a solution that we are trying to measure. The light either is absorbed by the dissolved substance or is transmitted through the solution and exits the sample tube. The spectrophotometer compares the amount of light exiting the tube (that is, the transmitted light) with the amount entering the tube and calculates transmittance—the more solute, the lower the transmittance. The spectrophotometer also calculates the amount of light absorbed—the more solute, the higher the absorbance. The basic parts of a spectrophotometer are shown in figure 8.2.

The **light source** of a spectrophotometer produces white light, a combination of all visible wavelengths. A mixture of all colors of light is white. Spectrophotometry may involve wavelengths outside the visible range, such as ultraviolet and infrared, but the spectrophotometer must have special light sources to produce these wavelengths. In this exercise you will work only with visible light.

The **filter** is adjusted to select the wavelength that you wish to pass through the sample. The filter may be a prism that separates white light into a rainbow of colors and focuses the desired wavelength (color) on the sample. Or the filter may be a series of colored glass plates that absorb all but the selected wavelength focused on the sample.

The **sample** is a solution contained in a clear test tube or cuvette made of glass or quartz. Light of the specific

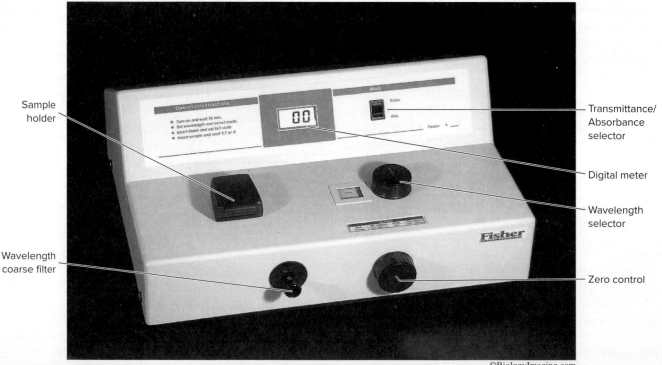

©BiologyImaging.com

**Figure 8.2** A spectrophotometer and its parts.

wavelength determined by the filter passes into the sample, where it may be completely or partially absorbed or transmitted. The amount of light absorbed depends on the amount and type of chemicals in the solution and the dimensions of the tube.

A **blank** is a test tube or cuvette containing only the solvent used to dissolve the chemical you are analyzing. A blank is used to calibrate the spectrophotometer for the solution used in your experiment. For most of your experiments, the solvent, and therefore the blank, is distilled water.

Any light transmitted through the sample exits the sample on the opposite side and is focused on a **photodetector** that converts light energy into electrical energy. The amount of electricity produced by the photodetector is proportional to the amount of transmitted light: The more light that is transmitted through the sample, the more electricity produced. The **meter** on the front of the spectrophotometer measures the electrical current produced by the photodetector and displays the results on a scale of absorbance or transmittance values. If a chemical is in solution, we usually refer to its transmittance rather than its reflectance, although both terms refer to nonabsorbed light.

The double scale on the spectrophotometer indicates that you can measure either absorbance or transmittance of radiation. **Absorbance** is the amount of radiation retained by the sample, and **transmittance** is the amount of radiation passing through the sample. In mathematical terms, transmittance is the intensity of light exiting the sample divided by the amount entering the sample. Transmittance is usually expressed as a percentage:

$$\textbf{Percentage transmittance} = (I_t/I_o) \times \textbf{100}$$

where

$I_t$ = **transmitted (exiting) light intensity**
$I_o$ = **original (entering) light intensity**

Absorbance is the logarithm of the reciprocal of transmittance and is expressed as a ratio with no units. Absorbance can be calculated with the Beer-Lambert Law:

$$\textbf{Absorbance} = \textbf{log}_{10}\,(I_o/I_t)$$

Absorbance is not a percentage and is not simply the opposite (reciprocal) of transmittance. Instead, it is a logarithmic function and has no units. This calculation makes absorbance values directly proportional to the concentration of the substance in solution. Thus, a twofold increase in absorbance indicates a twofold increase in concentration. This convenient and direct relationship between concentration and absorbance helps scientists measure an unknown concentration of a chemical.

To measure unknown concentrations of dissolved chemicals, two procedures must be done—determine a chemical's absorption spectrum and build a standard curve. As you proceed with this exercise, be sure that you understand the differences between these procedures.

# ABSORPTION SPECTRUM OF COBALT CHLORIDE

Your first task is to learn to operate a spectrophotometer while deriving the **absorption spectrum** of a common chemical, cobalt chloride ($CoCl_2$). The pattern of wavelengths absorbed by $CoCl_2$ is its "fingerprint" because it is unique to that chemical. This fingerprint is the **absorption spectrum** of the chemical and is represented as a graph relating absorbance to wavelength (fig. 8.3). Your instructor may choose to use red dye to simulate $CoCl_2$.

 **SAFETY FIRST** Before coming to lab, you were asked to read this exercise so you would know what to do and be aware of safety issues. In the space below, briefly list the safety issues associated with today's procedures. If you have questions about these issues, contact your laboratory assistant before starting work.

## *Procedure 8.1*  Determine the absorption spectrum of $CoCl_2$

1.  **When you get to the lab, turn on the spectrophotometer at your work-station. Let it warm up for 10–15 min before you begin work.**

2.  Check with your instructor or manufacturer's directions for any special instructions for using the spectrophotometer in your lab.

3.  Prepare seven solutions in spectrophotometer tubes (test tubes or cuvettes) with the mixtures of distilled water and stock solution of $CoCl_2$ (100 mg/mL) listed in table 8.1. Only tubes 0 and 6 are needed to determine the absorption spectrum. You will use the others later in the lab period.

4.  After you have prepared the dilutions, clean the outside of all the tubes with a cloth or paper towel.

5.  Verify that the solutions in each of your tubes are free of particles (dust, chalk, etc.) that might scatter the light from the spectrophotometer and produce false absorbance values. If necessary, centrifuge the tubes at 2000 rpm for 5–10 min.

### Question 2
Why is it important to clean the sample tubes?

6.  Cap the tubes and label the caps 0–6. If you label the tube rather than the caps, be sure that the labels don't interfere with light entering the tube while it is in the spectrophotometer.

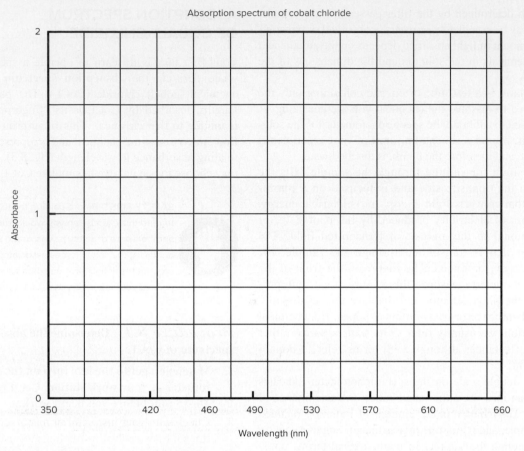

Absorption spectrum of cobalt chloride

**Figure 8.3** Absorption spectrum of cobalt chloride ($CoCl_2$).

| Table 8.1 | | | |
|---|---|---|---|
| **Volumes of Cobalt Chloride Stock Solution (100 mg/mL) and Water Used to Prepare Seven Known Dilutions** | | | |
| **Tube Number** | **Concentration of $CoCl_2$ (mg/mL)** | **$CoCl_2$ Stock (mL)** | **Distilled $H_2O$ (mL)** |
| 0 | 0 | 0 | 10.0 |
| 1 | 1 | 0.1 | 9.9 |
| 2 | 10 | 1.0 | 9.0 |
| 3 | 20 | 2.0 | 8.0 |
| 4 | 30 | 3.0 | 7.0 |
| 5 | 40 | 4.0 | 6.0 |
| 6 | 50 | 5.0 | 5.0 |

7. Place the blank (tube 0) in the sample holder of the spectrophotometer.

8. Adjust the filter to the lowest wavelength (350 nm) and read the absorbance value indicated on the meter. The distilled water blank has no color and should not absorb any visible light.

9. If the absorbance is not zero, use the zero adjust knob to calibrate the meter to zero on the absorbance scale.

10. Remove the blank and replace it with tube 6 (50 mg/mL). This is the sample you will use to determine the absorption spectrum of cobalt chloride.

11. After the meter has stabilized (5–10 sec) read the absorbance value and record the wavelength and absorbance value in table 8.2.

12. Remove tube 6 and adjust the filter to 420 nm.

## Table 8.2

### Absorbance for CoCl$_2$ (50 mg/mL)

| Wavelength | Absorbance | Wavelength | Absorbance |
|---|---|---|---|
| 350 nm | _____ | 530 nm | _____ |
| 420 nm | _____ | 570 nm | _____ |
| 460 nm | _____ | 610 nm | _____ |
| 490 nm | _____ | 660 nm | _____ |

## Table 8.3

### Absorbance Values for Six Solutions of Known Concentration (Standards) of CoCl$_2$ at the Peak Absorbance Wavelength

| Concentration of Standards (mg CoCl$_2$/mL) | Absorbance | Peak Wavelength = _____ |
|---|---|---|
| 1 | _____ | _____ |
| 10 | _____ | _____ |
| 20 | _____ | _____ |
| 30 | _____ | _____ |
| 40 | _____ | _____ |
| 50 | _____ | _____ |

13. Put the blank back into the spectrophotometer and readjust for zero absorbance at the new wavelength. **The spectrophotometer should be recalibrated with the blank often, especially when you change the wavelength.**

14. Insert tube 6 and measure its absorbance at the new wavelength. Record the absorbance in table 8.2.

15. Complete table 8.2 for the other wavelengths by repeating steps 5–12 and measuring the absorbance of the contents of tube 6.

The absorbance values in table 8.2 represent the absorption spectrum for CoCl$_2$ and are expressed best with a graph. Plot on figure 8.3 the *Absorbance* versus *Wavelength* values recorded in table 8.2. Connect the points with straight lines.

### Question 3

a. What wavelength is the peak absorbance of CoCl$_2$?

b. Why is it important to recalibrate with your blank sample often?

c. Would you expect a curve of the same shape for another molecule such as chlorophyll? Why or why not?

## THE STANDARD CURVE

A graph showing a chemical's concentration versus its absorbance of a wavelength of light is called a **standard curve,** and the relationship is a straight line. In this exercise you will construct a standard curve and then use it to determine some unknown concentrations of solutions of CoCl$_2$ prepared by your instructor. Use the six dilutions that you prepared previously to construct your standard curve for CoCl$_2$. These solutions are **standards** because their concentrations are known, and they are used to determine the concentration of an unknown solution. The absorbance of each standard is measured at the peak wavelength of the absorption spectrum for CoCl$_2$.

### *Procedure 8.2* Construct a standard curve for cobalt chloride

1. Refer to your data in table 8.2. Determine the wavelength of peak absorbance for CoCl$_2$ and set the filter of your spectrophotometer to this wavelength.

2. Insert the solvent blank (tube 0) and adjust the spectrophotometer for zero absorbance.

## Table 8.4

**Measurements of Absorbance and Concentration for Four Unknown Solutions of CoCl₂**

| Tube Number | Absorbance | Concentration (mg/mL) |
|---|---|---|
| Unknown _____ | _____ | _____ |
| Unknown _____ | _____ | _____ |
| Unknown _____ | _____ | _____ |
| Unknown _____ | _____ | _____ |

3. Replace the blank with tube 1 (1 mg CoCl₂/mL), measure its absorbance, and then record the absorbance value in table 8.3.

4. Repeat steps 2–3 for the other five tubes (standards). Be sure and check the zero-absorbance calibration with the blank before each standard measurement.

5. To construct your standard curve for CoCl₂, plot the data in table 8.3 with *Concentration (mg/mL)* on the horizontal axis and *Absorbance* on the vertical axis on the graph paper at the end of this exercise.

6. Because the relationship between concentration and absorbance is linear (directly proportional), you should draw a straight line that lies as close as possible to each data point. Do not merely connect the dots. Extremely high concentrations of a solute can produce a nonlinear segment of the standard curve. However, the concentrations used in this lab exercise are not high enough to produce such "saturation" effects.

7. If a computer and software are available, calculate and plot the line of best fit.

### Question 4
Do the plotted data points of your standard curve lie on a straight line?

### Using the Standard Curve to Measure the Unknown Concentration of a Solution

After you have created a standard curve, measuring the unknown concentration of a CoCl₂ solution is easy. Your instructor has prepared a series of numbered tubes containing unknown concentrations of CoCl₂.

### *Procedure 8.3*  **Determine unknown concentrations**

1. Obtain a tube with an unknown solution and record the tube number in table 8.4.

2. Use the blank tube (tube 0) to zero the spectrophotometer at the wavelength of peak absorbance for CoCl₂.

3. Measure the absorbance of the unknown solution and record this value in table 8.4.

4. Find this absorbance value on the vertical axis of your standard curve and draw a line from this point parallel to the horizontal axis until the line intersects the standard curve (see the example in fig. 8.4).

5. Draw a line from the intersection straight down until it intersects the horizontal axis. This point on the horizontal axis marks the concentration of the unknown solution.

6. Record the concentration of the unknown solution in table 8.4.

7. Obtain three more tubes with unknown solutions, determine their absorbance and concentration, and record the values in table 8.4. Ask your instructor to check your results.

## ABSORPTION SPECTRUM OF CHLOROPHYLL

To give you more experience with absorption spectra, your instructor has prepared a plant extract containing chlorophyll, a photosynthetic pigment (fig. 8.5). The extract was made by grinding leaves in acetone.

Acetone is flammable. Keep all solvents away from hotplates and flames at all times.

### *Procedure 8.4*  **Determine the absorption spectrum of chlorophyll**

1. Obtain a tube of the extract and prepare a blank. Your instructor will provide the solvent used for the blank.

2. Using procedure 8.3 for determining an absorption spectrum, measure the absorbance of chlorophyll for at least eight wavelengths available on your spectrophotometer as listed in table 8.5.

3. Record your results in table 8.5.

4. Graph your results (in fig. 8.6) as you did for the absorption spectrum of CoCl₂.

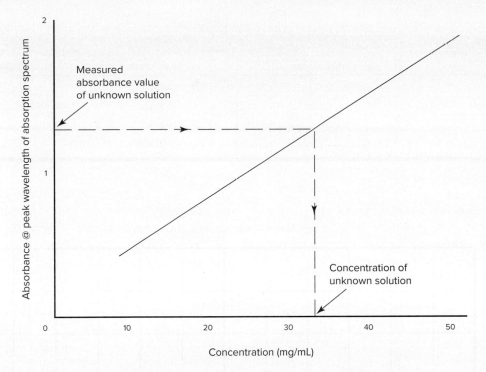

**Figure 8.4**   A standard curve showing the graphical determination of the concentration of an unknown solution of $CoCl_2$.

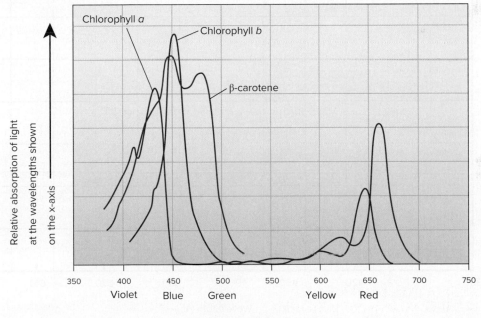

**Figure 8.5**   The absorption spectrum of chlorophyll and carotenoids. The peaks represent wavelengths of sunlight absorbed by the two common forms of photosynthetic pigment, chlorophylls *a* and *b*, and by the carotenoids. Chlorophylls absorb predominantly violet-blue and red light in two narrow bands of the spectrum and reflect green light in the middle of the spectrum; this is why chlorophyll appears green. Carotenoids absorb mostly blue and green light and reflect orange and yellow light; this is why carotenoids appear orange and yellow.

**Question 5**

*a.* What is the proper blank for determining the absorption of chlorophyll in a plant extract?

*b.* Which wavelengths are least absorbed by chlorophyll?

Table 8.5

**Absorbance Values for a Plant Extract Containing Chlorophyll**

| Wavelength | Absorbance | Wavelength | Absorbance |
|---|---|---|---|
| 350 nm | _____ | 530 nm | _____ |
| 420 nm | _____ | 570 nm | _____ |
| 460 nm | _____ | 610 nm | _____ |
| 490 nm | _____ | 660 nm | _____ |

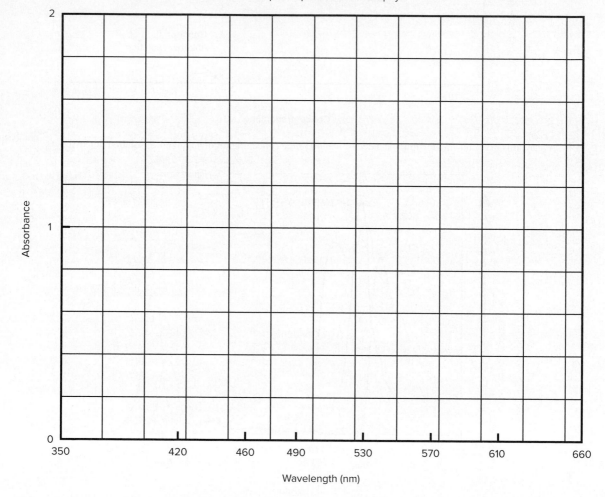

Absorption spectrum of chlorophyll

**Figure 8.6** Absorption spectrum of chlorophyll.

*c.* Which wavelengths are most absorbed by chlorophyll?

## INQUIRY-BASED LEARNING

*Biological samples are not always pure. How do contaminants impact spectrophotometry?*

Observation: Spectrophotometry uses the differential absorption of light to identify and measure the concentrations of various molecules. The molecules of interest are not the only molecules that absorb light; contaminants do also and, therefore, affect measurements.

Question: How does a contaminant such as salt affect the absorption of light by a spectrophotometric sample?

a. Establish a working lab group and obtain Inquiry-Based Learning Worksheet 8 from your instructor.

b. Discuss with your group well-defined questions relevant to the preceding observation and question. Choose and record your group's best question for investigation.

c. Translate your question into a testable hypothesis and record it.

d. Outline on Worksheet 8 your experimental design and supplies needed to test your hypothesis. Ask your instructor to review your proposed investigation.

e. Conduct your procedures, record your data, answer your question, and make relevant comments.

f. Discuss with your instructor any revisions to your questions, hypothesis, or procedures. Repeat your work as needed.

## Questions for Further Study and Inquiry

*1.* What is the difference between an absorption spectrum and a standard curve?

*2.* Can spectrophotometry be used to determine the concentration of "colorless" solutes such as salt or sugar? Explain your answer.

*3.* Why is it important to use standards or to develop a standard curve in spectrophotometry?

*4.* Why do leaves of plants appear green? Would plants grow well in greenish-yellow light? Explain your answer.

*5.* How might the basic techniques that you learned about today be used to solve crimes?

*6.* Would you expect the absorption spectra to look the same for green leaves collected during the summer and red leaves collected in autumn? Explain your answer.

## Standard Curve for CoCl₂ (Procedure 8.2)

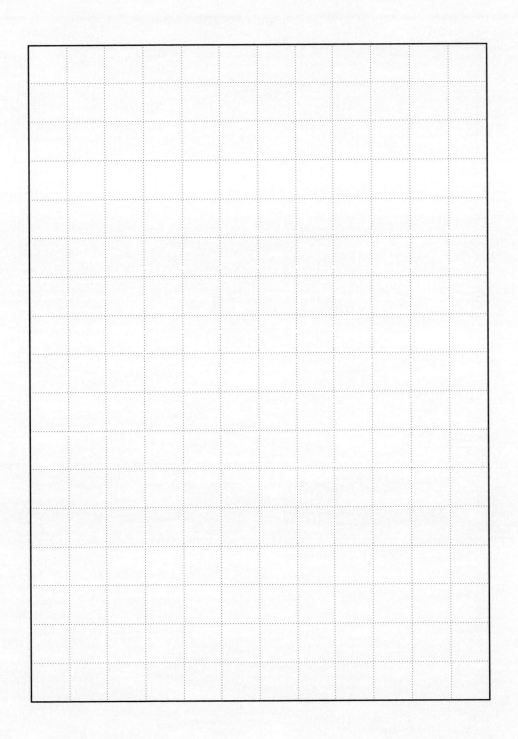

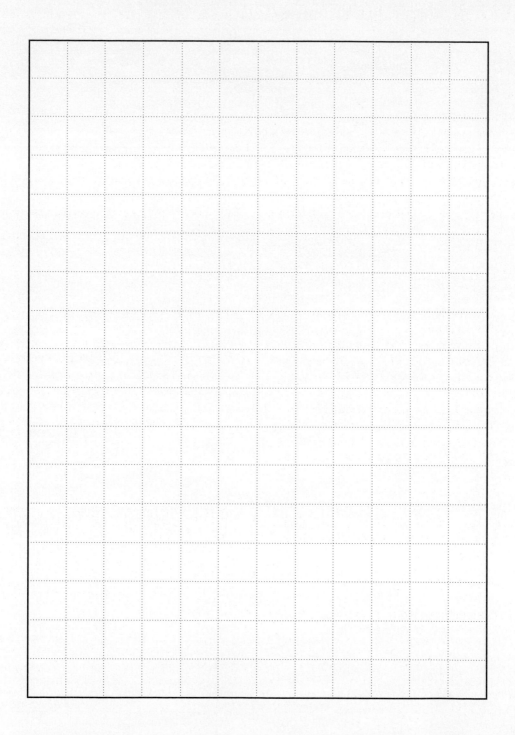

EXERCISE 8

# Diffusion and Osmosis
## Passive Movement of Molecules in Biological Systems

## Learning Objectives

By the end of this exercise you should be able to:
1.  Understand Brownian movement and its relationship to molecular movement.
2.  Explain the factors controlling a substance's direction and rate of diffusion.
3.  Determine the direction and relative rates of diffusion of molecules of different sizes.
4.  Predict the direction and rate of osmosis into and out of cells surrounded by hypotonic, hypertonic, and isotonic environments.
5.  Describe how hypotonic, hypertonic, and isotonic solutions affect the volume and integrity of blood cells and plant cells.

Please visit **connect.mheducation.com** to review online resources tailored to this lab.

All molecules display random thermal motion, or kinetic energy; this is why a dissolved molecule tends to move around in a solution. This random movement is constant, but the *net* movement of molecules from areas of high concentration to areas of low concentration continues until the distribution of molecules is homogenous throughout the solution. For example, when a dye dissolves in a container of water, the dye disperses. The rate of dispersal depends on the temperature and concentration of the dye, the size of the dye molecules, and the temperature and density of the solvent. Regardless of this rate, the dye will eventually disperse itself uniformly throughout the solution. This is easily illustrated by placing a drop or crystal of dye into a glass of water (fig. 9.1). Diffusion is critical to oxygen uptake during breathing and the distribution of nutrients and salts into and out of our cells.

**SAFETY FIRST** Before coming to lab, you were asked to read this exercise so you would know what to do and be aware of safety issues. In the space below, briefly list the safety issues associated with today's procedures. If you have questions about these issues, contact your laboratory assistant before starting work. Check with your instructor about the availability and appropriate use of safety googles for these experiments.

Charles D. Winters/McGraw Hill

**Figure 9.1** Beakers of water before and after diffusion of a dye. Random movements of water and dye molecules drive diffusion, eventually resulting in a uniform distribution of the dye. Convection currents caused by heat may also help distribute the dye in these solutions.

## BROWNIAN MOVEMENT

Heat causes **random motion** of molecules that passively moves molecules in biological systems. Although we cannot directly see molecules move, we can see small particles move after they collide with moving molecules. This motion was originally described in 1827 by Robert Browning after he prepared dead pollen grains in water and viewed them with a microscope. The dead pollen grains were moving. They were being jostled by collisions from water molecules. **Brownian movement** is visible using your microscope's high magnification. Carmine red dye mixed with soap is a good suspension of small particles. The red dye particles are small enough to vibrate when water molecules bump into them.

## Procedure 9.1  Observe Brownian movement

1. Place a small drop of a carmine red suspension on a microscope slide and cover the drop with a coverslip.

2. Focus first at low magnification; then rotate to higher power (40×). Be careful not to get dye on the objective lens.

3. Fine focus the image. At first the field of view will appear uniformly reddish gray. But with sharp focus, you will see thousands of small particles vibrating rapidly.

4. Check with your instructor to determine if your microscope has oil immersion magnification and if you need this to easily view the particles. If needed, follow his or her instructions for using this objective.

5. Leave the microscope light on. Observe any changes in motion caused by increased heat.

### Question 1

*a.* Briefly describe your observation of the moving pigment particles.

*b.* Does the movement of particles change visibly with heat? If so, how?

## DIFFUSION

In biological systems, substances often move through solutions and across membranes in a predictable direction. This passive, directional movement of molecules is **diffusion** (fig. 9.2). Each breath you take effectively balances oxygen intake and carbon dioxide removal according to the laws of diffusion (fig. 9.3). The *direction* of diffusion depends on the concentration gradient, heat, and pressure. Specifically, molecules diffuse from an area of high concentration, heat, and pressure to an area of low concentration, heat, and pressure. The *rate* of diffusion is determined by the steepness of the gradient and other characteristics of the specific molecule in question, such as its size, polarity, or solubility.

Temperature, pressure, and concentration all affect diffusion, but temperature and pressure are relatively constant in most biological systems. Therefore, concentration is usually the best predictor of a substance's direction of diffusion. But remember that temperature and pressure gradients may also affect diffusion. Nowhere is the interplay among diffusion, osmosis, and solute concentration gradients more elegant and critical than in kidney function. Review in your textbook how your kidneys help maintain your body's water and salt balance.

### Diffusion and Molecular Weight

Before your class meeting, your instructor inoculated some petri plates containing agar with either potassium permanganate (molecular weight = 158 g mole$^{-1}$), malachite green (molecular weight = 929 g mole$^{-1}$), or methylene blue (molecular weight = 374 g mole$^{-1}$).

### Question 2

Which would you predict would diffuse faster: a substance having a high molecular weight or a substance having a low molecular weight? Why?

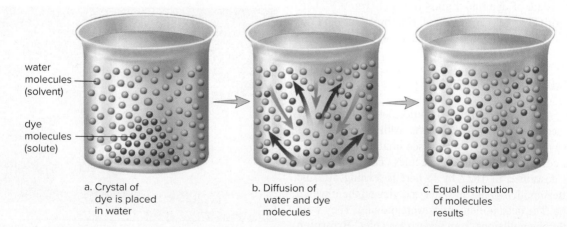

water molecules (solvent)

dye molecules (solute)

a. Crystal of dye is placed in water

b. Diffusion of water and dye molecules

c. Equal distribution of molecules results

**Figure 9.2**  Process of diffusion. Diffusion is spontaneous, and no chemical energy is required to bring it about. (*a*) When a dye crystal is placed in water, it is concentrated in one area. (*b*) The dye dissolves in the water, and there is a net movement of dye molecules from higher to lower concentration. There is also a net movement of water molecules from a higher to a lower concentration. (*c*) Eventually the water and dye molecules are equally distributed throughout the container.

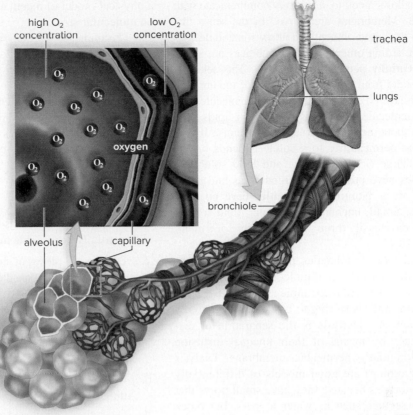

high O₂ concentration

low O₂ concentration

trachea

lungs

oxygen

bronchiole

alveolus

capillary

McGraw Hill

**Figure 9.3** Gas exchange in lungs. Oxygen ($O_2$) diffuses into the capillaries of the lungs because there is a higher concentration of oxygen in the alveoli (air sacs) than in the capillaries.

---

*Procedure 9.2* **Observe diffusion as affected by molecular weight**

1.  Examine one of the prepared agar plates and note the three halos of color. These halos indicate that the chemicals have diffused away from the two original spots and moved through the agar.

2.  Measure the halos with a ruler.

3.  Record within the outline of a petri dish your observations of the size of each halo.

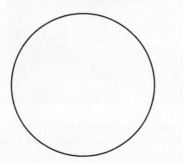

**Question 3**

*a.* Considering the different molecular weights of potassium permanganate, malachite green, and methylene blue, which should have the larger halo after the same amount of time? Why?

*b.* Do molecules stop moving when diffusion stops? Explain your answer.

## DIFFUSION AND DIFFERENTIALLY PERMEABLE MEMBRANES

Membranes surround cells and organelles and organize an immense number of simultaneous reactions. However, the barrier imposed by a cellular membrane does not isolate a

9–3

*Diffusion and Osmosis* **97**

cell. Instead, it allows a cell to *selectively* communicate with its environment. Membranes are "alive" in the sense that they respond to their environment and allow some molecules to pass while retarding others. Thus, membranes are selective and **differentially permeable** (fig. 9.4). This selective permeability results from the basic structure of membranes. They have a two-layered core of nonpolar lipid molecules that selects against molecules not readily soluble in lipids. You will learn more about membrane structure in Exercise 10.

Membrane permeability to a solute depends on the solute's size, charge (ions), polarity, and lipid solubility. **Polar molecules** have positively charged areas and negatively charged areas. **Nonpolar molecules** have no local areas of charge. Small, uncharged, nonpolar, lipid-soluble molecules pass most easily through the lipid core of a membrane (see fig. 10.4).

In general, small molecules pass through a membrane more easily than do large molecules. We can demonstrate membrane selection for molecular size by using a bag made from **dialysis tubing** to model a differentially permeable membrane. **Dialysis** is the separation of dissolved substances by means of their unequal diffusion through a differentially permeable membrane. Dialysis membranes (or tubing) are good models of differentially permeable membranes because they have small pores that allow small molecules such as water to pass, but block large molecules such as glucose. However, remember that living cell membranes also discriminate among molecules based on charge and solubility, whereas dialysis tubing used in these procedures does not. Dialysis tubing is only a physical model of a cell, and its selectivity is based only on molecular size.

Examine some dialysis tubing. Although the dried material looks like a narrow sheet of cellophane, it is a flattened, open-ended tube.

In procedure 9.3 you will use two indicators: **phenolphthalein** and **iodine.** Phenolphthalein is a pH indicator that turns red in basic solutions (see Exercise 5). Iodine is a starch indicator that changes from yellow to dark blue in the presence of starch (see Exercise 6).

---

### Procedure 9.3   Observe diffusion across a differentially permeable membrane

1. Obtain four pieces of string or dialysis clips and two pieces of water-soaked dialysis tubing approximately 15 cm long.
2. Seal one end of each bag by folding over 1–2 cm of the end. Then accordion-fold this end and tie it tightly with monofilament line or string (fig. 9.5). The ends of the tube must be sealed tightly to prevent leaks.
3. Roll the untied end of each tube between your thumb and finger to open it and form a bag.
4. Use either a graduated cylinder or pipet to fill one tube with 10 mL of water and add three drops of phenolphthalein. Seal the open end of the bag by folding the end and tying it securely.

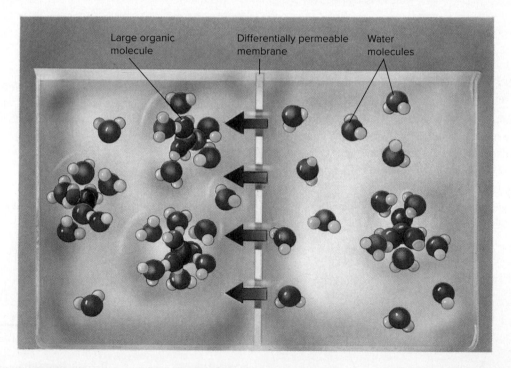

**Figure 9.4**   A differentially permeable membrane prevents the movements of some molecules but not others. Arrows indicate the movement of small molecules, such as water, from an area of high concentration to an area of lower concentration. The large molecules cannot pass through the membrane.

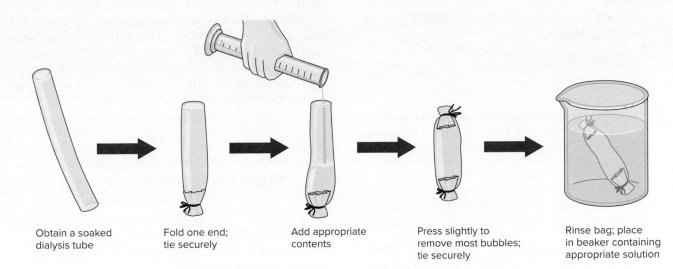

Obtain a soaked
dialysis tube

Fold one end;
tie securely

Add appropriate
contents

Press slightly to
remove most bubbles;
tie securely

Rinse bag; place
in beaker containing
appropriate solution

**Figure 9.5** Preparation of dialysis tubing as a model of a cell surrounded by a differentially permeable membrane.

5. Fill the other bag with 10 mL of starch suspension. Seal the open end of the bag by folding the end and tying it securely.

6. Gently rinse the outside of each bag in tap water.

7. Fill a beaker with 200 mL of tap water and add 10 drops of 1 M sodium hydroxide (NaOH). Submerge the dialysis bag containing phenolphthalein in the beaker.

 Do not spill the NaOH. It is extremely caustic. If it spills on your skin, wash it off immediately and clean tabletops thoroughly.

8. Fill a beaker with 200 mL of tap water and add 20–40 drops of iodine. Submerge the dialysis bag containing starch in the beaker.

9. Observe color changes in the two bags' contents and the surrounding solutions.

10. In this experiment some of the solutes can move through the membrane and some cannot. Water can freely move through the membrane, but the movement of water is not of interest in this experiment.

11. Record in figure 9.6 the color inside and outside the bags. Label the contents inside and outside the bags.

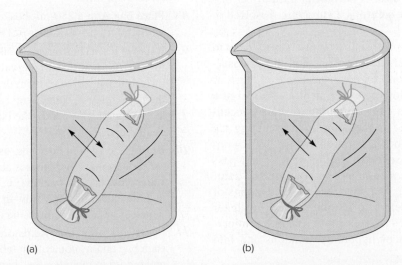

(a)

(b)

**Figure 9.6** (a) Movements and reaction of sodium hydroxide and phenolphthalein through a differentially permeable membrane. (b) Movements and reaction of iodine and starch through a differentially permeable membrane. Record the results of your experiment on this diagram.

## Question 4

*a.* Describe color changes in the two bags and their surrounding solutions.

*b.* For which molecules and ions (phenolphthalein, iodine, starch, Na⁺, OH⁻) does your experiment give evidence for passage through the semipermeable membrane?

*c.* What characteristic distinguishes those molecules and ions passing through the membrane from those that do not pass through the membrane?

## OSMOSIS AND THE RATE OF DIFFUSION ALONG A CONCENTRATION GRADIENT

The speed at which a substance diffuses from one area to another depends primarily on the concentration gradient between those areas. For example, if concentrations of a diffusing substance at the two areas differ greatly, then diffusion is rapid. Conversely, when the concentration of a substance at the two areas is equal, the diffusion rate is zero and there is no net movement of the substance.

**Osmosis** is diffusion of water across a differentially permeable membrane. Osmosis follows the same laws as diffusion but always refers to water, the principal solvent in cells. A **solution** is a homogenous, liquid mixture of two or more kinds of molecules. A **solvent** is a fluid that dissolves substances, and a **solute** is a substance dissolved in a solution.

We can simulate osmosis by using dialysis bags to model cells under different conditions and then measuring the direction and rate of osmosis. Each of the four dialysis bags in the following experiment is a model of a cell. Bag A simulates a cell containing a solute concentration that is hypotonic to the solution surrounding the bag. **Hypotonic** describes a solution with a lower concentration of solutes, especially solutes that do not pass across the surrounding membrane. In contrast, the solution surrounding the bag is hypertonic to the contents of the bag. **Hypertonic** refers to a solution with a higher concentration of solutes. Water moves across semipermeable membranes out of hypotonic solutions and into hypertonic solutions.

Bag B represents a cell whose solute concentration equals the concentration of the surrounding solution; this cell (bag B) is isotonic to its environment. **Isotonic** refers to two solutions that have equal concentrations of solutes. The solutions in bags C and D are both hypertonic to the solutions surrounding the cells—that is, the cells (bags C and D) contain higher solute concentrations than their surrounding solutions. Remember that the solute (sugar) does not pass through the membrane—only the water does.

## NOTE

*Start this experiment at the beginning of the lab period so that you'll have enough time to see results.*

---

### *Procedure 9.4*  Observe osmosis across a concentration gradient

1. Obtain eight pieces of string and four pieces of water-soaked dialysis tubing each 15 cm long. Seal one end of each tube by folding and tying it tightly.

2. Open the other end of the tube by rolling it between your thumb and finger.

3. Fill the bags with the contents shown in figure 9.7. To label each bag, insert a small piece of paper with the appropriate letter (A, B, C, or D written on it in pencil).

4. For each bag, loosely fold the open end and press on the sides to push the fluid up slightly and remove most of the air bubbles. Tie the folded ends securely, rinse the bags, and check for leaks.

5. Gently blot excess water from the outside of the bags and weigh each bag to the nearest 0.1 g.

6. Record these initial weights in table 9.1 in the first column.

7. Place bags B, C, and D in three individual beakers or one large bowl filled with 1% sucrose (fig. 9.6). Record the time.

8. Place bag A in a 250-mL beaker and fill the beaker with 150 mL of 10% sucrose. Record the time.

9. Remove the bags from the beakers at 15-min intervals for the next hour (or at intervals indicated by your instructor), gently blot them dry, and weigh them to the nearest 0.1 g. Handle the bags delicately to avoid leaks, and quickly return the bags to their respective containers.

10. During the 15-min intervals, use your knowledge of osmosis to make hypotheses about the direction of water flow in each system (i.e., into or out of bag), and the extent of water flow in each system (i.e., in which system will osmosis be most rapid?).

11. For each 15-min interval record the total weight of each bag and its contents in table 9.1. Then calculate and record in table 9.1 the change in weight since the previous weighing.

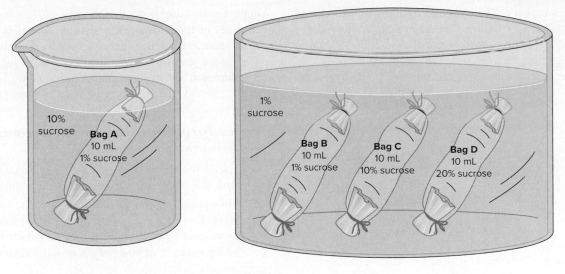

**Figure 9.7** Experimental setup for four cellular models used to measure the rate of osmosis.

## Procedure 9.5 Graph osmosis

1. Use the graph paper at the end of this exercise to construct a graph with *Total Weight (g)* versus *Time (min)*. *Total Weight* changed in response to differences in the independent variable, so *Total Weight* is the **dependent variable.** The dependent variable is always graphed on the vertical axis. *Time* is the variable that you established and actively controlled and, therefore, is the **independent variable.** The independent variable is always graphed on the horizontal axis.

2. Graphs must have a title (e.g., Relationship between Time and Weight Gain), correctly labeled axes (e.g., *Total Weight*, *Time*), a label showing measurement units (e.g., g and min), and values along each axis (e.g., 0, 15, 30, 45, 60). Include these in your graph.

3. Plot the data for total weight at each time interval from table 9.1.

4. Include the data for all four bags as four separate curves on the same graph.

**Question 5**

*a.* Did water move across the membrane in all bags containing solutions of sugar?

*b.* In which bags did osmosis occur?

*c.* A concentration gradient for water must be present in cells for osmosis to occur. Which bag represented the steepest concentration gradient relative to its surrounding environment?

| Table 9.1 | | | | | | | | | | |
|---|---|---|---|---|---|---|---|---|---|---|
| **Changes in Weight of Dialysis Bags Used as Cellular Models*** | | | | | | | | | | |
| | 0 Min | 15 Min | | 30 Min | | 45 Min | | 60 Min | | |
| | Initial Weight | Total Weight | Change in Weight | Total Weight | Change in Weight | Total Weight | Change in Weight | Total Weight | Change in Weight | |
| Bag A | | | | | | | | | | |
| Bag B | | | | | | | | | | |
| Bag C | | | | | | | | | | |
| Bag D | | | | | | | | | | |

*Each change in weight is only for the previous 15-min interval.

*d.* The steepest gradient should result in the highest rate of diffusion. Examine the data in table 9.1 for Change in Weight during the 15- and 30-min intervals. Did the greatest changes in weight occur in cells with the steepest concentration gradients? Why or why not?

*e.* If a cell (bag) containing 36% sucrose solution was placed in a 57% sucrose environment, would the water move into or out of the cell? Why? Would the bag expand?

**Question 6**
*a.* Refer to your graph. How does the slope of a segment of a curve relate to the rate of diffusion?

*b.* What influence on diffusion (i.e., temperature, pressure, concentration) causes the curves for bags C and D eventually to become horizontal (i.e., have a slope = 0)?

## WATER POTENTIAL

Plants must balance the uptake and loss of water as it moves from one part of a plant to another and in and out of cells by osmosis. However, the concentration gradient of water and solutes doesn't solely determine the direction and rate of water movement. Physical pressure influenced by cell walls and evaporation is also important. Plant physiologists refer to the combined effects of these factors as **water potential;** water flows from an area of high water potential to an area of low potential. Both high water concentration (low solute concentration) and high pressure increase water potential. Similarly, high solutes and low pressure decrease water potential. In simple terms, water flows through a plant from higher water potentials of roots toward lower water potentials of leaves. These lower potentials in leaves are created by their loss of water to the atmosphere (see Exercise 33). In the following procedure you will measure the concentration of solutes in potato cells and relate this concentration to water potential.

### *Procedure 9.6* Determine the concentration of solutes in living plant cells

1. Locate the five beakers prepared by your instructor with five concentrations of salt (NaCl) solution.
2. The cylinders of potato that you see in the solutions were all originally the same size (i.e., the same length or weight). Check the beaker labels to determine which measure of size (length or weight) you will be using as your data.
3. Record the initial values in table 9.2.
4. Carefully remove three of the potato cylinders from each solution and measure their size.
5. Record your data in table 9.2.
6. Calculate the mean change in size and record the data in table 9.2.
7. Your instructor may ask you to graph your data (see Question 7*f*). Follow his or her instructions.

**Question 7**
*a.* Which potato cylinders increased in size or weight? Why?

*b.* Which solution(s) contained a higher concentration of solutes and therefore a lower water potential than in the potato cells? Explain your answer.

**Table 9.2**

**Change in Length of Potato Cylinders Surrounded by Different Salt Concentrations**

| Concentration of Salt Solution (%) | Initial Size of Cylinders (Millimeters or Grams) | Changes in Size of Three Sample Cylinders | | Mean Change in Size |
|---|---|---|---|---|
| 0 | _____ | _____ | _____ | _____ |
| 0.9 | _____ | _____ | _____ | _____ |
| 5 | _____ | _____ | _____ | _____ |
| 10 | _____ | _____ | _____ | _____ |
| 15 | _____ | _____ | _____ | _____ |

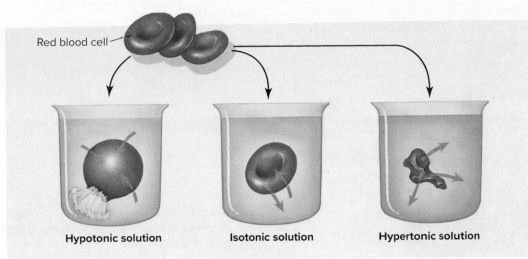

Red blood cell

**Hypotonic solution**   **Isotonic solution**   **Hypertonic solution**

**Figure 9.8**  Osmosis of water surrounding animal cells. Blue arrows indicate the movement of water. When the outer solution is hypotonic with respect to the cell, water will move into the cells and the cells will lyse; when it is hypertonic, water will move out of the cells and the cells will shrink (i.e., become crenate).

*c.* Which salt solution best approximated the water potential in the potato cells? How do you know this?

*d.* For a growing potato plant, what would you predict as the water potential of the potato relative to the soil? Relative to the leaves?

*e.* What might be some sources of error in this experiment?

*f.* How could a graph of your data help you estimate the solute concentration of potato cells?

## HEMOLYSIS OF BLOOD CELLS

Living red blood cells (erythrocytes) are good models for studying osmosis and diffusion in hypotonic, hypertonic, and isotonic solutions. Osmosis occurs when living cells are placed in a hypotonic or hypertonic environment and water diffuses into or out of the cell (fig. 9.8). For example, in the previous experiment, water moved into cells toward the low concentration of water. However, osmosis into animal cells increases the hydrostatic (i.e., water) pressure and may burst the cells because they lack cell walls. This destruction of a cell by the influx of water (causing the cell to burst) is called **lysis.** Such destruction of a red blood cell is called **hemolysis.** If water flows out of a cell into a hypertonic solution, the cell will shrivel and become crenate.

### *Procedure 9.7*  Observe hemolysis

*1.* Obtain and label three test tubes and fill them with the solutions listed in table 9.3.

*2.* Add four drops of fresh sheep's blood to each tube.

| Table 9.3 | | | |
|---|---|---|---|
| **Hemolysis of Red Blood Cells Exposed to Three Solutions with Different Solute Concentrations** | | | |
| Tube | Contents | Readable Print (yes/no) | Cell Condition (crenate/normal/lysed) |
| 1 | 5 mL 10% NaCl | _____ | _____ |
| 2 | 5 mL 0.9% NaCl | _____ | _____ |
| 3 | 5 mL distilled water | _____ | _____ |

Wash your hands thoroughly after working with blood products. Always handle sheep blood with caution and avoid skin contact.

**Question 8**

*a.* Through which test tubes could you read the printed page? Why?

*b.* Which concentration of NaCl lysed the cells?

*c.* Which of the three solutions most closely approximates the solute concentration in a red blood cell? How do you know?

**Figure 9.9** Experimental setup for determining hemolysis. Hypotonic solutions will hemolyze cells.

3. Cover each tube with Parafilm and invert the tubes to mix the contents.

4. Hold each tube in front of a printed page and determine if you can read the print through the solution (fig. 9.9). Record your results in table 9.3.

5. Obtain a microscope, slide, and coverslip.

6. Use an eyedropper or pipet to obtain one drop from each tube. Make a wet mount and examine the blood cells. Use low magnification first and then higher magnification.

7. Record in table 9.3 the cell's condition as crenate, normal, or lysed.

## PLASMOLYSIS OF PLANT CELLS

Plasmolysis is the shrinking of the cytoplasm of a plant cell in response to diffusion of water out of the cell and into a hypertonic solution (high salt concentration) surrounding the cell (fig. 9.10). During plasmolysis the cellular membrane pulls away from the cell wall (fig. 9.11).

### *Procedure 9.8* Observe plasmolysis

1. Prepare a wet mount of a thin layer of onion epidermis or *Elodea* leaf. Examine the cells.

2. Add two or three drops of 30% NaCl to one edge of the coverslip.

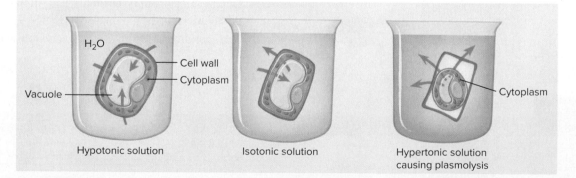

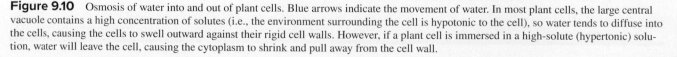

**Figure 9.10** Osmosis of water into and out of plant cells. Blue arrows indicate the movement of water. In most plant cells, the large central vacuole contains a high concentration of solutes (i.e., the environment surrounding the cell is hypotonic to the cell), so water tends to diffuse into the cells, causing the cells to swell outward against their rigid cell walls. However, if a plant cell is immersed in a high-solute (hypertonic) solution, water will leave the cell, causing the cytoplasm to shrink and pull away from the cell wall.

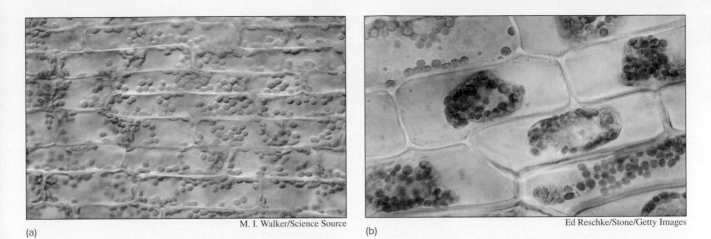

(a)                                M. I. Walker/Science Source    (b)                                Ed Reschke/Stone/Getty Images

**Figure 9.11**   (*a*) Turgid *Elodea* cells (100×). (*b*) Plasmolyzed *Elodea* cells (200×) showing the effects of exposure to a hypertonic solution.

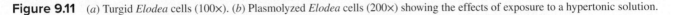

*3.*  Wick this salt solution under the coverslip by touching a piece of absorbent paper towel to the fluid at the opposite edge of the coverslip.

*4.*  Examine the cells. The cytoplasm is no longer pressed against the cell wall. This shrinkage is **plasmolysis.**

**Question 9**

*a.*  Why did the plant cells plasmolyze when immersed in a hypertonic solution?

*b.*  What can you conclude about the permeability of the cell membrane (i.e., the membrane surrounding the

cytoplasm) and vacuolar membrane (the membrane surrounding the vacuole) to water?

To observe the effects of cellular plasmolysis on a larger scale, compare petioles of celery that have been immersed overnight in distilled water or in a salt solution.

**Question 10**

What causes crispness (i.e., firmness, crunchiness) in celery?

## INQUIRY-BASED LEARNING

### *How concentrated are the solutes of plant cells?*

Observation: Water moves into and out of cells along a concentration gradient. The more solutes present in cells, the greater the tendency for water to move into the cells.

Question: What is the approximate concentration of solutes in a piece of apple?

**a.** Establish a working lab group and obtain Inquiry-Based Learning Worksheet 9 from your instructor.

**b.** Discuss with your group well-defined questions relevant to the preceding observation and question. Choose and record your group's best question for investigation.

**c.** Translate your question into a testable hypothesis and record it.

**d.** Outline on Worksheet 9 your experimental design and supplies needed to test your hypothesis. Ask your instructor to review your proposed investigation.

**e.** Conduct your procedures, record your data, answer your question, and make relevant comments.

**f.** Discuss with your instructor any revisions to your questions, hypothesis, or procedures. Repeat your work as needed.

1.  Why must particles be extremely small to demonstrate Brownian movement?

2.  What is the difference between molecular motion and diffusion?

3.  If you immerse your hand in distilled water for 15 min, will your cells lyse? Why or why not?

4.  Your data for diffusion of water across a differentially permeable membrane in response to a sucrose gradient could be graphed with *Change in Weight* on the vertical axis rather than *Total Weight*. How would you interpret the slope of the curves produced when you do this?

5.  How do cells such as algae and protists avoid lysis in fresh water?

**WRITING TO LEARN BIOLOGY**

Where in an animal might pressure affect diffusion of a substance?

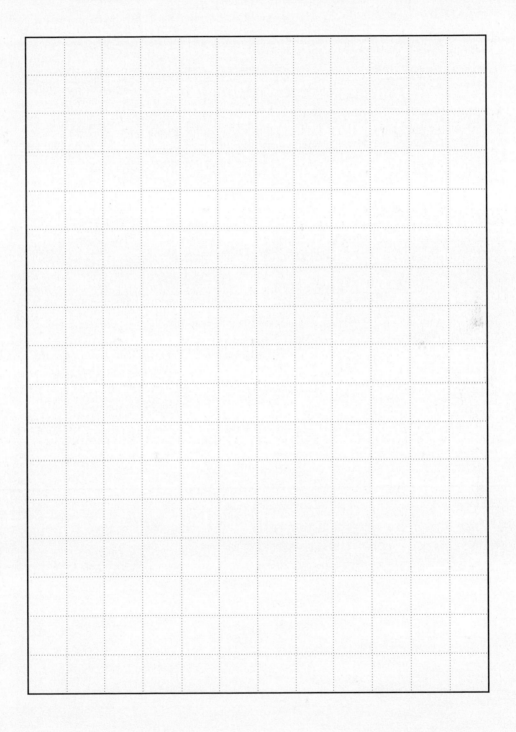

# Cellular Membranes
## Effects of Physical and Chemical Stress

Membranes separate and organize the myriad of reactions within cells and allow communication with the surrounding environment. Although they are only a few molecules thick (6–10 nm), membranes (1) retard diffusion of selected molecules; (2) house receptor molecules that detect other cells, organelles, and hormones; (3) provide sites for active and passive transport of selected molecules; (4) organize life processes by providing surfaces to accommodate chemical reactions; and (5) help maintain the integrity of cells.

As with all biological entities, the structure of a membrane reflects its function. Membranes consist of a phospholipid bilayer; attached to or embedded within this bilayer are thousands of proteins with a variety of functions. A phospholipid molecule consists of a phosphate group and two fatty acids bonded to a three-carbon, glycerol chain (fig. 10.1). Phospholipids have an unevenly distributed charge; that is, they have charged (polar) and uncharged (nonpolar) areas. In phospholipids, the phosphate group and glycerol are polar and **hydrophilic** ("water-loving"), whereas the fatty-acid chains are nonpolar and **hydrophobic** ("water-fearing"). Such molecules with two different affinities are **amphipathic,** and amphipathic phospholipids have a natural tendency to self-assemble into a double-layered sheet (fig. 10.2). In this double layer, the hydrophobic tails of lipids form the core of the membrane, and the hydrophilic phosphate groups line both surfaces. This elegant assembly is stable, is self-repairing, and resists penetration by most hydrophilic molecules.

Membranes also include proteins dispersed throughout the fluid bilayer of lipids (fig. 10.3). These proteins are not fixed in position; they move about freely and may be densely packed in some membranes and sparse in others. Carbohydrate

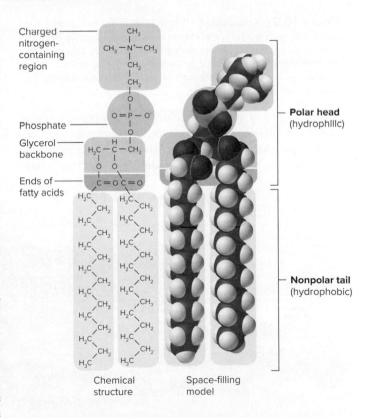

**Figure 10.1** The structure of a phospholipid. The backbone of a phospholipid is glycerol, a three-carbon alcohol. Glycerol is bonded to two fatty acids, both hydrophobic, and to one phosphate group, which is hydrophilic. Phospholipids vary by their fatty acids and by the side chains attached to the phosphate. The polar head may include glycerol, sugars, and nitrogen-containing groups as shown here.

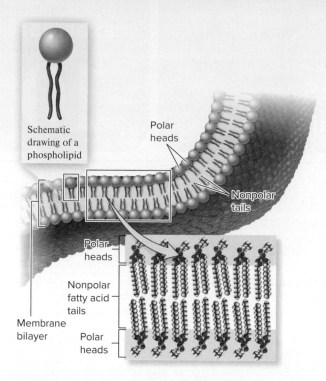

Schematic
drawing of a
phospholipid

Polar
heads

Nonpolar
tails

Polar
heads

Nonpolar
fatty acid
tails

Membrane
bilayer

Polar
heads

**Figure 10.2**   Arrangement of phospholipids in a biological membrane, such as the plasma membrane that encloses cells. The hydrophilic regions of the phospholipid face the watery environments on either side of the membrane, while the hydrophobic regions associate with each other in the interior of the membrane, thereby forming a bilayer.

To effectively regulate the flow of materials moving into and out of cells, membranes must be **selectively permeable**. The proteins embedded in the phospholipid bilayer can selectively take up or expel molecules that otherwise could not penetrate the membrane. In doing so, these proteins function as pores, permitting and often facilitating the passage of specific ions and polar molecules. In addition to forming pores and sites for active transport, membrane-bound proteins also function as enzymes and receptors that detect signals from the environment or from other cells.

The physical and chemical integrity of a membrane is crucial for the proper functioning of the cell or organelle that it surrounds. As a stable sheet of interlocking molecules, the membrane functions as a barrier to simple diffusion. In general, the permeability of a membrane to a solute depends on a combination of the solute's size, charge (ions), polarity, and lipid solubility. Small, uncharged, nonpolar, lipid-soluble molecules pass most easily through the lipid core of a membrane (fig. 10.4).

### Question 1

*a.*   What ions must routinely move across cell membranes?

*b.*   How could membranes promote the movement of ions out of or into cells? How could membranes restrict the movement of ions out of or into cells?

chains (strings of sugar molecules) are often bound to these proteins and to lipids. These chains serve as distinctive tags that identify particular types of cells. These elaborate molecular elements form the **fluid mosaic model** of membrane structure.

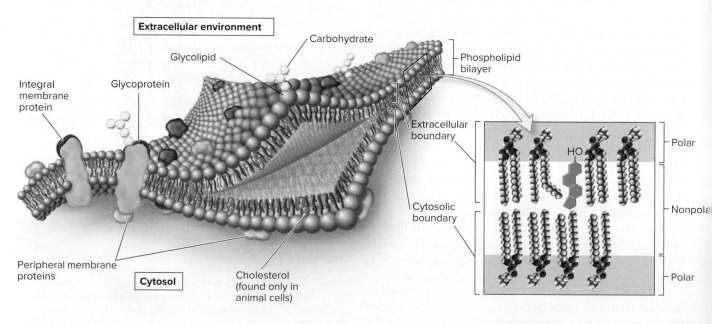

Extracellular environment

Carbohydrate

Glycolipid

Phospholipid
bilayer

Glycoprotein

Integral
membrane
protein

Extracellular
boundary

Cytosolic
boundary

Polar

Nonpola

Polar

Peripheral membrane
proteins

Cytosol

Cholesterol
(found only in
animal cells)

HO

**Figure 10.3**   Fluid-mosaic model of membrane structure. The basic framework of a plasma membrane is a phospholipid bilayer. Proteins may span the membrane and may be bound on the surface to other proteins or lipids. Proteins and lipids, which have covalently bound carbohydrates, are called glycoproteins and glycolipids, respectively.

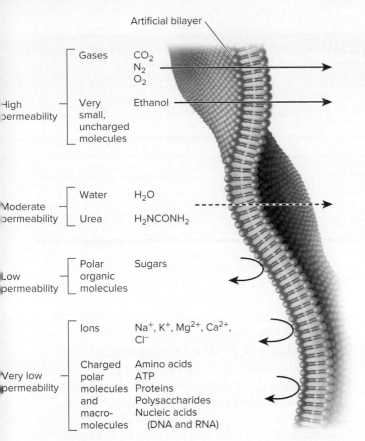

**Figure 10.4** Relative permeability of an artificial phospholipid bilayer to a variety of solutes. Solutes that easily penetrate the membrane are shown with a straight arrow that passes through the bilayer. The dotted line indicates that the membrane has moderate permeability to those solutes. For the remaining solutes shown at the bottom, the membrane is relatively impermeable.

*c.* Could a cell survive without an intact cell membrane? Explain.

## BEET CELLS AS AN EXPERIMENTAL SYSTEM

Beet tissue will be your model to investigate membrane integrity. Roots of beet (*Beta vulgaris*) contain large amounts of a reddish pigment called **betacyanin** localized almost entirely in the large central vacuoles of cells. In healthy cells, betacyanin remains inside the vacuoles, surrounded by a vacuolar membrane called the **tonoplast.** The entire cell (including the vacuole, tonoplast, and cytoplasm) is surrounded by a cell membrane and cell wall.

In two procedures you will subject beet cells to a range of temperatures and organic solvents and determine which treatments stress and damage the membranes the most. If stress damages the membranes, betacyanin will leak through the tonoplast and plasma membrane. This leakage from the stressed beet will color the surrounding water red. Thus, you can measure membrane damage by measuring the intensity of color resulting from a treatment.

**SAFETY FIRST** Before coming to lab you were asked to read this exercise so you would know what to do, and be aware of safety issues. In the space below, briefly list the safety issues associated with today's procedures. If you have questions about these issues, contact your laboratory assistant before starting work.

## THE EFFECT OF TEMPERATURE STRESS ON MEMBRANES

Membranes are sensitive to extreme temperatures. High temperatures cause violent molecular collisions that degrade a membrane as a physical barrier to diffusion. Conversely, freezing temperatures cause water to crystallize as ice and expand because of hydrogen bond alignment. This expansion and formation of ice often rupture membranes.

---

### *Procedure 10.1* Observe the effect of temperature stress on cellular membranes

1. Examine the treatments listed in table 10.1. Hypothesize which treatments will cause the most and least damage.

2. Record your rankings alongside the tube numbers in the column marked Tube Number.

3. Cut six uniform cylinders of beet using a cork borer with a 5-mm inside diameter. Trim each cylinder to exactly 15 mm in length. All of the cylinders must be the same size.

4. Place these cylinders of beet tissue in a beaker and rinse them with tap water for 2 min to wash betacyanin from the injured cells on the surface. Be sure that all of the cylinders are washed in the same way. Discard the colored rinse-water.

5. Gently place one of the six beet sections into each of six dry test tubes. Do not crush, stab, or otherwise damage the cylinders when moving them to the test tubes.

6. Label the tubes 1–6 and write the temperature treatment on each tube as listed in table 10.1.

**Table 10.1**

**The Color Intensity of Betacyanin Leaked from Damaged Cells Treated at Six Temperatures**

| Tube Number | Treatment (°C) | Color Intensity (0–10) | | Absorbance (460 nm) | |
|---|---|---|---|---|---|
| | | Working Group | Class Average | Working Group | Class Average |
| 1 | 70 | | | | |
| 2 | 55 | | | | |
| 3 | 40 | | | | |
| 4 | 20 | | | | |
| 5 | 5 | | | | |
| 6 | −5 | | | | |

**7. FOR COLD TREATMENTS:**

*a.* Place tube 5 in a refrigerator (5°C) and tube 6 in a freezer (−5°C). If a refrigerator and freezer are not available in your lab, give your labeled tubes to an assistant who will take them to another facility.

*b.* Leave tubes 5 and 6 in the cold for 30 min. While waiting, proceed with hot treatments (step 8). However, watch your time and return to steps 6c and 6d after 30 min.

*c.* After 30 min, remove the beets from the freezer and refrigerator and add 10.0 mL of distilled water at room temperature to each of the tubes.

*d.* Let the cold-treated beets soak in distilled water for 20 min. Then remove and discard the beets from tubes 5 and 6.

**8. FOR HOT TREATMENTS:**

*a.* Take the beet section out of tube 1 and immerse it in a beaker of hot water at 70°C for 1 min. If a 70°C water-bath is not available, hot tap water should be adequate, but carefully adjust the temperature to 70°C. Handle the beet gently with forceps, and don't squeeze it tightly because you may rupture the beet's cells.

*b.* After 1 min at 70°C, return the beet to tube 1 and add 10.0 mL of distilled water at room temperature.

*c.* If a 55°C water-bath is not available, slowly add ice chips or cold water to cool the beaker of hot water to 55°C. Then immerse the beet from tube 2 for 1 min. Return the beet to tube 2 and add 10.0 mL of distilled water at room temperature.

*d.* If a 40°C water-bath is not available, cool the beaker of hot water to 40°C. Then immerse the beet from tube 3 for 1 min. Return the beet to tube 3 and add 10.0 mL of distilled water at room temperature.

*e.* If a 20°C water-bath is not available, cool the beaker of hot water to 20°C. Then immerse the beet from tube 4 for 1 min. Return the beet to tube 4 and add 10.0 mL of distilled water at room temperature.

*f.* Allow the treated beets in tubes 1–4 to soak in distilled water at room temperature for 20 min. Then remove and discard the beets and measure the extent of membrane injury according to the amount of betacyanin that diffused into the water.

**9. FOR ALL SIX TEMPERATURE TREATMENTS:**

Quantify the relative color of each solution between 0 (colorless) and 10 (darkest red). If color standards are available in the lab, use them to determine relative values for the colors of your samples. Record the results for your work group in table 10.1. You may be asked to provide your results to the instructor to calculate the class averages.

Use the graph paper at the end of this exercise to graph *Temperatures* versus *Relative Color* for the class averages according to a demonstration graph provided by your instructor. Your instructor may also ask you to quantify your results further using a spectrophotometer. If so, see Exercise 8 for instructions for using a spectrophotometer. Read the absorbance of the solutions at 460 nm and record your results and the class average results in table 10.1. Then graph *Temperature* versus *Absorbance* for the class averages.

**Question 2**

*a.* Which temperature damaged membranes the most? Which the least? How do you know?

*b.* In general, which is more damaging to membranes, extreme heat or extreme cold? Why?

**Table 10.2**

| | | Color Intensity (0–10) | | Absorbance (460 nm) | |
| | | Working Group | Class Average | Working Group | Class Average |
|---|---|---|---|---|---|

**The Color Intensity of Betacyanin Leaked from Damaged Cells Treated with Various Concentrations of Two Organic Solvents**

| Tube Number | Treatment | Working Group | Class Average | Working Group | Class Average |
|---|---|---|---|---|---|
| 1 | 1% acetone | | | | |
| 2 | 25% acetone | | | | |
| 3 | 50% acetone | | | | |
| 4 | 1% methanol | | | | |
| 5 | 25% methanol | | | | |
| 6 | 50% methanol | | | | |
| 7 | Isotonic saline | | | | |

*c.* If the results of this experiment are easily observed with the unaided eye, why use a spectrophotometer?

*d.* The beets were subjected to cold temperatures longer than to hot temperatures to make sure that the beet sections were thoroughly treated. Why does the freezing treatment require more time?

*e.* How accurate were your hypothesized rankings for the temperature treatments?

## THE EFFECT OF ORGANIC SOLVENT STRESS ON MEMBRANES

Organic solvents, such as acetone and methanol, dissolve lipids. If the lipids in membranes dissolve when exposed to acetone and/or methanol, then the colored cell contents of beets will leak out of the damaged cells and color the surrounding solution.

 Organic solvents are flammable. Extinguish all open flames and heating elements before doing the following procedure. Do not pour organic solvents down the drain. Dispose of them properly. Use a fume hood as directed by your instructor.

**Question 3**

Which of the organic liquids (acetone or methanol) do you predict will damage membranes the most? Why?

### *Procedure 10.2*   Observe the effect of organic solvents on cellular membranes

*1.* Examine the treatments listed in table 10.2. Hypothesize which treatments will cause the most and least damage.

*2.* Note your rankings alongside the tube numbers in the column marked Tube Number.

*3.* Cut seven uniform cylinders of beet using a cork borer with a 5-mm inside diameter. Trim each cylinder to exactly 15 mm in length. All the cylinders must be the same size.

*4.* Place these cylinders of beet tissue in a beaker and rinse them with tap water for 2 min to wash betacyanin from the injured cells on the surface. Be sure that all of the cylinders are the same size. Discard the colored rinse-water.

*5.* Place one of the seven beet sections into each of seven dry test tubes. Do not crush, stab, or otherwise damage the cylinders when moving them to the test tubes.

*6.* Label the tubes 1–7 and write the organic-solvent treatment on each tube as listed in table 10.2.

*7.* Add 10.0 mL of the appropriate solvent (see table 10.2) to each of the seven tubes.

*8.* Keep all beets at room temperature for 20 min and shake them occasionally. Then remove and discard the beet sections and measure the extent of membrane damage according to the amount of betacyanin that diffused into the water.

9. Quantify the relative color of each solution between 0 (colorless) and 10 (darkest red). If color standards are available in the lab, use them to determine relative values for the colors of your samples. Record the results for your work group in table 10.2. You may be asked to provide your results to the instructor to calculate the class averages.

 Be sure to dispose of the organic solvents as directed by your instructor.

Graph *Concentration of Organic Solvent* versus *Relative Color* for the class averages according to a demonstration graph provided by your instructor. Your instructor may also ask you to further quantify your results using a spectrophotometer. If so, see Exercise 8 for instructions for using a spectrophotometer. Read the absorbance of the solutions at 460 nm and record your results and the class average results in table 10.2. Then graph *Concentration of Organic Solvent* versus *Absorbance* for the class averages.

**Question 4**

*a.* Based on your results, are lipids soluble in both acetone and methanol?

*b.* Based on your results, which damages membranes more: 50% methanol or 25% acetone?

*c.* In which solvent are lipids most soluble?

*d.* The concentration of solvent affects its ability to dissolve lipids. Based on your results, did the highest concentration of both solvents cause the most damage?

*e.* What other solvents might be interesting to test in this experiment?

*f.* What was the purpose of tube 7?

*g.* How accurate were your hypothesized rankings for the treatments involving organic solvents?

## INQUIRY-BASED LEARNING

### *How sensitive are cellular membranes to their environment?*

Observation: Cellular membranes are the interface between cells and their environment. The integrity of cellular membranes, which is critical for the proper functioning of the membranes and cells, is affected by environmental stimuli.

Question: How do solvents or temperature affect membrane permeability?

**a.** Establish a working lab group and obtain Inquiry-Based Learning Worksheet 10 from your instructor.

**b.** Discuss with your group well-defined questions relevant to the preceding observation and question. Choose and record your group's best question for investigation.

**c.** Translate your question into a testable hypothesis and record it.

**d.** Outline on Worksheet 10 your experimental design and supplies needed to test your hypothesis. Ask your instructor to review your proposed investigation.

**e.** Conduct your procedures, record your data, answer your question, and make relevant comments.

**f.** Discuss with your instructor any revisions to your questions, hypothesis, or procedures. Repeat your work as needed.

# Questions for Further Study and Inquiry

1.  Are your conclusions about membrane structure and stress valid only for beet cells? Why or why not?

2.  What characteristics of beets make them useful as experimental models for studying cellular membranes?

3.  Explain why phospholipids have a natural tendency to self-assemble into a bilayer. Why is this biologically important?

4.  Freezing temperatures are often used to preserve food. Considering the results of this experiment, which qualities of food are preserved and which are not?

5.  Movement of water through membranes has long puzzled scientists. Why would you not expect water to move easily through a membrane?

6.  What suggestions would you make to improve the experimental designs in these procedures?

 **Doing Biology Yourself**
How would you design an experiment to determine the relative lipid solubilities of various organic solvents?

 **WRITING TO LEARN BIOLOGY**
What role did the stability and tendency for self-assembly play in the early evolution of life?

# Enzymes
## Factors Affecting the Rate of Activity

Please visit **connect.mheducation.com** to review online resources tailored to this lab.

Fortunately, not all chemical reactions within our cells occur spontaneously. If they did, our metabolism would be chaotic. Instead, most reactions in cells are controlled by proteins called **enzymes.** Enzymes are **biocatalysts,** meaning that they accelerate metabolic reactions to biologically useful rates. Specifically, enzymes catalyze (accelerate)

reactions by lowering the activation energy needed for the reaction to occur (fig. 11.1).

Enzymes bind to reacting molecules called the **substrate** to form an **enzyme–substrate** complex. This complex stresses or distorts chemical bonds and forms a **transition state** in which the substrate becomes more

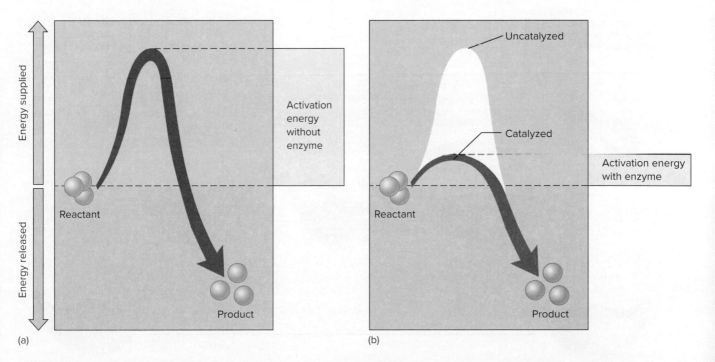

**Figure 11.1**    Activation energy and catalysis. (*a*) Exergonic reactions (those that release energy) do not necessarily proceed rapidly because energy must be supplied to destabilize existing chemical bonds. This extra energy is the activation energy for the reaction. (*b*) Catalysts such as enzymes accelerate particular reactions by lowering the amount of activation energy required to initiate the reaction.

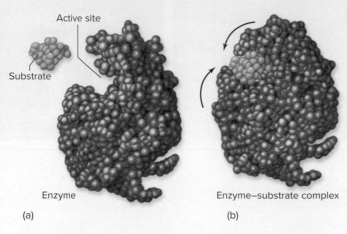

Active site

Substrate

Enzyme

(a)

Enzyme–substrate complex

(b)

**Figure 11.2** Enzyme binding its substrate. (*a*) The active site of an enzyme fits the shape of its substrate. (*b*) When the substrate, indicated in yellow, slides into the active site, the protein binds the substrate tightly and slightly alters the substrate's shape.

$$\text{Enzyme} + \text{substrate} \rightarrow \text{enzyme–substrate} \rightarrow \text{enzyme} + \text{product}$$
$$\text{complex}$$

Enzymes are proteins made of long chains of amino acids that form complex shapes. Although cells contain many enzymes, each enzyme has a precise structure and function, and catalyzes a specific reaction. This specificity results from an enzyme's unique structure and shape. The complex shape of the active site on the enzyme's surface usually couples with only one type of substrate.

Any structural change in an enzyme may **denature** or destroy its effectiveness by altering the active site and slowing the reaction. A denatured enzyme is often permanently ineffective. Denatured enzymes may result from extreme temperature, extreme pH, or any environmental condition that fundamentally alters a protein's structure. Therefore, the rate of an enzymatic reaction depends on conditions in the immediate environment. These conditions affect the shape of the enzyme and modify the active site and precise fit of an enzyme and its substrate.

The range of values for environmental factors such as temperature and pH within which an enzyme functions best represents that enzyme's **optimal conditions.** The optimal conditions for the enzymes of an organism are usually adaptive for the environment of the organism. Other factors such as the amount of substrate or concentration of enzyme also affect the reaction rate.

In this exercise you will learn that environmental factors such as temperature and pH affect enzymatic reactions

reactive. The energy needed to form the transition state is called **energy of activation** and is lowered by the enzyme. The site of attachment and the surrounding parts of the enzyme that stress the substrate's bonds constitute the enzyme's **active site** (fig. 11.2).

The reaction is complete when the **product** forms and the enzyme is released in its original condition. The enzyme then repeats the process with other molecules of substrate (fig. 11.3). Enzymes are reusable.

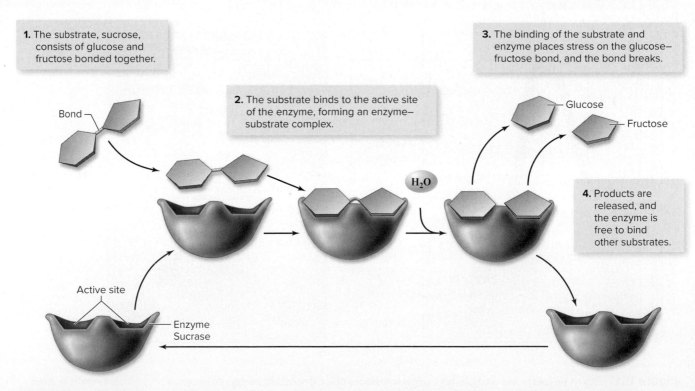

**1.** The substrate, sucrose, consists of glucose and fructose bonded together.

**2.** The substrate binds to the active site of the enzyme, forming an enzyme–substrate complex.

**3.** The binding of the substrate and enzyme places stress on the glucose–fructose bond, and the bond breaks.

Bond

Glucose

Fructose

$H_2O$

**4.** Products are released, and the enzyme is free to bind other substrates.

Active site

Enzyme
Sucrase

**Figure 11.3** The catalytic cycle of an enzyme. Enzymes increase the speed of chemical reactions but are not themselves permanently altered by the process. Here, the enzyme sucrase splits the disaccharide sucrose (steps 1, 2, 3, and 4) into its two parts, the monosaccharides glucose and fructose. After the enzyme releases the glucose and fructose, it can bind another molecule of sucrose and begin the catalytic cycle again.

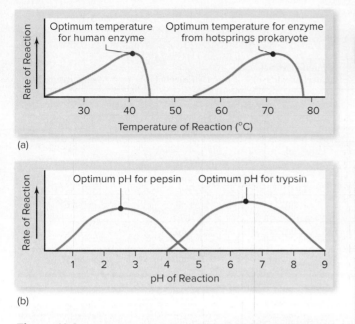

(a)

(b)

**Figure 11.4** Enzymes are sensitive to their environment. The activity of an enzyme is influenced by both (*a*) temperature and (*b*) pH. Most enzymes in humans, such as the protein-degrading enzyme trypsin, work best at temperatures about 40°C and within a pH range of 6 to 8. As you can see, however, pepsin works best at a much lower pH than does trypsin.

(fig. 11.4). You will also investigate how inhibitors affect enzymatic activity.

## TEMPERATURE AFFECTS THE ACTIVITY OF ENZYMES

Heat increases the rate of most chemical reactions. During enzymatic reactions, faster molecular motion caused by heat increases the probability that enzyme molecules will contact substrate molecules. The rate of chemical reactions generally doubles with a 10°C rise in temperature. However, higher temperatures do not always accelerate enzymatic reactions; enzymatic reactions have an optimal range of temperatures. Temperatures above or below this range decrease the reaction rate. Extreme temperatures often denature enzymes.

The effects of temperature on enzyme activity can be investigated with **catechol oxidase,** a plant enzyme that converts catechol to benzoquinone. When fruit is bruised, injured cells release catechol and catechol oxidase, which react to form a brownish product, benzoquinone. Toxic to bacteria, benzoquinone prevents decay in damaged cells. Your source of catechol oxidase will be potato extract.

$$\text{Catechol} \xrightarrow[\text{oxidase}]{\textit{catechol}} \text{benzoquinone}$$

**Catechol**        **benzoquinone**
**(substrate)**    **(enzyme)**        **(product)**

Catechol is toxic. Wash well with soap and water after skin contact.

### *Procedure 11.1* Determine the effect of temperature on catechol oxidase activity

1. Read all steps in this procedure and state the null hypothesis that your experiments will test.

2. Prepare water-baths at 40°C and 80°C. Locate a refrigerator or ice bath at or below 4°C. Place a test-tube rack in each bath and in the refrigerator.

3. Obtain seven test tubes and number them at the top 1–7.

4. Obtain a tube of potato extract and a tube of 1% catechol from your instructor.

5. Add distilled water, pH buffer, and potato extract to the tubes as listed in table 11.1. Shake or swirl to resuspend the potato extract.

6. Place the tubes in the appropriate bath or refrigerator. Allow each tube to stand undisturbed for 5 min at its respective temperature. Put tubes 1–4 in a test-tube rack at room temperature (approximately 22°C).

7. Add 1% catechol solution to tubes 2 and 4–7 as listed in table 11.1. For each tube immediately record in table 11.2 any color changes for 0 min. Record qualitative color changes on a scale between 0 (no change) and 5 (drastic change).

8. Every 5 min observe and note color changes in the seven tubes over the next 20 min. Always return the tubes to their original temperature locations (e.g., refrigerator, water-bath).

9. If your instructor asks you to further quantify your data, then measure the absorbance of the solution in each tube using a spectrophotometer set to 470 nm with tube 3 as a blank. Refer to Exercise 8 and the videos tailored to that exercise for instructions on how to use the spectrophotometer.

10. Clean your work area and materials. **Catechol must be disposed into waste containers, not down the sink drain.**

**Question 1**

*a.* Do your data support or refute your hypotheses?

Table 11.1

**Experimental Conditions to Test the Effect of Temperature on Catechol Oxidase Activity**

| Tube | Distilled Water | pH 6 Buffer | Potato Extract (catechol oxidase) | 1% Catechol | Temperature |
|------|-----------------|-------------|-----------------------------------|-------------|-------------|
| 1 | 2 mL | 1 mL | | | 22°C |
| 2 | 1 mL | 1 mL | | 1 mL | 22°C |
| 3 | 1 mL | 1 mL | 1 mL | | 22°C |
| 4 | | 1 mL | 1 mL | 1 mL | 22°C |
| 5 | | 1 mL | 1 mL | 1 mL | 4°C |
| 6 | | 1 mL | 1 mL | 1 mL | 40°C |
| 7 | | 1 mL | 1 mL | 1 mL | 80°C |

**Table 11.2**

**Qualitative and Quantitative Color Changes as Catechol Oxidase Activity Produces Brown Benzoquinone**

| Tube | Qualitative Color Change Results | | | | | Quantitative Absorbance Results | | | | |
|------|-------|-------|--------|--------|--------|-------|-------|--------|--------|--------|
| | 0 min | 5 min | 10 min | 15 min | 20 min | 0 min | 5 min | 10 min | 15 min | 20 min |
| 1 | 0 | 0 | 0 | 0 | 0 | 0 | 0 | 0 | 0 | 0 |
| 2 | | | | | | | | | | |
| 3 | | | | | | | | | | |
| 4 | | | | | | | | | | |
| 5 | | | | | | | | | | |
| 6 | | | | | | | | | | |
| 7 | | | | | | | | | | |

**b.** Write a hypothesis and a null hypothesis for the effect of temperature on catechol oxidase activity.

**c.** What were the enzyme, substrate, and product of the enzymatic reaction?

**d.** Why was each tube left undisturbed for 5 min in step 6 of procedure 11.1?

**e.** Explain the results observed for tubes 1–3. What was the purpose of these tubes?

**f.** Use your results for tubes 4–7 to construct a line graph of *Enzyme Activity* versus *Time* on graph paper provided at the end of the exercise. Your graph will include four curves.

*g.* Use your results to argue for or against the statement, "Catechol oxidase functions equally and efficiently at various temperatures."

*h.* Over what range of temperatures tested was catechol oxidase active? Should other temperatures be tested to more accurately determine the range of activity?

*i.* At which temperature was catechol oxidase activity greatest? Should more temperatures be tested to determine its optimum?

*j.* At what temperature was catechol oxidase denatured? How do you know?

*k.* What is the effect of denaturing an enzyme?

*l.* If an enzyme has a single optimal temperature, then an organism might have difficulty dealing with an environment with wide temperature variation. What adaptive advantage is there in having repetitive enzyme systems (i.e., more than one enzyme to catalyze the same reaction) that we know many organisms have?

## pH AFFECTS THE ACTIVITY OF ENZYMES

Enzymatic activity is sensitive to pH. Acidic and basic solutions are rich in $H^+$ and $OH^-$ ions (see Exercise 5), respectively, and they readily react with the charged side groups of the enzyme molecules. As the pH is lowered, side groups gain $H^+$ ions; as the pH is raised, side groups lose $H^+$ ions. In this way, solutions having an extreme pH can change an enzyme's shape enough to alter its active site. Extreme pH can denature an enzyme just as drastically as can high temperatures. Many enzymes function optimally in the neutral pH range, while others (such as pepsin, an

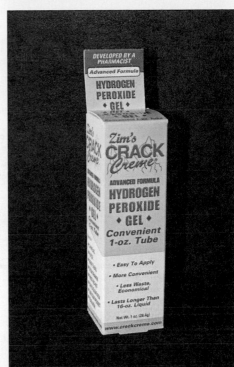

Evelyn Jo Johnson/McGraw Hill

**Figure 11.5** Ointments with high concentrations of hydrogen peroxide can be effective antibiotics. Hydrogen peroxide is a naturally produced toxin in cells and is degraded by the enzyme catalase.

enzyme in your digestive tract) function best at pH as low as 1.6 (fig. 11.4*b*).

The effects of pH can be investigated with **catalase,** an enzyme in plants and animals that speeds the breakdown of hydrogen peroxide, toxic to cells. It may surprise you that cells can produce a toxin naturally. Could that ability be useful to cells? Occasionally we take advantage of hydrogen peroxide as a powerful oxidizer that denatures macromolecules and kills "germs" (fig. 11.5). Hydrogen peroxide is broken down by catalase to water and oxygen.

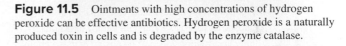

$$2 \ H_2O_2 \xrightarrow{\ catalase\ } 2 \ H_2O + O_2$$

*Procedure 11.2* **Observe the effects of pH on catalase activity**

1. Read all steps in this procedure and state the null hypothesis that your experiments will test. Review Exercise 1 for the meaning and significance of a null hypothesis.

2. Prepare catalase solution.

   *a.* Use a mortar and pestle to macerate a marble-size portion of fresh, raw ground meat in 10 mL of distilled water.

   *b.* Filter the solution through cheesecloth into a test tube and add an equal volume of distilled water.

3. Obtain 10 test tubes and number them at the top 1–10.

Table 11.3

**Experimental Conditions to Test the Effect of pH on Catalase Activity**

| Tube | Distilled Water | Buffer | Hydrogen Peroxide | HCl | NaOH | pH | Catalase Solution |
|------|-----------------|--------|-------------------|-----|------|-----|-------------------|
| 1 | 5 mL | 1 mL, pH 7 | | | | | |
| 2 | 4 mL | 1 mL, pH 7 | | | | | 1 mL |
| 3 | 2 mL | 1 mL, pH 7 | 3 mL | | | | |
| 4 | 1 mL | | 3 mL | 1 mL | | | 1 mL |
| 5 | 1 mL | 1 mL, pH 5 | 3 mL | | | | 1 mL |
| 6 | 1 mL | 1 mL, pH 7 | 3 mL | | | | 1 mL |
| 7 | 1 mL | 1 mL, pH 9 | 3 mL | | | | 1 mL |
| 8 | 1 mL | | 3 mL | | 1 mL | | 1 mL |
| 9 | 1 mL | 1 mL, pH 7 | 3 mL | 1 mL | | | |
| 10 | 1 mL | 1 mL, pH 7 | 3 mL | | 1 mL | | |

**4.** Obtain stock solutions of distilled water, hydrogen peroxide, buffer pH 5, buffer pH 7, buffer pH 9, 0.1 M HCl, and 0.1 M NaOH.

HCl is a strong caustic acid, and NaOH is a strong caustic base. Follow your instructor's directions for handling, dispensing, and disposing of these chemicals. Rinse immediately with water if you spill any acid or base on your skin.

**5.** Add distilled water and hydrogen peroxide to each tube as listed in table 11.3. If you are measuring by drops, then 1 mL equals about 20 medium-sized drops. Wait 2 min before proceeding to step 6.

**6.** Add 1 mL of HCl to tubes 4 and 9. Verify that the pH is approximately 3 or lower.

**7.** Add 1 mL of NaOH to tubes 8 and 10. Verify that the pH is approximately 11 or higher.

**8.** Add 1 mL of the buffer solutions as indicated in table 11.3.

**9.** Your instructor may ask you to verify that the buffers produce the indicated pH. If so, use pH paper to measure the values for each solution and record them in table 11.3.

**10.** No catalase is added to tubes 1, 3, 9, or 10.

**11.** Add catalase to tube 2 according to table 11.3. After adding catalase, swirl the solution gently and immediately record in table 11.4 qualitative changes in the bubbling intensity of oxygen production on a scale of 0 (no bubbling) to 5 (vigorous bubbling).

**12.** If your instructor asks you to more rigorously quantify your results, then immediately after adding the catalase place a stopper with tubing over each tube to collect and measure the volume of gases produced in

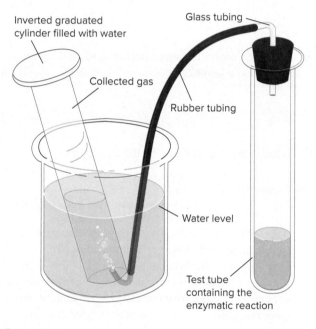

**Figure 11.6** A method to capture oxygen released by catalase activity.

a water-filled graduated cylinder inverted in a beaker of water (fig. 11.6). Be sure that the graduated cylinder does not pinch off the rubber tubing. Also be sure the cylinder is vertical when you measure volume. Record these results in table 11.4.

**13.** Repeat step 11 for each remaining solution.

**14.** After you have gathered your data for all 10 tubes, record in table 11.4 your explanation for the results of the catalase activity in each of the tubes.

**15.** Clean your work area and materials. Follow your instructor's directions concerning the disposal of waste solutions containing HCl and NaOH.

## Table 11.4

**Production of Oxygen by Catalase Activity. Qualitative Data Are Observations of Intensity of Oxygen Effervescence Ranging from 1–5. Quantitative Data Are Milliliters of Oxygen Produced.**

| Tube | Oxygen Production | | Explanation |
|------|-------------------|---|-------------|
| | Qualitative (0–5) | Quantitative (mL $O_2$) | |
| 1 | | | |
| 2 | | | |
| 3 | | | |
| 4 | | | |
| 5 | | | |
| 6 | | | |
| 7 | | | |
| 8 | | | |
| 9 | | | |
| 10 | | | |

## Question 2

**a.** Do your data support or refute your hypothesis?

**b.** What were the enzyme, substrate, and product of the enzymatic reaction?

**c.** What was the purpose of completing steps 2–9 for all tubes before adding the catalase in step 11?

**d.** What was the purpose of tubes 1, 2, 3, 9, and 10?

**e.** Use your data for tubes 4–8 to construct a line graph of *Enzyme Activity* versus *pH*.

**f.** Over what pH range was catalase active?

**g.** What pH levels denatured catalase? Specifically how do solutions of high or low pH change an enzyme's reactivity?

**h.** At which of the tested pH values did catalase react most rapidly? Should more values be tested to accurately determine its optimum?

**i.** After experimenting with the effects of pH on enzymes, would you suspect that human blood has a constant pH? Why? What would be the adaptive advantage of this?

## INHIBITORS AFFECT THE ACTIVITY OF ENZYMES

**Peroxidase** is an enzyme in plants (such as turnips) and some bacteria that converts toxic hydrogen peroxide to $H_2O$ and $O_2$ in a reaction similar to that of catalase. Peroxidase is a large protein with a reactive iron atom at its active site.

$$2\ H_2O_2 \xrightarrow{\ \ peroxidase\ \ } 2\ H_2O + O_2$$

Enzymes such as peroxidase can be inhibited by chemicals in various ways. One mechanism is **competitive inhibition.** Competitive inhibitors are molecules structurally similar to the substrate and therefore competitive for positions at the active sites of enzymes. This ties up the enzyme and makes it unable to bind with the substrate. For example, hydroxylamine ($HONH_2$) is structurally similar to hydrogen peroxide ($H_2O_2$) and binds to the iron atom at the active site of peroxidase. Thus, hydroxylamine competes with hydrogen peroxide for the active sites on peroxidase molecules and reduces the frequency of hydrogen peroxide binding with peroxidase molecules. However, a high enough concentration of enzyme with a constant concentration of inhibitor can reduce the inhibition.

The production of oxygen by peroxidase provides a method to measure the ongoing reaction rate. One method would be to capture liberated bubbles of oxygen and measure their total volume. But in the following procedure you will measure oxygen by combining it with a dye that changes color when it is oxidized. Guaiacol is a convenient dye that turns from colorless to brown as it is oxidized by oxygen. The amount of brown color in the final product is proportional to the amount of oxygen formed by the reaction. You can measure the color change qualitatively by rating the color of a reacting solution on an arbitrary scale from 0 to 5, or quantitatively with a spectrophotometer measuring the solution's absorbance of 470 nm light. Review Exercise 8 and the associated video for instructions on using a spectrophotometer.

$$O_2 + \textbf{guaiacol} \longrightarrow \textbf{oxidized guaiacol}$$
$$\textbf{(colorless)} \qquad\qquad \textbf{(brown)}$$

### *Procedure 11.3*   Observe the effects of an inhibitor on enzymatic activity

1. Read all steps in this procedure and state the null hypothesis that your experiments will test.
2. Prepare turnip extract.
   a. Thoroughly blend 6 g of the inner portion of a peeled turnip in a blender with 200 mL of cold water.
   b. Filter the turnip slurry through cheesecloth into a beaker.
   c. Pour about 7 mL of the extract into a test tube and determine its absorbance in a spectrophotometer. Refer to Exercise 8 for instructions on spectrophotometry. The absorbance for the turnip

extract should be between 0.1 and 0.2 at 470 nm to give a reasonable concentration of enzyme.
   d. Dilute or concentrate the suspension as necessary. Your instructor may provide directions for standardizing this enzyme solution more precisely.
3. Obtain nine test tubes and number them at the top 1–9.
4. Add distilled water to each tube as listed in table 11.5.
5. To tube 1, add 0.1 mL (2 drops) of guaiacol as listed in table 11.5 and swirl the contents.
6. Immediately determine the solution's absorbance at 470 nm using a spectrophotometer. Record the absorbance value in table 11.6. Measure the absorbance every 30 sec for 5 min and record the value each time.
7. To tube 2, add 0.2 mL (4 drops) of hydrogen peroxide as listed in table 11.5. Swirl the contents. Repeat step 6 quickly.
8. To tube 3, add 0.1 mL (2 drops) of guaiacol and 0.2 mL of hydrogen peroxide as listed in table 11.5. Swirl the contents. Repeat step 6 quickly.
9. To tube 4, add 0.1 mL (2 drops) of guaiacol and 1.0 mL of turnip extract as listed in table 11.5. Swirl the contents. Repeat step 6 quickly.
10. Complete all of the measurements for steps 5–9 before proceeding to step 11.
11. For each of tubes 5–9, add 0.1 mL (2 drops) of guaiacol and 0.2 mL (4 drops) of hydrogen peroxide as listed in table 11.5.
12. For tube 5, add 1.0 mL of turnip extract and swirl the contents. Repeat step 6 quickly.
13. For tube 6, add 1.0 mL of turnip extract and 0.5 mL (10 drops) of hydroxylamine and swirl the contents. Repeat step 6 quickly.
14. For tube 7, add 1.5 mL of turnip extract and 0.5 mL (10 drops) of hydroxylamine and swirl the contents. Repeat step 6 quickly.
15. For tube 8, add 2.0 mL of turnip extract and 0.5 mL (10 drops) of hydroxylamine and swirl the contents. Repeat step 6 quickly.
16. For tube 9, add 3.0 mL of turnip extract and 0.5 mL (10 drops) of hydroxylamine and swirl the contents. Repeat step 6 quickly.
17. Clean your work area and materials.

### Question 3
a. Do your data support or refute your hypothesis?



b. What were the enzyme, substrate, and product of this enzymatic reaction?

Table 11.5

**Experimental Conditions to Test the Inhibition of Hydroxylamine on Peroxidase Activity**

| Tube | Distilled Water | Guaiacol (25 mM) | Hydrogen Peroxide (3%) | Turnip Extract | Hydroxylamine (10%) |
|------|-----------------|------------------|------------------------|----------------|---------------------|
| 1 | 5.9 mL | 0.1 mL | | | |
| 2 | 5.8 mL | | 0.2 mL | | |
| 3 | 5.7 mL | 0.1 mL | 0.2 mL | | |
| 4 | 4.9 mL | 0.1 mL | | 1.0 mL | |
| 5 | 4.7 mL | 0.1 mL | 0.2 mL | 1.0 mL | |
| 6 | 4.2 mL | 0.1 mL | 0.2 mL | 1.0 mL | 0.5 mL |
| 7 | 3.7 mL | 0.1 mL | 0.2 mL | 1.5 mL | 0.5 mL |
| 8 | 3.2 mL | 0.1 mL | 0.2 mL | 2.0 mL | 0.5 mL |
| 9 | 2.2 mL | 0.1 mL | 0.2 mL | 3.0 mL | 0.5 mL |

**Table 11.6**

**Absorbance at 470 nm of Peroxide/Peroxidase Solutions**

| Tube | 0.0 min | 0.5 min | 1.0 min | 1.5 min | 2.0 min | 2.5 min | 3.0 min | 3.5 min | 4.0 min | 4.5 min | 5.0 min |
|------|---------|---------|---------|---------|---------|---------|---------|---------|---------|---------|---------|
| 1 | | | | | | | | | | | |
| 2 | | | | | | | | | | | |
| 3 | | | | | | | | | | | |
| 4 | | | | | | | | | | | |
| 5 | | | | | | | | | | | |
| 6 | | | | | | | | | | | |
| 7 | | | | | | | | | | | |
| 8 | | | | | | | | | | | |
| 9 | | | | | | | | | | | |

*c.* Explain the results you observed for tubes 1, 2, 3, and 4. What was the purpose of these tubes?

*d.* Use your data for tubes 5–9 to construct a line graph of *Enzyme Activity (Absorbance)* versus *Time*. There will be five curves on the graph. You will not graph the values for tubes 1–4.

*e.* In which tubes was peroxidase still active after 5 min?

*f.* How does hydroxylamine affect peroxidase activity?

*g.* Was it possible to detect peroxidase activity in the presence of the inhibitor by increasing enzyme concentration? Why or why not?

*h.* Inhibitors are common in biological systems. Why might some organisms release enzyme inhibitors into their surrounding environment?

## INQUIRY-BASED LEARNING

### *What factors speed up or slow down enzymatic activity?*

Observation: Numerous factors affect enzyme reaction rates. Some products inhibit activity, whereas others can stimulate activity. You learned in an earlier lab that acids, bases, and buffers affect pH, and you learned in this lab that pH affects enzymatic activity.

Question: How do common antacids affect enzymatic activity?

**a.** Establish a working lab group and obtain Inquiry-Based Learning Worksheet 11 from your instructor.

**b.** Discuss with your group well-defined questions relevant to the preceding observation and question. Choose and record your group's best question for investigation.

**c.** Translate your question into a testable hypothesis and record it.

**d.** Outline on Worksheet 11 your experimental design and supplies needed to test your hypothesis. Ask your instructor to review your proposed investigation.

**e.** Conduct your procedures, record your data, answer your question, and make relevant comments.

**f.** Discuss with your instructor any revisions to your questions, hypothesis, or procedures. Repeat your work as needed.

## Questions for Further Study and Inquiry

*1.* More substrate increases the probability that an enzyme will contact the substrate and should increase the enzymatic reaction rate. How do you explain the increase in time to complete hydrolysis when more substrate was present?

*2.* What term describes a change in an enzyme's structure that renders the proteinaceous enzyme nonfunctional? What factors in addition to temperature influence a protein's structure?

*3.* What happens when an enzyme is denatured?

*4.* Can a denatured enzyme be "re-natured"? Explain your answer.

*5.* Commercial meat tenderizers contain papain (extracted from papaya) and/or bromelain (extracted from pineapple), both of which are enzymes. Because these enzymes "tenderize" meat, what group of organic compounds that you studied in Exercise 6 do you suspect that these enzymes react with? How could you test your answer?

*6.* Enzymes are proteins, and therefore are structurally linked to DNA sequences. How could natural selection alter the metabolism of an organism?

# DOING BIOLOGY YOURSELF

Review the structure of starch and the action of the enzyme amylase. Design an experiment that uses a spectrophotometer to detect the progress and completion of hydrolysis of starch by amylase.

# WRITING TO LEARN BIOLOGY

Propose a mechanism involving enzyme production by which a cell could counteract a sudden increase in the amount of substrate.

EXERCISE 11

# Respiration

## Aerobic and Anaerobic Oxidation of Organic Molecules

## Learning Objectives

By the end of this exercise you should be able to:

1. Measure carbon dioxide production during anaerobic respiration.
2. Understand the effects of inhibitors, intermediate compounds, and cofactors in anaerobic respiration.
3. Determine oxygen consumption during aerobic respiration.
4. Use a pH-indicator to measure the relative production of carbon dioxide by plants and animals.
5. Use a respirometer to determine the metabolic rate of an animal.
6. Demonstrate practical applications of anaerobic respiration, such as making wine and kimchee.

Please visit **connect.mheducation.com** to review online resources tailored to this lab.

**A**ll living organisms respire, meaning that they have metabolic pathways that release energy from organic (rarely inorganic) molecules and capture it in ATP. Some need oxygen to do it, some don't, but they all respire because all organisms need usable chemical energy to fuel their life processes. Respiration is the chemistry that provides that energy. Usually, organic carbon molecules are the energy source, and $CO_2$ and $H_2O$ are released as waste. Humans release the waste as they exhale. Respiring yeasts don't exhale, but they can "pump up" rising bread by liberating $CO_2$ as the yeasts break down sugar during respiration (fig. 12.1).

**Cellular respiration** involves oxidation of organic molecules and a concomitant release of energy. Some of this energy is stored in chemical bonds of **adenosine triphosphate (ATP),** which is used later as a direct source of energy for cellular metabolism. Organisms use the energy stored in ATP to do work such as transport materials, synthesize new compounds, reproduce, contract muscles, and remove wastes.

Photosynthesis, the topic of Exercise 13, uses light energy to split $H_2O$ and harvest high-energy electrons. These energetic electrons (and accompanying $H^+$) are passed to $CO_2$, thereby reducing $CO_2$ to energy-storing sugars. Respiration removes electrons from (i.e., oxidizes) glucose, captures some of the energy in ATP, and ultimately passes the electrons to oxygen to form $H_2O$.

In most cells, respiration begins with the oxidation of glucose to pyruvate via a set of chemical reactions called **glycolysis** (fig. 12.2a). During glycolysis, some of the

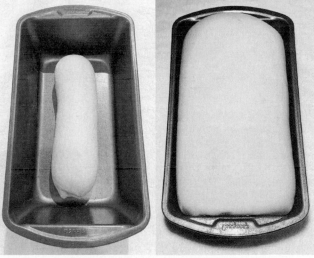

Darrell Vodopich

**Figure 12.1** Bread dough rises because respiring yeasts break down sugars to obtain their energy for growth and liberate $CO_2$, thereby forming small bubbles that cause the dough to rise. The loaf on the right has been rising 4 hours longer than the loaf on the left.

energy released from each glucose molecule is stored in ATP. Glycolysis occurs with or without oxygen. If oxygen is present, most organisms continue respiration by oxidizing pyruvate to $CO_2$ via chemical reactions of the **citric acid cycle.** Organisms that use oxygen for respiration beyond glycolysis are called **aerobes.**

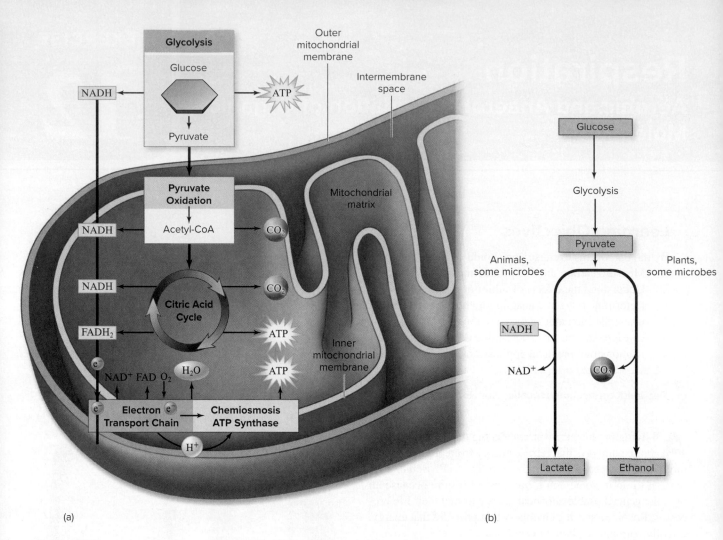

**Figure 12.2**   (*a*) An overview of aerobic respiration. Glycolysis occurs in the cytoplasm, and the  citric acid cycle and electron transport chain occur in mitochondria. (*b*) During anaerobic fermentation, pyruvate is reduced with electrons extracted during glycolysis and carried by NADH. In organisms that reduce pyruvate directly, as in muscle cells, the product is lactate. In organisms that first remove carbon dioxide, as in yeast cells, the product is ethanol.

As aerobes oxidize the acetyl group from the pyruvate in the citric acid cycle, they store energy in electron carriers such as NAD+ (nicotinamide adenine dinucleotide). Specifically, aerobes store energy by reducing (adding high-energy electrons to) NAD+ and FAD+. These compounds later transfer their high-energy electrons to a series of compounds collectively called the electron transport chain. The **electron transport chain** generates proton gradients from energy stored in reduced NAD and related compounds that lead to formation of approximately 18-times more ATP than that formed in glycolysis. Oxygen, the final electron-acceptor in the electron transport chain, is reduced to form $H_2O$ (fig. 12.2*a*). Without oxygen to accept electrons passed through the electron transport chain, the chain is not functional and an aerobic organism will quickly die. We can summarize aerobic respiration as follows:

### Summary Equation for Aerobic Respiration

$$C_6H_{12}O_6 + 6\ O_2 \rightarrow 6\ CO_2 + 6\ H_2O + ATP + Heat$$

Glucose   Oxygen   Carbon   Water
Dioxide

**Question 1**
Why must aerobic organisms such as yourself inhale oxygen and exhale $CO_2$?

Other organisms called **anaerobes** live without oxygen and may even be killed by oxygen in the atmosphere. Some of these anaerobes are primitive bacteria that gather their energy with a pathway of anaerobic respiration that uses inorganic electron acceptors other than oxygen. For example, many bacteria use nitrate, sulfate, or other inorganic compounds as the electron acceptor instead of oxygen. Other anaerobes use glycolysis, but the pyruvate from glycolysis is reduced via anaerobic **fermentation** to either $CO_2$ and ethanol (in plants and some microbes such as yeast) or lactic acid (in other microbes and oxygen-stressed muscles of animals; fig. 12.2*b*). We can summarize anaerobic fermentation in figure 12.2*b* in the following equations:

## Anaerobic Fermentation in Plants and Some Microbes

$$C_6H_{12}O_6 \rightarrow 2\ C_2H_5OH + 2\ CO_2 + ATP + Heat$$

Glucose      Ethanol     Carbon
Dioxide

## Anaerobic Fermentation in Animals and Some Microbes

$$C_6H_{12}O_6 \rightarrow 2\ CH_3CHOHCOOH + ATP + Heat$$

Glucose        Lactic Acid

Notice from these equations that plants (as well as prokaryotes and other eukaryotes such as yeasts) can temporarily conduct anaerobic fermentation that reduces pyruvate from glycolysis to ethanol and carbon dioxide. This occurs, for example, in roots that penetrate anaerobic soils and sediments.

Anaerobic fermentation does not involve or benefit from the additional ATP produced by the citric acid cycle or electron transport chain. Thus, the ability of an organism to live in the absence of oxygen comes at a price: Anaerobic fermentation produces 18-fold less ATP per glucose molecule than does aerobic respiration.

### Question 2

What are the advantages and disadvantages of anaerobic fermentation?

In today's exercise, you will study the major features of cellular respiration. Let's begin with a type of anaerobic fermentation with which you are already familiar: alcoholic fermentation by yeast.

 **SAFETY FIRST** Before coming to lab, you were asked to read this exercise so you would know what to do and be aware of safety issues. In the space below, briefly list the safety issues associated with today's procedures. If you have questions about these issues, contact your laboratory assistant before starting work.

## PRODUCTION OF CO$_2$ DURING ANAEROBIC FERMENTATION

Yeast are fungi used in baking and producing alcoholic beverages. They can respire in the absence of O$_2$ and can oxidize glucose to ethanol and CO$_2$. To demonstrate CO$_2$ production during anaerobic fermentation by yeast (fermentation), follow procedure 12.1. In this procedure you will observe the effects of these compounds on respiration:

**Pyruvate**—a product of glycolysis; pyruvate is reduced to ethanol or lactic acid during anaerobic fermentation

**Magnesium sulfate (MgSO$_4$)**—provides Mg$^{2+}$, a cofactor that activates some enzymes of glycolysis

**Sodium fluoride (NaF)**—an inhibitor of some enzymes of glycolysis

**Glucose**—a common organic molecule used as an energy source for respiration

## *Procedure 12.1*   Measure CO$_2$ production during anaerobic fermentation

1. Label seven test-tubes and add the solutions listed in table 12.1.
2. Completely fill the remaining volume in tubes 1–6 with the yeast suspension that is provided. Fill the remaining volume in tube 7 completely with distilled water.

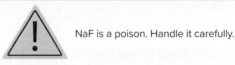 NaF is a poison. Handle it carefully.

3. For each tube, slide an inverted, flat-bottomed test tube down over the yeast-filled tube (see fig. 1.3). Hold the yeast-filled tube firmly against the inside bottom of the cover tube and invert the assembly. Your instructor will demonstrate how to slide this slightly larger empty tube over the top of each yeast tube and invert the assembly. If done properly, air will not be trapped at the top of the tube of yeast after inversion.
4. Incubate the tubes at 37°C for 40 min. While you are waiting, write your predictions for each tube:

| Tube | Predicted Results and Brief Explanation |
|------|------------------------------------------|
| 1. | |
| 2. | |
| 3. | |
| 4. | |
| 5. | |
| 6. | |
| 7. | |

5. After 40 min, measure the height (in millimeters) of the bubble of accumulated CO$_2$. Record your results in table 12.1.
6. The effects of pyruvate, MgSO$_4$, NaF, and glucose on CO$_2$ production are best determined by comparing each tube to the control rather than by ranking all of

the treatment tubes. For each variable in table 12.2, record the number of the tube containing the compound being tested and the number of the tube serving as the control for that compound.

## Question 3

*a.* What was the purpose of tube 7?

*b.* How was the effect of concentration of inhibitor tested in this experiment? How did the concentration of NaF affect anaerobic fermentation in your experiment? Why?

*c.* Which compounds listed in step 6 are intermediates in the respiratory pathway?

*d.* Why did tube 6 produce $CO_2$ even though an inhibitor of glycolysis was present?

*e.* Compare tubes 4 and 5. How was $CO_2$ production affected by the 10-fold increase in the amount of NaF? For example, was it also changed 10-fold?

*f.* Did magnesium (a cofactor that activates many enzymes) promote respiration? If not, what are some possible reasons?

*g.* Smell the contents of the tube containing the most $CO_2$. What compound do you smell?

### Table 12.1

**Experimental Treatments and $CO_2$ Production during Anaerobic Fermentation**

| Tube | 3 M Na Pyruvate (Activator) | 0.1 M MgSO$_4$ (Activator) | 0.1 M NaF (Inhibitor) | 5.0% Glucose (Activator) | Water | Fill With | $CO_2$ Produced After 40 min (mm) |
|---|---|---|---|---|---|---|---|
| 1 | — | — | — | — | 7.5 mL | Yeast suspension | |
| 2 | — | — | — | 2.5 mL | 5.0 mL | Yeast suspension | |
| 3 | — | 5.0 mL | — | 2.5 mL | — | Yeast suspension | |
| 4 | — | — | 0.5 mL | 2.5 mL | 4.5 mL | Yeast suspension | |
| 5 | — | — | 5.0 mL | 2.5 mL | — | Yeast suspension | |
| 6 | 2.5 mL | — | 2.5 mL | 2.5 mL | — | Yeast suspension | |
| 7 | — | — | — | 2.5 mL | 2.5 mL | Water | |

### Table 12.2

**Effects of Four Chemical Variables on $CO_2$ Production during Anaerobic Fermentation**

| Variable | Tube # with Variable | Tube # Control | Effect of Variable on Respiration Rate | Mechanism for the Effect |
|---|---|---|---|---|
| Yeast | | | | |
| Glucose | | | | |
| NaF | | | | |
| Na Pyruvate | | | | |
| MgSO$_4$ | | | | |

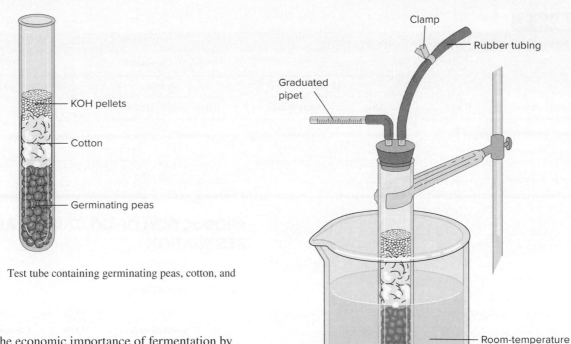

**Figure 12.3** Test tube containing germinating peas, cotton, and KOH pellets.

*h.* What is the economic importance of fermentation by yeast?

*i.* What gas is responsible for the holes in baked bread?

**Figure 12.4** Test tube with stopper having capillary tubes attached. The tube will be covered in foil.

If time and facilities are available, repeat procedure 12.1 and incubate the tubes at 4°C (refrigerator), 20°C (incubator), and/or 55°C (incubator). Use your data to explain the effect of temperature on fermentation by yeast.

## OXYGEN CONSUMPTION DURING AEROBIC RESPIRATION

Aerobic respiration uses oxygen as the terminal electron acceptor in the electron transport chain. Because this oxygen is reduced to water, you can measure aerobic respiration by measuring the consumption of oxygen. During respiration, $CO_2$ is produced, while $O_2$ is consumed. Review the summary equation for aerobic respiration in the introduction of this exercise. In the following experiment, KOH is used to absorb the $CO_2$. Therefore, the net change in gas volume is a measure of oxygen consumption.

### Procedure 12.2 Determine oxygen consumption during aerobic respiration (may be done as a demonstration)

*1.* Fill a test tube or flask half-full with germinating peas and another half-full with heat-killed peas. The germinating peas have been soaked in water in the dark for three to four days.

*2.* Cover the contents of each tube with a loose-fitting plug of cotton.

*3.* Cover the cotton with approximately 1 cm of loosely packed pellets of potassium hydroxide (KOH) (fig. 12.3).

 Potassium hydroxide (KOH) is a strong, caustic base. Handle it carefully. If you get any KOH on your skin, rinse immediately with water.

*4.* Place a stopper containing a capillary tube or graduated pipet with an attached outlet tube into both tubes containing peas (fig. 12.4). The capillary tube or graduated pipet should be oriented horizontally.

*5.* Cover the tube with foil to prevent light and photosynthesis.

*6.* Vertically clamp the tubes to a ring stand so that the bottom of each tube is submerged in a room-temperature water-bath. The water-bath will minimize temperature fluctuations in the tube.

*7.* Use a Pasteur pipet to inject enough dye into each capillary tube so that approximately 1 cm of dye is drawn into each capillary tube. The capillary tube or graduated pipet should be oriented horizontally.

**Table 12.3**

**Oxygen Consumption by Seeds at Three Temperatures**

| Treatment | 0 min | 10 min | | 20 min | | 30 min | |
|---|---|---|---|---|---|---|---|
| | | Alive | Heat-Killed | Alive | Heat-Killed | Alive | Heat-Killed |
| Room temperature | 0 | _____ | _____ | _____ | _____ | _____ | _____ |
| Ice bath | 0 | _____ | _____ | _____ | _____ | _____ | _____ |
| Warm water-bath | 0 | _____ | _____ | _____ | _____ | _____ | _____ |

mL $O_2$ Consumed

8. After waiting 1 min for equilibration, attach a pinch clamp to the outlet tube and mark the position of the dye with a wax pencil. Write your predicted results and a brief explanation here: _____
_____
_____
_____
_____

9. Use a wax pencil to mark the position of the dye every 10 min for the next 30 min.

10. After each time interval, measure the distance the dye moved from its starting point; record your data in table 12.3.

11. Remove the pinch clamp from the outlet valve and return the dye to the end of the capillary tube by tilting the capillary tube.

12. Repeat steps 1–11 using tubes incubated in an ice bath and warm (35°C) water-bath. You can save time by running all of these treatments simultaneously. Record your results in table 12.3.

**Question 4**

*a.* What was the purpose of adding heat-killed peas to a tube?

*b.* In which direction did the dye move? Why?

*c.* What does this experiment tell you about the influence of temperature on oxygen consumption during cellular respiration?

## PRODUCTION OF $CO_2$ DURING AEROBIC RESPIRATION

$CO_2$ produced during cellular respiration can combine with water to form carbonic acid:

$$CO_2 \quad + \quad H_2O \quad \leftrightarrow \quad H_2CO_3$$

**Carbon dioxide**      **Water**      **Carbonic acid**

In this procedure (fig. 12.5), you will use phenolphthalein to detect changes in pH resulting from the production of $CO_2$ (and, therefore, carbonic acid) during cellular respiration. Phenolphthalein is red in basic solutions and colorless in acidic solutions. Thus, you can monitor cellular respiration by measuring acid production as change in pH. **pH** is a measure of the acidic or basic properties of a solution; pH 7 is neutral. Solutions having a pH < 7 are acidic, and solutions having a pH > 7 are basic (see Exercise 5).

In procedure 12.3, you will not directly measure the volume of $CO_2$ produced by a respiring organism. Instead, you will measure the volume of NaOH used to neutralize the carbonic acid produced by the $CO_2$, and thereby calculate a relative measure of respiration.

**Question 5**

The organisms you will study include an animal (snail) and a plant (*Elodea*). Which do you think will respire more? Write your hypothesis here:

*Procedure 12.3*    **Measure relative $CO_2$ production by aerobic organisms**

**Experimental Setup**

1. Obtain 225 mL of culture solution provided by your instructor. This solution has been dechlorinated and adjusted to be slightly acidic.

2. Place 75 mL of this solution in each of three labeled beakers (fig. 12.5).

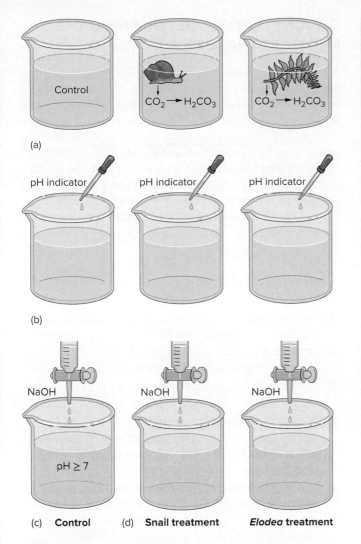

(a)

(b)

pH indicator        pH indicator        pH indicator

pH ≥ 7

(c)  **Control**    (d)  **Snail treatment**    ***Elodea* treatment**

**Figure 12.5**  The procedure to determine the relative respiration rates of a plant and animal. (*a*) During respiration, organisms release $CO_2$, which combines with water to form carbonic acid ($H_2CO_3$). (*b*) The acidic solutions remain colorless after addition of phenolphthalein, a pH indicator. (*c*) Titration of the control with NaOH (a base) will make the solution basic and pink when the pH reaches the end-point of phenolphthalein. (*d*) The treatment solutions are then titrated to the pink end-point matching the control. The volume of NaOH needed to reach the end-point indicates the relative amounts of dissolved $CO_2$ produced during respiration.

3. Obtain the organisms listed in table 12.4 from your instructor and determine the volume of each organism by following steps 4–6. Your instructor may substitute a small fish for the snail. Note that the control beaker contains no living organisms.

## Determine Volume by Water Displacement

4. Place exactly 25 mL of water in a 50-mL graduated cylinder.

5. Place the organism in the cylinder and note the increase in volume above the original 25 mL. This increase equals the volume of the organism.

6. Record the volumes in table 12.4. Gently place similar masses of each plant or animal in the appropriate beaker.

## Incubate Experimental Treatments

7. Cover each beaker with a plastic film or petri dish top and set them aside on your lab bench. Place the beaker containing the *Elodea* in the dark by covering it with a coffee can or aluminum foil.

8. Allow the organisms to respire for 15 min.

9. Gently remove the organisms from the beakers and return them to their original culture bowls.

## Titrate to Gather Your Raw Data

10. Add four drops of phenolphthalein to the contents of each beaker. The solutions should remain clear because the solutions are acidic.

11. Obtain a burette or dropper bottle to dispense NaOH (2.5 mM). Add NaOH drop by drop to the contents of the control beaker. Thoroughly mix the contents of the beaker after adding each drop. Record in table 12.4 the milliliters of NaOH required to reach the end-point of phenolphthalein. The end-point is when you first notice that the solution is pink.

12. Repeat step 11 for beaker 1; be sure to add NaOH only until the solution is the same shade of pink as the control beaker. Record the number of milliliters of NaOH added to beaker 1 in table 12.4.

13. Repeat step 11 for beaker 2.

| Table 12.4 | | | | |
| --- | --- | --- | --- | --- |
| **Data for Measuring $CO_2$ Production during Respiration** | | | | |
| (Organisms) | Total Volume of Organisms (mL) | Milliliters of NaOH to Reach End-Point (mL NaOH) | Relative Respiration Rate of Organisms (mL NaOH) | Respiration Rate per Milliliter of Organism (mL NaOH/mL Organisms) |
| Beaker 1: 4 snails | ___ | ___ | ___ | ___ |
| Beaker 2: *Elodea* | ___ | ___ | ___ | ___ |
| Control beaker | 0 | ___ | 0 | 0 |

## Calculate Your Results

**14.** For beaker 1, determine the relative respiration rate for organisms by subtracting the milliliters NaOH added to the control beaker from the milliliters NaOH added to beaker 1. Record this value in table 12.4.

**15.** Repeat step 14 for beaker 2.

**16.** For beakers 1 and 2, determine the respiration rate per milliliter of organism by dividing the relative respiration rate for organisms by the volume of the organism(s). Record these values in table 12.4.

### Question 6

*a.* In this exercise you measured the relative respiration rates of an animal and a plant. Why should you be cautious about having no algae in the control beaker?

*b.* Before you gathered your raw data, you formulated a hypothesis about the expected results. After considering your data, do you accept or reject your hypothesis? Why?

*c.* What is your major conclusion from the results of this procedure?

*d.* What features of the biology of the organisms that you used most likely contributed to the observed differences in respiration rate?

*e.* Do you feel justified in drawing conclusions from your work about all plants and animals? Or only about snails and *Elodea?* Why?

*f.* How would you expand this experiment to further test your conclusions about other plants and other animals?

*g.* What other organisms might you include in an expanded experiment? Why did you choose these organisms?

## DEMONSTRATION: DETERMINING THE METABOLIC RATE OF A MOUSE

The rate of $O_2$ uptake during cellular respiration indicates the metabolic rate of an organism. In procedure 12.4 you will measure $O_2$ uptake by measuring changes in air pressure as $O_2$ is removed from the air by a respiring mouse. Changes in air pressure can be attributed primarily to $O_2$ consumption (rather than $CO_2$ production or exhalation of water vapor) only if exhaled $CO_2$ and $H_2O$ are removed from the air. This is accomplished by adding ascarite (which adsorbs $CO_2$) and drierite (which adsorbs $H_2O$) to the experimental setup (fig. 12.6). Use procedure 12.4 to estimate the metabolic rate of a mouse.

### *Procedure 12.4* Estimate the metabolic rate of a mouse

**1.** Weigh a mouse to the nearest 0.1 g. Record this weight in table 12.5 and place the mouse in the jar of a respirometer (fig. 12.6). Use a fan to circulate air in the jar and allow the mouse to get accustomed to the jar.

**2.** Attach a 10-mL syringe filled with air to the respirometer.

**3.** Seal the respirometer jar with a lid. Then close the air escape line with a clamp and record the position of the dye solution in the right column of the curved capillary tube. This tube is called a manometer.

**4.** Inject 10 mL of air into the respirometer. The level of dye in the right side of the manometer will rise because of the increased presence of air. After injecting the 10 mL of air into the respirometer, record the position of the dye solution in table 12.5.

**5.** Allow the mouse to respire. The air pressure in the respirometer should decrease as $O_2$ is consumed, and the dye level in the right column of the manometer should decrease.

**6.** Record in table 12.5 the elapsed time for the dye level to return to its original position. This is the time for the mouse to consume 10 mL of $O_2$. Record this time as "A" in the Calculations section of table 12.5.

**7.** Gently return the mouse to its cage.

**8.** Calculate the number of liters consumed by the mouse per day by using the following formula:

Liters of $O_2$ consumed per day =

$$\frac{1440 \text{ minutes per day} \div \text{ minutes}}{\text{to consume 1 liter of } O_2}$$

Record this as "B" in the Calculations section of table 12.5.

**9.** Calculate and record in table 12.5 the mouse's metabolic rate in kcal/day, assuming that 4.8 kcal of energy are used for each liter of $O_2$ consumed. Record this as "C" in the Calculations section of table 12.5.

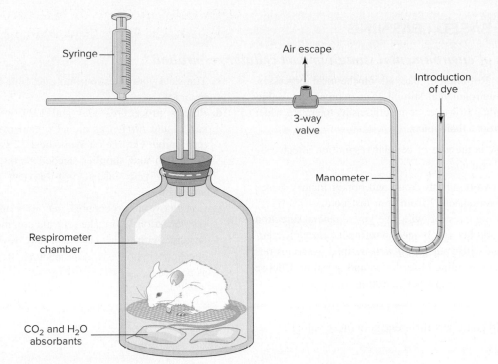

**Figure 12.6** Respirometer with mouse.

| Table 12.5 | | |
|---|---|---|
| **Data for Determination of Metabolic Rate of a Respiring Mouse** | | |
| **Observations and Data** | | |
| Weight of mouse: _____ grams | | |
| Initial position of dye solution: _____ | | |
| Position of dye solution after injection of 10 mL of air: _____ | | |
| Minutes for dye level to return to initial position: (minutes per 10 mL oxygen): _____ min | | |
| **Calculations** | | |
| A | Minutes to consume 1 liter of $O_2$ = (minutes per 10 mL oxygen) × 100 = _____ min | |
| B | Liters of $O_2$ consumed per day = 1440 minutes per day ÷ A = _____ liters per day | |
| C | Experimental metabolic rate as kcal per day = B × 4.8 kcal per liter $O_2$ = _____ kcal per day | |
| D | Predicted metabolic rate = 70 × (weight of mouse)$^{3/4}$ = _____ kcal per day | |

*10.* Calculate and record the predicted metabolic rate obtained from the following general equation for metabolic rate of small mammals:

**Predicted metabolic rate = 70 × (body weight in kg)$^{3/4}$**

Record this rate as "D" in the Calculations section of table 12.5.

*11.* Compare your experimental value with the predicted value for metabolic rate.

**Question 7**

*a.* Is the predicted metabolic rate similar to that which you determined experimentally?

## INQUIRY-BASED LEARNING

### *What sorts of environmental stimuli impact cellular respiration?*

Observations: Respiration, like all biochemical processes, responds to environmental stimuli (e.g., temperature, salinity, acidity, light). However, some organisms tolerate a wider range of conditions than others.

Question: How is the rate of cellular respiration affected by environmental stimuli?

**a.** Establish a working lab group and obtain Inquiry-Based Learning Worksheet 12 from your instructor.

**b.** The preceding question will give you a general direction for your work, but you'll need to refine it before proceeding. Discuss with your group well-defined questions relevant to the preceding observation and question. Choose and record your group's best question for investigation.

**c.** Translate your question into a testable hypothesis and record it.

**d.** Review procedures 12.1 and 12.3, which use yeast, snails, and *Elodea* as model organisms to investigate respiration. Outline on Worksheet 12 your experimental design and supplies needed to test your hypothesis. Ask your instructor to review your proposed investigation.

**e.** Conduct your procedures, record your data, answer your question, and make relevant comments.

**f.** Discuss with your instructor any revisions to your questions, hypothesis, or procedures. Repeat your work as needed.

---

***b.*** What could cause any differences in these values?

***c.*** Determine the metabolic rate of other organisms available in the lab. How do their metabolic rates compare with that of a mouse?

## APPLICATIONS OF ANAEROBIC RESPIRATION

### Making Wine

In this exercise, you've seen how easy it is to demonstrate alcoholic fermentation by yeast. Many biologists as well as nonbiologists use this reaction to make their own wine. If you're game for an introduction to home wine-making, try the following procedure.

### Procedure 12.5   Making wine

*1.* Thoroughly clean and sterilize all glassware.

*2.* Combine a cake of yeast with either bottled grape juice or cranberry juice. Mix the yeast and juice in a ratio of approximately 5 liters of juice to 1 gram of yeast.

*3.* Add approximately 650 mL of the juice-yeast mix to each of four 1-liter Erlenmeyer flasks (or use 1- to 2-liter recycled plastic pop bottles).

*4.* Dissolve the following amounts of sucrose in each flask:

   Flask 1: 75 g      Flask 3: 300 g

   Flask 2: 150 g     Flask 4: no sucrose

**Figure 12.7**   Experimental setup for making wine.

*5.* For each flask, set up the fermentation apparatus as shown in figure 12.7.

*6.* Be sure to keep the procedure anaerobic by keeping the end of the exit tube under water in the adjacent flasks. This will prevent contamination by airborne bacteria and yeast.

*7.* Incubate the flasks at temperatures between 15°C and 22°C. Although fermentation will continue for a month or so, most fermentation will occur within the first 14 days. Fermentation is complete when bubbling stops.

*8.* To test your wine, remove the stopper and use a piece of tubing to siphon off the wine solution without disturbing the sediment in the bottom of the flasks. You may then want to filter the solution to remove any remaining yeast cells from the wine.

*9.* Taste your wine. If your wine has been contaminated by bacteria that produce acetic acid, vinegar may have been formed, so take your first sip cautiously.

## Question 8

***a.*** What differences are there in wines produced with different amounts of sugar?

***b.*** What would happen if oxygen were present (i.e., if conditions were not anaerobic)? If you have time, test your hypothesis.

1. How would you modify the experimental setup to introduce oxygen?

2. What results would you predict?

3. Based on your results, was your hypothesis accurate? Explain.

If you're interested in the finer points of wine making, visit your local bookstore or library. There may also be a local society of amateur wine-makers in your area who will be glad to give you some pointers on creating "a simply delightful bouquet."

## Making Kimchee

Pickling is an ancient way of preserving food. Pickling involves the anaerobic fermentation of sugars to lactic acid; this acid lowers the pH of the medium, thereby creating an environment in which other food-spoiling organisms cannot grow. Common foods preserved with pickling include sauerkraut, yogurt, and dill pickles. The ancient Chinese cabbage product kimchee, still a major part of the Korean diet, is also made with pickling. Here's how to make kimchee.

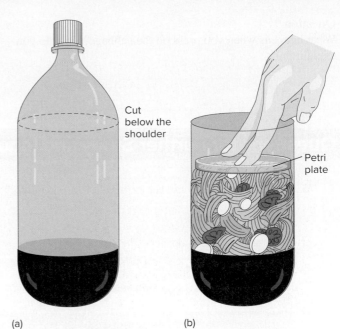

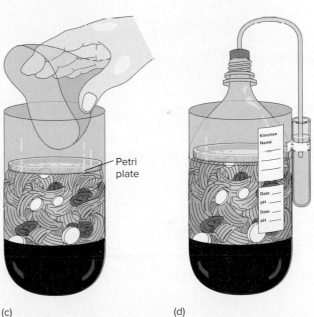

**Figure 12.8** Experimental setup for making kimchee. See the text for the recipe and procedure.

---

### Procedure 12.6 Making kimchee

1. Coarsely shred a head of cabbage. Place it in a mixing bowl with salt and allow it to wilt. This will draw some of the liquid out and prevent the finished kimchee from being watery.

2. Cut a 2-liter bottle just below the shoulder, as shown in figure 12.8*a*.

3. Add alternating layers of cabbage, garlic, pepper, and a sprinkling of salt in the bottle, pressing each layer down until the bottle is full. If you're using chilies or pepper, do not touch your eyes or mouth.

4. Place the petri plate lid, rim side up, atop the ingredients. Press down (fig. 12.8*b*). Within a few minutes, the salt will draw liquid from the cabbage; that liquid will begin to accumulate in the bottle.

5. For the next hour or so, continue to press the cabbage. You should then be able to fit the bottle top inside the bottle bottom, forming a sliding seal (fig. 12.8*c*). When you press with the sliding seal, cabbage juice will rise above the petri plate and air will bubble out around the edge of the plate.

6. The cabbage will pack half to two-thirds of the bottle's volume (fig. 12.8*d*). Every day, press on the sliding seal to keep the cabbage covered by a layer of juice.

**Question 9**

What happens when you press on the cabbage? How do you explain this?

7. Use pH-indicator paper to measure and record the pH of the juice each day (see Exercise 5).

8. After 4 to 7 days (depending on the temperature), the pH will have dropped from about 6.5 to about 3.5. Enjoy your kimchee!

## Questions for Further Study and Inquiry

1. What is the difference between respiration and breathing?

2. Does cellular respiration occur simultaneously with photosynthesis in plants? How could you determine the relative rates of each?

3. What role does cellular respiration play in an organism's metabolism?

4. What modifications of cellular respiration might you expect to find in dormant seeds?

5. In procedure 12.3, why did you subtract the control value from the titrant in beaker 1 and beaker 2?

6. Why is the volume of $CO_2$ production rather than $O_2$ uptake an adequate measure of respiration for the study of respiration rate?

### Doing Biology Yourself

Repeat the procedure to measure relative $CO_2$ production by aerobic organisms and include in your design an animal such as a fish. Would you expect greater $CO_2$ production from a fish or a snail? Why?

### Doing Biology Yourself

Repeat procedure 12.1 to measure $CO_2$ production in yeast and incubate the tubes at 4°C (refrigerator), 20°C (incubator), and/or 50°C (incubator). How does temperature affect the rate of fermentation by yeast?

# Photosynthesis
## Pigment Separation, Starch Production, and CO₂ Uptake

## Learning Objectives

By the end of this exercise you should be able to:

1. Relate each part of the summary equation for photosynthesis to the synthesis of sugar.
2. Describe the differences between the light-dependent and light-independent reactions involved in photosynthesis.
3. Separate the photosynthetic pigments using paper chromatography and calculate their $R_f$ numbers.
4. Use a spectroscope to describe the absorption of visible light by chlorophyll.
5. Describe fluorescence.
6. Describe the process of electron transport in chloroplasts and its role in photosynthesis.
7. Describe the change of pH that occurs as plants take up $CO_2$ from their environment during photosynthesis.
8. Describe the distribution of starch in leaves resulting from photosynthesis relative to the amount of light they receive and the distribution of pigments.

Please visit **connect.mheducation.com** to review online resources tailored to this lab.

Photosynthesis is the most important series of chemical reactions that occurs on earth (fig. 13.1). Indeed, virtually all life depends on photosynthesis for food and oxygen. **Photosynthesis** is a complex chemical process that converts radiant energy (light) to chemical energy (sugar). The following equation summarizes photosynthesis:

$$6\ CO_2 + 12\ H_2O \xrightarrow[chlorophyll]{light} C_6H_{12}O_6 + 6\ H_2O + 6\ O_2$$

Carbon   Water                  Sugar   Water  Oxygen
dioxide

Thus, photosynthesis is the light-dependent and chlorophyll-dependent conversion of carbon dioxide and water to sugar, water, and oxygen. Oxygen is released to the environment, and sugar is used to fuel growth or is stored as starch, a polysaccharide. Although water is present on both sides of the summary equation, these are not the same water molecules. The "reactant" water molecules (i.e., those on the left side of the equation) are split to release electrons during the photochemical (i.e., light-dependent) reactions. The "product" water molecules (i.e., those on the right side of the equation) are assembled from hydrogen and oxygen released during the photochemical and biochemical (i.e., light-independent) reactions. The photochemical reactions of photosynthesis are often referred to as the "light reactions." The biochemical

Nigel Cattlin/Science Source

**Figure 13.1** Sunlight powers photosynthesis, an energy-dependent process that fixes $CO_2$ into sugars. Less than 1% of the energy that reaches Earth is captured by photosynthesis, but this energy fuels virtually all life on the planet. Photosynthesis releases $O_2$, which is visible as bubbles on the aquatic plant *Elodea*.

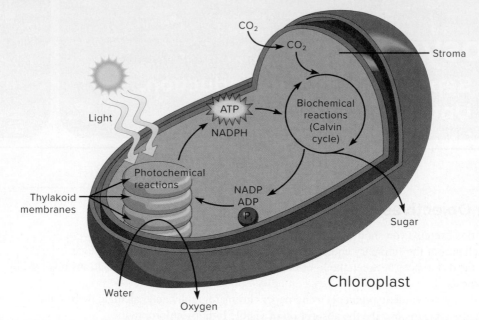

**Figure 13.2** Photosynthesis occurs in chloroplasts and consists of photochemical (the light-dependent "light reactions") and biochemical (the light-independent "dark reactions" including the Calvin cycle) reactions. The photochemical (i.e., light) reactions convert light-energy to chemical energy captured in ATP and NADPH. The biochemical (i.e., dark) reactions use the ATP and NADPH produced by the photochemical reactions to reduce $CO_2$ to sugars. The photochemical reactions occur on thylakoid membranes, whereas the biochemical reactions occur in the stroma. Chloroplasts perform the chemical reactions linking the inorganic world ($CO_2$) to the organic world (carbohydrates).

reactions are often referred to as the "dark reactions" or the Calvin cycle, in honor of Melvin Calvin, the botanist who described the reactions.

As mentioned, photosynthesis can be divided into two sets of reactions (fig. 13.2). Some characteristics of these reactions are compared here:

| *Photochemical "Light" Reactions* | *Biochemical "Dark" Reactions* |
|---|---|
| Fast (practically instantaneous) | Slower, but still extremely fast |
| Light-dependent | Light-independent |
| Splits water to release oxygen, electrons, and protons | Reduces (fixes) carbon dioxide to sugar |

In today's exercise, you'll investigate some of the major aspects of photosynthesis, beginning with the isolation and identification of photosynthetic pigments.

Before you begin studying photosynthesis, we should remind you that *all* organisms (including plants) carry out respiration in one form or another, but chlorophyll-containing organisms can *also* photosynthesize.

## PAPER CHROMATOGRAPHY OF PHOTOSYNTHETIC PIGMENTS

Light must be absorbed before its energy can be used. A substance that absorbs light is a **pigment.** The primary photosynthetic pigments that absorb light for photosynthesis are **chlorophylls *a* and *b*.** However, chlorophylls are not the only photosynthetic pigments; **accessory pigments** such as

**carotenoids** and **xanthophylls** also absorb light and transfer energy to chlorophyll *a*. During spring and summer, chlorophylls in leaves mask the colors of accessory pigments, and leaves appear green. In the fall, however, the production of chlorophyll stops, and the accessory pigments become prominent as beautiful "fall colors" of trees such as maples.

**Paper chromatography** is a technique for separating dissolved compounds such as chlorophyll, carotene, and xanthophyll. When a solution of these pigments is applied to strips of paper, the pigments adsorb onto the fibers of the paper. When the tip of the paper is immersed in a solvent, the solvent is absorbed and moves up through the paper. As the solvent moves through the spot of applied pigments, the pigments dissolve in the moving solvent. However, the pigments do not always keep up with the moving solvent—some pigments move almost as fast as the solvent, whereas others move more slowly. This differential movement of pigments results from each pigment's solubility and characteristic tendency to stick (i.e., be adsorbed) to the cellulose fibers of the paper. A pigment's molecular size, polarity, and solubility determine the strength of this tendency; pigments adsorbed strongly move slowly, whereas those adsorbed weakly move fastest. Thus, each pigment has a characteristic rate of movement, and the pigments can be separated from each other. In procedure 13.1, four bands of color will appear on the strip—a yellow band of xanthophylls, a yellow-orange band of carotenes, a blue-green band of chlorophyll *a*, and a yellow-green band of chlorophyll *b*.

The relationship of the distance moved by a pigment to the distance moved by the solvent front is specific for a given set of conditions. We call this relationship the $R_f$ number and define it as follows:

$$R_f = \frac{\text{Distance moved by pigment}}{\text{Distance from pigment origin to solvent front}}$$

Paper chromatography can be used to identify each pigment by its characteristic $R_f$. This $R_f$ is constant for a given pigment in a particular solvent-matrix system.

 **SAFETY FIRST** Before coming to lab, you were asked to read this exercise so you would know what to do and be aware of safety issues. In the space below, briefly list the safety issues associated with today's procedures. If you have questions about these issues, contact your laboratory assistant before starting work.

## Procedure 13.1 Separate plant pigments by paper chromatography

1. Observe the contents of the container labeled "Plant Extract." You'll use paper chromatography to separate its pigments.

 Extinguish all hotplates and flames before you do this experiment. Keep all solvents away from hotplates and flames at all times.

**Question 1**
What color is the plant extract, and why is it this color?

2. Obtain a strip of chromatography paper from your lab instructor. Handle the paper by its edges so that oil on your fingers does not contaminate the paper.

3. Use a Pasteur pipet or a fine-tipped brush to apply a stripe of plant extract approximately 2 cm from the tip of the paper (fig. 13.3). Blow the stripe dry and repeat this application at least 15 times. For this separation to work well, you must start with an extremely concentrated application of extract on the paper.

4. An alternate procedure is to place a fresh leaf directly on the paper, then press and roll the edge of a coin (quarter) over the leaf to crush the cells and form a stripe of pigment.

5. Place the chromatography strip in a test tube containing 2 mL of chromatography solvent (9 parts petroleum ether : 1 part acetone). Position the chromatography strip so that the tip of the strip (but not the stripe of plant extract) is submerged in the solvent. You can do this by hooking the strip of paper with a pin inserted in the tube's stopper (fig. 13.4).

6. Place the tube in a test-tube rack and watch as the solvent moves up the paper. Keep the tubes capped and undisturbed during solvent movement.

7. Remove the chromatography strip when the solvent front is within 1 cm of the top of the strip (i.e., after 2–3 min). Mark the position of the solvent front with a pencil and set the strip aside to dry. Observe the bands of color, then draw your results on figure 13.5. Use your textbook or other materials in lab to identify the different bands of pigments according to their position and color. For example, xanthophylls appear yellow.

8. Use a ruler to measure the distance from the pigment origin to the solvent front and from the

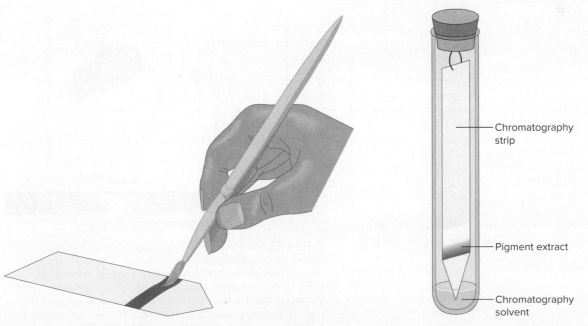

**Figure 13.3** Application of pigment extract to a chromatography strip.

**Figure 13.4** Chromatography setup.

— Chromatography strip

— Pigment extract

— Chromatography solvent

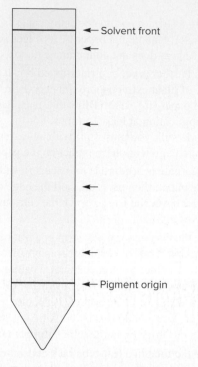

**Figure 13.5** Completed chromatogram. On the chromatogram, indicate the color of the band of pigment to the left of the arrows. To the right of the arrows, write the name of the pigment.

**Table 13.1**

$R_f$ Numbers for Four Plant Pigments

| Pigment | $R_f$ |
|---|---|
| Carotene | |
| Xanthophyll | |
| Chlorophyll $a$ | |
| Chlorophyll $b$ | |

origin to each pigment band. Calculate the $R_f$ number for each pigment; record your data in table 13.1.

**Question 2**

*a.* What does a small $R_f$ number tell you about the characteristics of the moving molecules?

*b.* Which are more soluble in the chromatography solvent, xanthophylls or chlorophyll $a$? How do you conclude this?

*c.* Would you expect the $R_f$ number of a pigment to change if you altered the composition of the solvent? Why or why not?

*d.* If yellow xanthophylls were present in the extract, why did the extract appear green?

*e.* Is it possible to have an $R_f$ number greater than 1? Why or why not?

## ABSORPTION OF LIGHT BY CHLOROPHYLL

A **spectroscope** is an instrument that separates white light into its component colors. These colors range from red to violet and appear as a spectrum when separated (fig. 13.6). Observe this spectrum by looking through the spectroscope provided in the lab. Now insert a chlorophyll sample between the light and spectroscope, and observe the resulting spectrum. Light not visible through the extract has been absorbed.

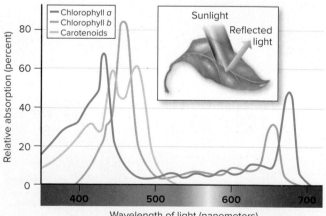

**Figure 13.6** The absorption spectrum of chlorophyll and carotenoids. The peaks represent wavelengths of sunlight absorbed by the two common forms of photosynthetic pigment, chlorophylls $a$ and $b$, and by the carotenoids. Chlorophylls absorb predominantly violet-blue and red light in two narrow bands of the spectrum and reflect green light in the middle of the spectrum. Carotenoids absorb mostly blue and green light and reflect orange and yellow light.

## Question 3
What colors are diminished or absent?

Based on this observation, complete the following absorption spectrum for chlorophyll. For each color, estimate the relative absorbance of that color by placing an X above the color name at the appropriate position along the y-axis. Connect the X's for all colors to complete the absorption spectrum.

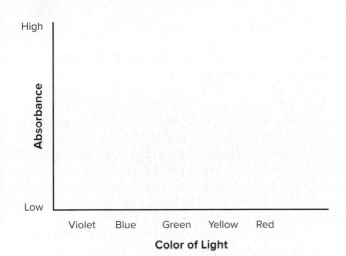

## Question 4
*a.* What color of light would be least effective for plant photosynthesis? Why?

*b.* If available, use an extract from red or orange peppers to plot an absorption spectrum for carotenoids. What colors of light are absorbed by carotenoids? How does the absorption spectrum for carotenoids differ from that of chlorophylls?

## FLUORESCENCE

Light produces reactions only if it is absorbed by a molecule. When sunlight strikes a plant, the chlorophyll absorbs some of the light and reflects some of the light. The green light is reflected and is responsible for the plant's green color. The absorbed light "excites" the chlorophyll by boosting electrons to a higher-energy orbital. During photosynthesis, the energy of these excited electrons from chlorophyll and chlorophyll's

central magnesium atom is passed efficiently to another pigment molecule and photosynthesis proceeds. However, to easily observe these energized electrons, we can disrupt the photosynthetic system by blending the cells during the preparation of the plant extract. As the electrons return to their original orbital, the energy is passed efficiently to another pigment molecule as light is harvested. Sometimes this efficient transfer of energy to another pigment molecule does not occur, and a photon of red light is emitted as the electron returns to its original orbital. This release of light energy is **fluorescence.** The wavelength of reemitted light is determined by the structure of the molecule reemitting the light.

To decrease the efficiency of energy transfer so that we can observe fluorescence, your instructor has disrupted the photosynthetic system by blending during the preparation of the extract you will use in the next procedure.

### *Procedure 13.2* Observe fluorescence by chlorophyll
Place a glass test tube containing chlorophyll extract in front of a bright light. View the extract from the side. (If a UV light is available, you can use the thin-layer chromatography strip from procedure 13.1 to observe fluorescence.)

## Question 5
What color light does the extract fluoresce?

## ELECTRON TRANSPORT IN CHLOROPLASTS

The photochemical reactions of photosynthesis transfer electrons among various compounds within chloroplasts (fig. 13.7). In 1937, Robin Hill demonstrated that isolated

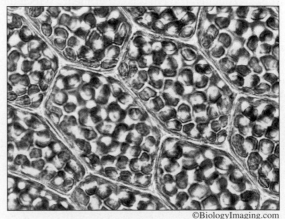

©BiologyImaging.com

**Figure 13.7** These photosynthetic cells of a moss are packed with bright-green chloroplasts (1000×).

chloroplasts could transfer electrons in the absence of $CO_2$ if provided with an alternate or artificial electron-acceptor. This observation indicated that electron transport does not require $CO_2$-fixation to occur. That is, electron transfer and $CO_2$-fixation involve separate sets of reactions.

You can detect electron transfer using a dye called 2,6-dichlorophenol-indolephenol (DCPIP). In its oxidized state, DCPIP is blue. After accepting electrons, DCPIP becomes reduced and colorless. DCPIP can accept electrons released in chloroplasts during photosynthesis. The rate of DCPIP decoloration depends on its concentration and the rate of electron flow. By measuring decoloration of DCPIP we can indirectly measure the rate of some reactions of photosynthesis. Because the rates of many chemical reactions are pH-dependent, a constant pH of approximately 6.5 is necessary for this experiment. The phosphate buffer used in this experiment maintains a constant pH of the incubation mixture.

### Procedure 13.3   Observe electron transport in chloroplasts

1.  Prepare test tubes according to table 13.2. Metabolically active chloroplasts will be provided by your instructor.

2.  Mix the contents of each tube well and place tubes 1–3 approximately 15 cm in front of a high-intensity light-bulb. Wrap tube 4 in aluminum foil and place it with the other three tubes. Do not position tubes behind each other. Keep all tubes directly in the path of the light.

3.  Observe the contents of the tubes intermittently; describe the changes in color that you see.

    _____

    _____

4.  If you have time, prepare a replicate of tube 2 in which water is replaced by 1 mL of 0.1 mM simizane or monuron, both herbicides. Handle all herbicides and pesticides carefully.

### Question 6

a.  What was the purpose of each of the different tubes (i.e., treatments) used in this experiment? Which tubes were controls?

b.  What happens when you illuminate the tube containing herbicide?

c.  Based on this result, what do you think is the mode of action of these herbicides?

## UPTAKE OF CARBON DIOXIDE DURING PHOTOSYNTHESIS

Phenol red (phenol-sulfonphthalein) is a pH-indicator that turns yellow in an acidic solution (pH < 7) and becomes red in a neutral to basic solution (pH > 7). (For more about pH and pH indicators, see Exercise 5.) In this experiment you will use the pH-indicator phenol red to detect the uptake of $CO_2$ by a photosynthesizing aquatic plant, _Elodea_ (see fig. 4.6). Recall that plants use $CO_2$ during the light-independent reactions of photosynthesis.

To detect $CO_2$ uptake you will put a plant into an environment that you have made slightly acidic with your breath. Carbon dioxide in your breath will dissolve in water to form carbonic acid, which lowers the pH of the solution:

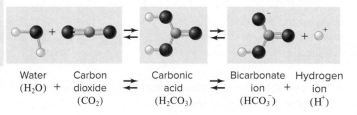

| Water ($H_2O$) + | Carbon dioxide ($CO_2$) | ⇌ | Carbonic acid ($H_2CO_3$) | ⇌ | Bicarbonate ion ($HCO_3^-$) + | Hydrogen ion ($H^+$) |

As the plant fixes $CO_2$ the pH rises. When the pH rises above 7, the solution turns red.

| Table 13.2 |
| --- |

**Solutions for Comparison of Photosynthetic Reaction Rates**

| Tube | Chloroplasts | 0.1 M PO$_4$ Buffer (pH 6.5) | H$_2$O | 0.2 mM DCPIP |
| --- | --- | --- | --- | --- |
| 1 | 0.5 mL | 3 mL | 1.5 mL | 0 |
| 2 | 0.5 mL | 3 mL | 0.5 mL | 1 mL |
| 3 | 0 | 3 mL | 1.0 mL | 1 mL |
| 4 | 0.5 mL | 3 mL | 0.5 mL | 1 mL |

## Procedure 13.4  Observe the uptake of $CO_2$ during photosynthesis

1. Fill two test tubes half full with a dilute solution of phenol red provided by your laboratory instructor (fig. 13.8). (Your instructor may have prepared this solution with carbonated water, which is acidic because it has been enriched with $CO_2$. If this has occurred, the solution will already be yellow, and you should skip to step 3 of this procedure.)

2. Use a straw to gently blow your breath into the phenol red solution. Because excess carbonic acid will lengthen this experiment, stop blowing in the tubes as soon as the color changes to yellow.

3. Add pieces of healthy *Elodea* totaling about 10 cm to one of the tubes. Pour off excess solution above the *Elodea*. The other tube, which is the control, will not include *Elodea*.

4. Cover the tops of the tubes with plastic film or foil to prevent gases from the atmosphere from diffusing into the tubes. Then place both tubes approximately 0.5 m in front of a 100-watt bulb for 30–60 min. What do you think will happen? Write your prediction and a brief explanation here: _____

   _____

   _____

5. Observe the tubes about every 10 min.

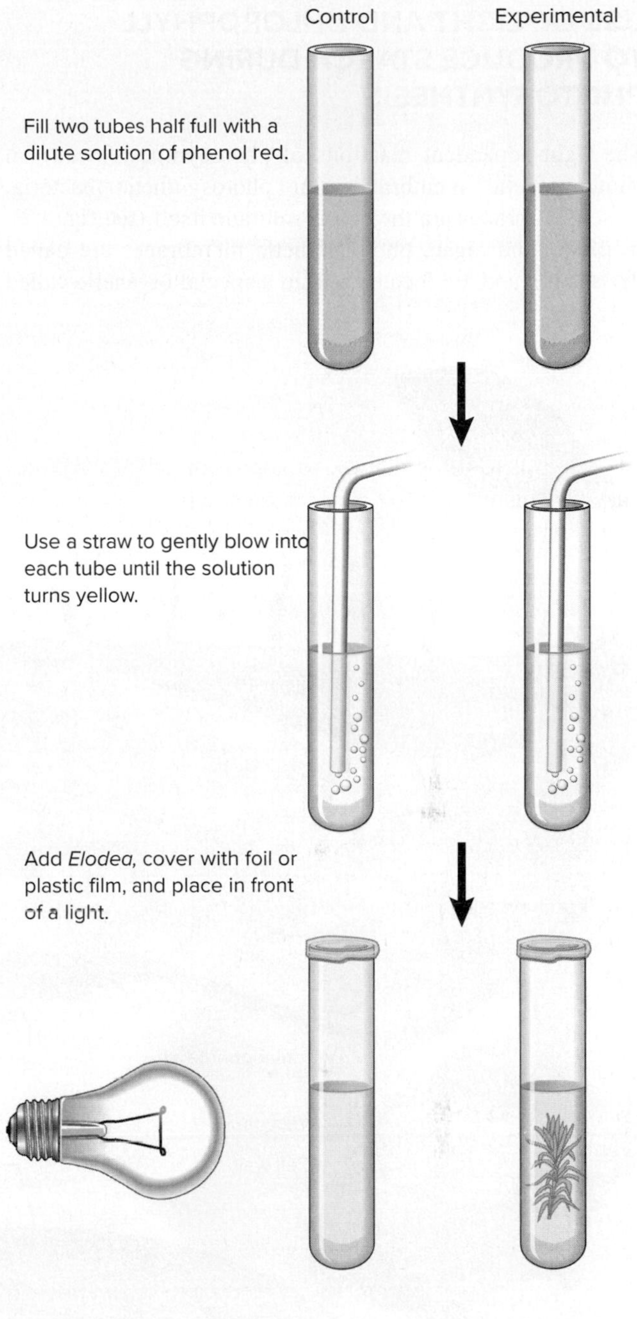

### Question 7

*a.* What happens to the color of the indicator?

*b.* What is the reason for the color change?

*c.* Did the solution in the control tube change color? Why or why not?

*d.* Considering the summary equation for photosynthesis, what is the basis for this change in color?

Control          Experimental

Fill two tubes half full with a dilute solution of phenol red.

Use a straw to gently blow into each tube until the solution turns yellow.

Add *Elodea*, cover with foil or plastic film, and place in front of a light.

**Figure 13.8**  Preparation of treatment and control test tubes to determine how photosynthesis affects the pH of a solution.

# USE OF LIGHT AND CHLOROPHYLL TO PRODUCE STARCH DURING PHOTOSYNTHESIS

The light-dependent reactions of photosynthesis occur on photosynthetic membranes. In photosynthetic bacteria, these membranes are the cell membrane itself (see fig. 4.2). In plants and algae, photosynthetic membranes are called **thylakoids** and are located within a special organelle called a **chloroplast**, an organelle specialized for photosynthesis (fig. 13.9; also see fig. 4.8). Thylakoids form stacks called **grana,** which are connected by **lamellae.** The semi-liquid stroma of the chloroplast interior contains the enzymes that catalyze the light-independent reactions of photosynthesis.

Sugars produced by photosynthesis are often stored as starch. Thus, starch production is another indirect measure of photosynthesis. To produce this starch, photosynthesis requires light as an energy source. In the absence of light, sugars and starch are not produced. Photosynthesis also requires chlorophyll to capture light energy. In the absence of chlorophyll, sugars and starch are not produced.

In the following procedures you will detect the presence of starch by staining it with a solution of iodine and observe the requirement of light and chlorophyll for photosynthesis.

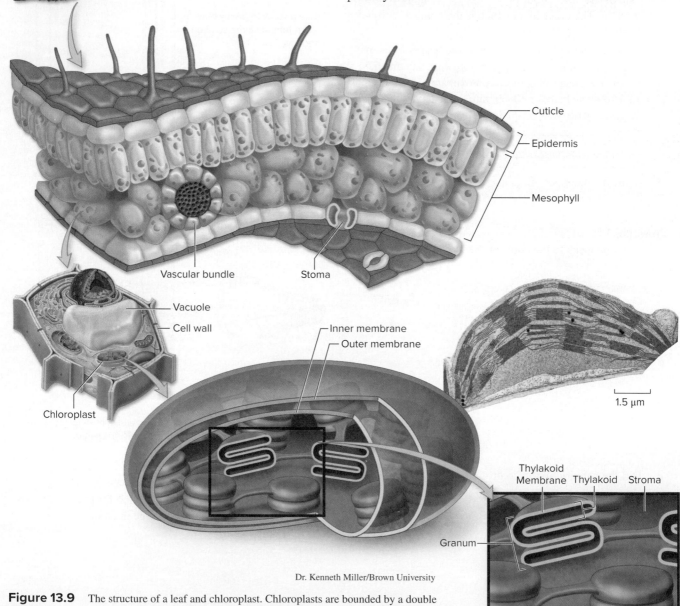

Dr. Kenneth Miller/Brown University

**Figure 13.9**    The structure of a leaf and chloroplast. Chloroplasts are bounded by a double membrane and contain photosynthetic membranes called thylakoids. Stacks of thylakoids are referred to as a granum (plural, grana), and the semi-liquid interior of the chloroplast is called the stroma. The openings that enable $CO_2$ to enter the leaf are stomata (singular, stoma).

## Procedure 13.5 Stain starch with iodine

1. Place separate drops of water, glucose, and starch solutions on a glass slide.
2. Add a drop of iodine to each and describe your results.

## Procedure 13.6 Observe starch production during photosynthesis

1. Remove a leaf from a *Geranium* plant that has been illuminated for several hours.
2. After immersing the leaf in boiling water for 1 min, remove the pigments from the leaf by boiling the leaf in methanol for 3–5 min. This part of the procedure (i.e., the boiling methanol) must be done in a fume hood. Boiling the leaf will remove pigments so that you can see the color changes of the iodine starch test.

 Exercise extreme caution when you heat methanol.

3. Place the leaf in a petri dish containing a small amount of water, and then add five to eight drops of iodine.
4. Observe any color change in the leaf.
5. Record in figure 13.10*a* the color of the leaves after each successive treatment.

### Question 8
*a.* Was starch stored in the leaf? How can you tell?

*b.* Would you expect leaves to be the primary organ for starch storage in plants? Why or why not?

## Procedure 13.7 Observe the requirement of light for photosynthesis

1. Obtain a *Geranium* leaf that has been half or completely covered with metal foil or thick paper for three or four days.

2. Repeat the bleaching and staining steps described in procedure 13.6.
3. Describe and explain any color change in the leaf.

_____

_____

4. Record in figure 13.10*b* the color of the leaves after each successive treatment.

### Question 9
Does a leaf produce starch if it has been deprived of light?

## Procedure 13.8 Observe the requirement of chlorophyll for photosynthesis

1. Obtain leaves of a variegated *Coleus* plant (fig. 13.11*a*) and a purple-leafed *Coleus* plant (fig. 13.11*b*). Make sketches of their original pigmentation patterns in figure 13.10*c, d.* Indicate which areas are green, red, green/red, and white.
2. Extract the pigments and stain for starch according to procedure 13.6. Boiling the leaf in water will remove the water-soluble pigments such as the red cyanins, and boiling the leaf in methanol will remove chlorophyll. These pigments must be removed for you to see the color changes of the iodine starch test.
3. Record in figure 13.10*c, d* the color of the leaves after each successive treatment.

### Question 10
*a.* How does the pattern of starch storage relate to the distribution of chlorophyll?

*b.* Photosynthesis requires chlorophyll (green), but some of the *Coleus* leaves that you tested were purple. How do you explain your results?

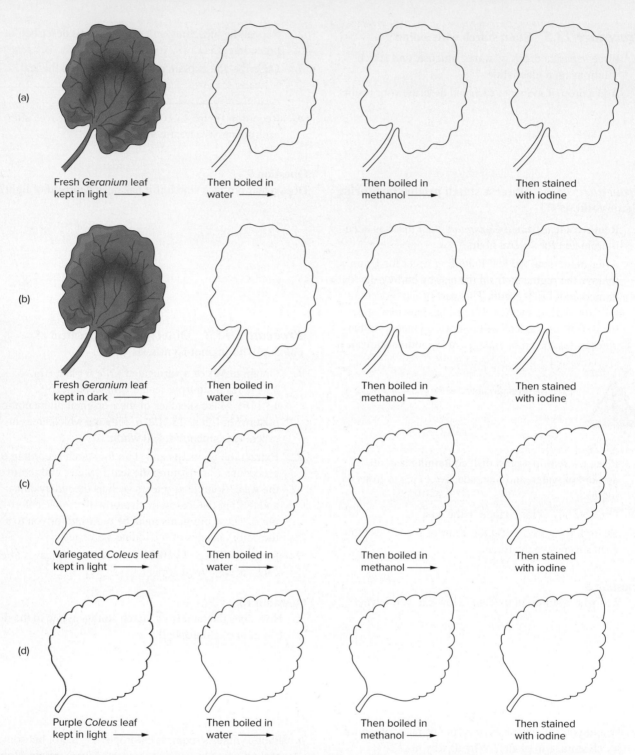

**Figure 13.10** The requirement of light and chlorophyll and the production of starch during photosynthesis. Within each diagram, record the color of the leaf following the treatments to indicate (*a*) the production of starch, (*b*) the need for light, and (*c, d*) the need for chlorophyll for photosynthesis. Record your results from the appropriate procedure by writing the resulting color of each treated leaf directly onto the outline of the leaf.

(a) ©BiologyImaging.com    (b) ©BiologyImaging.com

**Figure 13.11** *Coleus* plants. (*a*) Leaves of this variegated plant have green, white, purple, and pink areas resulting from combinations of chlorophylls and anthocyanin (red) pigments. (*b*) Leaves of this purple *Coleus* have the same pigment combination throughout the leaf.

## INQUIRY-BASED LEARNING

### *Does a plant's respiration produce as much carbon dioxide as photosynthesis captures?*

Observations: Recall from Exercise 12 that aerobic cellular respiration releases $CO_2$, which can combine with water to form carbonic acid and lower the pH (see procedure 12.3). *Elodea* growing in light respires and photosynthesizes. *Elodea* in darkness only respires. Design an experiment to measure *Elodea's* relative uptake and production of $CO_2$.

Question: What is the relative uptake versus production of $CO_2$ during photosynthesis and respiration?

a. Establish a working lab group and obtain Inquiry-Based Learning Worksheet 13 from your instructor.

b. Discuss with your group a well-defined question relevant to the preceding observation and question. Record it on Worksheet 13.

c. Translate your question into a testable hypothesis and record it.

d. Review procedure 12.3 that provides a method to quantify $CO_2$ production. Outline on Worksheet 13 your experimental design and supplies needed to test your hypothesis. Ask your instructor to review your proposed investigation.

e. Conduct your procedures, record your data, answer your question, and make relevant comments.

f. Discuss with your instructor any revisions to your questions, hypothesis, or procedures. Repeat your work as needed.

## DOING BIOLOGY YOURSELF

Recall that respiration produces $CO_2$, which combines with water to form carbonic acid that lowers the pH of a surrounding solution. Design an experiment to measure the relative dynamics (mass balance) of photosynthesis versus respiration for *Elodea*.

## WRITING TO LEARN BIOLOGY

Use a reference to determine the relative penetration of different wavelengths of light through water. Describe how this could affect the existence and distribution of submerged plants.

*1.* Why does chlorophyll appear green?

*2.* Is starch produced when a leaf is kept in the dark? Why or why not?

*3.* What causes leaves to turn from green to yellow and red in autumn?

*4.* Of what value to plants is starch? Of what value to animals is starch?

*5.* What is the significance of electron transport in the photochemical (i.e., light-dependent) reactions of photosynthesis?

*6.* Design an experiment to determine if plants respire. Be sure to explain how you would measure respiration and the controls you would include in the experimental design.

# Mitosis
## Replication of Eukaryotic Cells

## Learning Objectives

By the end of this exercise you should be able to:
1. Describe events associated with the cell cycle.
2. Describe events associated with mitosis.
3. Distinguish the stages of mitosis on prepared slides of mitotic cells.
4. Stain and examine chromosomes in mitotic cells.
5. Estimate the duration of various stages of mitosis from experimental observations.

Please visit **connect.mheducation.com** to review online resources tailored to this lab.

The growth of organisms includes cell enlargement, cell replication, and cell specialization to sustain life. All of these activities are part of a repeating set of events called the **cell cycle.** A major feature of the cell cycle is cellular replication, and a major feature of cellular replication is mitosis. **Mitosis** is the replication and division of the nucleus of a eukaryotic cell in preparation for cytokinesis. During mitosis, chromosomes within the cell are replicated by enzymes and then separated into two identical sets—each set is then surrounded by a nuclear membrane. Each of the two new nuclei has a full set of chromosomes containing a copy of all of the genetic information for the organism. Prokaryotic cells lack nuclei and do not undergo mitosis. Instead, they replicate their chromosome and then divide in half during a process called binary fission (described in Exercise 24).

Mitosis is usually associated with **cytokinesis,** the division of the cell and cytoplasm into halves that each contains a nucleus. In some tissues, cytokinesis is delayed or does not occur at all, and the cells are multinucleate. Mitosis and cytokinesis are important because they provide a mechanism for orderly growth of living organisms.

### Question 1
Consider the surface-to-volume ratios of large versus small cells. Is it adaptive for cells of a growing organism to remain small? Explain your answer.

## THE CELL CYCLE

This exercise emphasizes events associated with mitosis, but mitosis is only part of the cell cycle (fig. 14.1). The remainder of the cycle is called **interphase** and is subdivided further into **cytokinesis (C), gap 1 ($G_1$), synthesis (S),** and **gap 2 ($G_2$)** phases.

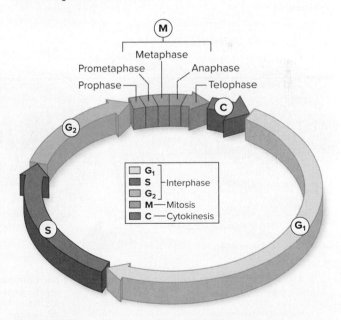

**Figure 14.1** The cell cycle is depicted as a circle. The first gap phase, $G_1$, involves growth and preparation for DNA synthesis. During the S phase, a copy of the genome is synthesized. The second gap phase, $G_2$, prepares the cell for mitosis. During mitosis (M), replicated chromosomes are partitioned. Cytokinesis divides the cell into two cells with identical genomes.

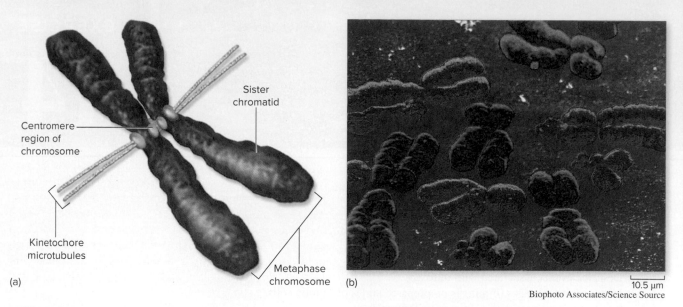

(a)

(b)

10.5 μm

**Figure 14.2** Chromosomes. (*a*) In a metaphase chromosome, kinetochore microtubules are anchored to proteins at the centromere. (*b*) This electron micrograph shows how human chromosomes appear during the early stages of nuclear division. Each strand of DNA has already been replicated and condensed to form discrete sister chromatids identical to each other and held together by a centromere.

The cell cycle begins with the formation of a new cell and ends with replication of that cell. The G₁ phase of the cell cycle occurs after mitosis and cytokinesis, and is when the majority of cellular activity for the functions of the cell occurs. Many cell-specific proteins and other molecules are produced for the metabolism of the cell during G₁. During the S phase, the elongate DNA molecules are replicated. The DNA copies condense to form compact structures called chromosomes that can be easily seen with a light microscope. At the end of the S phase each chromosome consists of an identical pair of chromosomal DNA strands, called sister **chromatids,** attached at a **centromere** (fig. 14.2). During the G₂ phase, molecules and structures necessary for mitosis are synthesized.

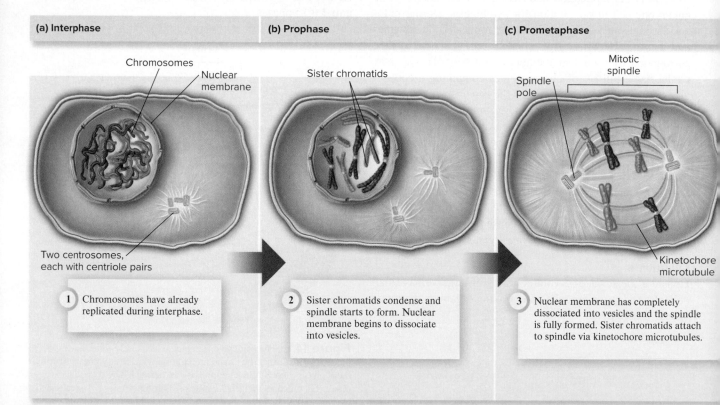

**Figure 14.3** Interphase and the stages of mitosis in an animal cell. The cleavage furrow signifying cytokinesis may first appear during anaphase or more typically during telophase.

Mitosis (M phase) usually lasts for less than 10% of the time of the cell cycle, which usually lasts 10 to 30 h. Actively dividing cells such as those in rapidly growing tissues may spend more than 10% of their time in mitosis, whereas static cells such as bone cells or neurons may rarely enter M phase. Cytokinesis may begin during mitosis but is highly variable in length and timing. Tissues such as striated muscle fibers, and some algal filaments, may undergo mitosis without cytokinesis and produce multinucleate cells.

### Question 2

*a.* Mitosis and cytokinesis are often referred to collectively as "cellular division." Why are they more accurately called cellular replication?

*b.* Does the cell cycle have a beginning and an end? Explain.

## STAGES AND EVENTS OF MITOSIS

Mitosis (1) separates the genetic material duplicated during interphase into two identical sets of chromosomes and (2) reconstitutes a nucleus to house each set. As a result, mitosis produces two identical nuclei from one. In animals, mitosis occurs in body (i.e., non-sex) cells.

Mitosis is traditionally divided into five stages: **prophase, prometaphase, metaphase, anaphase,** and **telophase** (fig. 14.3). The actual events of mitosis are not discrete but occur in a continuous sequence; separation of mitosis into five stages is merely convenient for our discussion and organization. During these stages, important cellular structures are synthesized and perform the mechanics of mitosis (fig. 14.3). For example, in animal cells, two microtubule-organizing centers called **centrosomes** contain cylinders of microtubules called **centrioles,** which replicate at the onset of mitosis. The pairs of centrioles move apart and form an axis of proteinaceous microtubules between them called **polar microtubules.** The centrioles continue to move apart until they reach opposite poles of the cell and have a bridge of microtubules called the **mitotic spindle** (or spindle apparatus) extending between them (fig. 14.6). **Kinetochore microtubules** (fig. 14.2) attach to each chromosome's **kinetochore,** which is a complex of proteins that binds to the centromere. The pairs of sister chromatids align at the midpoint of the cell on an imaginary plane called the **metaphase plate.** At the poles of the cell, the

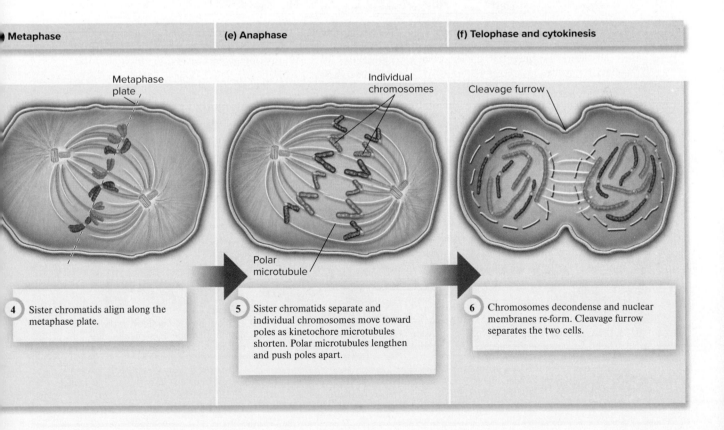

**Metaphase**      **(e) Anaphase**      **(f) Telophase and cytokinesis**

Metaphase plate

Individual chromosomes

Cleavage furrow

Polar microtubule

4   Sister chromatids align along the metaphase plate.

5   Sister chromatids separate and individual chromosomes move toward poles as kinetochore microtubules shorten. Polar microtubules lengthen and push poles apart.

6   Chromosomes decondense and nuclear membranes re-form. Cleavage furrow separates the two cells.

# Cancer—A Corrupt Cell Cycle

Cancer is unrestrained cell proliferation caused by damage to genes that regulate the cell division cycle. Cancer produces a cluster of cells called a *tumor* that constantly expands. Tumors from cells in connective tissue, bone, or muscle are known as **sarcomas,** while those from epithelial tissue, such as skin, are called **carcinomas** (fig. 14.4). In the United States, the four deadliest human cancers (about 55% of all cancer deaths) are lung cancer, colon cancer, breast cancer, and leukemia/lymphomas. Recent work has identified one of the culprits in cancer. Officially dubbed p53, this gene plays a key role in the $G_1$ checkpoint of cell division (fig. 14.5). The gene's product—the p53 protein—monitors the integrity of DNA, checking that it is undamaged. If the p53 protein detects damaged DNA, it halts cell division and stimulates enzymes to repair the damage. After repair, p53 allows cell division to continue. If the DNA damage is irreparable, then the p53 directs the cell to kill itself.

The p53 gene prevents the development of many mutated cells and is therefore a **tumor-suppressor gene.** Researchers have found that p53 is absent or damaged in the majority of cancerous cells they have examined. It is precisely because p53 is nonfunctional that cancer cells are able to repeatedly undergo cell division without being halted at the $G_1$ phase of the cell cycle.

**Figure 14.4** Portrait of a cancer. This carcinoma is developing from epithelial cells that line the interior of a human lung. As the mass of cells grows, it invades the surrounding tissues, eventually penetrating lymphatic and blood vessels, both of which are plentiful in the lung. These vessels carry metastatic cancer cells throughout the body, where they lodge and grow, forming new masses of cancerous tissue.

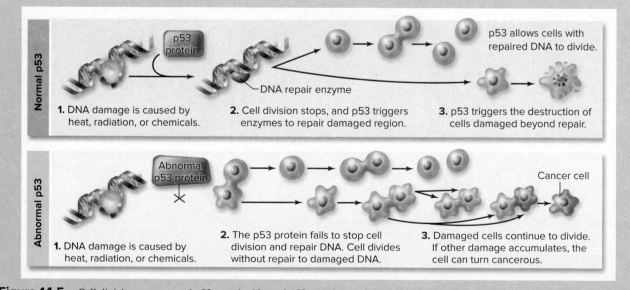

**Figure 14.5** Cell division, cancer, and p53 protein. Normal p53 protein monitors DNA, destroying cells that have irreparable damage to their DNA. Abnormal p53 protein fails to stop cell division or repair DNA. As damaged cells proliferate, cancer develops.

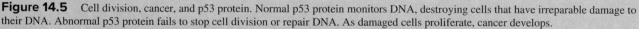

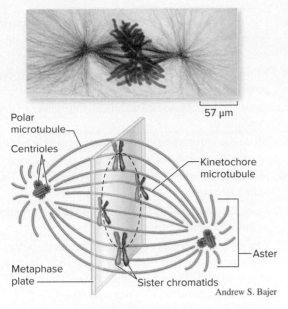

Polar microtubule
Centrioles
Kinetochore microtubule
Aster
Metaphase plate
Sister chromatids

57 μm

Andrew S. Bajer

**Figure 14.6** The structure of the mitotic spindle. The mitotic spindle is formed by the centrosomes from three types of microtubules. The astral microtubules emanate away from the region between the poles. The polar microtubules project into the region between the two poles. The kinetochore microtubules are attached to the kinetochores of sister chromatids. The spindle is a three-dimensional structure. The chromosomes are arrayed in a rough circle on the metaphase plate.

centrioles radiate an array of microtubules outward in addition to the spindle apparatus. These microtubules brace the centrioles against the cell membrane. This arrangement of microtubules radiating from a centriole is called an **aster** with **astral microtubules.** Plant cells lack centrioles and asters, but spindle fibers still form between opposite poles of the cell.

Interestingly, animal cells deprived of centrioles will still form a spindle apparatus. Chromosomes will eventually distribute themselves on the spindle apparatus and are moved and separated to opposite poles. The distribution of chromosomes will also occur if the cell is haploid (i.e., has a single set of chromosomes). The vegetative cells of many organisms such as fungi are haploid rather than diploid (have a double set of chromosomes). However, the steps of mitosis are the same as for diploid cells.

**SAFETY FIRST** Before coming to lab, you were asked to read this exercise so you would know what to do and be aware of safety issues. In the space below, briefly list the safety issues associated with today's procedures. If you have questions about these issues, contact your laboratory assistant before starting work.

| Table 14.1 | |
| --- | --- |
| **Events of Mitosis and Interphase** | |
| Interphase | Although interphase is not part of nuclear replication, understanding its events is essential to understanding mitosis. |
| Prophase | |
| Prometaphase | |
| Metaphase | |
| Anaphase | |
| Telophase | |

## Procedure 14.1 Describe the specific events of mitosis

Before your lab meeting, review in your textbook the events associated with each stage of mitosis in addition to the preparatory stage, interphase. List these events in table 14.1. Some events and structures occur only in plant cells and some occur only in animal cells. Mark these events in your list with an asterisk. This list can serve as an excellent study guide, so be as complete as possible. One event for each stage is provided in figure 14.3.

### Question 3

***a.*** If a nucleus has eight chromosomes during interphase, how many chromosomes does it have during metaphase?

***b.*** How many does it have after mitosis is complete?

Understanding the movements of chromosomes is crucial to understanding mitosis. You can simulate these movements easily with chromosome models made of pipe cleaners or popsicle sticks. This is a simple procedure but a valuable one. It will be especially helpful when you are comparing the events of mitosis to the events of meiosis, which you will simulate in the next exercise.

Note that one chromosome consists of all DNA strands attached to the centromere. A replicated chromosome (fig. 14.2*a*) has two strands of DNA, but it is still considered one chromosome. When the strands (chromatids) separate during anaphase, the replicated chromosome becomes two chromosomes.

## Procedure 14.2 Simulate chromosomal replication and movement during mitosis

1. Examine the materials to be used as chromosome models provided by your instructor.

2. Identify the differences in chromosomes represented by various colors, lengths, or shapes of materials. Also identify materials representing centromeres.

3. Place a sheet of notebook paper on your lab table to use in representing the boundaries of the mitotic cell.

4. Assemble the chromosomes needed to represent nuclear material in a cell of a diploid organism with a total of six chromosomes. Place the chromosomes in the cell.

5. Arrange the chromosomes to depict the position and status of chromosomes during interphase $G_1$. (During $G_1$ the chromosomes are usually not condensed, as the chromosome models imply, but the models are an adequate representation.)

6. Depict the status of chromosomes after completing interphase S. Use additional "nuclear material" if needed.

7. Move the chromosome models appropriately to depict prophase.

8. Move the chromosome models appropriately to depict metaphase.

9. Move the chromosome models appropriately to depict anaphase.

10. Move the chromosome models appropriately to depict telophase.

11. Draw the results of cytokinesis and the re-formation of nuclear membranes.

12. Chromosomal events occur as a continuous process of movements rather than in distinct steps. Therefore, repeat steps 4–11 as a continuous process and ask your instructor to verify your simulation.

## MITOSIS IN ANIMAL CELLS

The most distinctive features of cellular replication in animal cells are the formation of asters with centrioles at their center (discussed earlier) and cytokinesis. Cytokinesis includes formation of a **cleavage furrow** that begins on the periphery of the cell, pinches inward, and eventually divides the cytoplasm into two cells (fig. 14.7). Cells of a whitefish blastula provide good examples of the stages of mitosis and cytokinesis. Whitefish are commonly cultured fish whose eggs and early developmental stages undergo rapid cell divisions (as do all embryonic cells). A blastula is an early embryonic stage of a vertebrate and consists of a sphere of 25–100 cells with a high frequency of different mitotic stages. Exercise 50 (Embryology) details the formation of a blastula during embryonic development.

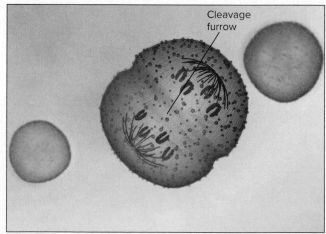

Cleavage furrow

MedicalRF.com/Alamy Stock Photo

**Figure 14.7** Cytokinesis in an animal cell. Cytokinesis, the physical division of the cell's cytoplasm, usually occurs after nuclear replication is complete. A cleavage furrow is forming around this dividing sea urchin egg.

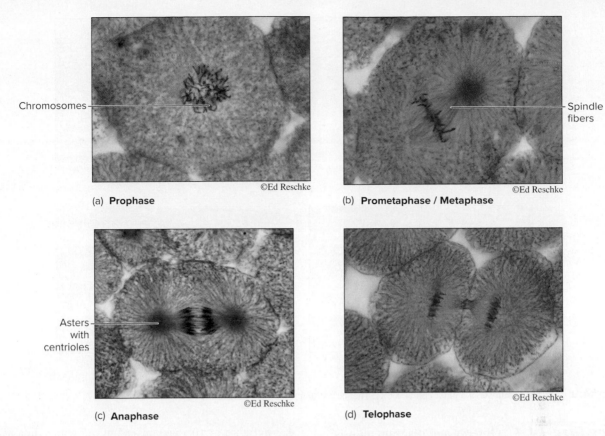

Chromosomes

(a) **Prophase** ©Ed Reschke

Spindle fibers

(b) **Prometaphase / Metaphase** ©Ed Reschke

Asters with centrioles

(c) **Anaphase** ©Ed Reschke

(d) **Telophase** ©Ed Reschke

**Figure 14.8** Stages of mitosis in cells of a whitefish embryonic blastula (400×). Prometaphase and metaphase may not always be distinguishable by light microscopy.

## *Procedure 14.3* **Observe and describe mitosis in animal cells**

1. Obtain a prepared slide of a cross section through the blastula of a whitefish.
2. Examine the cells first on low (10×) then high (40× or 100×) magnification. Some of the cells contain condensed and stained chromosomes.
3. Refer to figure 14.3 for a summary of the stages of mitosis. Identify examples of each stage on your prepared slide (fig. 14.8). Verify these stages with your lab partner or teaching assistant.
4. Also identify cells that you believe are between stages.
5. Examine the whitefish cells for signs of cytokinesis.
6. Prepared cross-sections of cells show only two dimensions, but mitosis is a three-dimensional process. In the following space, draw two cells in metaphase: one in which the cross section is parallel to the axis of the spindle apparatus and one in which the cross section is perpendicular to the spindle apparatus.

## Question 4

**a.** Why would we choose an embryonic mass of cells for procedure 14.3 in which to study the stages of mitosis?

**b.** Which stage of mitosis most often is associated with the beginning of cytokinesis?

## MITOSIS IN PLANT CELLS

Our model to study cellular replication in plants is the root tip of *Allium* (onion). Root tips of plants contain **meristems,** which are localized areas of rapid cell division due to active growth at the root tips. In plant cells, cytokinesis includes formation of a partition called a **cell plate** perpendicular to the axis of the spindle apparatus. The cell plate forms in the middle of the cell and grows out to the periphery. It will separate the two new cells.

Interestingly, the formation of the spindle apparatus and other microtubule systems in plant and fungal cells is organized by centrosomes, as in animal cells. But plant and fungal cells have no centrioles within the centrosomes. Thus, the function and necessity of centrioles remain somewhat of a mystery.

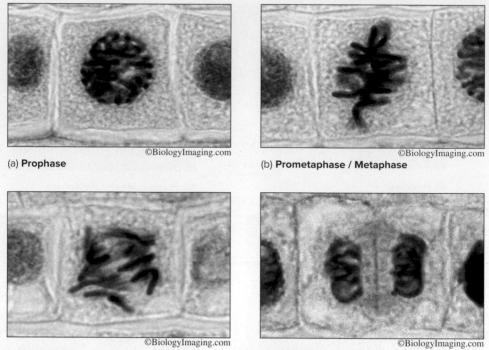

(a) **Prophase**                    ©BiologyImaging.com

(b) **Prometaphase / Metaphase**    ©BiologyImaging.com

(c) **Anaphase**                    ©BiologyImaging.com

(d) **Telophase**                   ©BiologyImaging.com

**Figure 14.9**   Stages of mitosis in a plant cell (1000×). The dark structures are chromosomes.

*Procedure 14.4*   **Observe and diagram mitosis in plant cells**

1.  Examine a prepared slide of a longitudinal section through an onion root tip.

2.  Search for examples of all stages of mitosis (fig. 14.9). Notice that most cells are in some part of interphase. Prometaphase may be difficult to distinguish from metaphase.

3.  Search for signs of cell plate formation.

4.  In figure 14.10, diagram a plant cell with a diploid number of three pairs of chromosomes in each of the stages of mitosis. **Diploid** refers to a nucleus with two of each type of chromosome. Be sure to label the cell wall and cell plate.

5.  Prepared cross sections of cells show only two dimensions, but mitosis is a three-dimensional process. In the following space draw two cells in metaphase: one in which the plane of section is parallel to the axis of

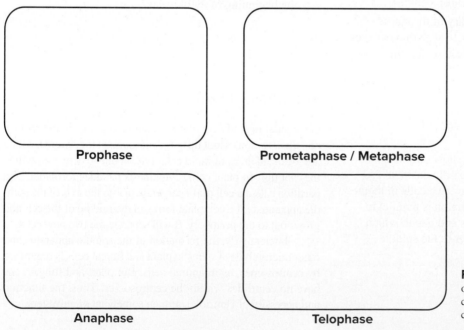

**Prophase**

**Prometaphase / Metaphase**

**Anaphase**

**Telophase**

**Figure 14.10**   Diagram the stages of mitosis in a plant cell with six chromosomes. The outlines represent the cell walls of each of four cells.

# INQUIRY-BASED LEARNING

## How much time elapses during the various stages of mitosis?

Observations: The cell cycle of actively dividing cells of root tips of *Allium* is approximately 24 h long. The phases of mitosis usually occupy only a small portion of that time.

Question: How long does mitosis take?

a. Establish a working lab group and obtain Inquiry-Based Learning Worksheet 14 from your instructor.

b. Discuss with your group a well-defined question relevant to the preceding observation and question. Record it on Worksheet 14.

c. Translate your question into a testable hypothesis and record it.

d. Each prepared slide of a root tip of *Allium* reveals a snapshot in time of all stages of mitosis.

e. The relative abundance of cells in a phase of mitosis is directly proportional to the length of time for that phase.

f. Outline on Worksheet 14 your experimental design and supplies needed to test your hypothesis. The table below offers insight to a reasonable design. Ask your instructor to review your proposed investigation.

g. Conduct your procedures, record your data, answer your question, and make relevant comments.

h. Discuss with your instructor any revisions to your questions, hypothesis, or procedures. Repeat your work as needed.

| Stage of Mitosis | Predicted Duration (hours) | Number of Cells in Each Stage | Total Number of Cells in Each Stage | Class Total Number for Each Stage | Calculated Duration (hours) |
|---|---|---|---|---|---|
| Interphase | _____ | ___ ___ ___ | _____ | _____ | _____ |
| Prophase | _____ | ___ ___ ___ | _____ | _____ | _____ |
| Prometaphase/ Metaphase | _____ | ___ ___ ___ | _____ | _____ | _____ |
| Anaphase | _____ | ___ ___ ___ | _____ | _____ | _____ |
| Telophase | _____ | ___ ___ ___ | _____ | _____ | _____ |
| **Totals** | _____ | | _____ | _____ | _____ |

the spindle apparatus and one in which the cross section is perpendicular to the spindle apparatus.

**Question 5**

*a.* What region of a root has the most mitotic activity?

*b.* How does cytokinesis differ in plant versus animal cells?

*c.* Why is pinching of the cytoplasm inadequate for cytokinesis in plant cells?

*d.* What is a cell plate, and in what stage of mitosis does it form?

*e.* Locate a plant cell in late telophase. What is the volume of the two new cells relative to a mature cell?

## PREPARING AND STAINING CHROMOSOMES

Your instructor has prepared some living onion root tips for you to process further and use to observe the stages of mitosis.

### Procedure 14.5   Stain chromosomes

1. Obtain an onion root tip and place it in a small vial with Schiff's reagent for 30 min. Handle Schiff's reagent carefully because it is a colorless liquid that becomes bright red after reaction.

   Keep the vial in the dark and at room temperature until the root tip becomes purple. Your instructor may have already stained some root tips for you.

2. Place the root tip in a drop of 45% acetic acid on a slide and cut away and remove all of the root tissue except the terminal 1 mm of the tip.

Acetic acid is corrosive. Do not spill it.

3. Crush the root tip with a blunt probe and cover the tissue with a coverslip.

4. Smash the tissue by pressing on the coverslip with the eraser of your pencil. Your instructor will demonstrate this procedure.

5. Scan your preparation at low magnification to locate stained chromosomes. Then switch to high magnification and locate formations of chromosomes that indicate each of the stages of mitosis.

6. Add a drop of acetic acid to the edge of the coverslip to avoid desiccation.

7. Locate as many stages of mitosis as you can. Be sure to look at preparations done by other students.

## Questions for Further Study and Inquiry

1. Interphase has sometimes been called a "resting stage." Why is this inaccurate?

2. Most general functions of a cell occur during $G_1$ of interphase. What events that occur during other phases of the cell cycle might inhibit general metabolism?

3. Read in your textbook about prokaryotic cellular replication; list the fundamental cellular/structural differences between it and eukaryotic cellular replication. What is the basis for these differences?

4. Some specialized cells such as neurons and red blood cells lose their ability to replicate when they mature. Which phase of the cell cycle do you suspect is terminal for these cells? Why?

5. Find a concise definition of "cancer." How might methods to treat cancer relate to what you learned in this lab exercise?

### WRITING TO LEARN BIOLOGY

Refer to your textbook to review the properties of a chemical called colchicine. Describe how colchicine affects dividing cells. What is the mechanism of this effect? How might colchicine be used as a tool in scientific research or medicine?

### WRITING TO LEARN BIOLOGY

Write a summary of the mechanism and consequences of using the drug Vincristine that blocks mitotic spindle formation by cancer (and other) cells. Reference the scientific literature.

# Meiosis
## Reduction Division and Gametogenesis

## Learning Objectives

By the end of this exercise you should be able to:
1. Describe the events of meiosis.
2. Compare and contrast meiosis and mitosis.
3. List the most significant events of meiosis.
4. Explain the relevance of meiosis to sexual reproduction and evolutionary change.
5. Explain the relationship of meiosis and gametogenesis.
6. Describe the events of spermatogenesis and oogenesis.

Please visit **connect.mheducation.com** to review online resources tailored to this lab.

**S**ex is one of the most experienced and scrutinized processes of life. Biologists know that the significance of sex, and meiosis in particular, is the recombination of a parent's genes and the packaging of these genes as a gamete. During sexual reproduction, a gamete and its genes are combined with another parent's gamete and genes to endow the new offspring with new genetic combinations. Production of new combinations allows populations to adapt to changing environmental conditions.

Chromosomes in typical, eukaryotic nuclei occur in pairs; that is, the nuclei are **diploid** ($2n$). The two chromosomes of a pair are called **homologous chromosomes,** and each homologue of a pair has the same sites, or **loci,** for the same genes, although the homologues may carry different alleles at homologous loci. A nucleus, such as that in a gamete, with only one chromosome of each homologous pair is **haploid** ($n$).

**Meiosis** produces haploid daughter nuclei and is sometimes called "reduction division." Reducing the number of chromosomes in the nucleus of a gamete to only one of each pair is important because such a haploid nucleus can fuse with another haploid nucleus during sexual reproduction and restore the original, diploid number of chromosomes to the new individual (fig. 15.1).

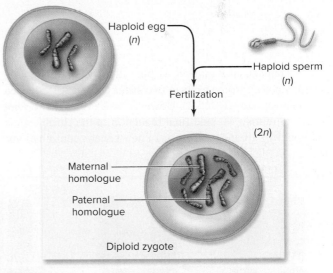

**Figure 15.1** In animals, meiosis produces eggs and sperm, both of which are haploid ($n$). Fusion of an egg and sperm during fertilization produces a zygote, which is diploid ($2n$). There are three homologous chromosomes in this zygote.

## Question 1
***a.*** Why would shuffling genetic material to produce new combinations of characteristics be advantageous to a species?

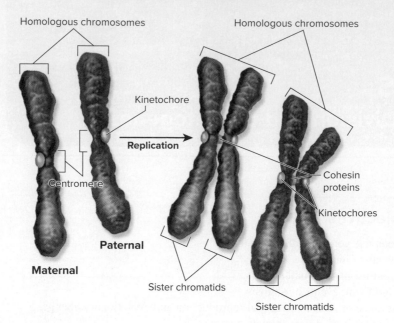

**Figure 15.2** The difference between homologous chromosomes and sister chromatids. Homologous chromosomes are the maternal and paternal copies of the same chromosome—say, chromosome number 16. Sister chromatids are the two replicas of a single chromosome held together at their centromeres by cohesin proteins after DNA replication. The kinetochore is composed of proteins found at the centromere that attach to microtubules during mitosis.

***b.*** When would it be deleterious?

Meiosis, like mitosis, is preceded by the replication of each chromosome to form two sister chromatids attached at a centromere (fig. 15.2). However, two events that do *not* occur in mitosis include final reduction of the chromosome number by half and production of new genetic combinations.

Meiosis reduces the chromosome number during *two* rounds of chromosome separation called **meiosis I** and **II.** Thus, the genetic material is replicated once just before meiosis but divided twice during meiosis. This allocates half the original number of chromosomes (one of each original pair) to each daughter cell. That is, the nuclei are haploid.

To produce new genetic combinations, each chromosome (composed of two sister chromatids) initially pairs along its length with its homologue to form a **bivalent** (fig. 15.3). This pairing of homologous chromosomes is

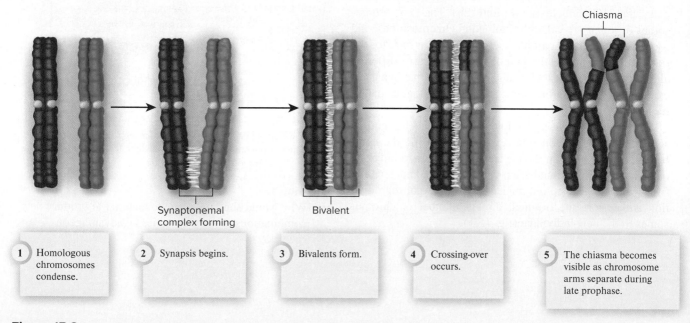

1. Homologous chromosomes condense.
2. Synapsis begins.
3. Bivalents form.
4. Crossing-over occurs.
5. The chiasma becomes visible as chromosome arms separate during late prophase.

**Figure 15.3** Formation of a bivalent and crossing-over during meiosis I. At the beginning of meiosis, homologous chromosomes pair with each other to form a bivalent, usually with a synaptonemal complex between them. Crossing-over then occurs between homologous chromatids within the bivalent. During this process, homologues exchange segments of chromosomes.

called **synapsis,** and the four chromatids exchange homologous segments of genetic material called **alleles.** Alleles are alternate states of a gene, such as a Type A allele, Type B allele, or Type O allele, which together determine a person's blood type. This exchange of genetic material among chromatids is called **crossing-over** and produces new genetic combinations. During crossing-over there is no gain or loss of genetic material. But afterward, each chromatid of the chromosomes contains different segments (alleles) that it exchanged with other chromatids. The temporary points of attachment of two chromatids at a point of genetic exchange are called **chiasmata** (fig. 15.3).

### Question 2

*a.* Synapsis occurs after chromosomal DNA has replicated. How many chromatids are involved in crossing-over of a homologous pair of chromosomes?

*b.* Suppose synapsis occurred between two homologous chromosomes, and one had alleles for blue eyes and brown hair where the other had alleles for green eyes and blonde hair. How many different combinations of these alleles would be possible?

## STAGES AND EVENTS OF MEIOSIS

Although meiosis is a continuous process, we can study it more easily by dividing it into stages just as we did for mitosis. Meiosis and mitosis are similar, and their corresponding stages of prophase, prometaphase, metaphase, anaphase, and telophase have much in common. However, meiosis takes longer than mitosis because meiosis involves two divisions instead of one. These two reductions are called meiosis I and meiosis II. Homologous chromosomes are separated at the end of meiosis I, and chromatids composing each chromosome are separated during meiosis II. Each reduction involves the events of prophase, prometaphase, metaphase, anaphase, and telophase (fig. 15.4).

Before your lab meeting, review in your textbook the events associated with each stage of meiosis, including the preparatory stage, interphase. List these events in table 15.1. This list can serve as an excellent study guide, so be as complete as possible. Ask your instructor to check for errors. One or two events for each stage are provided in figure 15.4.

### Premeiotic Interphase

Meiosis I is preceded by an interphase similar to the $G_1$, S, and $G_2$ of mitotic interphase, including replication of the chromosomes. Each chromosome is replicated.

Compare the outline in table 15.1 with your outline in table 14.1 on mitosis.

### Question 3

*a.* If a nucleus has eight chromosomes when it begins meiosis, how many chromosomes does it have after telophase I? Telophase II?

*b.* What are the major differences between the events of meiosis and mitosis?

*c.* What are some minor differences, and why do you consider them minor?

Understanding the movements of chromosomes is crucial to understanding meiosis. You can simulate these movements easily with chromosome models made of pipe cleaners or popsicle sticks. This simple but valuable exercise is especially instructive if you compare your simulation of meiosis with your simulation of mitosis from the previous exercise. There are important differences between them.

### *Procedure 15.1* Simulate chromosomal replication and movement during meiosis

*1.* Examine the materials for chromosome models provided by your instructor.

*2.* Identify the differences in chromosomes represented by various colors, lengths, or shapes of the materials. Also identify materials representing centromeres.

*3.* Place a clean sheet of notebook paper on your lab table and arrange the materials to represent the boundaries of the meiotic cell.

*4.* Assemble the chromosomes needed to represent the nuclear material in a cell of a diploid organism with a total of six chromosomes (three homologous pairs). Place the chromosomes in the cell.

*5.* Arrange the chromosomes to depict the position and status of chromosomes during interphase $G_1$. (During $G_1$ the chromosomes are usually not condensed as the chromosome models imply; nevertheless, the models are an adequate representation.)

*6.* Depict the chromosomes after completing interphase S. Use additional "nuclear material" if needed.

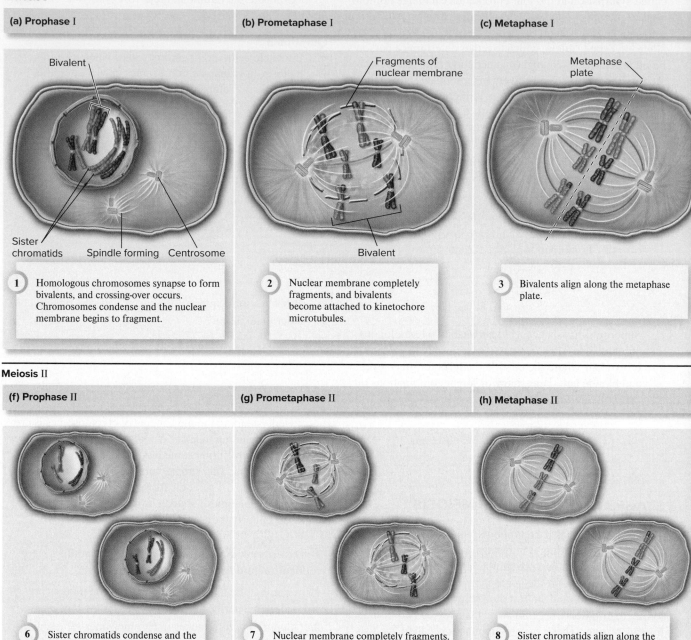

**Meiosis I**

**(a) Prophase I**

Bivalent

Sister chromatids    Spindle forming    Centrosome

1    Homologous chromosomes synapse to form bivalents, and crossing-over occurs. Chromosomes condense and the nuclear membrane begins to fragment.

**(b) Prometaphase I**

Fragments of nuclear membrane

Bivalent

2    Nuclear membrane completely fragments, and bivalents become attached to kinetochore microtubules.

**(c) Metaphase I**

Metaphase plate

3    Bivalents align along the metaphase plate.

**Meiosis II**

**(f) Prophase II**

6    Sister chromatids condense and the spindle starts to form. Nuclear membrane begins to fragment.

**(g) Prometaphase II**

7    Nuclear membrane completely fragments. Sister chromatids attach to spindle via kinetochore microtubules.

**(h) Metaphase II**

8    Sister chromatids align along the metaphase plate.

**Figure 15.4**    Stages of meiosis.

7. Depict the chromosomes during prophase I, metaphase I, anaphase I, and telophase I.

8. The interval between meiosis I and meiosis II is called **interkinesis.** Draw the results of cytokinesis, which occurs at this stage in some organisms.

9. Depict the status of chromosomes during prophase II for both daughter nuclei. Repeat this for metaphase II, anaphase II, and telophase II.

10. Draw the results of cytokinesis and the re-formation of nuclear membranes.

11. Chromosomal events are a continuous process rather than distinct steps. Therefore, repeat steps 4–10 as a continuous process and ask your instructor to verify your simulation.

## GAMETOGENESIS

Meiosis occurs in all sexually reproducing eukaryotes and produces haploid nuclei. However, organisms vary in the timing and structures associated with producing functional gametes. **Gametes** are reproductive cells with haploid nuclei resulting

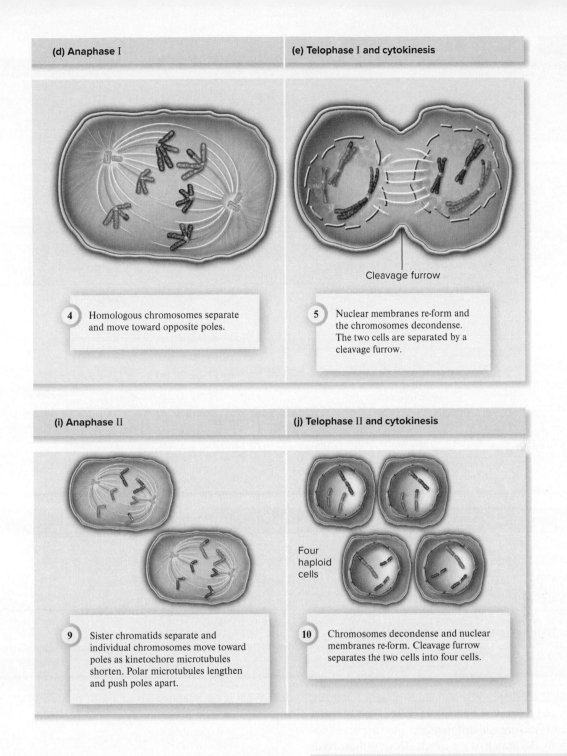

**(d) Anaphase I**

**4** Homologous chromosomes separate and move toward opposite poles.

**(e) Telophase I and cytokinesis**

Cleavage furrow

**5** Nuclear membranes re-form and the chromosomes decondense. The two cells are separated by a cleavage furrow.

**(i) Anaphase II**

**9** Sister chromatids separate and individual chromosomes move toward poles as kinetochore microtubules shorten. Polar microtubules lengthen and push poles apart.

**(j) Telophase II and cytokinesis**

Four haploid cells

**10** Chromosomes decondense and nuclear membranes re-form. Cleavage furrow separates the two cells into four cells.

from meiosis, and the formation of gametes is called **gametogenesis.** Meiosis is the primary element of gametogenesis in animals, but after meiosis the cells must mature and usually change their morphology before becoming a functional gamete.

In this exercise, you will examine mammalian gametogenesis. Gametogenesis includes **spermatogenesis,** the formation of sperm cells, and **oogenesis,** the formation of egg cells (fig. 15.5).

## Mammalian Spermatogenesis

Spermatogenesis occurs in male testes made of tightly coiled tubes called **seminiferous tubules** (fig. 15.6). Examine a prepared slide of a cross-section through the seminiferous tubules of a monkey, rat, or grasshopper. Packed against the inner walls of the tubules are diploid cells called **spermatogonia.** These cells constantly replicate mitotically during the life of males. They are assisted by nongerminal cells called **sertoli cells.** Some of the daughter cells from spermatogonia mitosis move inward toward the lumen of the tubule and begin meiosis. These cells are called **primary spermatocytes.** Meiosis I of a primary spermatocyte produces two **secondary spermatocytes,** each with a haploid set of replicated chromosomes. Review figure 15.2 to see the structural difference between a replicated chromosome and an unreplicated chromosome.

Table 15.1

**Events of Meiosis**

| Prophase I | |
|---|---|
| Prometaphase/Metaphase I | |
| Anaphase I | |
| Telophase I | |
| Prophase II | |
| Prometaphase/Metaphase II | |
| Anaphase II | |
| Telophase II | |

## Human Sperm Cells by the Numbers

- Healthy males produce 100–200 million sperm cells per day.

- There are an estimated 250 million sperm cells per ejaculation.

- There are 200–300 million sperm cells for every one egg.

- There are about 20 million sperm/mL of semen.

- About 20% of all sperm are deformed.

- Sperm can live inside a female reproductive system for up to 6 or 7 days.

- Conception can occur several days after sexual intercourse.

Meiosis II separates the chromatids of each replicated chromosome and produces two haploid cells called **spermatids.** Spermatids mature and differentiate into **sperm** cells as they move along the length of the tubule. Review these basic stages of spermatogenesis in figure 15.5. Then examine some prepared slides of sperm cells from vertebrates such as guinea pig, rat, and human.

### Question 4

*a.* During gametogenesis a sperm cell undergoes considerable structural change. What are the basics of sperm structure and how do these features relate to function?

*b.* What is the adaptive advantage of producing sperm in a system of tubes rather than in solid tissue?

*c.* What is each strand of a double-stranded chromosome called?

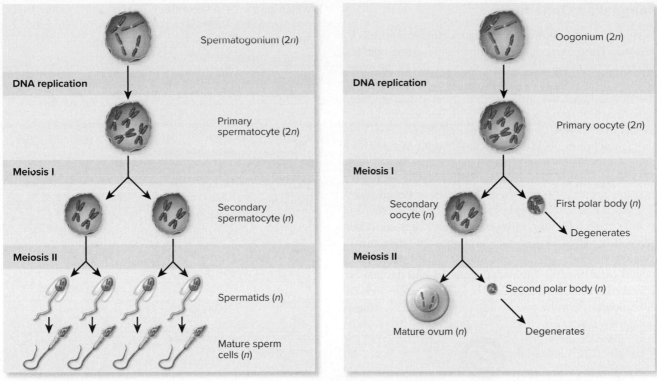

**(a) Spermatogenesis (males)**

**(b) Oogenesis (females)**

**Figure 15.5** Gametogenesis in (*a*) males and (*b*) females. Both male and female germ cells are diploid (2*n*) cells that undergo two meiotic divisions to produce mature haploid (*n*) gametes.

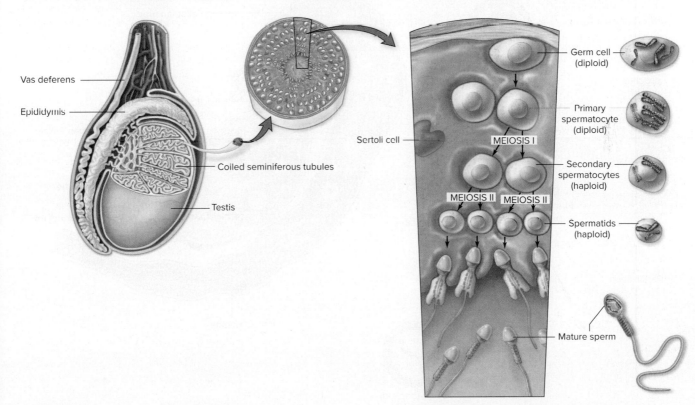

**Figure 15.6** The testis and spermatogenesis. Spermatogenesis occurs in the seminiferous tubules, shown on the left. Events begin on the outside of the tubule, progressing inward to release mature spermatozoa in the tubule. The first meiotic division separates and distributes homologous chromosomes into the daughter cells, which mature into haploid secondary spermatocytes. The second meiotic division distributes sister chromatids into four haploid spermatids, which mature into spermatozoa.

## Mammalian Oogenesis

Oogenesis occurs in ovaries of females (fig. 15.7). Cells of the ovary that produce female gametes are called **oocytes.** However, oocytes are not produced continually by the ovary, as spermatocytes are produced by the testes. During early fetal development, **oogonia** (germinal cells) are produced in the ovaries. These oogonia replicate mitotically to produce as many as two million **primary follicles,** each containing a **primary oocyte.** In humans, ovaries of a newborn female contain all of the primary oocytes that she will ever have (i.e., oogonia produce no more primary oocytes). At birth the primary oocytes in a female have begun meiosis I but are arrested in prophase I. They are surrounded by supportive **follicular cells,** and together they are called **follicles.** At puberty, circulating hormones stimulate growth of one or two of these dormant follicles (and their primary oocytes) each month. The oocyte enlarges and the number of follicular cells increases. Just before **ovulation** (release of the oocyte from the ovary), the oocyte completes meiosis

I, which produces a **secondary oocyte** and a **polar body.** This mature follicle is called a **Graafian follicle** and contains a secondary oocyte (fig. 15.8). Each secondary oocyte contains a haploid set of double-stranded chromosomes (two chromatids), but cytoplasmic cleavage is unequal. The secondary oocyte retains most of the cytoplasm and the first polar body usually disintegrates.

Examine a prepared slide of a mammalian ovary cross-section. In the following space, sketch a Graafian follicle and two or three less mature stages.

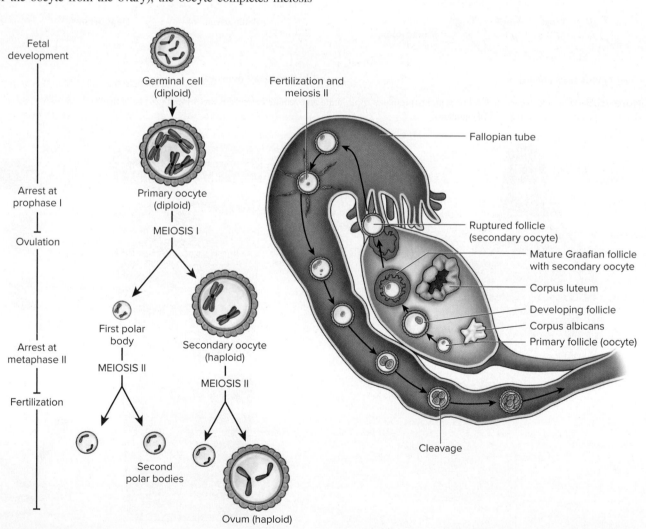

**Figure 15.7** Oogenesis. A primary oocyte is diploid ($2n$). After its first meiotic division, one product is eliminated as a polar body. The other product, the secondary oocyte, is released during ovulation. Sperm penetration stimulates the second meiotic division, and a second polar body and a haploid ovum are produced. Fusion of the haploid ($n$) ovum nucleus with a haploid ($n$) sperm nucleus produces a diploid ($2n$) zygote that subsequently forms an embryo.

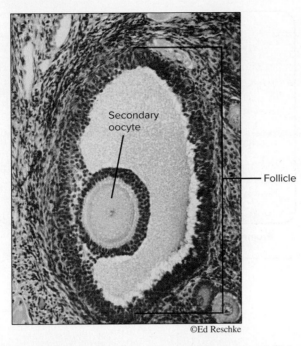

Secondary oocyte

Follicle

©Ed Reschke

**Figure 15.8** A mature secondary oocyte in an ovarian follicle of a cat (400×). This secondary oocyte awaits ovulation.

**Question 5**
How would retaining extra cytoplasm enhance survival of a developing oocyte?

Mciosis II procccds but is not completed until after a sperm cell penetrates the egg. Completion of meiosis II produces another polar body and a haploid cgg ccll rcady for **fertilization** (fusion of nuclei). Review these basic stages of oogenesis in figure 15.5. Then examine a cross-section of a cat ovary.

**Question 6**
*a.* What are the relative sizes of oocytes in a dormant follicle, a growing follicle, and a Graafian follicle?

*b.* Are polar bodies visible in your prepared slide of a cat ovary? Why or why not?

After ovulation the remaining follicle cells form the **corpus luteum** on the surface of the ovary. The corpus luteum produces hormones that prepare the uterus for the potential arrival of a fertilized egg. The corpus luteum declines after 8–9 days and becomes a white body of connective tissue called the **corpus albicans.**

## Plant Gametogenesis

The formation of gametes in plants is somewhat different because their sexual life cycle includes an alternation of generations between haploid and diploid forms. However, meiosis is still the critical process by which plants reduce the number of chromosomes by half to prepare for gamete production.

In flowering plants, meiosis occurs in the anthers and ovary of the flowers. In the anther, the spores resulting from meiosis produce a stage of the life cycle (pollen) that will eventually produce male gametes. In the ovary, the spores resulting from meiosis produce a stage of the life cycle (ovule) that will eventually produce female gametes. You'll learn more about these events in Exercise 31. In this procedure, you will observe prepared slides showing stages of the beginning, middle, and end of meiosis I and II in a representative plant.

*Procedure 15.2* **Diagram and observe stages of meiosis**

*1.* In figure 15.9, diagram a plant cell with three pairs of chromosomes in each of the stages of meiosis. Be sure to label the cell wall and cell plate.

*2.* Examine the following prepared slides of stages of meiosis in a *Lilium* anther (see figs. 31.10, 31.11).

   *a.* *Lilium* anther—early prophase I

   *b.* *Lilium* anther—late prophase I

   *c.* *Lilium* anther—first meiotic division

   *d.* *Lilium* anther—second meiotic division

   *e.* *Lilium* anther—pollen tetrads. Each of these cells will produce a pollen grain.

*3.* Examine the following prepared slides of stages of meiosis in a *Lilium* ovary.

   *a.* *Lilium* ovary—"mother cell," prophase I

   *b.* *Lilium* ovary—binucleate stage, end of meiosis I

   *c.* *Lilium* ovary—four nucleate stage, end of meiosis II

**Interphase**

**Prophase I**

**Prometaphase I**

**Metaphase I**

**Anaphase I**

**Telophase I**
**(label cell plate)**

**Prophase II**

**Prometaphase II**

**Metaphase II**

**Anaphase II**

**Telophase II**
**(label cell plates)**

**Figure 15.9** Meiosis in plants. Use these cellular outlines to diagram a plant cell with three pairs of chromosomes in each of the stages of meiosis.

# MITOSIS VERSUS MEIOSIS

Mitosis and meiosis are both forms of cellular replication, but they play different roles in the life cycle of animals and plants. Mitosis may occur in either haploid or diploid cells and is necessary for cell production and growth. Meiosis occurs in diploid cells. Its role is to produce cells with a reduced number of chromosomes and shuffle the genetic material so an organism can reproduce sexually. To compare mitosis and meiosis, review table 15.2 and complete the column with the contrasting features of meiosis.

# INQUIRY-BASED LEARNING

*How much does the anatomy of sperm cells vary among the animal species that you examined?*

Observation: The morphology of sperm cells directly relates to their function. Sperm of vertebrates such as guinea pigs, rats, and humans vary in size and shape.

Question: How does sperm cell morphology vary among species of vertebrates?

a. Establish a working lab group and obtain Inquiry-Based Learning Worksheet 15 from your instructor.

b. Discuss with your group the measurements and observations you might make to reveal variation in sperm morphology. Pose a well-defined question relevant to the preceding observation and question. Record it on Worksheet 15.

c. Translate your question into a testable hypothesis and record it.

d. Outline on Worksheet 15 the procedures and supplies needed to test your hypothesis. Ask your instructor to review your proposed investigation.

e. Conduct your procedures, record your data, answer your question, and make relevant comments.

f. Discuss with your instructor any revisions to your questions, hypothesis, or procedures. Repeat your work as needed.

## Table 15.2

**A Comparison of the Major Features of Mitosis and Meiosis**

|  | Mitosis | Meiosis |
|---|---|---|
| Purpose of process |  |  |
| Number of cells produced |  |  |
| Number of nuclear divisions |  |  |
| Haploidy or diploidy of resulting cells |  |  |
| Genetically identical cells (yes or no) |  |  |
| Pairing of homologues (yes or no) |  |  |
| Occurrence of crossing-over (yes or no) |  |  |

*1.* How would you diagram the chromosomal arrangement for transitional stages such as late prophase/early metaphase I? Or late anaphase/early telophase I?

*2.* Would evolution occur without the events of meiosis and sexual reproduction? Why or why not?

*3.* What are the general characteristics of sexual reproduction in humans and other vertebrates that are associated with continuous production of many sperm cells but intermittent, finite production of egg cells?

*4.* Which process is most accurately referred to as nuclear division: meiosis or mitosis?

*5.* What special event does interkinesis lack compared to premeiotic interphase?

*6.* How are mammalian sperm cells produced and incubated at a lower temperature than body temperature?

*7.* How old is an ovulated oocyte of a 35-year-old woman? What consequences does this have?

## WRITING TO LEARN BIOLOGY

Wouldn't it be easier for a cell simply to divide the chromosomes once rather than duplicating them and then dividing them twice during meiosis? Why do you suppose this isn't done?

# Molecular Biology and Biotechnology

## DNA Isolation and Genetic Transformation

## Learning Objectives

By the end of this exercise you should be able to:

1. Isolate DNA from a bacterium.
2. Understand how temperature and pH affect DNA.
3. Insert a gene for resistance to ampicillin into a bacterium.

Please visit **connect.mheducation.com** to review online resources tailored to this lab.

**B**iotechnology is the manipulation of organisms to do practical things and to provide useful products. Biotechnology has been around for centuries: Humans have selectively bred livestock for meat products, controlled pollination to produce more food, and used bacteria and fungi to make wine and cheese. But recent progress in molecular biology has revolutionized biotechnology, and its products include vaccines, detergents, drugs, biodegradable plastics, alcoholic beverages, industrial chemicals (e.g., ethanol, acetone), antibiotics, hybrid crops, livestock food, cooking oils, improved textiles and fabrics, lower-cholesterol meats and eggs, and many common foods (e.g., corn, watermelon, chicken, cheese) and beverages (e.g., milk, beer, wine). The revolution stems from new molecular techniques that have made genetic engineering possible.

**Genetic engineering** is the direct manipulation of genes for practical purposes (fig. 16.1). Genetic engineers can intervene *directly* in the genetic fate of organisms. We can isolate genes, move them from one organism to the next, and even move genes from one species to the next. The most common goals of this engineering are to harvest the valuable proteins made by engineered genes and to benefit from the new characteristics the genes provide to the target organisms, including humans. Indeed, the impact on society of the current revolution in genetic engineering may soon surpass that of such historical changes as the industrial revolution of the past two centuries.

Genetic engineering got its start in 1973 when Stan Cohen and Herb Boyer transplanted a gene for antibiotic resistance from a frog into a bacterium, and thus "engineered" an antibiotic-resistant organism. In 1980, molecular biologists successfully inserted a human gene for interferon, an antiviral drug, into a bacterium. When the "transformed" bacterium reproduced, it generated billions of progeny, each a miniature drug factory. As a result of this engineering, biologists could cheaply harvest a drug that was previously

MyLoupe/Getty Images

**Figure 16.1** The United States is the world's top per-capita consumer of genetically modified crops, and the production of these crops demonstrates the importance of genetic engineering in our lives. For example, in 1996, when the first biotech crops were commercialized, only 8% of the cropland devoted to corn was used to grow genetically engineered corn. By 2017, this percentage had increased to more than 90%, and in 2020 was 92%. Yields also increased, from 127 bushels per acre in 1996 to more than 170 bushels per acre in 2020. The more than 13 billion bushels of corn harvested in 2020 were eaten and used to make ethanol (a renewable additive to gasoline), beverages, dyes, adhesives, tires, drugs, and countless other products. The 113-fold increase in land devoted to genetically engineered crops since 1996 makes biotechnology the fastest adopted crop technology in the history of modern agriculture.

expensive and generally unavailable. Genetic engineering has since been applied to medicine (gene therapy, drug production) and the production of new foods and environmentally benign pesticides. Today, millions of diabetics worldwide use synthetic human-insulin to regulate their blood-sugar levels. Insulin is made by genetically engineered bacteria and yeast.

At the heart of genetic engineering is the science of **molecular biology** (i.e., the study of molecules critical to life). Molecular biologists recently have concentrated their efforts on manipulating "information molecules" such as DNA and proteins because all outward characteristics of organisms have their basis in proteins from genes made of DNA. In addition to providing techniques for genetic engineering, molecular biology has also impacted fields such as forensics (e.g., linking suspects to crimes, settling paternity disputes), hiring practices (pinpointing employees at high risk for cancer), and agriculture (e.g., in 2020, more than 17 million farmers in 26 countries planted biotech crops on more than 470 million acres of farmland). In the United States, genetically modified crops constituted more than 90% of the nation's sugar beets, feed corn, cotton, and soybeans in 2017. To introduce yourself to the core information of molecular biology, you should review DNA structure in your textbook and in Exercise 6, "Biologically Important Molecules."

In this exercise you'll learn two techniques used routinely by molecular biologists: (1) isolation of DNA and (2) genetic transformation.

## ISOLATION OF DNA

Isolation of DNA is a routine and important procedure for molecular biologists. Once isolated, the DNA and the sequence of its subunits can be determined, manipulated, or altered. Bacteria each contains only about $10^{-14}$ g of DNA, which accounts for approximately 5% of the organism's dry weight. However, molecules of this DNA can be very long; for example, the DNA in an *E. coli,* if strung out, would be approximately 1 mm long. (By analogy, if the bacterium were the size of a grapefruit, then its DNA would be more than 80 km long.) The ability to pack this much DNA into a tiny cell is impressive, especially because DNA is a rather stiff molecule (fig. 16.2).

When DNA or other large molecules are isolated from cells, the surrounding solution becomes viscous (i.e., thick, syrupy, and resistant to flow). This is because DNA molecules are long and tend to stick to each other due to cohesion among molecules and hydrogen bonding (recall that hydrogen bonding also holds the two strands of DNA's double helix together; see fig. 6.9). Harsh chemicals will nonspecifically disrupt hydrogen bonds, including those between corresponding nitrogenous bases of DNA. Molecular collisions produced by high heat can tear molecules apart, and any breakdown in molecular structure will reduce viscosity.

In this exercise, you will isolate DNA from *Halobacterium salinarum,* a halophilic ("salt-loving") archaea of domain Archaea that grows only in salty environments (4–5 M NaCl). It's especially easy to isolate DNA from this organism because its cell walls disintegrate when placed in low-salt environments (0–2 M NaCl).

Don W. Fawcett/Science Source

**Figure 16.2** A human chromosome contains an enormous amount of DNA (35,000×). The dark element at the bottom of the photograph is part of the protein matrix of a single chromosome. All of the surrounding material is the DNA of the chromosome.

**SAFETY FIRST** Before coming to lab, you were asked to read this exercise so you would know what to do and be aware of safety issues. In the space below, briefly list the safety issues associated with today's procedures. If you have questions about these issues, contact your laboratory assistant before starting work.

## *Procedure 16.1* Isolate DNA from *Halobacterium salinarum*

1. Several days ago your laboratory instructor inoculated petri dishes of culture media with *Halobacterium salinarum.* Obtain one of these cultures from your instructor.

2. Use a flexible plastic ruler to scrape the bacterial growth from the surface of the agar and collect it in a test tube.

3. Add 1 mL of distilled water to the tube. One milliliter equals about 20 drops.

4. Place a small piece of plastic film over the top of the test tube and hold it securely in place with your thumb. Invert the tube several times to mix the contents. This mixing with water lyses the cells, and the liquid will become viscous.

5. Use an eyedropper to slowly and gently pour 1 mL of ice-cold 95% ethanol down the side of the tube. Do not shake or disturb the tube. DNA will precipitate at the interface between the layers of water and ethanol.

6. Insert a wooden splint (e.g., coffee stirrer) or glass rod into the liquid. Rotate the rod. DNA will adhere to the rod as it is twirled.

**Question 1**
What is the texture of the DNA you've isolated?

The DNA that you've isolated is not pure; rather, it is contaminated with small amounts of protein and RNA.

## ISOLATE DNA FROM YOUR CELLS

If time permits, your instructor may also ask you to extract DNA from your cheek cells. These procedures will work best if you have not eaten or chewed gum lately.
Examine the cells from which you will extract your DNA.

### *Procedure 16.2*

1. Place a drop of methylene blue on a clean microscope slide. Be careful; methylene blue will stain your skin and clothes.

2. Gently scrape the inside of your cheek with the flat side of a toothpick. Swirl this end of the toothpick in the drop of methylene blue. Then throw the toothpick away.

3. Place a coverslip on the drop of methylene blue and examine the slide with your light microscope.

At low power, cells will appear as small, purplish objects. After finding some of these cells, examine the cells with a higher-power objective. These are the cells from which you will extract your DNA.

### Extracting your DNA

You'll extract your DNA by (1) collecting your cells, (2) releasing the DNA by using detergent to lyse the cell and nuclear membranes of your cells, and (3) precipitating your DNA by adding alcohol.

### *Procedure 16.3*

1. Pour 10 mL of a lightly colored sports-drink (or a 0.9% solution of sodium chloride) into a small, disposable cup.

2. Collect your cheek cells by *vigorously* swirling the sports drink or salt solution in your mouth for 1 min. While swirling the solution in your mouth, rub your cheek with your tongue and teeth. Cells lining your mouth are easily loosened, and this swirling will dislodge cells, which will be the source of your DNA. The longer and more vigorously you swirl the solution in your mouth, the more cells and DNA you will collect.

(This solution will also contain bacteria from your mouth, from which you will also isolate the DNA.)

3. Spit the sports-drink solution back into the disposable cup.

4. Pour the sports-drink solution from the cup into a test tube containing 5 mL of a 25% solution of dishwashing detergent. The detergent will break the lipid-based cell and nuclear membranes, thereby releasing the cells' DNA into solution. (This is why we use detergents to remove fats—that is, lipids—from dirty dishes.)

5. Add a pinch of meat tenderizer to the test tube. Each of our cells contains enough DNA that, if stretched end-to-end, would span 2–3 m. To fit this much DNA into a tiny cell, the DNA is wrapped tightly around proteins. Meat tenderizer is a protease, which is an enzyme that digests protein and, in doing so, releases DNA from the protein.

6. Cap or cork the test tube and *gently* rock the tube from side to side for 3 min. Also *gently* invert the tube four to five times. Do not vigorously shake or tilt the tube; doing so will shear the DNA into smaller pieces that will be hard to see later. If caps or corks are not available, you can also cover the end of the test tube with your gloved thumb or a piece of Parafilm.

7. Let the tube stand for 1 min.

8. After uncapping the tube, gently tilt the tube to a 45° angle and use a pipette to gently add 10 mL of ice-cold ethanol down the side of the tube. (The colder the alcohol, the more DNA will precipitate.) Alcohol is less dense than the detergent solution, so the alcohol will form a layer atop the detergent.

9. Let the tube stand for 10 min. Do not shake, tip, mix, or agitate the tube. Watch what happens at the interface between the alcohol and detergent.

10. Use a glass rod to slowly move some of the alcohol into the detergent. When you do this, your DNA—which is insoluble in alcohol—will precipitate as white, cottony strands at the alcohol-detergent interface. The lipids and proteins will remain dissolved. Bubbles displaced from the sports drink may get trapped in the DNA and make it easier to see.

11. Spool your DNA on the glass rod by slowly twirling the rod in one direction.

## The Influence of Heat and pH on DNA

Heat and pH strongly affect the properties of DNA. For example, DNA typically denatures at alkaline pH and at 80–97°C. Test these effects with the following procedures.

### *Procedure 16.4*   **Test the influence of heat on DNA**

1. Precipitate DNA in a test tube following steps 1–5 in procedure 16.1 for isolating DNA.

2. Place the tube into a boiling water-bath for 10 min.

3. Place the tube in an ice bath.

4. Insert and twirl a glass rod in the tube.

5. Compare the viscosity of the heat-treated DNA with untreated DNA.

**Question 2**

*a.* What effect does heat have on the viscosity of DNA?

*b.* What do you think is the mechanism for this change in viscosity?

---

*Procedure 16.5*   **Test the influence of pH on DNA**

*1.* Precipitate DNA in a test tube following steps 1–5 in procedure 16.1 for isolating DNA.

*2.* Add 2 mL of 1.0 N NaOH to the tube.

NaOH is caustic. Don't spill it on yourself or your clothes.

*3.* Insert and twirl a glass rod in the tube.

*4.* Compare the viscosity of the alkali-treated DNA with untreated DNA and with heat-treated DNA.

**Question 3**

*a.* What effect does alkaline pH have on the viscosity of DNA?

*b.* What do you think is the mechanism for this pH-induced change in viscosity?

## GENETIC TRANSFORMATION

Much of biotechnology is based on **genetic transformation,** which is the uptake and expression of DNA by a living cell. A successful transformation, summarized in figure 16.3, requires three conditions: (1) a host into which DNA can be inserted, (2) a means of carrying the DNA into the host, and (3) a method for selecting and isolating the successfully transformed organisms.

### The Host: *Escherichia coli*

The **host organism** you'll use is the bacterium *Escherichia coli,* one of the most intensively studied organisms in the world. *E. coli* has the following properties that make it ideally suited for transformation:

- It contains only one chromosome made of five million base-pairs. This is less than 0.2% of that of the human genome.

- *E. coli* grows rapidly. Transformations in bacteria are rare and occur in only about 0.1% of cells. Therefore, transformations are observed most easily in large, rapidly growing populations. *E. coli* is ideal for transformation studies because, in ideal conditions, it divides every 20 min. As a result, in 10 h a bacterium can produce a billion progeny (30 generations) in only 1 mL of nutrient broth.

Only a small percentage of bacterial cells in a culture can be transformed. Also, small lengths of DNA are taken up more readily than long lengths. However, competence of the bacteria (i.e., the chances for successful transformation) increases during the early and middle stages of its growth.

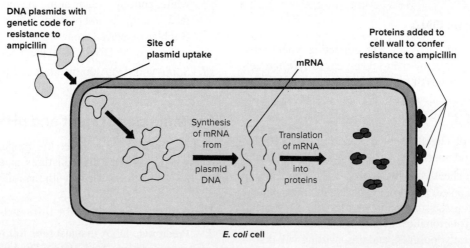

**Figure 16.3**   Transformation of an *E. coli* cell with plasmid DNA. In this example, the DNA plasmid contains the genetic code for resistance to the antibiotic ampicillin. After uptake of the plasmid, the code is transcribed to messenger RNA, which is translated during protein synthesis. Addition of these proteins to the cell wall will retard attack by ampicillin.

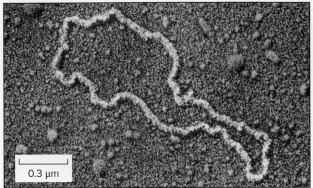

0.3 μm

Professor Stanley N. Cohen/Science Source

**Figure 16.4**    A famous plasmid. The circular molecule in this electron micrograph (70,000×) was the first plasmid used successfully to clone a vertebrate gene, pSC101. Its name refers to it being the 101st plasmid isolated by Stanley Cohen.

Competence also increases when cells suspended in a cold solution of $CaCl_2$ are heat-shocked. Yield is usually about $10^6$ transformants per milligram of DNA available for insertion.

## A Vector to Move DNA into the Host

A biological **vector** is a DNA molecule that carries DNA sequences into a host. The simplest bacterial vectors are **plasmids,** circular pieces of DNA made of 1000 to 200,000 base-pairs (fig. 16.4). Plasmids exist separately from the bacterial chromosome, and they must contain a gene that confers some selective advantage (e.g., resistance to an antibiotic) to remain in the host. We don't completely understand how plasmids enter host cells, but they seem to enter consistently.

## Selecting Transformed Organisms

You'll insert into *E. coli* a plasmid (pAMP) containing a gene for resistance to ampicillin, an antibiotic lethal to many bacteria (fig. 16.5). (Refer to Exercise 24 for information on bacterial cell wall structure and how ampicillin might kill bacteria.) Then you'll select transformed bacteria based on their resistance to ampicillin by spreading the transformed organisms onto nutrient medium containing ampicillin. Organisms that grow on this medium have been transformed. Because *E. coli* grows so fast, you can check for transformed organisms only 12–24 h after completing the experiment.

### Procedure 16.6    Transform *E. coli*

**Preparation**

1. Carefully read this procedure and review figure 16.5 before beginning. Wash your hands. Your instructor will demonstrate and discuss **sterile technique,** which will eliminate contamination of your cultures by other organisms (e.g., bacteria, fungi). Use sterile technique when doing this procedure.

2. Fill a 250-mL beaker with about 50 mL of ethanol and place a glass-rod bacterial spreader in the ethanol to soak.

**Preincubation to Increase Competency**

3. Obtain a test tube with 1 mL of sterile, yellow nutrient broth and a tube with 1 mL of clear, colorless 50 mM $CaCl_2$. Label these tubes NB and $CaCl_2$, respectively.

4. Fill a 250-mL beaker half full with crushed ice. Place the tube of $CaCl_2$ in the beaker of ice.

5. Obtain two sterile, plastic transformation tubes and label one of them (+)P, meaning with plasmid, and the other (−)P, meaning without plasmid. Place them in the beaker of ice.

6. Obtain a packaged, sterile, plastic pipet, and locate the graduation indicating a volume of 0.25 mL.

7. Open the packaged pipet without touching and contaminating the pipet's open end. Use this pipet to add 0.25 mL of a sterile, ice-cold solution of 50 mM $CaCl_2$ to each of the two transformation tubes. Use sterile technique. This is shown in step 1 of figure 16.5.

8. Several days ago your lab instructor streaked starter plates of nutrient agar with *E. coli*. Scrape a colony of *E. coli* (3-mm diameter) from one of these plates with a sterile, plastic inoculating loop. The bacteria are growing as a thin film on the surface of the agar; when you scrape a colony off the surface, be sure not to take up any of the agar. Use sterile technique. This is shown in step 2 of figure 16.5.

9. Place the loopful of bacteria into the transformation tube labeled (−)P. Rinse the bacteria from the loop by gently twirling the loop handle between your fingers.

10. To mix the bacterial suspension, open a sterile, packaged pipet. Insert the pipet into the suspension in the bottom of the tube and gently use a rubber bulb to suck the fluid in and out of the tip three or four times. Be sure the tip is empty before withdrawing the pipet.

11. For tube (+)P, repeat steps 8–10. Use a fresh loop and pipet to inoculate and mix the bacteria. Try to get the same amount of bacteria into each tube. Replace the two tubes in the ice bath and chill the tubes for at least 5 min.

**Incubation**

12. Use a sterile loop to obtain one loopful (10 μL) of an ice-cold solution of DNA plasmids from a vial kept by your instructor. Add this loopful of plasmids to tube (+)P and gently rotate the loop to rinse the plasmids from the loop. These plasmids contain the gene for resistance to ampicillin. Do **not** add plasmids to tube (−)P. This is shown in step 3 of figure 16.5.

13. Place both tubes in ice for 15 min.

14. While the tubes are cooling, obtain two agar plates labeled (−)AM, meaning nutrient agar without ampicillin, and two plates labeled (+)AM, meaning nutrient agar with ampicillin. Label one of each pair of plates as (−)P and the other two plates as (+)P.

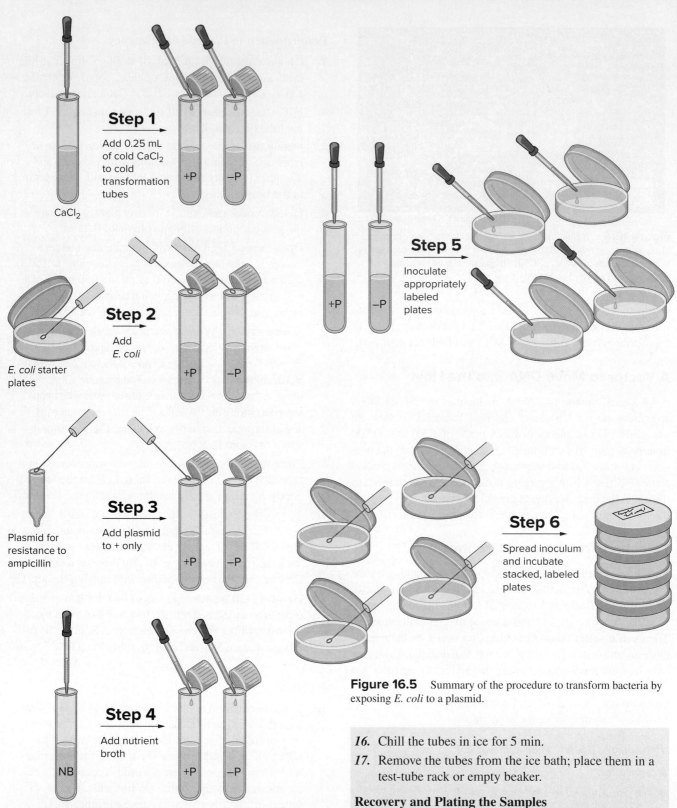

**Step 1**
Add 0.25 mL of cold CaCl₂ to cold transformation tubes

CaCl₂

+P   −P

**Step 2**
Add *E. coli*

*E. coli* starter plates

+P   −P

**Step 3**
Add plasmid to + only

Plasmid for resistance to ampicillin

+P   −P

**Step 4**
Add nutrient broth

NB

Nutrient broth (NB)

+P   −P

**Step 5**
Inoculate appropriately labeled plates

+P   −P

**Step 6**
Spread inoculum and incubate stacked, labeled plates

**Figure 16.5**   Summary of the procedure to transform bacteria by exposing *E. coli* to a plasmid.

### Heat Shock

*15.* Heat-shock the transformation tubes (−)P and (+)P by placing them in a 42°C water-bath for 2 min. Shake the tubes while they are in the water-bath. Heat shock increases the uptake of the plasmid by the bacterial cells.

*16.* Chill the tubes in ice for 5 min.

*17.* Remove the tubes from the ice bath; place them in a test-tube rack or empty beaker.

### Recovery and Plating the Samples

*18.* Using a sterile pipet, add 0.25 mL nutrient broth to each tube and tap gently to mix the contents. This is shown in step 4 of figure 16.5.

*19.* Using sterile technique and a sterile pipet, transfer 0.10 mL of the (−)P cell suspension onto the agar's surface in the middle of the plate labeled (+) AM/(−)P. Transfer another 0.10 mL onto the plate labeled (−)AM/(−)P. Close the plates. This is shown in step 5 of figure 16.5.

20. Using another sterile pipet, transfer 0.10 mL of the (+)P cell suspension onto the middle of the plate labeled (+)AM/(+)P. Transfer another 0.10 mL onto the plate labeled (−)AM/(+)P. Close the plates.

21. Light an alcohol lamp. Dip the bacteria spreader into the ethanol and then into the flame of the alcohol lamp. Let the ethanol burn away; then count to 10 to let the spreader cool. Before spreading the bacteria in the first of your four plates, further cool the spreader by touching it to the agar at the edge of the plate. Then touch the spreader to the cell suspension in the middle of the plate and gently drag it back and forth three times. Rotate the plate 90° and repeat. This is shown in step 6 of figure 16.5. Remember to sterilize the bacteria spreader between each plate. Repeat this procedure to spread the bacteria on the other three plates.

## Selecting the Transformed Organisms

22. Put your name and date on each of the four plates, and tape the plates together. Incubate the plates upside down at 37°C.

23. Place all the tubes, loops, and other such materials, in a central location for disposal. Wipe your work area with a weak bleach solution and wash your hands before leaving the laboratory.

24. In the space below indicate your predictions for growth (+) or no growth (−). Explain your reasoning for each prediction in the provided space.

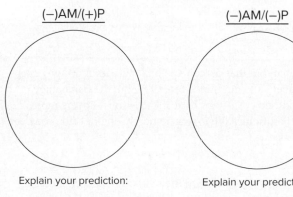

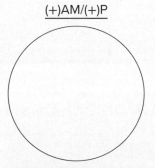
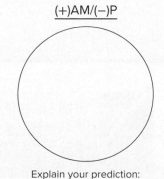

25. In the space below indicate with a (+) which plates had bacterial growth after 24 h. Draw the appearance and coverage of bacterial colonies and explain possible reasons for growth and possible reasons for no growth.

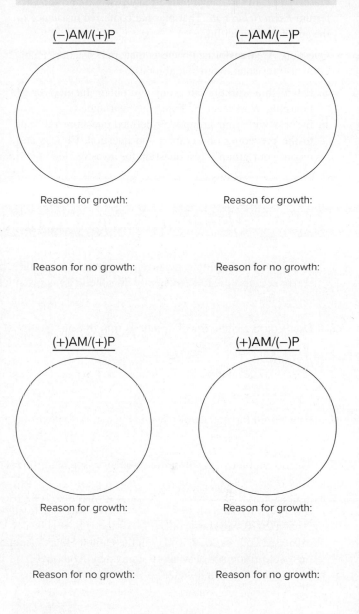

## Question 4

a. Which treatment produced transformed bacteria?

b. How many transformed colonies grew on each plate?

c. What was the purpose of tube 2 without plasmid?

## INQUIRY-BASED LEARNING

### *How narrow is the range of antibiotic resistance for transformed bacteria?*

Observation: In this lab you inserted a plasmid into the bacterium *Escherichia coli*. That plasmid conferred resistance to the antibiotic ampicillin.

Question: Are transformed bacteria that are resistant to ampicillin also resistant to other antibiotics?

**a.** Establish a working lab group and obtain Inquiry-Based Learning Worksheet 16 from your instructor.

**b.** Discuss with your group well-defined questions relevant to the preceding observation and question. Choose and record your group's best question for investigation.

**c.** Translate your question into a testable hypothesis and record it.

**d.** Outline on Worksheet 16 your experimental design and supplies needed to test your hypothesis. Ask your instructor to review your proposed investigation.

**e.** Conduct your procedures, record your data, answer your question, and make relevant comments.

**f.** Discuss with your instructor any revisions to your questions, hypothesis, or procedures. Repeat your work as needed.

## Questions for Further Study and Inquiry

*1.* Consider the ubiquitous occurrence of bacteria in nature, along with the constant fragmentation and release of DNA as cells decompose. How frequently do genetic transformations occur in nature? Explain your answer.

*2.* How could genetic transformations improve our quality of life? How could they decrease our quality of life?

*3.* Why is molecular biology often referred to as genetic engineering or biotechnology?

*4.* How could the uptake of plasmids in natural systems be important?

*5.* The development of antibiotic resistance is a major threat to our health. Why? How extensive is this problem?

*6.* In 1997, only 10% of the U.S. cotton crop was genetically engineered, but that percentage had increased to more than 90% in 2020. Similarly, only 17% of the U.S. soybean crop was genetically engineered in 1997, but the percentage exceeded 85% in 2020. Today in the United States, genetically modified corn—the largest U.S. crop—covers more than 84 million acres, which is equivalent to about 65 million football fields. Why are genetically engineered crops so popular in the United States?

*7.* Most of the member states of the European Union (EU) have partially or fully banned the cultivation of genetically modified organisms (GMOs). Indeed, the only widely adopted GMO crop grown in the EU (primarily in Spain and Portugal) is corn. What concerns have limited the impact of genetically engineered crops in Europe? Do you consider these concerns to be valid? Why or why not?

 **DOING BIOLOGY YOURSELF**

Design and conduct an experiment to test the effects of acid pH on the integrity of isolated DNA. How do the results compare to the effects of basic solutions on DNA?

 **WRITING TO LEARN BIOLOGY**

Many people resist the use of genetic engineering to alter organisms. What are their arguments? Do you agree?

# Genetics
## The Principles of Mendel

---

## Learning Objectives

By the end of this exercise you should be able to:
1. Describe simple genetic dominance, incomplete dominance, and lethal inheritance.
2. Describe possible genotypes for some of your personal traits inherited as dominant and recessive genes.
3. Explain the importance of Mendel's Law of Segregation and Law of Independent Assortment.
4. Distinguish between an organism's phenotype and genotype.

---

Please visit **connect.mheducation.com** to review online resources tailored to this lab.

---

**P**ublished papers are the primary means of communicating scientific discoveries. One of the most famous of these papers, titled "Experiments in Plant Hybridization," was written in 1866 by Gregor Mendel, an Austrian monk (fig. 17.1). Although Mendel's paper later became the basis for genetics and inheritance, it went largely unnoticed until it was rediscovered independently by several European scientists in 1900. The experiments and conclusions in Mendel's paper now form the foundation of **Mendelian genetics,** the topic of today's exercise.

Mendel's greatest contribution was to replace the blending theory of inheritance, which stated that all traits

blend with each other, with the **particulate theory.** Mendel's particulate theory states that (1) inherited characters are determined by particular factors (now called genes), (2) these factors occur in pairs (i.e., genes occur on maternal and paternal homologous chromosomes), and (3) when gametes form, these genes segregate so that only one of the homologous pair is contained in a particular gamete. Recall from Exercise 15 (Meiosis) that each gamete has an equal chance of possessing either member of a pair of homologous chromosomes. This part of the particulate theory is collectively known as Mendel's First Law, or the **Law of Segregation.** Mendel's Second Law, or the **Law of Independent Assortment,** states that genes on nonhomologous or different chromosomes will be distributed randomly into gametes (figs. 17.2, 17.3).

Mendel's laws describe the inheritance of traits linked to single genes on chromosomes in the nucleus of cells. In today's lab, you'll learn about some of these traits and their inheritance. Remember, however, that not all traits are inherited according to Mendel's laws. For example, several diseases that affect eyes and muscles are inherited from DNA in mitochondria (i.e., not the nucleus). Your textbook discusses this and other types of non-Mendelian inheritance.

Before you start this exercise, briefly review in your textbook some principles and terms pertinent to today's exercise. A **gene** is a unit of heredity on a chromosome. A gene has alternate states called **alleles,** contributed to an organism by its parents. Alleles for a particular gene occur in pairs. Alleles that mask expression of other alleles but are themselves expressed are **dominant;** these alleles are usually designated by a capital letter (for example, *P*). Alleles whose expression is masked by dominant alleles are **recessive** and designated by a lowercase letter (for example, *p*). The **genotype** of an organism includes all the alleles present in

**Figure 17.1** Gregor Mendel (1822–1884) grew and tended pea plants (*Pisum sativum*) like these for his experiments. For each experiment, he observed and counted as many offspring as possible. Pea plants are easy to grow and have many distinct traits; this made it easy for Mendel to analyze many crosses involving lots of offspring. In one set of crosses, he observed and counted a total of 7324 peas!

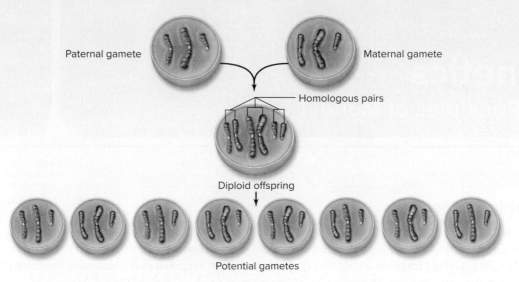

**Figure 17.2** Independent assortment increases genetic variability. Independent assortment contributes new gene combinations to the next generation because the orientation of chromosomes on the metaphase plate is random. For example, in cells with three pairs of chromosomes, eight different gametes can result, each with different combinations of parental chromosomes.

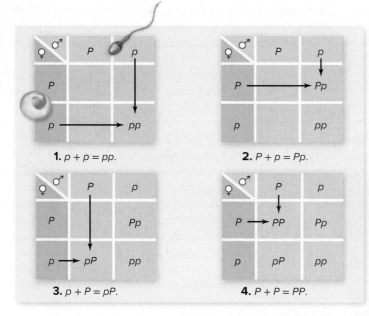

**1.** $p + p = pp$.

**2.** $P + p = Pp$.

**3.** $p + P = pP$.

**4.** $P + P = PP$.

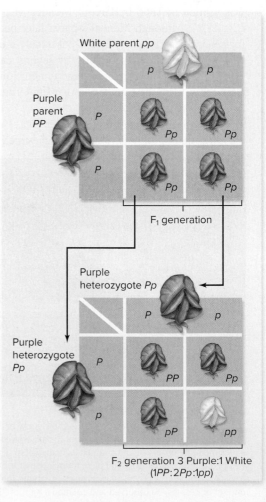

White parent $pp$

Purple parent $PP$

$F_1$ generation

Purple heterozygote $Pp$

Purple heterozygote $Pp$

F₂ generation 3 Purple:1 White
($1PP$:$2Pp$:$1pp$)

**Figure 17.3** In Mendel's crosses of pea plants having purple flowers with plants having white flowers, the original parents each make only one type of gamete. Offspring in the resulting $F_1$ generation are all $Pp$ heterozygotes with purple flowers. These $F_1$ offspring then each makes two types of gametes that can be combined to produce three kinds of $F_2$ offspring: $PP$ homozygotes (purple flowers); $Pp$ heterozygotes (also purple flowers); and $pp$ homozygotes (white flowers). The ratio of dominant to recessive phenotypes is 3:1. The ratio of genotypes is 1:2:1 ($1PP$:$2Pp$:$1pp$).

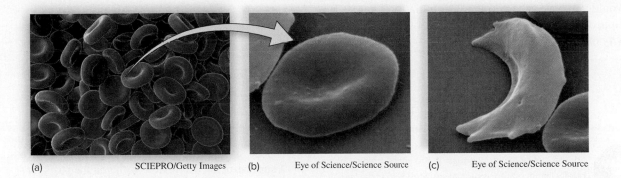

(a)      SCIEPRO/Getty Images      (b)      Eye of Science/Science Source      (c)      Eye of Science/Science Source

**Figure 17.4**    Sickle cell anemia. (*a, b*) Normal red blood cells (4175×, 7400×). (*c*) Individuals homozygous for the sickle cell allele have many red blood cells with irregular and sickle shapes (7400×).

the cell, whether they are dominant or recessive. The physical appearance of the trait is the **phenotype.** Thus, if purple flowers (*P*) are dominant to white flowers (*p*), a plant with purple flowers can have a genotype *PP* or *Pp*. A plant with white flowers can only have a genotype *pp* (fig. 17.3). When the paired alleles are identical (*PP* or *pp*), the genotype is **homozygous** (fig. 17.4). When the paired alleles are different (*Pp*), the genotype is **heterozygous.** With this minimal review, you're prepared to apply this information to solve some genetics problems.

## SIMPLE DOMINANCE

Assume that purple flowers are dominant to white flowers. If a homozygous purple-flowered plant is crossed (mated) with a homozygous white-flowered plant, what will be the phenotype (physical appearance) and genotype of the offspring?

| | |
|---|---|
| Parents: | *PP* (homozygous dominant = purple flowers) × *pp* (homozygous recessive = white flowers) |
| Gametes: | *P* from the purple-flowered parent *p* from the white-flowered parent |
| Offspring: | genotype = *Pp* phenotype = purple flowers |

This first generation of offspring is called the **first filial** or **F₁ generation** (fig. 17.3).

Each of the F₁ offspring can produce two possible gametes, *P* and *p*. Mendel noted that the gametes from each of the parents combine with each other randomly. Thus, you can simulate the random mating of gametes from the F₁ generation by flipping two coins simultaneously. Assume that heads designates the purple-flower allele (*P*) and tails designates the white-flower allele (*p*). Flipping one coin will determine the type of gamete from one parent and flipping the other will determine the gamete from the other parent. To demonstrate this technique, flip two coins simultaneously 64 times and record the occurrence of each of the three possible combinations in table 17.1.

| Table 17.1 | |
|---|---|
| **Results of Coin-Flipping Experiment Simulating Random Mating of Heterozygous (*Pp*) Individuals** | |
| **Response** | **Number** |
| Heads-heads = *PP* = purple flowers | |
| Heads-tails = *Pp* = purple flowers | |
| Tails-tails = *pp* = white flowers | |

### Question 1

What is the ratio of purple-flowered (*PP* or *Pp*) to white-flowered (*pp*) offspring?

Keep these results in mind and return to the original problem: What are the genotypes and phenotypes of the offspring of the F₁ generation?

| | |
|---|---|
| Parents: | *Pp* × *Pp* |
| Gametes: | (*P* or *p*) × (*P* or *p*) |
| Offspring genotypes: | *PP Pp pP pp* |
| Offspring phenotypes: | 3 purple    1 white |

Thus, the theoretical genotypic ratio for the offspring of the F₁ generation is 1 *PP* : 2 *Pp* : 1 *pp*, and the phenotypic ratio is 3 purple : 1 white.

### Question 2

*a.*   How do these ratios compare with your data derived from coin flipping?

**b.** Would you have expected a closer similarity if you had flipped the coins 64,000 times instead of 64 times? Why or why not?

 **SAFETY FIRST** Before coming to lab, you were asked to read this exercise so you would know what to do and be aware of safety issues. In the space below, briefly list the safety issues associated with today's procedures. If you have questions about these issues, contact your laboratory assistant before starting work.

## Procedure 17.1  Determine genotypic and phenotypic ratios for albinism

Albinos are homozygous recessive for the pair of alleles that produce pigments of skin, hair, and eyes. Suppose a woman having normal colored skin and an albino mother marries an albino man. Record the genotypic and phenotypic ratios of their children.

Genotype of children's mother  _____

Genotype of children's father  _____

Possible gametes of mother  _____

Possible gametes of father  _____

Possible offspring  _____

Genotypic ratio of children  _____

Phenotypic ratio of children  _____

## Procedure 17.2  Determine color and height ratios for corn plants

An ear of corn provides a large family of siblings in which we can study how traits are passed from one generation to the next. The color of grains (karyopses) and the height of *Zea mays* (corn) plants are often determined by a single gene.

**1.** Examine (a) the ears of corn having red and yellow grains, and (b) the tray of tall and dwarf plants on demonstration.

**2.** Record your observations here and determine the probable genotypes of the parents of each cross. Probable genotypes of parents:

### Color of Corn Grains

Number of red grains  _____

Number of white grains  _____

Ratio of red : white grains  _____

Probable genotypes of parents  _____ × _____

### Height of Plants

Number of tall plants  _____

Number of dwarf plants  _____

Ratio of tall : dwarf plants  _____

Probable genotypes of parents  _____ × _____

**3.** The preceding crosses involved only one trait and thus are termed **monohybrid crosses.** Let's now examine a cross involving two traits; that is, a **dihybrid cross.** Your instructor will review with you the basis for working genetics problems involving dihybrid crosses.

In corn, red (*R*) seed color is dominant to white (*r*) seed color, and smoothness (*S*) is dominant to wrinkled (*s*) seed. Observe the cobs of corn derived from a cross between parents having genotypes *RRSS* and *rrss*.

## Question 3
**a.** What is the expected genotype for the $F_1$ generation?

**b.** Will all $F_1$ offspring have the same genotype?

**c.** Will all $F_1$ offspring have the same phenotype?

## Question 4
**a.** What are the predicted phenotypes for the $F_2$ (i.e., second) generation that is produced by the cross *RrSs* × *RrSs*?

**b.** In what ratio will they occur?

**4.** To test your prediction in Question 4, count the number of kernels having each of the following phenotypes in five rows of kernels on the cobs of corn available in lab. These kernels are the $F_1$ generation produced by the cross *RrSs* × *RrSs*.

Red, smooth  _____

Red, wrinkled  _____

White, smooth  _____

White, wrinkled  _____

## Question 5
*a.* How do your data compare with those that you predicted?

*b.* What are the genotypes of the F$_2$ generation that is produced by the cross $RrSs \times RrSs$?

*c.* In what ratio will they occur?

## INCOMPLETE DOMINANCE

Some traits such as flower color are controlled by incomplete dominance. In this type of inheritance, the heterozygous genotype results in an intermediate characteristic. For example, if a plant with red flowers ($RR$) is crossed with a plant having white flowers ($rr$), all of the offspring in the first filial (F$_1$) generation will have pink flowers ($Rr$).

| | |
|---|---|
| Parents: | $RR$ (red) $\times$ $rr$ (white) |
| Gametes: | $R \times r$ |
| Offspring: | $Rr$ (pink) |

## Question 6
What are the expected ratios of red, pink, and white flowers in a cross involving two pink-flowered parents?

## LETHAL INHERITANCE

Lethal inheritance involves inheriting a gene that kills the offspring. Observe the tray of green and albino seedlings of corn. (Your instructor may substitute tobacco seedlings for the corn seedlings.) The albino plants cannot photosynthesize and therefore die as soon as their food reserves are exhausted.

## Question 7
*a.* What is the ratio of green to albino seedlings?

*b.* Based on this ratio, what might you expect were the genotypes of the parents?

## Question 8
Why is it impossible to cross a green and an albino plant?

## OTHER SOURCES OF GENETIC DIVERSITY

Genetic diversity can also result from multiple alleles, gene interactions (epistasis), continuous variation, pleiotropy, environmental effects, linkage, and sex linkage. Although time limitations prohibit exercises about these topics, be sure to review them in your textbook.

## BLOOD TYPE

Blood type of humans provides an excellent example of **codominance,** another type of Mendelian inheritance. In codominance, both alleles contribute to the phenotype of a heterozygote. For example, all people have one of four blood types: A, B, AB, and O (fig. 17.5). These blood groups are determined by the presence of compounds called **antigens** on the surfaces of their red blood cells. If antigen A or B is present, no antibodies against this antigen are produced. Thus, if a person has antigen-A on his or her blood cells, then the person has type A blood and possesses blood **antibodies** (proteins) that agglutinate type B blood cells. Similarly, a person having antigen-B on his or her blood cells has type B blood and has antibodies that agglutinate type A blood cells. If a person has antigen-A and antigen-B on his or her blood cells, then the person has type AB blood and lacks A and B antibodies. If a person has no A or B antigens on his or her blood cells, the blood type is O and the person possesses antibodies against both A and B antigens (table 17.2). This system is rather unusual in that individuals have antibodies against the blood antigens that they do not possess.

Blood typing is often important for establishing the possible identity of an individual in forensic work and paternity suits. For example, assume that a woman with type O blood has a child having type O blood. The suspected father has type AB blood.

Could the suspected father with type AB blood be the child's father? The answer is no because the cross would have the following results:

| | |
|---|---|
| Parents: | $ii$ (type O) $\times$ $I^A I^B$ (type AB) |
| Gametes: | $i$ and $i$, $I^A$ and $I^B$ |
| Offspring: | $I^A i$ or $I^B i$ |

Half of the offspring from the mother and the suspected father would have type A blood (genotype $= I^A i$), and the other half would have type B blood (genotype $= I^B i$).

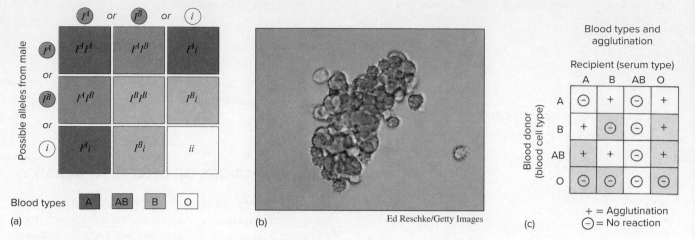

**Figure 17.5** ABO blood groups. (*a*) Multiple alleles control the ABO blood groups. Different combinations of the three *I* gene alleles result in four different blood type phenotypes: type A (either $I^AI^A$ homozygotes or $I^Ai$ heterozygotes), type B (either $I^BI^B$ homozygotes or $I^Bi$ heterozygotes), type AB ($I^AI^B$ heterozygotes), and type O (*ii* homozygotes). (*b*) The blood agglutination reaction. Agglutination occurs when blood cells stick and clump together. (*c*) Agglutination will occur when donor blood cells are incompatible with recipient serum, as designated by a +.

Thus, the suspected father with blood type AB could not have fathered a child with type O blood with this mother.

ABO blood typing can be used to eliminate a person as a potential parent but not to prove paternity. To appreciate this, suppose there is a mix-up of children in the maternity ward of a hospital after the genotypes of the children are determined from the parents' blood types. The following unidentified children have these blood types:

Child 1: type A (genotype $I^AI^A$ or $I^Ai$)

Child 2: type B (genotype $I^BI^B$ or $I^Bi$)

Child 3: type AB (genotype $I^AI^B$)

Child 4: type O (genotype *ii*)

**Question 9**

Which child or children could belong to a couple having AB and O blood types?

Blood typing is also important for determining the safety of blood transfusions. Your body automatically produces antibodies for antigens you do not carry (fig. 17.5).

For example, people with type A blood have antibodies against B antigens, and people with type B blood produce antibodies against A antigens. If someone having type B blood received blood from someone having type A blood, the recipient's antibodies would react with and agglutinate the red blood cells received from the donor (fig. 17.5*b*). As a result, the recipient would die.

**Question 10**

*a.* Can a person with type O blood safely donate blood to a person having type A blood? Why or why not?

*b.* Which blood type would be a universal donor?

*c.* Which blood type would be a universal recipient?

**Table 17.2**

**Characteristics of Individuals with the Four Major Blood Types**

| Blood Type | Antigen on Red Blood Cell | Antibody in Plasma | Genotype | % of U.S. Population |
|---|---|---|---|---|
| A | A | anti-B | $I^AI^A$ or $I^Ai$ | 42 |
| B | B | anti-A | $I^BI^B$ or $I^Bi$ | 10 |
| AB | A and B | none | $I^AI^B$ | 4 |
| O | O (none) | anti-A and anti-B | *ii* | 44 |

## Procedure 17.3 Determine blood type for ABO system

1. You will be provided with various samples of synthetic blood. This material simulates the blood type characteristics of human blood, and it is safe. Also obtain two bottles of antisera.

2. Obtain a clean slide and label the ends A and B. Near one end of the slide place a drop of antiserum A (containing antibodies against antigen-A), and near the other end of the slide place a drop of antiserum B (containing antibodies against antigen-B).

3. Place drops of blood near (but not touching) the two drops of antisera.

4. Mix one of the drops of blood with antiserum A and one with antiserum B. Use a different toothpick to mix each antiserum.

5. Dispose of all used materials properly.

6. Observe any agglutination of blood cells in either of the two antisera.

Agglutination of blood mixed with an antiserum is indicated by a grainy appearance. Agglutination indicates the presence of the respective antigen on red blood cells (fig. 17.5). Determine and record the blood type from your sample based on the presence of antigens.

### Question 11
*a.* What antigens are present on the artificial red blood cells that you tested?

### Rh Incompatibility

You've probably heard of the incompatibility (agglutination) problems that Rh-negative women may have with their Rh-positive babies (the Rh-positive trait is inherited from the child's father). This problem usually occurs with the second and subsequent children because women with the Rh blood system must be sensitized to the antigen before antibody production begins. This sensitization usually occurs during birth of the first child.

If you are a woman having Rh-negative blood, you should be concerned but not alarmed. Rh incompatibility is handled routinely by injections of anti-Rh antibodies. These antibodies destroy Rh-positive red cells and thus eliminate the Rh-associated risk of subsequent childbirth.

*b.* What is the blood type of your sample?

You are probably familiar with another characteristic of blood called **Rh factor.** Although more than two alleles determine Rh, we'll use "positive" and "negative" for simplicity and convenience.

### Procedure 17.4 Determine Rh

1. Place a drop of anti-Rh serum on a clean slide.

2. Using the procedure just described, mix a drop of blood from the synthetic blood sample provided with the antiserum.

3. Label the slide with your initials and place it on the warming plate in the lab.

4. The blood sample will agglutinate within a few minutes if it is Rh-positive. The absence of agglutination indicates the blood is Rh-negative.

5. Dispose of all materials properly.

## OTHER HUMAN TRAITS

The following traits are determined by a single gene. List your phenotype for each trait in table 17.3 and, if possible, list your genotype. If you have the recessive trait for gene *G,* for example, your genotype is homozygous recessive (*gg*). If you have the dominant trait, your genotype could be *GG* or *Gg,* in which case you should enter *G* in table 17.3. If you have the dominant trait and one of your parents shows the recessive trait, you must be heterozygous (*Gg*) for that trait. Give your results to your instructor so that she or he can provide you with the phenotypic results for your class.

**Widow's peak**—The *W* allele for widow's peak (i.e., a pointed hairline) is dominant to the *w* allele for a straight hairline (fig. 17.6).

**Bent little finger**—Lay your hands flat on the table and relax them. If the last joint of your little finger bends toward the fourth finger, you have the dominant allele *B* (fig. 17.7).

**Albinism**—The *A* allele is dominant and leads to production of melanin, a pigment. Individuals with an *aa* genotype lack pigment in their skin, hair, and iris.

**Pigmented iris**—If you are homozygous for the recessive allele *p,* you do not produce pigment in the front layer of your iris, and your eyes are either blue or gray (i.e., your eyes are the color of the back layer of the iris). The *P* allele produces pigment in the front layer of the iris (green, hazel, brown, or black), which masks the blue or gray color of the back layer of the iris.

Table 17.3

**Phenotypes and Genotypes of Human Traits**

| Characteristic | Your Phenotype | Your Genotype* | Phenotypes of Class | |
| --- | --- | --- | --- | --- |
| | | | Dominant | Recessive |
| Widow's peak | | | | |
| Bent little finger | | | | |
| Albinism | | | | |
| Pigmented iris | | | | |
| Attached earlobes | | | | |
| Hitchhiker's thumb | | | | |
| Interlacing fingers | | | | |
| PTC tasting | | | | |
| Middigital hair | | | | |
| Dimpled chin | | | | |
| Six fingers | | | | |

*Homozygous dominant, heterozygous, or homozygous recessive

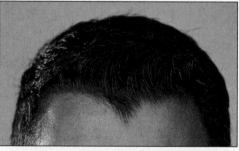

Dzmitry Kliapitski/123RF

Bent little        Straight little
finger             finger

**Figure 17.7**  Bent little finger.

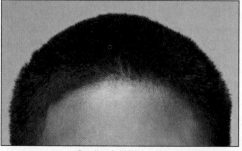

Caroline Schiff/Blend Images/Getty Images

**Figure 17.6**  Widow's peak hairline (*top*). People lacking a widow's peak have a relatively straight hairline (*bottom*).

**Free earlobes**—If part of your earlobe is unattached (i.e., free) below where it attaches to your head, you have the dominant *E* allele for a free earlobe (fig. 17.8).

**Hitchhiker's thumb**—Bend your thumb backward as far as possible. If you can bend the last joint of the thumb back at an angle of 60° or more, you are showing the recessive allele *h* (fig. 17.9).

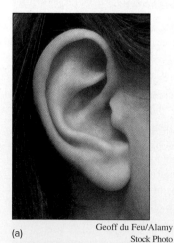

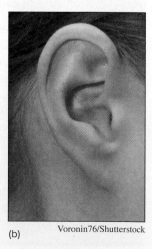

(a)        Geoff du Feu/Alamy        (b)        Voronin76/Shutterstock
           Stock Photo

**Figure 17.8**  (*a*) Free earlobe. (*b*) Attached earlobe.

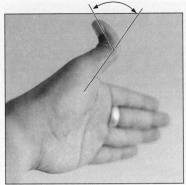

**Figure 17.9** Hitchhiker's thumb.

**Interlacing fingers**—Casually fold your hands together so that your fingers interlace. The *C* allele for crossing the left thumb over the right thumb when you interlace your fingers is dominant over the *c* allele for crossing your right thumb over your left.

**PTC tasting**—Obtain a piece of paper impregnated with phenylthiocarbamide (PTC). Taste the paper by chewing on it for a few seconds. If you detect a bitter taste, you have the dominant allele *T*.

**Middigital hair**—The allele *M* for hair on the middle segment of your fingers is dominant to the *m* allele for no middigital hair. If hair is present on the middigit of any finger, you have the dominant allele.

**Dimpled chin**—A dimpled chin is caused by a dominant allele *M*. People who have a dimpled chin are either homozygous dominant (*MM*) or heterozygous (*Mm*) for this trait. Homozygous recessive (*mm*) individuals do not have a dimpled chin.

**Six fingers**—In humans, the occurrence of six fingers results from a dominant allele *S*. People who have six fingers are either homozygous dominant (*SS*) or heterozygous (*Ss*). People who have only five fingers are homozygous recessive (*ss*) for this trait.

Several diseases are inherited as single-gene traits. These diseases include

**Cystic fibrosis,** a disease characterized by chronic bronchial obstruction and growth reduction. This disease is inherited as a recessive trait; people who are heterozygous or homozygous dominant do not have this disease. Medications can help manage the symptoms of cystic fibrosis, but there is no cure.

**Galactosemia,** an inability to metabolize galactose, a sugar in human milk. Galactosemia is inherited as an autosomal recessive trait. Approximately five cases occur per million births. Prenatal diagnosis can be performed on cells obtained through amniocentesis or chorionic villi sampling. Galactosemia is inherited as a recessive trait; people who are heterozygous or homozygous dominant

do not have this disease. The treatment for galactosemia is a galactose-free diet.

**Phenylketonuria (PKU),** an inability to metabolize the amino acid phenylalanine. Approximately 100 cases occur per million births. The treatment for PKU is a phenylalanine-free diet; this is why people who have phenylketonuria avoid artificial sweeteners such as "Equal," whose packets and containers include a warning that they contain phenylalanine. Artificial sweeteners such as "Splenda" lack phenylalanine and therefore are not accompanied by a warning. If untreated, phenylketonuria produces mental impairment. This disease is inherited as a recessive trait; people who are heterozygous or homozygous dominant do not have this disease.

**Juvenile retinoblastoma,** a cancer of the retina. The allele is located on chromosome 13. This disease is inherited as a recessive trait; people who are heterozygous or homozygous dominant do not have this disease.

**Huntington's disease,** a neurological disorder that leads to progressive degeneration of brain cells. The disease usually appears in middle age, so many affected individuals bear children before they realize that they are carriers. Approximately 100 cases occur per million births. Unlike most other genetic diseases, Huntington's disease is inherited as a dominant trait; people who are homozygous recessive (*hh*) do not have the disease, and people who are heterozygous (*Hh*) or homozygous dominant (*HH*) have the disease. Death comes 10–15 years after the onset of symptoms. Medications can help manage the symptoms of Huntington's disease, but treatment cannot prevent the behavioral, mental, and physical decline associated with the condition.

## Question 12

What conclusion about your genotype is evident if one of your siblings, but neither parent, shows the recessive trait?

All of the traits listed above are produced by alleles on **autosomes,** which are chromosomes that are not sex chromosomes. As a result, these traits are not linked to gender and occur in equal frequencies in males and females. This is why autosomal traits such as galactosemia, Tay Sachs, and Huntington's disease are equally frequent in males and females. Several other traits, however, are produced by alleles on **sex chromosomes,** and are referred to as **sex-linked traits.**

## SEX-LINKED INHERITANCE

Humans have two kinds of sex chromosomes: X and Y. Sex-linked traits are produced by alleles on these chromosomes and include hemophilia, Duchenne muscular dystrophy, and red-green colorblindness.

Inheritance patterns among affected and unaffected individuals have revealed 59 common diseases specific to segments of the X chromosome, indicated here by brackets (fig. 17.10). Only a partial map for the human X chromosome is shown here; a more detailed map would require a much larger figure. The black bands represent staining patterns that are specific to alleles involved in the indicated disease. The constriction represents the position of the centromere. The X chromosome sequence has 1098 gene loci, many of which may have mutant alleles that produce diseases.

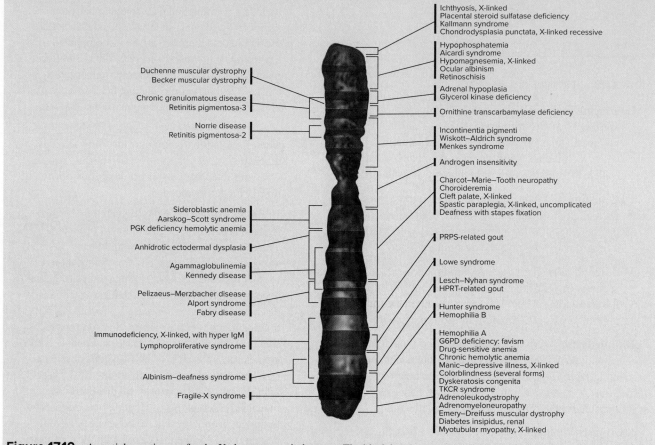

**Figure 17.10** A partial genetic map for the X chromosome in humans. The black bands represent alleles involved in the listed diseases.

Females have two X chromosomes (i.e., they are XX), and males have one X and one Y chromosome (i.e., they are XY). Because females and males each inherits an X chromosome, both can have X-linked traits. If the X chromosome in a male carries the recessive allele, that allele is expressed because there is no chance for the allele to be masked by its counterpart on another X chromosome.[1] For example, consider a cross between a male who is not colorblind (i.e., $X^AY$) and a female who is not colorblind but is a carrier of the recessive allele (i.e., $X^AX^a$):

|       | $X^A$    | Y      |
|-------|----------|--------|
| $X^A$ | $X^AX^A$ | $X^AY$ |
| $X^a$ | $X^AX^a$ | $X^aY$ |

In this cross, half of the boys (and none of the girls) would be colorblind.

If a male has an X-linked recessive disorder ($X^aY$) and his female partner does not carry the allele ($X^AX^A$), all of their girls will be carriers of the allele, and none of their boys will inherit the disorder.

|       | $X^a$    | Y      |
|-------|----------|--------|
| $X^A$ | $X^AX^a$ | $X^AY$ |
| $X^A$ | $X^AX^a$ | $X^AY$ |

## Question 13

*a.* Suppose that Donna is colorblind, but Darrell is not. If Donna and Darrell have a family, what percentage of their boys will be colorblind?

---

1   *The X chromosome carries more than 1000 alleles, but the much smaller Y chromosome carries fewer than 30 alleles. Because only males inherit Y chromosomes, they are the only ones to inherit Y-linked traits. Nine of the alleles on the Y chromosome are involved with sperm production. If any of these alleles are missing, the result is low sperm count and infertility (almost one-third of infertile couples in the United States are infertile because of Y-linked alleles). Only one gene on the Y chromosome (the SRY gene) produces male anatomical traits.*

## Table 17.4

### Some Dominant and Recessive Traits in Humans

| Recessive Traits | Phenotypes | Dominant Traits | Phenotypes |
|---|---|---|---|
| Albinism | Lack of melanin pigmentation | Middigital hair | Presence of hair on middle segment of fingers |
| Alkaptonuria | Inability to metabolize homogentisic acid | Brachydactyly | Short fingers |
| Red-green color blindness | Inability to distinguish red or green wavelengths of light | Huntington's disease | Degeneration of nervous system, starting in middle age |
| Cystic fibrosis | Abnormal gland secretion, leading to liver degeneration and lung failure | Phenylthiocarbamide (PTC) sensitivity | Ability to taste PTC as bitter |
| Duchenne muscular dystrophy | Wasting away of muscles during childhood | Camptodactyly | Inability to straighten the little finger |
| Hemophilia | Inability of blood to clot properly; some clots form, but the process is delayed | Familial hypercholesterolemia (the most common human Mendelian disorder) | Elevated levels of blood cholesterol and risk of heart attack |
| Sickle cell anemia | Defective hemoglobin that causes red blood cells to curve and stick together | Polydactyly | Extra fingers and toes |

*b.* What percentage of their girls will be colorblind?

## ANALYZING PEDIGREES

Many human traits display both dominant and recessive inheritance (table 17.4). Researchers cannot control crosses in humans the way Mendel did with pea plants, so to analyze human inheritance, geneticists study crosses that have been performed already—in other words, family histories. This involves a **pedigree,** which is a consistent graphical presentation of matings and offspring over multiple generations for a particular trait. Information in a pedigree allows geneticists to deduce how the trait is inherited.

If you understand the simple patterns of inheritance presented in this lab, you can trace a trait in a pedigree (i.e., family tree) to determine if it is inherited in a dominant or recessive pattern of inheritance.

### Question 14

*a.* What features would characterize pedigrees of dominant traits?

*b.* What features would characterize pedigrees of recessive traits?

Biologists use the following symbols in pedigrees:

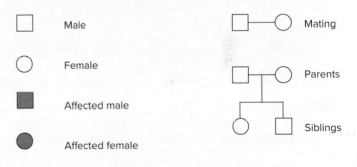

### Procedure 17.5  Analyze a pedigree of inheritance of cystic fibrosis

1. Among Caucasians, about 1 of every 2500 newborn infants is born with cystic fibrosis. In these individuals, a defective membrane protein results in the production of unusually thick and dry mucus that lines organs such as the tubes in the respiratory system. People having cystic fibrosis often have recurrent and serious infections, and most die in their 20s or 30s.

2. Use the following pedigree to determine whether the allele for cystic fibrosis is inherited as a dominant or recessive allele.

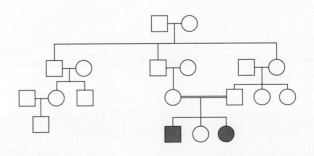

## Question 15

*a.* What is the inheritance pattern for the cystic fibrosis allele? What is your reasoning for this conclusion?

*b.* Can you determine the genotypes of any individuals in the pedigree? If so, which ones? Explain your reasoning.

### *Procedure 17.6* Analyze a pedigree of inheritance of Huntington's disease

*1.* Huntington's disease is a severe disorder of the nervous system that usually causes death.

*2.* Use the following pedigree to determine whether the allele for Huntington's disease is inherited as a dominant or recessive allele.

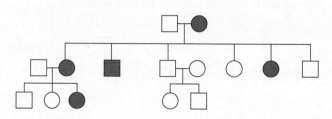

## Question 16

*a.* What is the inheritance pattern for the Huntington's disease allele? What is your reasoning for this conclusion?

*b.* Can you determine the genotypes of any individuals in the pedigree? If so, which ones? Explain your reasoning.

*c.* Examine figure 17.11. Shana's mother has Huntington's disease, and Shana has a 50–50 chance of developing Huntington's disease. Explain the genetic basis for Shana's chances of inheriting Huntington's disease.

### *Procedure 17.7* Analyze a pedigree of inheritance of phenylketonuria

*1.* Phenylketonuria, or PKU, results from an inability to metabolize the amino acid phenylalanine. If untreated, PKU leads to mental retardation.

*2.* Use the following pedigree to determine whether the allele for phenylketonuria is inherited as a dominant or recessive allele.

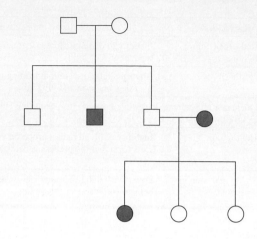

## Question 17

*a.* What is the inheritance pattern for the phenylketonuria allele? What is your reasoning for this conclusion?

*b.* Can you determine the genotypes of any individuals in the pedigree? If so, which ones? Explain your reasoning.

## TRANSPOSONS

For many decades, geneticists thought that genes do not move in cells. However, in 1947 Barbara McClintock (1902–1992) proposed that genes could move within and between chromosomes. McClintock based her conclusion on a series of experiments involving genetic crosses in corn. Specifically, McClintock showed that there is a fragment of DNA that can move to and be inserted at the locus for the production of pigments in corn kernels. Because this insertion renders the cell unable to make the purple pigment, the resulting kernel is yellow or white. However, subsequent removal of the DNA fragment results in the cell resuming production of the purple pigment; therefore, the resulting kernel is purple. Thus, Indian corn often has kernels with varying pigmentation, depending on when the DNA fragment was inserted or removed. This colorful pigmentation makes Indian corn a symbol of the fall harvest season and favorite decoration at Halloween and Thanksgiving. Indian corn is one of the earliest varieties of corn that Native Americans taught early colonists how to grow.

Huntington's Disease Society of America

**Figure 17.11** Huntington's disease is a degenerative disease inherited as a dominant trait. What is the genetic basis for Shana's statement that she has a 50–50 chance of getting Huntington's disease?

A similar phenomenon occurs with the production of other pigments in corn kernels. The translocation to and from the locus for production of these pigments several times during kernel development produces the red-orange swirls characteristic of many kernels of Indian corn.

The fragments of DNA that McClintock studied are now called **transposons** (fig. 17.12). Transposons are a useful tool for genetic engineering because they provide a way of inserting foreign DNA into a host cell's chromosome. For her work, McClintock received a Nobel Prize for Medicine in 1983; she was the first American woman to win an unshared Nobel Prize.

### Procedure 17.8 Observe corn kernels to understand the effects of transposons

*1.* Work in a group of two to four people. Obtain an ear of Indian corn for your group.

*2.* Look for examples of kernels with (a) purple or white spots, (b) red-orange swirls, and (c) other unusual color patterns. Sketch the pigmentation patterns in 2–3 kernels.

*3.* Use the information presented in this exercise and in your textbook to determine how transposons could produce such unusual patterns of pigmentation.

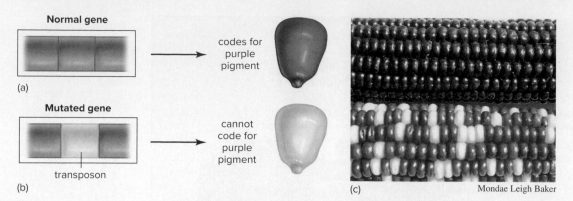

(a) Normal gene → codes for purple pigment

(b) Mutated gene → cannot code for purple pigment

transposon

(c) Mondae Leigh Baker

**Figure 17.12** Transposons may disrupt gene expression. (*a*) A purple-coding gene ordinarily codes for a purple pigment. (*b*) A transposon "jumps" into the purple-coding gene. This mutated gene is unable to code for purple pigment, and a white kernel results. (*c*) Indian corn displays a variety of colors and patterns due to transposon activity.

## INQUIRY-BASED LEARNING

### *Are homozygous recessive traits rare among humans?*

Observation: In humans, traits such as widow's peak, attached earlobes, and a dimpled chin are homozygous recessive traits. In many people, these traits are easily observed.

Question: How common are homozygous recessive traits such as widow's peak, attached earlobes, and a dimpled chin among your classmates?

a. Establish a working lab group and obtain Inquiry-Based Learning Worksheet 17 from your instructor.

b. Discuss with your group well-defined questions relevant to the preceding observation and question. Choose and record your group's best question for investigation.

c. Translate your question into a testable hypothesis and record it.

d. Outline on Worksheet 17 your experimental design and supplies needed to test your hypothesis. Ask your instructor to review your proposed investigation.

e. Conduct your procedures, record your data, answer your question, and make relevant comments.

f. Discuss with your instructor any revisions to your questions, hypothesis, or procedures. Repeat your work as needed.

## Testing for Understanding: Solving Genetics Problems

1. When reclusive billionaire Howard Hughes died in 1976, a variety of people claimed that they were entitled to Hughes' estate because they were his children. Hughes had type AB blood. One man who claimed that Hughes was his father had type O blood, and the man's mother had type A blood. If you were the judge in the case, what would you rule? Explain your answer.

2. Suppose that flower color is inherited by simple dominance and that purple flowers are dominant to white flowers. If a homozygous recessive individual is crossed with a homozygous dominant individual, what is the probability of obtaining a purple-flowered offspring?
   a. 100%   b. 75%  c. 50%  d. 25%  e. 0%

3. Bob is heterozygous for phenylketonuria, and Loretta is homozygous recessive for phenylketonuria. What is the probability that their first child will have phenylketonuria?
   a. 100%   b. 75%  c. 50%  d. 25%  e. 0%

4. Suppose that (1) Randy is heterozygous for the allele that causes Huntington's disease, (2) Susan is homozygous recessive for the allele that causes Huntington's disease, and (3) Randy and Susan decide to have a child. What is the probability that their child will get Huntington's disease?

   a. 100%   b. 75%  c. 50%  d. 25%  e. 0%

5. In question 4, what is the probability that their first daughter will get Huntington's disease?
   a. 100%   b. 75%  c. 50%  d. 25%  e. 0%

6. Suppose that someone having type AB blood has a child with someone having type O blood. What is the probability that their child will have type A blood?
   a. 100%   b. 75%  c. 50%  d. 25%  e. 0%

7. Suppose that you cross a red-flowered carnation with a white-flowered carnation. All of the offspring have pink flowers. What can you conclude?
   a. Flower color in carnations is inherited by incomplete dominance, and the red-flowered carnation is homozygous dominant for the trait.
   b. At least one of the parents is heterozygous for flower color.
   c. Flower color is inherited by simple dominance.
   d. Half of the offspring are heterozygous and half are homozygous for flower color.
   e. None of the above statements are true.

8. Suppose that a trait is inherited by simple dominance. If two heterozygotes are mated, what is the probability of having a homozygous recessive offspring?
   **a.** 100%   **b.** 75%  **c.** 50%  **d.** 25%  **e.** 0%
9. Tay-Sachs disease is characterized by the inability to produce an enzyme needed to metabolize lipids in brain cells. If this enzyme is not present, lipids accumulate in the brain and gradually destroy its ability to function (homozygous recessive children usually die by the age of four or five). Suppose that you are a carrier for Tay-Sachs disease and that your partner is not. What is the probability that you and your partner will have a child with Tay-Sachs disease?
   **a.** 100%   **b.** 75%  **c.** 50%  **d.** 25%  **e.** 0%
10. A normally pigmented man marries a normally pigmented woman. Their second child is an albino.
   **a.** What is the genotype of the man? _____
   **b.** What is the genotype of the woman? _____
   **c.** What is the genotype of the albino child?
   _____
   **d.** What is the probability that their next child will be an albino? _____
11. Darrell and Matilda each has type O blood. If they start a family, the probability that they will have a child having type A blood is _____.
   **a.** 100%   **b.** 75%  **c.** 50%  **d.** 25%  **e.** 0%
12. Is it possible for someone having type B blood and someone having type A blood to have a child having type O blood? Explain your answer.
13. Wanda, who has type O blood, gives birth to a baby having type O blood. The woman then claims that the child's father is Randy, who has type A blood.
   **a.** Could Randy be the father?
   **b.** Can this information alone prove that Randy is the father? Explain your answer.
14. Suppose that two people having free earlobes start a family. Their first child has free earlobes and their second has attached earlobes.
   **a.** What are the genotypes of the parents? _____
   **b.** What is the genotype of their first child? _____
   **c.** What is the genotype of their second child?
   _____
15. Suppose that a trait is inherited by simple dominance. If two heterozygotes are mated, what is the probability of having an offspring that has the same phenotype as the parents?
   **a.** 100%   **b.** 75%  **c.** 50%  **d.** 25%  **e.** 0%
16. Suppose that Donna is a carrier for hemophilia and that Darrell does not have hemophilia. If Donna and Darrell have a family, what percentage of their girls will have hemophilia? What percentage of their boys will have hemophilia?
17. Darrell has type B blood. To people having what blood types can he safely donate blood?
18. Albinism, which is inherited as an autosomal recessive allele, is a genetic condition characterized by the absence of the pigment melanin in skin, hair, and eyes.

The lack of melanin results from being homozygous recessive (*aa*). Approximately one in 17,000 people worldwide is an albino. The ability to taste the chemical phenylthiocarbamide (PTC) is also inherited as an autosomal recessive allele; not being able to taste PTC results from being homozygous recessive (*pp*). Approximately 70% of Americans can taste PTC.

Randy is a normally pigmented man who cannot taste PTC, and Randy's father is an albino who can taste PTC. Matilda is a homozygous, normally pigmented woman who can taste PTC, but her mother cannot taste PTC.
   **a.** If Randy and Matilda have a family, what are the possible genotypes of their children?
   **b.** What percentage of their children will be albino?
   **c.** What percentage of their children will be able to taste PTC?
19. In certain species of flies, eye color is controlled by simple dominance by a single pair of alleles. A red-eyed fly was crossed with a white-eyed fly, both of whose parents had white eyes. All of their offspring (both female and male) had red eyes.
   **a.** Which is dominant, the allele for red eyes or the allele for white eyes?
   **b.** What is the genotype of the white-eyed parents?
   **c.** If the white-eyed parent was mated with one of the red-eyed offspring, what phenotypic ratio would you expect regarding eye color?
20. In cattle, hair color is controlled by incomplete dominance. Red results from homozygous dominant alleles (*RR*), roan from heterozygous alleles (*Rr*), and white from homozygous recessive alleles (*rr*). What are the predicted phenotypes and their frequencies of the following crosses:
   **a.** a white cow and a red bull
   **b.** a roan cow and a red bull
   **c.** a roan cow and a roan bull
21. In each of the following cases of disputed paternity, identify the probable father:

| Blood type of mother | Blood type of child | Blood type of father #1 | Blood type of father #2 |
|---|---|---|---|
| B | O | A | AB |
| B | AB | A | B |

22. A man with type B blood starts a family with a woman having type A blood. The woman's parents both have type AB blood. What is the probability that the couple's first child will be a boy having type O blood?
23. Suppose that a woman having normal vision and a man having normal vision have a son who is colorblind. What is the probability that their second child will be a colorblind girl?
24. Albinism results from an autosomal recessive gene. Two parents with normal pigmentation have an albino child.
   **a.** What is the probability that their next child will be an albino?
   **b.** What is the probability that their next two children will both be albino?

1. The man could not be Hughes' son. Hughes had type AB blood. Regardless of the blood type of the mother, a child of Hughes could not have type O blood

2. a

3. c

4. c

5. c

6. c

7. a

8. d

9. e

10a. *Aa*    10b. *Aa*    10c. *aa*    10d. 25%

11. e

12. Yes, but only if the person having type B blood has a BO genotype, and if the person having type A blood has an AO genotype.

13a. Yes, if he has an AO genotype.

13b. No, information about blood type cannot prove that anyone is the parent of a child; it can only eliminate people who are not parents of the child.

14a. *Ee*    14b. *EE* or *Ee*    14c. *ee*

15. b

16. None of their girls will have hemophilia, and 50% of their boys will have hemophilia.

17. B and AB

18a. *AApp, AAPp, Aapp, AaPp*    18b. 0%    18c. 50%

19a. red    19b. homozygous recessive

19c. 50% white-eyed, 50% red-eyed (i.e., a 1:1 ratio)

20a. all roan    20b. 1 red : 1 roan    20c. 1 red : 2 roan : 1 white

21. In both cases, the probable father is father #1 (type A blood).

22. 0%

23. 0%

24a. 25%

24b. 6.25%

# Questions for Further Study and Inquiry

*1.* What determines how often a phenotype occurs in a population?

*2.* Are dominant characteristics always more frequent in a population than recessive characteristics? Why or why not?

*3.* Is it possible to determine the genotype of an individual having a dominant phenotype? How?

*4.* Why is hybrid seed so expensive to produce?

*5.* What blood types are not expected for children to have if their parents have AB blood? O blood?

## WRITING TO LEARN BIOLOGY

Organisms heterozygous for a recessive trait are often called carriers of that trait. What does this mean?

# Evolution
## Natural Selection and Morphological Change in Green Algae

## Learning Objectives

By the end of this exercise you should be able to:

1. Give a working definition of evolution, fitness, selection pressure, and natural selection.
2. Determine the genotypic and phenotypic frequency of a population while properly using the terms allele, dominant, recessive, homozygous, and heterozygous.
3. Explain the Hardy-Weinberg Principle and use it to demonstrate negative selection pressures on a population.
4. Describe the significance of the Volvocine line, particularly in the areas of cellular specialization and colonial complexity.

Please visit **connect.mheducation.com** to review online resources tailored to this lab.

**E**volution is the change in heritable characteristics of a population over successive generations. The theory of evolution broadly describes genetic change in populations. The existence of genetic change (and therefore evolution) is universally accepted by biologists. We know that many mechanisms can change the genetic makeup of populations, but the relative importance of each mechanism remains to be fully described. Events such as **mutations** (changes in the genetic message of a cell, fig. 18.1) and catastrophes (e.g., meteor showers, ice ages) can produce genetic change. However, Charles Darwin (fig. 18.2) formulated a theory that explains a major force behind genetic change that produces adaptation: natural selection.

Darwin postulated that organisms that survive and reproduce successfully have genetic traits aiding survival and reproduction. These traits enhance an organism's **fitness,** which is its tendency to produce more offspring than competing individuals, and therefore contribute more genes to the next generation. Darwin noticed that fit individuals (that is, ones that reproduce the most) produce more offspring because their traits are better adapted for survival and reproduction than the traits of their competitors. He further reasoned that if the traits of the more fit individuals are transmitted to the next generation more often, then more of these traits will be found in the next generation. After many generations, the frequency of these traits will increase in the population, and the nature of the population will gradually change. Darwin called this process **natural selection** and proposed it as a major force that guides genetic change while producing adaptations and forming new species. Review in your textbook the theories of evolution and the mechanism of natural selection.

©Otero/gtphoto

**Figure 18.1** Mutations produce new alleles and new genetic combinations. This child has a streak of white hair, caused by a somatic mutation in a single cell during embryonic development. This cell continued to divide to produce a streak of white hair.

Showing the effects of natural selection in living populations is usually time-consuming and tedious. Therefore, in this exercise you will simulate reproducing populations with nonliving, colored beads representing organisms and their gametes. With this artificial population

**Figure 18.2** Darwin greets his "monkey ancestor." In his time, Darwin was often portrayed unsympathetically, as in this drawing from an 1874 publication.

Science source

you can quickly follow genetic change over many generations. Before you begin work, review the previous exercise on genetics, especially the terms **gene, allele, dominant alleles, recessive alleles, homozygous,** and **heterozygous.**

You will begin your experiments using a "stock population" of organisms consisting of a container of beads. Each bead represents a haploid (having one set of chromosomes) gamete, and the color of the "gamete" (colored or white) represents an allele it is carrying. Individual organisms from this population are diploid (having two sets of chromosomes) and therefore are represented by two beads.

## UNDERSTANDING ALLELIC AND GENOTYPIC FREQUENCIES

**Frequency** is the proportion of individuals in a certain category relative to the total number of individuals considered. The frequency of an allele or genotype is expressed as a decimal proportion of the total alleles or genotypes in a population. For example, if 1/4 of the individuals of a population are genotype *Bb*, then the frequency of *Bb* is 0.25. If 3/4 of all alleles in a population are *B*, then the frequency of *B* is 0.75.

In this exercise you will simulate evolutionary changes in allelic and genotypic frequencies in an artificial population. The trait you will work with is fur color. A colored bead is a gamete with a dominant allele (complete dominance) for black fur (*B*), and a white bead is a gamete with a recessive allele for white fur (*b*). An individual is represented by two gametes (beads). Individuals with genotypes *BB* and *Bb* have black fur, and those with *bb* have white fur.

### *Procedure 18.1* Establish a parental population

1. Obtain a "stock population" of organisms consisting of a container of colored and white beads.

2. Obtain an empty container marked "Parental Population."

3. From the stock population, select 25 homozygous dominant individuals (*BB*) and place them in the container marked "Parental Population." Each individual is represented by two colored beads.

4. From the stock population, select 50 heterozygous individuals (*Bb*) and place them in the container marked "Parental Population." Each individual is represented by a colored and a white bead.

5. From the stock population, select 25 homozygous recessive individuals (*bb*) and place them in the container marked "Parental Population." Each individual is represented by two white beads.

6. Calculate the total number of individuals and the total number of alleles in your newly established parental population. Use this information to calculate and record in table 18.1 the correct genotypic frequencies for your parental population.

7. Complete table 18.1 with the number and frequency of each of the two alleles.

### Question 1

*a.* How many alleles are present for this particular trait?

*b.* How many of the total beads are colored and how many are white?

| Table 18.1 | | | | |
|---|---|---|---|---|
| **Frequencies of Genotypes and Alleles of the Parental Population** | | | | |
| **Genotypes** | **Frequency** | | **Alleles** | **Frequency** |
| *BB* ●● | _____ | | *B* ● | _____ |
| *Bb* ●○ | _____ | | *b* ○ | _____ |
| *bb* ○○ | 0.25 | | | |

*c.* What color fur do *Bb* individuals have?

*d.* How many beads represent the population?

## THE HARDY-WEINBERG PRINCIPLE

The **Hardy-Weinberg Principle** enables us to calculate and predict allelic and genotypic frequencies. We can compare these predictions with actual changes that we observe in natural populations and learn about factors that influence gene frequencies. Deviations of observed frequencies from frequencies predicted by the Hardy-Weinberg Principle indicate evolution.

This predictive model includes two simple equations first described for stable populations by Godfrey Hardy and Wilhelm Weinberg. Hardy-Weinberg equations (1) predict allelic and genotypic frequencies based on data for only one or two frequencies and (2) provide a set of theoretical frequencies that we can compare to frequencies from natural populations. For example, if we know the frequency of *B* or *BB*, we can calculate the frequency of *b, Bb,* and *bb.* Then we can compare these frequencies with those of a natural population that we might be studying. If our observed data vary from our predictions, we can study the reasons for this genetic change. This comparison is important because biological characteristics of natural populations rarely correspond exactly to theoretical calculations. Furthermore, deviations are important because they often reveal unknown factors influencing the population being studied.

According to the Hardy-Weinberg Principle, the frequency of the dominant allele of a pair is represented by the letter *p,* and that of the recessive allele by the letter *q.* Also, the genotypic frequencies of *BB* (homozygous dominant), *Bb* (heterozygous), and *bb* (homozygous recessive) are represented by $p^2$, $2pq$, and $q^2$, respectively. Examine the frequencies in table 18.1 and verify the Hardy-Weinberg equations:

$$p = \text{frequency of dominant allele}$$
$$q = \text{frequency of recessive allele}$$
$$p + q = 1$$
$$p^2 + 2pq + q^2 = 1$$

For example, consider a population of plants in which flower color is inherited by incomplete dominance (Exercise 17). Homozygous individuals (*RR*) are red, heterozygous individuals (*Rr*) are pink, and homozygous recessives (*rr*) are white. If you know that the frequency of the dominant allele (*R*) is 0.8, then the frequency of the recessive allele (*r*) must be $1.0 - 0.8 = 0.2$. This means that the probability of producing a red plant is $0.8 \times 0.8 = 0.64 = 64\%$, and the probability of producing a white plant is $0.2 \times 0.2 = 0.04 = 4\%$. The probability of producing a pink plant is $2 \times 0.8 \times 0.2 = 0.32 = 32\%$. The Hardy-Weinberg Principle and its equations predict that frequencies of alleles and genotypes will remain constant from generation to generation in stable populations. Therefore, these equations can be used to predict genetic frequencies through time. However, the Hardy-Weinberg prediction assumes that

- The population is large enough to overcome random events.
- Choice of mates is random.
- Mutation does not occur.
- Individuals do not migrate into or out of the population.
- There is no selection pressure.

### Question 2
*a.* Because all of the conditions listed above never occur simultaneously in nature, Hardy-Weinberg equilibrium never occurs in real populations. However, the Hardy-Weinberg Principle can be used to calculate and predict genotypic frequencies. Consider the Hardy-Weinberg equations. If the frequency of a recessive allele is 0.3, what is the frequency of the dominant allele?

*b.* If the frequency of the homozygous dominant genotype is 0.49, what is the frequency of the dominant allele?

*c.* If the frequency of the homozygous recessive genotype is 0.36, what is the frequency of the dominant allele?

*d.* If the frequency of the homozygous dominant genotype is 0.49, what is the frequency of the homozygous recessive genotype?

*e.* Which Hardy-Weinberg equation relates the frequencies of the alleles at a particular gene locus?

*f.* Which Hardy-Weinberg equation relates the frequencies of the genotypes for a particular gene locus?

*g.* Which Hardy-Weinberg equation relates the frequencies of the phenotypes for a gene?

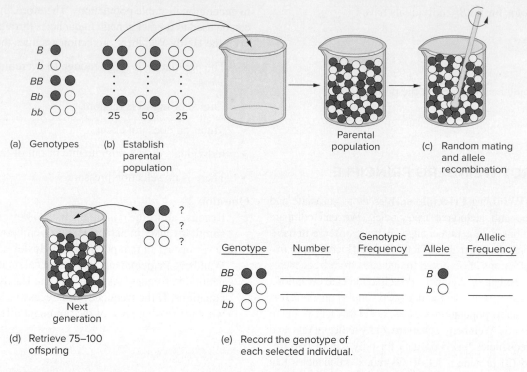

(a) Genotypes  (b) Establish parental population

Parental population

(c) Random mating and allele recombination

Next generation

(d) Retrieve 75–100 offspring

| Genotype | Number | Genotypic Frequency | Allele | Allelic Frequency |
|----------|--------|---------------------|--------|-------------------|
| BB ●● | _____ | _____ | B ● | _____ |
| Bb ●○ | _____ | _____ | b ○ | _____ |
| bb ○○ | _____ | _____ | | |

(e) Record the genotype of each selected individual.

**Figure 18.3** Verification of the Hardy-Weinberg Principle. See procedure 18.2 for an explanation of each step.

Additional problems to test your understanding of the Hardy-Weinberg Principle are at the end of this exercise.

To verify the predictions of the Hardy-Weinberg Principle, use the following procedure to produce a generation of offspring from the parental population you created in the previous procedure.

## *Procedure 18.2* **Verify the Hardy-Weinberg Principle**

1. Examine figure 18.3 for an overview of the steps of this procedure.

2. Establish the parental population from procedure 18.1 (fig. 18.3*a*, 18.3*b*).

3. Simulate the random mating of individuals by mixing the population (fig. 18.3*c*).

4. Reach into the parental container (without looking) and randomly select two gametes. Determine their genotype (fig. 18.3*d*). This pair of gametes with colored or white alleles represents an individual offspring.

5. Record the occurrence of the genotype in figure 18.3*e* as a mark under the heading "Number" or temporarily on a second sheet of paper and return the beads to the container.

6. Repeat steps 4 and 5 (100 times) to simulate the production of 100 offspring.

7. Calculate the frequency of each genotype and allele, and record the frequencies in figure 18.3*e*. Beside each of these new-generation frequencies write (in parentheses) the original frequency of that specific genotype or allele from table 18.1.

**Question 3**

*a.* The Hardy-Weinberg Principle predicts that genotypic frequencies of offspring will be the same as those of the parental generation. Were they the same in your simulation?

*b.* If the frequencies were different, then one of the assumptions of the Hardy-Weinberg Principle was probably violated. Which one?

## EFFECT OF A SELECTION PRESSURE

**Selection** is the differential reproduction of phenotypes (fig. 18.4); that is, some phenotypes (and their associated genotypes) are passed to the next generation more often than others. In positive selection, genotypes representing adaptive traits in an environment increase in frequency because their bearers are more likely to survive and reproduce. In negative selection, genotypes representing nonadaptive traits in an environment decrease in frequency because their bearers are less likely to survive and reproduce.

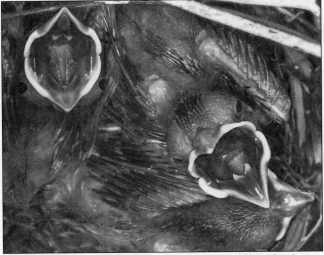

©BiologyImaging.com

**Figure 18.4** Reproduction. These birds hatching from eggs may or may not survive to reproduce. On average, birds with characteristics best adapted to their environment will survive and reproduce more than those with less adaptive characteristics. As a result, the frequencies of adaptive traits (and their alleles) in the population will increase from generation to generation. This change in frequencies of alleles over time is evolution.

**Selection pressures** are factors such as temperature and predation that affect organisms and result in selective reproduction of phenotypes. Some pressures may elicit 100% negative selection against a characteristic and eliminate any successful reproduction of individuals having that characteristic. For example, mice with white fur may be easy prey for a fox if they live on a black lava field. This dark environment is a negative selection pressure against white fur. If survival and reproduction of mice with white fur were eliminated (i.e., if there is 100% negative selection), would the frequency of white mice in the population decrease with subsequent generations? To test this, use the following procedure to randomly mate members of the original parental population to produce 100 offspring (fig. 18.5).

**Procedure 18.3** Simulate 100% negative selection pressure

1. Establish the same parental population that you used to test the Hardy-Weinberg prediction.

2. Simulate the production of an offspring from this population by randomly withdrawing two gametes to represent an individual offspring.

3. If the offspring is *BB* or *Bb*, place it in a container for the accumulation of the "Next Generation." Record the occurrence of this genotype on a separate sheet of paper.

4. If the offspring is *bb*, place this individual in a container for those that "Cannot Reproduce." Individuals in this container should not be used to produce subsequent generations. Record the occurrence of this genotype on a sheet of paper.

5. Repeat steps 2–4 until the parental population is depleted, thus completing the first generation.

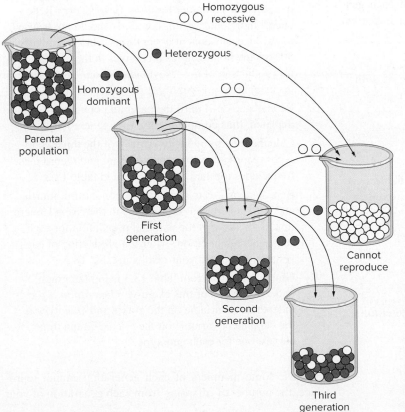

**Figure 18.5** Demonstrating the effect of 100% selection pressure on genotypic and phenotypic frequencies across three generations. In this example, selection is against the homozygous recessive genotype. Random mating within the parental population is simulated by mixing the gametes (beads), and the parental population is sampled by removing two alleles (i.e., one individual) and placing them in the next generation. Homozygous recessive individuals are removed (selected against) from the population. The genotypic and phenotypic frequencies are recorded after the production of each generation. The production of each generation depletes the beads in the previous generation in this simulation.

**Table 18.2**

**Genotypic Frequencies for 100% Negative Selection**

| Genotype | Generation | | | | |
|---|---|---|---|---|---|
| | First | Second | Third | Fourth | Fifth |
| BB ●● | _____ | _____ | _____ | _____ | _____ |
| Bb ●○ | _____ | _____ | _____ | _____ | _____ |
| bb ○○ | _____ | _____ | _____ | _____ | _____ |
| Total* | 1.0 | 1.0 | 1.0 | 1.0 | 1.0 |

*Note: The total of frequencies for each generation must equal 1.0.

6. Calculate the frequencies of each of the three genotypes recorded on the separate sheet and record these frequencies for the first generation in table 18.2. Individuals in the next generation will serve as the parental population for each subsequent generation.

7. Repeat steps 2–5 to produce a second, third, fourth, and fifth generation. After the production of each generation, record your results in table 18.2.

8. Graph your data from table 18.2 using the graph paper at the end of this exercise. *Generation* is the independent variable on the *x*-axis and *Genotype* is the dependent variable on the *y*-axis. Graph three curves, one for each genotype.

Because some members (i.e., the *bb* individuals that you removed) of each generation cannot reproduce, the number of offspring from each successive generation of your population will decrease. However, the frequency of each genotype, not the number of offspring, is the most important value.

**Question 4**

a. Did the frequency of white individuals decrease with successive generations? Explain your answer.

b. Was the decrease of white individuals from the first to second generation the same as the decrease from the second to the third generation? From the third to the fourth generation? Why or why not?

c. How many generations would be necessary to eliminate the allele for white fur?

Most naturally occurring selective pressures do not eliminate reproduction by the affected individuals. Instead, their reproductive capacity is reduced by a small proportion. To show this, use procedure 18.4 to eliminate only 20% of the *bb* offspring from the reproducing population.

### Procedure 18.4 Simulate 20% negative selection pressure

1. Establish the same parental population that you used to test the Hardy-Weinberg prediction.

2. Simulate the production of an offspring from this population by randomly withdrawing two gametes to represent an individual offspring.

3. If the offspring is *BB* or *Bb,* place it in a container for production of the "Next Generation." Record the occurrence of this genotype on a separate sheet of paper.

4. If the offspring is *bb,* place every fifth individual (20%) in a separate container for those that "Cannot Reproduce." Individuals in this container should not be used to produce subsequent generations. Place the other 80% of the homozygous recessives in the container for production of the "Next Generation." Record the occurrence of this genotype on a sheet of paper.

5. Repeat steps 2–4 until the parental population is depleted, thus completing the first generation.

6. Calculate the frequencies of each of the three genotypes recorded on the separate sheet and record these frequencies for the first generation in table 18.3.

7. Repeat steps 2–5 to produce a second, third, fourth, and fifth generation. Individuals in the "Next Generation" will serve as the parental population for each subsequent generation. After the production of each generation, record your results in table 18.3.

8. Graph your data from table 18.3 using the graph paper at the end of this exercise. *Generation* is the independent variable on the *x*-axis and *Genotype* is the dependent variable on the *y*-axis. Graph three curves, one for each genotype.

Because some members of each generation cannot reproduce, the number of offspring from each generation of your

Table 18.3

**Genotypic Frequencies for 20% Negative Selection**

| Genotype | Generation | | | | |
|---|---|---|---|---|---|
| | First | Second | Third | Fourth | Fifth |
| BB ●● | ___ | ___ | ___ | ___ | ___ |
| Bb ●○ | ___ | ___ | ___ | ___ | ___ |
| bb ○○ | ___ | ___ | ___ | ___ | ___ |
| Total* | 1.0 | 1.0 | 1.0 | 1.0 | 1.0 |

*Note: The total of frequencies for each generation must equal 1.0.

population will decrease. However, the frequency of each genotype, not the number of offspring, is the most important value.

**Question 5**

*a.* Did the frequency of white individuals decrease with successive generations?

*b.* Consult your graphs and compare the rate of selection for procedures 18.3 and 18.4. Was the rate of decrease for 20% negative selection similar to the rate for 100% negative selection? If not, how did the rates differ?

*c.* How many generations would be necessary to eliminate the allele for white fur?

## AN EXAMPLE OF EVOLUTION: THE VOLVOCINE LINE

The evolution of most species is too slow to witness in the lab, but we can examine modern species to learn about changes that likely occurred over evolutionary time. Researchers might ask which characteristics are conserved throughout an evolutionary lineage of species and which ones evolve rapidly and consistently. Are more complex species always more successful?

The **Volvocine line** of algae is a group of modern species that reflects an easily recognized sequence of changes

## INQUIRY-BASED LEARNING

*Heterozygotes can "hide" traits. How effective is selection against heterozygotes?*

Observations: Natural selection can change allelic frequencies in populations. Negative selection pressure (i.e., an environment that reduces reproduction by a particular phenotype) against a homozygous genotype can reduce allelic frequencies in only a few generations. The results of selection against heterozygotes may differ.

Question: How would selection against heterozygous individuals over many generations affect allelic frequencies in a population?

a. Establish a working lab group and obtain Inquiry-Based Learning Worksheet 18 from your instructor.

b. Discuss with your group a well-defined question relevant to the preceding observation and question. Record it on Worksheet 18.

c. Translate your question into a testable hypothesis and record it.

d. Review procedures 18.3 and 18.4. Outline on Worksheet 18 your experimental design and supplies needed to test your hypothesis. Ask your instructor to review your proposed investigation.

e. Conduct your procedures, record your data, answer your question, and make relevant comments.

f. Discuss with your instructor any revisions to your questions, hypothesis, or procedures. Repeat your work as needed.

as their common ancestors evolved. In this example, the changes were in colony complexity.

Studies of morphology and molecular genetics indicate that an ancient species similar to today's flagellated *Chlamydomonas* (fig. 18.6) was probably the original and most ancient common ancestor to the Volvocine line. The probable sequence of events was that an ancestor of unicellular *Chlamydomonas* evolved a novel colonial morphology that was successful and gave rise to *Gonium* (figs. 18.7 and 18.8). In turn that ancestor evolved greater colonial complexity to give rise to today's *Pandorina* (fig. 18.9) and then *Eudorina* (fig. 18.10). That colonial ancestor later gave rise to *Volvox* (fig. 18.11), the most complex alga of the Volvocine line. These five genera are modern representatives of a lineage of species that evolved along a path of colonial complexity.

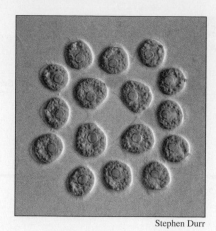

Stephen Durr

**Figure 18.8** *Gonium,* a colonial green alga composed of 16 cells (400×).

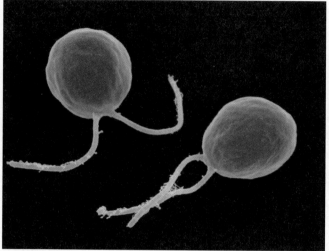

Aaron J. Bell/Science Source

**Figure 18.6** *Chlamydomonas,* a unicellular green alga (1700×). *Chlamydomonas* has two flagella.

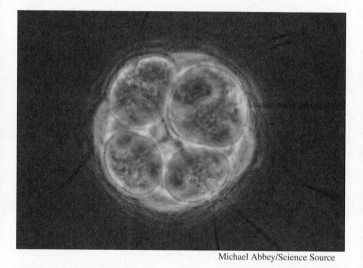

Michael Abbey/Science Source

**Figure 18.9** *Pandorina,* a colony of 16 or 32 flagellated green algal cells (890×).

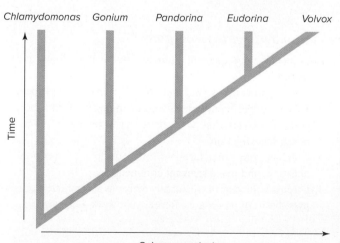

**Figure 18.7** A cladogram representing the simplified phylogeny (family tree) of the Volvocine line. Proposed common ancestors are represented by the branching points called nodes.

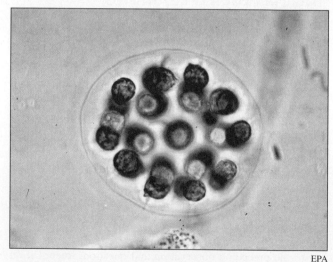

EPA

**Figure 18.10** *Eudorina,* a colony of 32 flagellated green algal cells (420×).

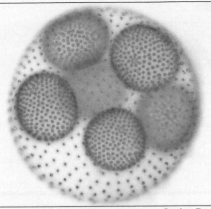

Stephen Durr

**Figure 18.11**  *Volvox,* a common green alga (200×). Colonies of *Volvox* often consist of hundreds of cells. Daughter colonies are visible within the larger parent colony.

### *Procedure 18.5*  **Examine members of the Volvocine line of algae**

1. Follow steps 2–6 to sequentially examine each of the organisms with your microscope. When preparing each of the colonial specimens, try both a standard microscope slide and a deep-well or depression slide. Determine which works best for colonies of cells.

2. *Chlamydomonas* is among the most primitive and widespread of the green algae. It is a unicellular biflagellate alga (fig. 18.6). All species of the Volvocine line consist of cells similar to *Chlamydomonas,* but the cells are in different configurations.

3. *Gonium* is the simplest colonial member of the Volvocine line (fig. 18.8). A *Gonium* colony consists of 4, 8, 16, or 32 *Chlamydomonas*-like cells held together in the shape of a disk by a gelatinous matrix. Each cell in the *Gonium* colony can divide to produce cells that produce new colonies. Like *Chlamydomonas, Gonium* is isogamous.

### Question 6
Why do colonies of *Gonium* consist of only 4, 8, 16, or 32 cells? Why are there no 23-cell colonies?

4. *Pandorina* consists of 16 or 32 *Chlamydomonas*-like cells held together by a gelatinous matrix (fig. 18.9). Examine how *Pandorina* moves. Flagella on *Pandorina* move the ellipsoidal alga through the water like

a ball. After attaining its maximum size, each cell of the colony divides to form a new colony. The parent matrix then breaks open like Pandora's box (hence the name *Pandorina*) and releases the newly formed colonies. *Pandorina* is isogamous.

### Question 7
What is the significance of a specialization at one end of the colony?

5. *Eudorina* is a spherical colony composed of 32, 64, or 128 cells (fig. 18.10). Cells in a colony of *Eudorina* differ in size; smaller cells are located at the anterior part of the colony. The anterior surface is determined by the direction of movement.

### Question 8
What is the significance of these structural and functional specializations of *Eudorina?*

6. *Volvox* is the largest and most spectacular organism of the Volvocine line. *Volvox* is a spherical colony made of thousands of vegetative cells and a few reproductive cells (fig. 18.11). Flagella spin the colony on its axis. In some species of *Volvox* and *Gonium,* cytoplasmic strands form a conspicuous network among the cells.

### Question 9
*a.* Does the *Volvox* colony spin clockwise or counterclockwise?

*b.* What is the significance of the cytoplasmic network in *Volvox?*

To organize your information and observations complete table 18.4.

## Table 18.4

### Evolutionary Specialization of Members of the Volvocine Line

| Characteristic | *Chlamydomonas* | *Gonium* | *Pandorina* | *Eudorina* | *Volvox* |
|---|---|---|---|---|---|
| Number of cells | | | | | |
| Colony size | | | | | |
| Structural and functional specializations of cells | | | | | |
| Reproductive specialization (isogamy versus oogamy) | | | | | |

## Testing for Understanding: Solving Hardy-Weinberg Problems

1. Galactosemia is inherited as a homozygous recessive trait (i.e., *gg*). You have sampled a population in which 36% of people have galactosemia.

   **a.** What is the frequency of the *g* allele?

   **b.** What is the frequency of the *G* allele?

   **c.** What is the frequency of the *GG* and *Gg* genotypes?

2. Suppose that in wasps, brown wings are dominant to white wings, and 40% of all wasps in a population you've sampled have white wings.

   **a.** What percentage of the wasps is heterozygous?

   **b.** What percentage of the wasps is homozygous dominant?

3. Suppose that you and 19 of your classmates (giving a final population of 10 males and 10 females) are on a cruise, and your ship sinks near a deserted island. You and all of your friends make it to shore and start a new population isolated from the rest of the world. Two of your friends carry the recessive allele (i.e., are heterozygous) for phenylketonuria. If the frequency of this allele does not change as the population on your island increases, what will be the incidence of phenylketonuria on your island?

4. Albinos produce very little of the pigment melanin in their skin and hair. Albinism is inherited as a homozygous recessive trait. In North America, about 1 in 20,000 people is an albino.

   **a.** What is the frequency of the dominant allele for albinism?

   **b.** What is the frequency of albinos?

   **c.** What is the frequency of heterozygotes?

5. People who are heterozygous recessive for the sickle-cell trait have some sickling of their blood cells, but not enough to cause death. Malarial parasites cannot infect these individuals' blood cells. People who are homozygous dominant for the sickle-cell trait have normal blood cells, but these cells are easily infected with malarial parasites. As a result, many of these individuals are killed by sickle-cell anemia. People who are homozygous recessive for the sickle-cell trait resist infections by malarial parasites, but their sickled blood cells collapse when oxygen levels drop, thereby killing the individuals. As a result, homozygous individuals—be they homozygous dominant or homozygous recessive—are less likely to survive than are heterozygous individuals. Suppose that 9% of a population in Africa is homozygous recessive for sickle-cell anemia. What percentage of the population will be heterozygous (i.e., more resistant to the disease)?

6. What would be the frequency of the recessive allele in a population that produces twice as many homozygous recessive individuals as heterozygotes?

## Answers to Hardy-Weinberg Problems

| | | |
|---|---|---|
| *1a.* 60% | *2b.* 14% | *4c.* 1.4% |
| *1b.* 40% | *3.* 0.25% | *5.* 42% |
| *1c.* *GG* = 16%; *Gg* = 48% | *4a.* 99.3% | *6.* $q = 0.8$ |
| *2a.* 47% | *4b.* 0.005% | |

1.  How would selection against heterozygous individuals over many generations affect the frequencies of homozygous individuals? Would the results of such selection depend on the initial frequencies of $p$ and $q$? Could you test this experimentally? How?

2.  How are genetic characteristics associated with nonreproductive activities such as feeding affected by natural selection?

3.  Although Charles Darwin wasn't the first person to suggest that populations evolve, he was the first to describe a credible mechanism for the process. That mechanism, which produces adaptations, is natural selection. What is natural selection, and how can it drive evolution?

4.  Does evolutionary change always leads to greater complexity? Why or why not?

5.  Is natural selection the only means of evolution? Explain.

6.  Is natural selection the only means of evolution that produces adaptations? Explain.

7.  What change in a population would you expect to see if a selection pressure was against the trait of the dominant allele?

8.  The application of evolution to understanding disease is widespread and productive. What is the benefit of applying Darwinian principles to medical practice?

9.  How could natural selection affect the frequency of sickle cell anemia (Fig. 17.4)?

**DOING BIOLOGY YOURSELF**

Design an experiment to determine the phylogenetic relationships among members of the Volvocine line of algae. What information about their DNA sequences would be useful?

**WRITING TO LEARN BIOLOGY**

The Hardy-Weinberg equilibrium assumes that pollination and subsequent fertilization must be random. Is that true for most wildflower populations? What characteristics of these plants influence pollination patterns?

**WRITING TO LEARN BIOLOGY**

Summarize the most recent books and publications that review the benefits of applying Darwinian principles to medical practice.

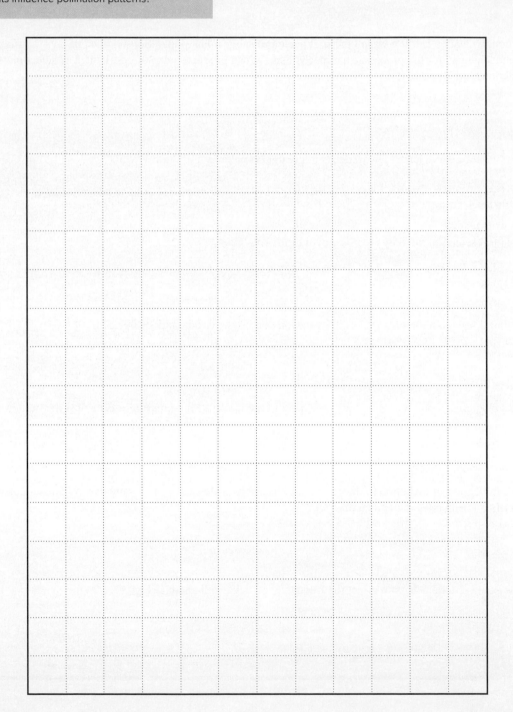

# Human Evolution
## Skull Examination

## Learning Objectives

By the end of this exercise you should be able to:
1. Describe the parts of a modern human skull.
2. Distinguish between skulls of males and females.
3. Distinguish between skulls of apes and modern humans.

Please visit **connect.mheducation.com** to review online resources tailored to this lab.

Throughout time, no issue has interested humans more than learning about our origins. Where did we come from? What did our ancestors look like? Where did they live? Today, we are beginning to understand those questions, thanks to evidence provided by biologists and anthropologists. Nevertheless, the topic remains controversial and often elicits strong responses from people.

In this exercise, you will examine some of the information underlying recent ideas about human evolution by examining the skulls of human and humanlike organisms. Specifically, you will examine skulls of

*Gorilla gorilla,* the modern gorilla. Gorillas and humans share a common ancestor.

*Australopithecus* (fig. 19.1), a relatively small, humanlike organism that had small, pointed canine teeth and an elongated face and was adapted for bipedalism and an upright stance. The most famous australopithecine is "Lucy," discovered in 1974 in Ethiopia. *Australopithecus,* the best known of the early humanlike organisms, lived 3.5–2.5 million years ago.

*Homo erectus,* a human ancestor characterized by an upright posture, a sloping forehead, a large brow ridge, a thick skull, and a larger braincase (900 cm³) than earlier humanlike organisms. *H. erectus* lived 1,800,000–27,000 years ago.

*Homo sapiens,* or modern humans, are characterized by a vertical forehead, small brow ridge, thin skulls, and a larger braincase (1400 cm³) than earlier humanlike organisms. *H. sapiens* appeared 200,000 years ago.

These species represent well-documented stages in the lineage of modern humans. As you examine these skulls, think

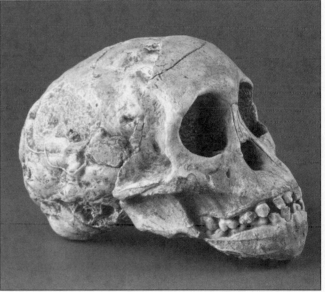

Dea/G Cigolini/age fotostock

**Figure 19.1** *Australopithecus africanus* was the first australopithecine to be discovered. It was unearthed in southern Africa by Raymond Dart in the 1920s.

of how their traits link modern humans with other primates. Also know that anthropologists are constantly making new discoveries of early human fossils and that the specimens you will analyze represent a small subset of these species (fig. 19.2).

## THE MODERN HUMAN SKULL

The human skull (fig. 19.3), including the lower jaw, consists of 22 bones, 8 of which are paired. All of the bones fit together at joints called **sutures,** which appear as wavy lines. Projections and raised lines are sites of muscle attachment.

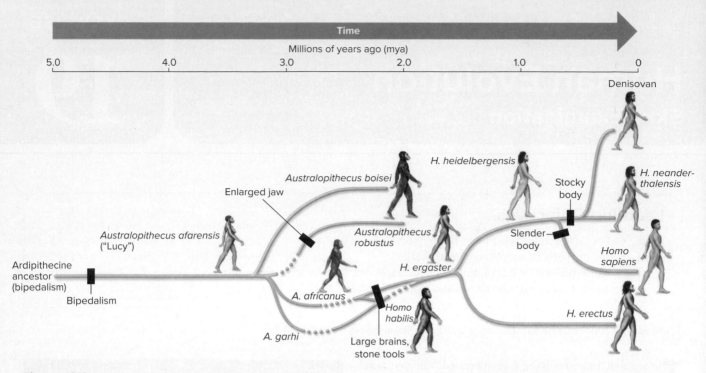

**Figure 19.2** A possible scenario for human evolution. In this human family tree, several species lived contemporaneously with one another, but only one lineage gave rise to modern humans (*Homo sapiens*). Dotted lines indicate controversial time lines and branching patterns. Molecular analysis distinguishes denisovans (known from fragmented fossils discovered in Denisova Cave in Siberia) from neanderthals. Note that four species (groups) of our genus *Homo* likely occupied the world simultaneously for thousands of years.

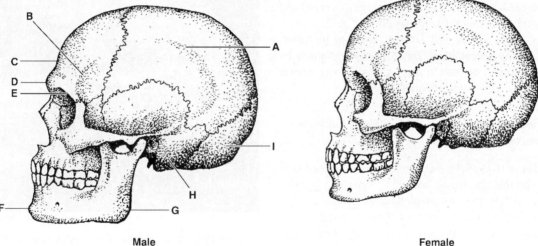

| Male Skull | Feature | Female Skull |
|---|---|---|
| Large | **A. Braincase** | Smaller than male |
| Marked | **B. Muscle lines** | Slight |
| Retreating | **C. Forehead** | Bulging |
| Developed | **D. Brow ridges** | Absent |
| Rounded | **E. Upper rim of eye socket** | Sharp |
| Square | **F. Chin** | Rounded |
| Nearly a right angle | **G. Angle of jaw** | Angle more obtuse (over 120°) |
| Large | **H. Mastoid process** | Small |
| Present | **I. External occipital protuberance** | Absent |

**Figure 19.3** A comparison of male and female skulls of modern *Homo sapiens*.

Although skulls from males and females share many features, they usually can be distinguished. However, such a diagnosis—even when done by experts—is only about 90% reliable (80% if the lower jaw is missing).

**Question 1**

*a.* How do skulls of females differ from those of males?

*b.* What is the biological significance of these differences?

## Face

**Prognathism** is the extent to which the face and jaws protrude forward when viewed from the side. Their larger teeth and jaws cause apes to exhibit more prognathism than do humans.

## Braincase

The **brow ridge,** the mass of bone over the eye sockets, supports the upper facial skeleton against forces produced by chewing. The brow ridge in apes is prominent. In humans, the brow ridge of modern humans is largely internalized because our frontal bone has expanded outward to a more vertical angle.

The **sagittal crest** is a thin ridge of bone atop and down the middle of the braincase. The sagittal crest is associated with having a small braincase and powerful jaws. In apes, the sagittal crest is an attachment site for the large temporalis muscle used for chewing.

The **foramen magnum** is the large opening in the base of the skull through which the spinal cord passes. The position of the foramen magnum reflects the posture of the body (and, indirectly, the pattern of movement) of **hominoids.** Humans stand erect and walk with the head directly over the vertical spinal column. Conversely, the knuckle-walking apes hold their heads forward, with the foramen magnum toward the rear. Thus, the foramen magnum is located in a more rear position in apes than in humans.

## Teeth and Jaws

Adult apes and humans have the same number and types of teeth: 4 canines, 8 premolars, 12 molars, and 8 incisors. Identify these teeth on the skulls and diagrams. In apes the canine teeth are longer and more pointed than others. Notice that nonhuman primates have large lower canines (fig. 19.4). Therefore, the upper jaw must have a **diastema** (space) on each side to receive those canines when the jaw is closed. The canine teeth seldom project above the others.

In humans, the four front teeth (incisors) are smaller, more vertical, and flatter than in apes. In nonhuman primates, the canine diastema is the gap in the teeth corresponding to the canines of the opposite jaw.

**Question 2**

*a.* Between which teeth does the gap occur? Why are these gaps essential in nonhuman primates?

*b.* Why are they usually absent in humans?

Humans have an outward projection on the lower part of their lower jaw (i.e., chin). Apes lack this feature; instead, they have a smooth, even slant to the front part of their jaw.

| Modern *Homo sapiens* | Gorilla (ape) | *Australopithecus* |
|---|---|---|

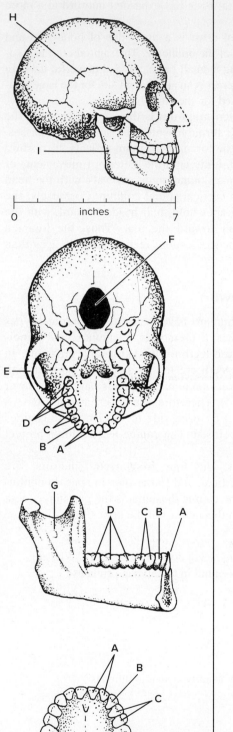
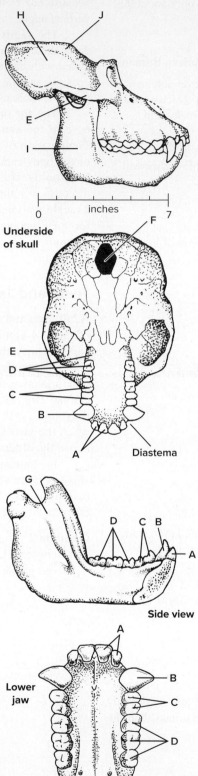
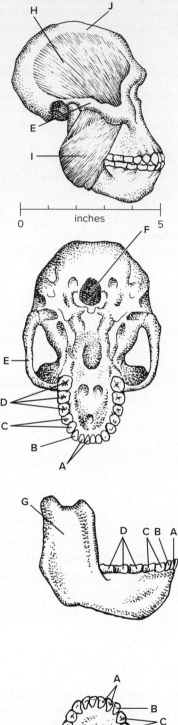
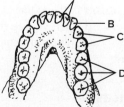

**Figure 19.4** Comparison of skull features of modern *Homo sapiens,* gorilla, and *Australopithecus* skull and jaws:

A. Incisors

B. Canines

C. Premolars

D. Molars

E. Zygomatic arch

F. Foramen magnum

G. Vertical ramus for muscle attachment

H. Skull surface for muscle attachment

I. Jaw surface for muscle attachment

J. Sagittal crest for muscle attachment

In humans, teeth are arranged in a relatively continuous curve from the third molar around to the other third molar. The arrangement of teeth in apes is straighter, with a slight curve in front. This is primarily because of the larger size of the incisors and canines.

To summarize the differences between skulls of modern humans and apes, complete table 19.1.

## Procedure 19.3   Study the skulls of fossil primates

Use the information and diagrams in figures 19.5–19.8 to learn about skulls of humans and ancient primates.

At first glance, full skeletons of humans and modern apes look quite similar. But closer examination reveals distinctive differences and adaptations, many of which are related to bipedalism, jaw shape, and brain size.

### Table 19.1

**Prominent Features of Skulls of Apes and Humans**

| Feature | Apes | Humans |
|---|---|---|
| Sagittal crest | | |
| Brow ridge | | |
| Foramen magnum | | |
| Prognathism | | |
| Canines | | |
| Canine diastema | | |
| Incisors | | |
| Chin | | |
| Arrangement of teeth | | |

## Procedure 19.4   Compare human and ape skeletal adaptations for standing

1. Compare the skeletons illustrated in figures 19.9 and 19.10.
2. For each of the following skeletal features, record the difference between human and ape anatomy. Then speculate on the adaptive significance of those features related to posture, bipedalism, and/or food habits. Summarize your work by completing table 19.2.

### Table 19.2

**Comparison of Skeletal Features of Apes and Humans**

| Feature | Ape | Human |
|---|---|---|
| Tooth number and size | | |
| Spine shape | | |
| Braincase size | | |
| Insertion point of head on the spinal column | | |
| Thumb and big toe size and angle | | |
| Arm length | | |
| Size of heel bone (calcaneus) | | |
| Thickness of knee joints (ends of tibia and femur) | | |
| Length of the angled head of the femur | | |
| Posterior exposure of sacrum | | |
| Risk of cervical vertebrae to compression | | |
| Size of birth canal relative to body size | | |

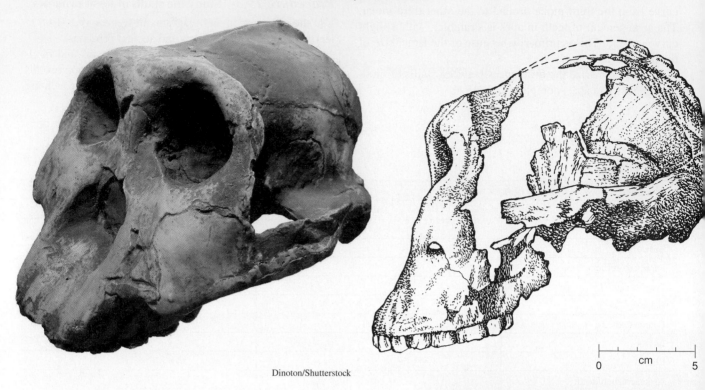

Dinoton/Shutterstock

0       cm       5

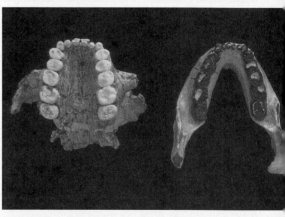

The Natural History Museum/Alamy Stock

**Figure 19.5**    Views of the skull and teeth (upper jaw, left; lower jaw, right) of *Australopithecus boisei*. Age: 1.8 million years. This skull includes m[...] sive molar and premolar teeth (similar in size to those of gorillas) and is nearly complete except for the lower jaw. It is commonly known as "Zinj," an a[...] viation of the original genus name *Zinjanthropus*. Zinj and the remains of many smaller hominins (humanlike organisms) were discovered in the Olduvai Gorge in Tanzania by Mary and Louis Leakey in 1959.

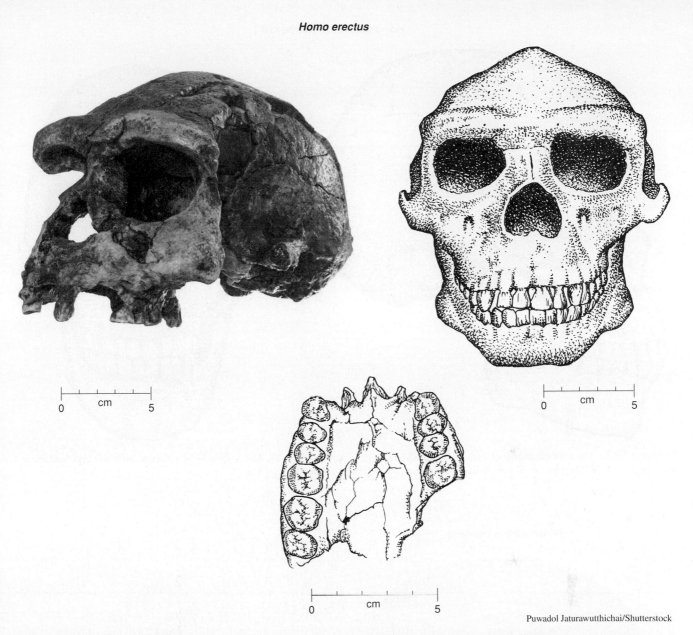

**Figure 19.6**  Views of the skull and teeth (upper jaw) of *Homo erectus*. Age: Less than 1 million years. This diagram shows a reconstruction that includes parts of skulls discovered in 1937 and 1939 in Java (Sangiran). The skull of *Homo erectus* differs from modern human skulls in that it is low vaulted and has a relatively small volume; *H. erectus* skulls have volumes of 900–1100 cm$^3$, whereas skulls of modern humans have volumes of approximately 1400 cm$^3$. Skulls of *H. erectus* also have small mastoid processes behind the ear openings, large jaws, small chins, and large molar teeth.

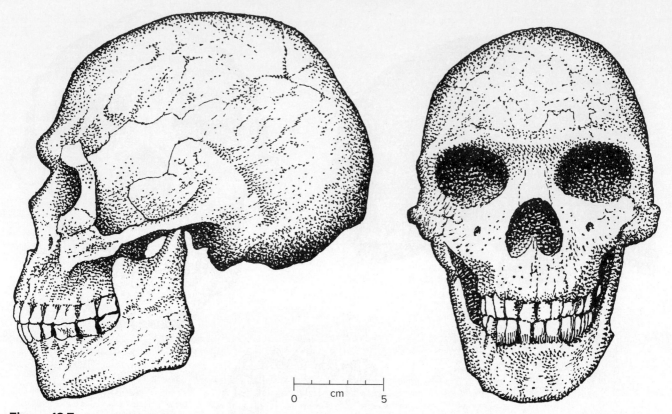

**Figure 19.7** Views of the skull and teeth of *Homo sapiens*. Age: About 32,000 years. Also known as Neanderthal Man, this type was recovered in 1932 from Mugharet-es-Skhull, Wadi el-Mughara, Israel. The skull is nearly complete.

*Homo sapiens* **(modern)**

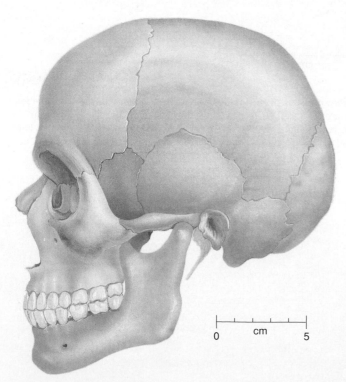

**Figure 19.8** Skull of modern *Homo sapiens*. Age: About 10,000 years. This skull is from one of at least 50 skeletons recovered at Oued Agrious, Algeria, during the late 1920s. See figure 19.4 for labeled features.

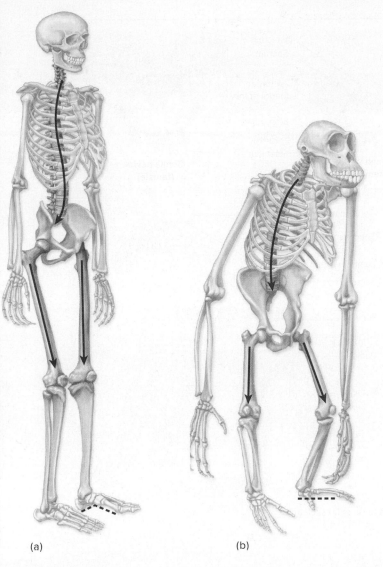

Human spine exits from the skull's center; ape spine exits from rear of skull.

Human spine is S-shaped; ape spine has a slight curve.

Human pelvis is bowl-shaped; ape pelvis is longer and more narrow.

Human femurs angle inward to the knees; ape femurs angle out a bit.

Human knee can support more weight than ape knee.

Human foot has an arch; ape foot has no arch.

(a)                    (b)

**Figure 19.9**   Adaptations for standing. (*a*) Human skeleton compared to (*b*) Ape (chimpanzee) skeleton.

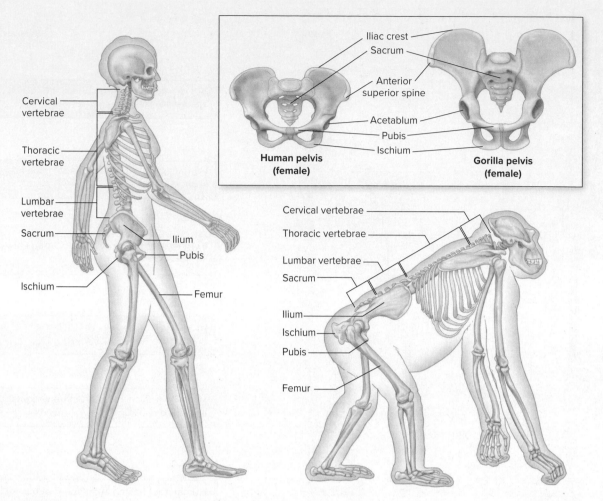

**Figure 19.10** Comparison of the major bones of Human and Ape (gorilla) skeletons.

Labels (human skeleton): Cervical vertebrae, Thoracic vertebrae, Lumbar vertebrae, Sacrum, Ilium, Pubis, Ischium, Femur

Labels (pelvis inset): Iliac crest, Sacrum, Anterior superior spine, Acetablum, Pubis, Ischium, Human pelvis (female), Gorilla pelvis (female)

Labels (gorilla skeleton): Cervical vertebrae, Thoracic vertebrae, Lumbar vertebrae, Sacrum, Ilium, Ischium, Pubis, Femur

## INQUIRY-BASED LEARNING

*Humans don't stand still. Along what paths did our common ancestors migrate during our evolution?*

Observations: Some of the oldest fossils of our human ancestors have been found in Africa. Current theories of human evolution involve the migration of ancestral primates and early hominins from Africa to Europe and back to Africa.

Question: What migration pattern of human ancestors is supported by fossil evidence?

**a.** Establish a working lab group and obtain Inquiry-Based Learning Worksheet 19 from your instructor.

**b.** Research the current literature and outline the patterns of migration by ancestral primates and early hominins away from and back to Africa.

**c.** Record your findings on Worksheet 19.

**d.** Discuss your findings with your instructor and with other students.

# Questions for Further Study and Inquiry

1. What could changes in our ancestors' teeth tell us about the evolutionary history of humans?

2. Is the skull shown in figure 19.7 that of a man or a woman? Explain your answer.

3. Neanderthals were once considered to be a separate species (*Homo neanderthalensis*). However, some anthropologists consider modern humans and Neanderthal Man to be the same species, *Homo sapiens*. Do you think that they should be the same species? Why or why not? What species-related criteria are you using as a basis for your claim? What does this judgment imply about the ability of these two groups of people to interbreed when they lived together in Europe 34,000 years ago?

## WRITING TO LEARN BIOLOGY

Describe what other evidence you would use to study the evolution of humans. Where would you get that evidence?

## WRITING TO LEARN BIOLOGY

Read an article on a more recent fossil discovery and describe how this new early human fossil is similar to and different from the species they measured in this lab. Describe distinctive features of the skeleton and describe what it may be adapted to.

# Ecology
## Diversity and Interaction in Plant Communities

## Learning Objectives

By the end of this exercise you should be able to:
1. Observe the physical factors, plant dominance, and interactions among organisms in a terrestrial community and characterize the community according to these features.
2. Quantify the distribution and abundance of plants in a community.
3. Detect the experimental effects of competition and allelopathy among plants.
4. Explain four different ways to quantify the relative abundances of different plant species in a plant community, using transect data.

Please visit **connect.mheducation.com** to review online resources tailored to this lab.

Ecological communities are extraordinarily complex. The assemblage of plants that you observe at any time results from interactions among plants and their physical surroundings, among different species of plants, and among plants and animals. All of these interactions are driven by a flow of energy captured by green plants and passed to herbivores (plant eaters), predators, and decomposers.

It is beyond the scope of this exercise to explain all of the processes occurring in a plant community, but we can assemble some basic observations that characterize and distinguish communities.

 **SAFETY FIRST** Before coming to lab, you were asked to read this exercise so you would know what to do and be aware of safety issues. In the space below, briefly list the safety issues associated with today's procedures. If you have questions about these issues, contact your laboratory assistant before starting work.

## QUALITATIVE COMMUNITY ASSESSMENT

*Procedure 20.1*   Observe and assess the ecological characteristics of a terrestrial community

1. Locate and visit a terrestrial community designated by your instructor.
2. Characterize the community according to the criteria and questions that follow. After you've answered the questions, discuss your observations with your instructor and other groups. Be prepared to use your observations as a basis for describing your assessment of energy flow through the community, diversity of the community, and interactions among organisms.

## Physical Factors

### Observations
1. What levels of light intensity occur throughout the community?

2. Does the community include shade-tolerant as well as shade-intolerant plants?

3. How does light differ among different vertical levels of vegetation? Why could this be important?

4. What is the temperature 2 m above ground?

5. What is the temperature at the soil surface?

6. How does the ground slope? Why is this important?

7. How would you characterize the soil? Loam (a fertile soil of clay, sand, and organic humus)? Clay (a sticky mix of ultrafine mineral particles)? Sand (a granular mix of hard mineral grains)?

8. What is the nature of the ground cover? Grasses? Bare soil?

9. Is there a layer of leaf litter on the ground?

10. Is the community generally a moist, moderate, or dry environment?

## Interpretations

1. How might shade affect the temperature of the community?

2. How might different amounts of light at different vertical levels within the community be important?

3. What parts of the community might be cooler than others? Why is this important?

4. Why would ground slope be important?

5. Based on your observations of slope and soil type, would you expect the soil to retain moisture?

6. How long has this community been left to develop naturally? That is, what is the age of the community?

# Plant Dominance

## Observations

*1.* Which plant species are most abundant in numbers?

*2.* Which plant species are most abundant in biomass?

*3.* What general categories of plant types (shrubs, trees, etc.) are apparent?

*4.* What is the vertical distribution of vegetation?

## Interpretations

*1.* Would you describe this community as diverse? Why or why not?

*2.* What comparison community in your local area would you consider to be more diverse? Less diverse?

*3.* What observations led you to your conclusion for the previous question?

*4.* What specific factors make your comparison community more or less diverse? Human impact? Stressful environmental factors? Geology?

# Interactions among Organisms

## Observations

*1.* What evidence do you see of resident vertebrates?

*2.* What evidence do you see of resident invertebrates?

3. What evidence do you see of plant–animal interactions?

4. What evidence do you see of plant–plant interactions?

5. What adaptations do the plants have to discourage herbivores?

6. Do you see any obvious or subtle evidence of competition by plants for available resources?

## Interpretations

1. If you don't see any vertebrates, does that mean they are not around? Explain your answer.

2. Reexamine the observations that you just listed. How would each observation affect the type and growth of plants in the community that you studied?

3. What types of competitive interactions are apparent in the community?

4. What types of mutually beneficial interactions are apparent in the community?

## QUANTITATIVE COMMUNITY ASSESSMENT

Many techniques have been developed to measure the numbers, densities, and distributions of organisms in terrestrial plant communities. One widely used technique is to count organisms within randomly distributed **quadrats** (sometimes called **plots**) of uniform size. Other techniques involve measuring distances between plants or the distance from randomly chosen points to nearest plants. In the **line-intercept method,** a **transect,** or line, is established and laid out within the community. Organisms that touch this line are counted and measured. Calculations based on these measurements of quadrats and transects reveal the relative abundances, frequencies, and distributions of the plant species that compose the community.

### Procedure 20.2   Assessing a community with the line-intercept method

1. With the help of your instructor, locate a suitable field site with a plant community to be examined.

2. Obtain a measuring tape 15–20 m long, a meterstick, and a notepad. If a measuring tape is unavailable, use a measured piece of string or rope.

3. Assess the general layout of the community to be sampled. With the aid of your instructor, decide on a reasonable set of criteria to govern the placement of a transect for each group of students.

## Table 20.1

**Summary of Raw Data for Species Occurring along a Transect**

Total transect length = _____          Total number of intervals _____

| Species $i$ | $f_i$ = Number of Intervals in Which Species $i$ Occurs | $n_i$ = Number of Individuals Encountered | $c_i$ = Length of Transect Intercepted |
|---|---|---|---|
| | | | |
| | | | |
| | | | |
| | | | |
| | | | |
| | | | |
| | | | |
| | | | |
| | | | |
| | | | |
| | | | |
| | F = Total of all frequencies = _____ | N = Total of all individuals = _____ | C = Total length of transect intercepted = _____ |

## Question 1

What concepts or ideas should govern the placement of your transect to obtain a representative sample of the community? Are there any "wrong" places to put a transect? Why or why not?

4. You and your lab partners will work on a single transect. Stretch the measuring tape on the ground to establish a transect.

5. Divide the transect into 1-, 2-, or 3-m intervals. Each interval represents a separate unit of the transect.

6. At one end of the transect begin counting plants that touch, overlie, or underlie the transect line. For each plant encountered record the type of plant (species) and the length of the line that the plant intercepts. For plants that overhang the line, record the length of the imaginary vertical plane of the line that the plant would intercept. Record these raw data for each interval in your field notepad. Also record any uncovered (bare) lengths within the transects.

7. When all plants from all intervals have been recorded, summarize your data in table 20.1.

8. Sum the values in each of the three data columns of table 20.1 to calculate F, N, and C. Record the calculations at the bottom of each column.

9. Use the data in table 20.1 to calculate the following four parameters for each species within the community. Record your results in table 20.2.

$$\text{Relative frequency} = f_i/F$$
$$\text{Relative density} = n_i/N$$
$$\text{Relative coverage} = c_i/C$$
$$\text{Importance value of species } i = \text{Relative frequency} + \text{Relative density} + \text{Relative coverage}$$

## Question 2

What is the meaning of an importance value? Why would we calculate this in addition to density, coverage, and frequency?

**Table 20.2**

| Relative Values of Each Species in a Selected Community Using Parameters of the Line-Intercept Method | | | | |
|---|---|---|---|---|
| Species | Relative Frequency | Relative Density | Relative Coverage | Importance Value |
| | | | | |
| | | | | |
| | | | | |
| | | | | |
| | | | | |
| | | | | |
| | | | | |
| | | | | |

## PLANT INTERACTIONS

### Competition

**Competition** is the interaction among individuals seeking a scarce common resource (fig. 20.1). The interaction is usually negative (disadvantageous) for both competitors. The intensity of competition for a resource such as light, food, water, space, and nutrients depends on the amount of resource, the number of individuals competing, and the needs of each individual for that resource. Competition among members of the same species is **intraspecific competition,** whereas competition among members of different species is **interspecific competition.**

*Procedure 20.3* **Examine competition by sunflower seedlings**

1. Obtain five pots containing enough potting soil to plant seeds. Measure the area at the surface of the potting soil.

2. Follow your lab instructor's directions to plant 2, 4, 8, 16, and 32 sunflower seeds in each of the five pots, respectively. In each of the pots plant at least 20% more seeds so you can later reduce the number of seedlings to the numbers previously listed. Label each pot with the number of seeds, the date, and your name.

3. Water each pot gently with a consistent amount of water.

4. Place the pots in the greenhouse or growth area so that each pot has the same environmental conditions of light, temperature, and so on. After the seedlings are established, remove excess seedlings so the treatments will have the correct number of seedlings listed in step 2. Use that number of plants in later calculations even if plants die before the experiment ends. For calculations of mean weight and mean height, use the treatment's original number of plants.

5. Examine the pots at regular intervals as directed by your instructor. Record measurements of the

Dom Hammond//PunchStock/Design Pics, Inc.

**Figure 20.1**   Competition in a forest. Each plant in this forest competes with all the individuals around it for light, soil nutrients, and moisture.

parameters called for in table 20.3. Your instructor may ask you to plot your results.

**Question 3**
*a.* In which pot was competition greatest?

*b.* Which parameters best showed the effects of competition?

## Table 20.3

**Effects of Competition on Sunflower Seedlings**

| | Plants per Pot | | | | |
|---|---|---|---|---|---|
| | **2** | **4** | **8** | **16** | **32** |
| **General observations:** | | | | | |
| Date _____ | | | | | |
| Date _____ | | | | | |
| Date _____ | | | | | |
| Date _____ | | | | | |
| **Mean height of individuals:** | | | | | |
| Date _____ | | | | | |
| Date _____ | | | | | |
| Date _____ | | | | | |
| Date _____ | | | | | |
| **Range of height of individuals:** | | | | | |
| Date _____ | | | | | |
| Date _____ | | | | | |
| Date _____ | | | | | |
| Date _____ | | | | | |
| **Mean width of 10 widest leaves:** | | | | | |
| Date _____ | | | | | |
| Date _____ | | | | | |
| Date _____ | | | | | |
| Date _____ | | | | | |
| **Mean fresh weight of aboveground biomass (measured at the end of the experiment):** | | | | | |
| Date _____ | | | | | |

*c.* What other characteristics of the competitors might you have measured?

*d.* Did competition more noticeably affect the number of individuals or the biomass of each individual?

*e.* In what environments would you expect that competition among sunflowers would be most intense?

*f.* Would you expect different results if a different type of potting soil was used? Why?

*g.* Would plants and animals compete for the same resources? What might be some differences?

## Allelopathy

**Allelopathy** is a form of competition involving the inhibition of a plant's germination or growth by exposure to compounds produced by another plant. These compounds may be airborne or leach from various plant parts. Rainfall, runoff, and diffusion typically distribute inhibitory compounds in the immediate area surrounding the producing plant (fig. 20.2).

### Question 4

*a.* What are the benefits of producing allelopathic compounds?

*b.* What are the possible disadvantages of producing allelopathic compounds?

### Procedure 20.4   Demonstrate allelopathy

*1.* Determine from your instructor the overall experimental design for the class—how many plants your group will test and how many replicates you will set up for each plant.

Arto Hakola/Shutterstock

**Figure 20.2**   Allelopathic inhibition of nearby plant competitors for nutrients and water often results in barren areas surrounding the established plant producing the inhibitors.

*2.* Obtain tissue (stems and leaves) from the variety of plants provided by your instructor. Some of these plants are suspected to produce allelopathic compounds.

*3.* For each plant: Mix 10 g of tissue with 100 mL of water in a blender and homogenize. Let the slurry soak for 5–10 min to leach chemicals from the disrupted tissue.

*4.* Filter or strain the slurry to remove large particulates. Drain the filtrate into a beaker.

*5.* Obtain a petri dish (7.6 cm diameter) and line its bottom with a circular piece of filter paper.

*6.* Label the petri dish appropriately for the plant extract and replicate being tested in that dish.

*7.* Saturate the filter paper with a measured amount (5–8 mL) of the extract.

*8.* Repeat steps 3–7 for each replicate and each plant being tested.

*9.* Obtain seeds of radish, lettuce, or oat. Distribute 50 seeds uniformly on the filter paper in each dish.

*10.* Your instructor may enhance the experimental design by asking you to set up replicate dishes for each extract and to test the effects on different kinds of seeds. Follow the directions given by your instructor.

*11.* You may also be asked to design and implement an experimental control of a water-only treatment that is blended and then filtered.

*12.* Incubate the covered dishes at room temperature in the laboratory or in the greenhouse. After 24 and 48 h, count the number of seeds that have germinated and calculate the percent germinated.

*13.* After 72 h (or the length of time specified by your instructor) measure the length of the radicle and make relevant observations about the apparent rate of growth of the emerged embryos. Extend your observations for as many days as specified by your instructor.

*14.* Record your results in table 20.4.

### Question 5

*a.* Did your observations reveal any differences in allelopathy among tested plants?

*b.* Which plant species demonstrated the most intense allelopathy?

**Table 20.4**

**Effects of Allelopathy on Germination and Growth Rate of Seeds**

Plant Extract _____     Seed Species _____     Total No. of Seeds _____

| | Time | % Germinated | Length of Radicle | Observations |
|---|---|---|---|---|
| Rep 1 | | | | |
| Rep 2 | | | | |
| Rep 3 | | | | |

Plant Extract _____     Seed Species _____     Total No. of Seeds _____

| | Time | % Germinated | Length of Radicle | Observations |
|---|---|---|---|---|
| Rep 1 | | | | |
| Rep 2 | | | | |
| Rep 3 | | | | |

Plant Extract _____     Seed Species _____     Total No. of Seeds _____

| | Time | % Germinated | Length of Radicle | Observations |
|---|---|---|---|---|
| Rep 1 | | | | |
| Rep 2 | | | | |
| Rep 3 | | | | |

Plant Extract _____     Seed Species _____     Total No. of Seeds _____

| | Time | % Germinated | Length of Radicle | Observations |
|---|---|---|---|---|
| Rep 1 | | | | |
| Rep 2 | | | | |
| Rep 3 | | | | |

## INQUIRY-BASED LEARNING

### *Do leaves, roots, and stems all produce allelopathic compounds?*

Observation: Many plants produce chemicals that inhibit the growth of other plants and/or the germination of other seeds. These allelopathic chemicals are adaptive because they reduce competition.

Question: Do all organs (e.g., roots, stems, leaves) of allelopathic plants produce the same amounts of allelopathic chemicals?

a. Establish a working lab group and obtain Inquiry-Based Learning Worksheet 20 from your instructor.

b. Discuss with your group well-defined questions relevant to the preceding observation and question. Choose and record your group's best question for investigation.

c. Translate your question into a testable hypothesis and record it.

d. Outline on Worksheet 20 your experimental design and supplies needed to test your hypothesis. Ask your instructor to review your proposed investigation.

e. Conduct your procedures, record your data, answer your question, and make relevant comments.

f. Discuss with your instructor any revisions to your questions, hypothesis, or procedures. Repeat your work as needed.

*1.* Diverse plant communities have many species representing a variety of plant types such as grasses, shrubs, succulents, hardwood trees, softwood trees, vines, and ferns. What factors increase a community's diversity? Age of the community? Energy input? Moisture? Nutrients? Disturbance? Human activity? How do they do so?

*2.* What characteristics of a community make it more resilient than other communities to a particular disturbance? Describe both the type of disturbance and the type of community.

*3.* What characteristics would indicate that a community has not been disturbed for a few years?

*4.* How does competition influence natural selection? Is the presence of competitors a selective force?

# Community Succession

## Learning Objectives

By the end of this exercise you should be able to:
1. Define community succession.
2. Describe how succession occurs.
3. Understand how the environment and resources influence succession.

As time passes, most environments are inhabited by a succession of different communities. **Succession** results from the differential growth of populations over time. A **community** includes all the organisms that live and interact in the same area at the same time. The growth of the bacterial community in a carton of milk produces a succession of changes in milk that, in turn, create conditions for other organisms to grow. This phenomenon—that is, the changes in community composition over time—is called **community succession.** During community succession, each community of organisms (i.e., each stage in succession) changes the environment. Ironically, during succession a community can change the environment such that the community inhibits its own long-term growth. When this happens, a different community (a new successional stage) takes its place. Thus, communities change over time.

**Primary succession** occurs in areas where there is little or no organic soil and no living organisms have become established. Examples of primary succession include the colonization of lava flows, as shown in figure 21.1, volcanic islands, and river deltas after a flood.

**Secondary succession** occurs when a disturbance restarts succession at a different point than that which occurs with primary succession. Examples of secondary succession include the changes caused to ecosystems by disease, forest fires, weather (e.g., hurricanes, drought), and human activities (e.g., cutting down a forest, plowing a field, mowing a lawn). Secondary succession proceeds faster than primary succession because organic soil is already present and conditioned for use by plants.

As you might guess, community succession occurs at multiple scales and levels, ranging from large-scale ecosystems (e.g., the forests surrounding Mt. St. Helens volcano, or the wildfires of the western states) to local areas affected by timber harvest, crop harvest, and development.

Don Hammond/PunchStock/Design Pics

**Figure 21.1** Plants colonizing a lava field along the coast of Hawaii.

Virtually everyone is familiar with the consequences of community succession in a container of milk. Milk contains carbohydrates (lactose, or milk sugar), protein (casein, or curd), and lipids (butterfat); all of these nutrients can support the growth of a variety of microbes. Pasteurization retards this growth and involves heating milk to about 170°C to kill pathogenic bacteria. Then the milk is rapidly cooled. However, pasteurization does not kill *all* nonpathogenic bacteria. Although these nonpathogenic bacteria divide slowly when they are refrigerated, they will ultimately "spoil" the milk, even if the milk is refrigerated. Leaving the milk at room temperature greatly speeds the spoilage.

### Question 1

What are some examples of community succession occurring on your campus or in nearby areas?

In large-scale, so-called old-field succession that occurs in places such as forests and fields, most of the organisms are multicellular, sexually reproducing, relatively long-lived species that colonize the site at various times. During this type of succession, the composition of the community changes and is dominated by autotrophs.

In contrast, small-scale communities such as carcasses or cartons of milk are dominated by unicellular, asexually reproducing, relatively short-lived individuals largely descended from individuals present at the start. (Carrion beetles and other insects that colonize carcasses are exceptions to this statement, for they are not present on living individuals.) This type of succession is "degradative" because the availability of energy and nutrients is highest at the early stages and declines over time.

## SUCCESSIONAL CHANGES IN MILK

In this exercise, you will study community succession in milk. This process of succession is often more complex than you might suspect. For example,

1. *Pseudomonas* and *Achromobacter* (both Gram-negative rods; see fig. 24.3 in Exercise 24) are common bacteria that digest butterfat and give milk a putrid smell.

2. *Lactobacillus* (a Gram-positive rod) and *Streptococcus* (a Gram-positive coccus; see Exercise 24) survive pasteurization. These bacteria ferment lactose to lactic acid and acetic acid.

3. Acidity sours the milk and converts the casein to curd.

4. Under acidic conditions, fungi such as yeast (which are much larger than bacteria) grow extremely well. These organisms often metabolize the acids into nonacidic compounds.

5. Finally, *Bacillus* (a Gram-positive rod) metabolize proteins into ammonia products and raise the milk's pH. The characteristic odor of spoiled milk becomes apparent when this occurs.

Succession in milk, under the right conditions and with the proper bacteria, can lead to the formation of cheese. Similarly, bacteria added to milk and the subsequent succession can produce buttermilk, yogurt, and sour cream. However, the most common instance of community succession in milk usually occurs in dairy cases at supermarkets and in our kitchen refrigerators. There, bacteria ferment lactose into acid, thereby spoiling the milk.

### Procedure 21.1   Compare community succession in different types of milk

1. Work in small groups as instructed by your lab instructor.

2. Each group will be given one of the following sets of samples:

   Treatment 1: Whole milk, kept at room temperature (25°C); 2, 5, and 8 days old.

   Treatment 2: Whole milk, kept in a refrigerator (4°C); 2, 5, and 8 days old.

   Treatment 3: Whole milk, incubated at 37°C; 2, 5, and 8 days old.

   Treatment 4: Whole milk, boiled, then cooled to room temperature (25°C) and sealed; 2, 5, and 8 days old.

   Treatment 5: Chocolate milk, kept at room temperature (25°C); 2, 5, and 8 days old.

   Treatment 6: Skim milk, kept at room temperature (25°C); 2, 5, and 8 days old.

   Treatment 7: Buttermilk, kept at room temperature (25°C); 2, 5, and 8 days old.

   Treatment 8: Optional treatment, determined by the instructor.

3. Use pH paper to measure the pH of each sample of each treatment. Record your data in table 21.1.

4. Note each sample's odor, color, and consistency. Pay particular attention to physical changes in the milk's composition and record any odors, particles, "growths," or new liquids forming in the flask.

5. Use the Gram stain procedure (see Exercise 24) to identify the shapes, staining properties, and relative amounts of the bacteria in the milk.

6. Record your observations in table 21.1.

7. Plot your pH data for Treatments 1–3 in figure 21.2, Treatments 1 and 4 in figure 21.3, and Treatments 1, 5, 6, and 7 in figure 21.4.

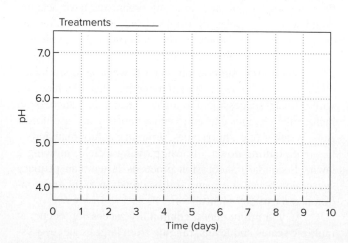

**Figure 21.2**   The effect of temperature on succession (as measured by changes in pH) in a biological community in whole milk.

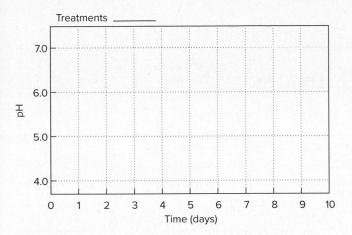

Treatments _____

**Figure 21.3** The effect of boiling and sealing on succession (as measured by changes in pH) in a biological community in whole milk.

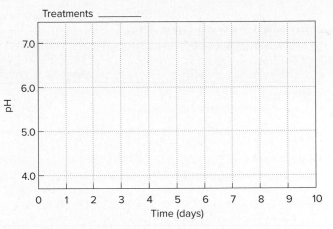

Treatments _____

**Figure 21.4** Community succession (as measured by changes in pH) in chocolate milk, skim milk, buttermilk, and whole milk.

## Table 21.1

### Community Succession in Milk

**Treatment 1: Whole milk kept at room temperature (25°C)**

| | pH | Odor | Color | Type of Bacteria Present | Comments |
|---|---|---|---|---|---|
| Day 0 | | | | | |
| Day 2 | | | | | |
| Day 5 | | | | | |
| Day 8 | | | | | |

**Treatment 2: Whole milk kept in a refrigerator (4°C)**

| | pH | Odor | Color | Type of Bacteria Present | Comments |
|---|---|---|---|---|---|
| Day 0 | | | | | |
| Day 2 | | | | | |
| Day 5 | | | | | |
| Day 8 | | | | | |

**Treatment 3: Whole milk kept at 37°C**

| | pH | Odor | Color | Type of Bacteria Present | Comments |
|---|---|---|---|---|---|
| Day 0 | | | | | |
| Day 2 | | | | | |
| Day 5 | | | | | |
| Day 8 | | | | | |

**Treatment 4: Whole milk, boiled, then cooled to room temperature (25°C)**

| | pH | Odor | Color | Type of Bacteria Present | Comments |
|---|---|---|---|---|---|
| Day 0 | | | | | |
| Day 2 | | | | | |
| Day 5 | | | | | |
| Day 8 | | | | | |

**Treatment 5: Chocolate milk kept at room temperature (25°C)**

| | pH | Odor | Color | Type of Bacteria Present | Comments |
|---|---|---|---|---|---|
| Day 0 | | | | | |
| Day 2 | | | | | |
| Day 5 | | | | | |
| Day 8 | | | | | |

**Treatment 6: Skim milk kept at room temperature (25°C)**

|  | pH | Odor | Color | Type of Bacteria Present | Comments |
|---|---|---|---|---|---|
| Day 0 |  |  |  |  |  |
| Day 2 |  |  |  |  |  |
| Day 5 |  |  |  |  |  |
| Day 8 |  |  |  |  |  |

**Treatment 7: Buttermilk kept at room temperature (25°C)**

|  | pH | Odor | Color | Type of Bacteria Present | Comments |
|---|---|---|---|---|---|
| Day 0 |  |  |  |  |  |
| Day 2 |  |  |  |  |  |
| Day 5 |  |  |  |  |  |
| Day 8 |  |  |  |  |  |

**Treatment 8: Optional treatment (to be determined by instructor)**

|  | pH | Odor | Color | Type of Bacteria Present | Comments |
|---|---|---|---|---|---|
| Day 0 |  |  |  |  |  |
| Day 2 |  |  |  |  |  |
| Day 5 |  |  |  |  |  |
| Day 8 |  |  |  |  |  |

**Question 2**

Which milk changed the slowest? Why? Which milk changed the quickest? Why?

**Question 3**

How did the pH of milk change over time in your different samples? Why do you think this happened?

**Question 4**

Of what value is pasteurization?

**Question 5**

How did the abundance and type of organisms in the milk change over time? Why is this important?

**Question 6**

Acidity sours the milk and converts the casein to curd. How long does it take for this change to occur at room temperature? When the milk is refrigerated?

**Question 7**

Why did you plot the pH data from Treatment 1 on all three of the graphs (figs. 21.2, 21.3, and 21.4)?

## EXAMINING LARGER ECOSYSTEMS

If time permits, your instructor will take you to a disturbed site on campus, an abandoned agricultural field, or a nature preserve. While you are there, examine the site closely. Describe the site in the following space.

**Question 8**

*a.* What caused the disturbance at the site?

*b.* How does modern agriculture affect succession?

*c.* How does the area you are examining differ from areas that surround it? What accounts for these differences?

## INQUIRY-BASED LEARNING

### *Community Succession and the "Spoilage" of Beverages*

Observation: In this lab you measured how dairy products "spoil" and how the changes associated with spoilage are associated with community succession. Many other kinds of beverages also "spoil."

Question: Is the "spoilage" of a beverage or product always associated with similar types of community succession?

a. Establish a working lab group and obtain Inquiry-Based Learning Worksheet 21 from your instructor.

b. Discuss with your group well-defined questions relevant to the preceding observation and question. Choose and record your group's best question for investigation.

c. Translate your question into a testable hypothesis and record it.

d. Outline on Worksheet 21 your experimental design and supplies needed to test your hypothesis. Ask your instructor to review your proposed investigation.

e. Conduct your procedures, record your data, answer your question, and make relevant comments.

f. Discuss with your instructor any revisions to your questions, hypotheses, or procedures. Repeat your work as needed.

## Questions for Further Study and Inquiry

*1.* Question 1 in this exercise asked you to describe local examples of community succession. How are these examples similar to the "milk community" model that you studied in this lab? How are they different?

*2.* What is biological succession, and why is it important?

*3.* How does succession occur? What factors influence the rate of succession?

*4.* What human activities try to prevent (or slow) biological succession?

*5.* In this exercise you studied how microbes affect a rather small ecosystem (i.e., a container of milk). In what ways do microbes affect larger ecosystems?

# Population Growth
## Limitations of the Environment

---

## Learning Objectives

By the end of this exercise you should be able to:
1. Describe how to quantify the growth of populations.
2. Show the effects of resources and environmental conditions on population growth.
3. Graphically analyze how the human population is increasing.

Please visit **connect.mheducation.com** to review online resources tailored to this lab.

To ecologists, the term **population** refers to a group of organisms of the same species living and reproducing together. Notice that all members of a population are of the same species and live in the same general area. A **species** is a group of organisms that can reproduce with each other but are reproductively incompatible with other organisms. Typically a species is distributed in multiple populations across its entire range. There may be little or no gene flow between populations. For example, the population of ground squirrels on the North Rim of the Grand Canyon might not reproduce with the population on the South Rim. They are the same species but constitute two populations.

In optimal conditions (e.g., plenty of food, water, space, etc.), populations grow in a predictable pattern. Early growth is slow (the so-called lag phase of growth), after which growth is extremely rapid (fig. 22.1). This rapid, logarithmic (i.e., "log") phase of growth represents the organism's **biotic potential,** which is its maximal reproductive capacity if given unlimited resources. The number of

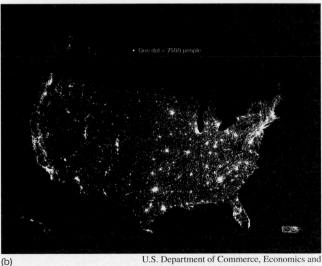

Thomas La Mela/Shutterstock

(b)   U.S. Department of Commerce, Economics and
Statistics Administration, U.S. Census Bureau

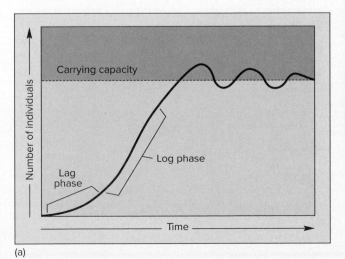

(a)

**Figure 22.1** Populations. (*a*, left) The theoretical growth of a population. The dotted line represents the ecosystem's carrying capacity. (*b*, above) The world's population already exceeds 7.9 billion people, and growth continues.

individuals at a given generation during logarithmic growth can be determined from the following formula:

$$N_t = a2^t$$

where

**$N_t$ = the number of individuals at time $t$**
**$a$ = the number of individuals present initially**
**$t$ = time, expressed as the number of generations**

This equation shows how fast populations can grow. For example, consider *Escherichia coli,* a common bacterium that can divide by way of binary fission every 20 min in ideal conditions. In only one day (1440 min), these bacteria can go through 72 (1440/20) generations. Therefore, if we start our experiment with one bacterium, and if we assume that no bacteria die, the number of bacteria present after one day would be:

**= (1 bacterium)($2^{72}$)**
**= 40,000,000,000,000,000,000,000 bacteria**
**= $4 \times 10^{22}$ bacteria**

This many bacteria would weigh about 3.2 million kg. Bacteria aren't the only organisms having such an incredible biotic potential. For example,

- Each oyster produces about 50 million eggs per year.

- In 6 years a single pair of Atlantic cod and their descendants reproducing without hindrance would completely fill the Atlantic Ocean.

- The 80 offspring produced every 6 months by a pair of cockroaches would produce 130,000 roaches in only 18 months—enough to overrun any apartment (fig. 22.2).

## ENVIRONMENTAL RESISTANCE AND CARRYING CAPACITY

Organisms in the "real world" do not always reproduce at maximum rates, and populations do not grow at ever-increasing rates. Maximum logarithmic growth cannot be sustained because **environmental resistance** includes factors such as disease, accumulation of waste products, and lack of

Chip Clark/National Museum of Natural History, Smithsonian Institution

**Figure 22.2** The consequences of exponential growth. All organisms have the potential to produce populations larger than those that actually occur in nature. The German cockroach (*Blatella germanica*), a major household pest, produces 80 offspring every 6 months. If every cockroach that hatched survived for three generations, kitchens might look like this culinary nightmare concocted by the Smithsonian Museum of Natural History.

food. Ultimately the size and growth of a population are balanced by the environment: environmental resistance prevents a population from continuing to reproduce at its biotic potential. The capacity of the environment to provide resources such as space, food, and energy is limited.

To understand the effects of environmental resistance, complete table 22.1. This table provides actual data for a growing but limited population of bacteria. For comparison, you must calculate the size of a theoretical population of *E. coli* if given unlimited resources and if no bacteria die. Every 20 min the population could potentially double. After doing these calculations, plot the growth of the theoretical and actual populations on figure 22.3.

### Table 22.1

**Theoretical and Actual Growth of *E. Coli* Bacteria**

| Generation | Time (minutes) | Size of Population ($10^3$ bacteria per mL) | |
| | | Theoretical | Actual |
| --- | --- | --- | --- |
| 1 | 0 | 8 | 8 |
| 2 | 20 | 16 | 15 |
| 3 | 40 | 32 | 28 |
| 4 | 60 | _____ | 48 |
| 5 | 80 | _____ | 120 |
| 6 | 100 | _____ | 220 |
| 7 | 120 | _____ | 289 |
| 8 | 140 | _____ | 794 |
| 9 | 160 | _____ | 1986 |
| 10 | 180 | _____ | 4381 |

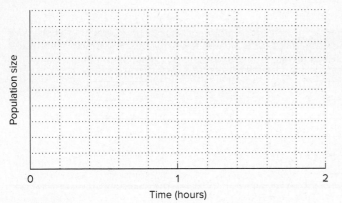

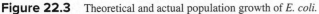

Time (hours)

**Figure 22.3** Theoretical and actual population growth of *E. coli.*

## Question 1

**a.** How did growth of the actual population compare with that of the theoretical population during early stages of the experiment? At later stages?

**b.** How long did it take the real populations to double during early stages of the experiment? Middle stages? Later stages?

**c.** When was growth of the actual population most rapid?

**d.** At what stage was growth slowest? Why?

The growth of most populations levels off after a logarithmic phase because of **limiting factors** such as disease, the presence of predators, competition for resources, and lack of food. This stability occurs when the birth rate equals the death rate and is referred to as the **carrying capacity** of the ecosystem (fig. 22.1). In most populations, population size remains near the carrying capacity as long as limiting factors are constant.

In the laboratory, you can measure the growth of real populations such as bacteria that reproduce quickly. As bacteria reproduce in a clear nutrient broth, the broth becomes turbid (cloudy). You can't accurately count individual bacteria

in this broth, but you can measure the increase in turbidity of a growing culture. More turbidity means more bacteria (living as well as dead). In most natural populations (excluding bacteria), immigration and emigration would also occur, but these factors are ignored for simplicity in the models of population growth that you will examine in this exercise.

There's a lag time before the limiting factors are realized. In these cases, the birth rate may not decline immediately as the carrying capacity is reached; this means that the population may overshoot the carrying capacity before settling down to reach the carrying capacity.

Your instructor has previously inoculated some test tubes of culture media with *E. coli,* a common bacterium. At regular time intervals some of the tubes were put into a refrigerator to stop growth. Examine the cultures by using procedure 22.1. Remember that the turbidity is due to living and dead bacteria.

**SAFETY FIRST** Before coming to lab, you were asked to read this exercise so you would know what to do and be aware of safety issues. In the space below, briefly list the safety issues associated with today's procedures. If you have questions about these issues, contact your laboratory assistant before starting work.

## *Procedure 22.1* Measure population growth of bacteria

**1.** Examine cultures of *E. coli* grown for 0, 4, 8, 12, 24, and 48 h. The greater the turbidity, the more dense the population of bacteria.

**2.** Estimate the relative turbidity of each culture between 0 (clear) and 10 (most turbid).

**3.** Record your results in table 22.2.

**4.** If turbidometers are available, measure the turbidity of the solutions according to procedures demonstrated by your instructor. Spectrophotometers may also be used at 600 or 750 nm. Record your results in table 22.2.

## *Procedure 22.2* Measure how resources and environmental conditions affect the size of a population

**1.** Examine cultures of *E. coli* that have grown for 10 days in the following environments:

Distilled water, pH 7

Nutrient broth, pH 3

Nutrient broth, pH 5

Nutrient broth, pH 7

Nutrient broth, pH 9

Nutrient broth, pH 11

**2.** Quantify the relative turbidity of each culture between 0 (clear) and 10 (most turbid).

**3.** Record your results in table 22.3.

## Table 22.2

**Growth of Bacteria in a Limited-Nutrient Medium**

| Time (hours) | Turbidity Intensity (0–10) | Absorbance Value |
|---|---|---|
| 0 | _____ | _____ |
| 4 | _____ | _____ |
| 8 | _____ | _____ |
| 12 | _____ | _____ |
| 24 | _____ | _____ |
| 48 | _____ | _____ |

## Table 22.3

**Growth of Bacteria in a Limited-Nutrient Medium**

| Media | Turbidity Intensity (0–10) | Absorbance Value |
|---|---|---|
| Distilled water, pH 7 | _____ | _____ |
| Nutrient broth, pH 3 | _____ | _____ |
| Nutrient broth, pH 5 | _____ | _____ |
| Nutrient broth, pH 7 | _____ | _____ |
| Nutrient broth, pH 9 | _____ | _____ |
| Nutrient broth, pH 11 | _____ | _____ |

**4.** If turbidometers are available, measure the turbidity of the solutions according to procedures demonstrated by your instructor. If spectrophotometers are available, measure the absorbance of the solution. Wear goggles if you handle any liquids. Record your results in table 22.3.

### Question 2

**a.** Compare your data for populations grown in nutrient broth and in distilled water. Does the presence of nutrients ensure rapid growth of bacteria? Why or why not?

**b.** What role does pH play in the growth of bacterial populations?

### Procedure 22.3 Measure population growth of duckweed (*Lemna*)

Duckweed (*Lemna;* fig. 22.5*a*) is a tiny (diameter = 1–4 mm) plant that forms mats on stagnant ponds. *Lemna* can reproduce sexually, but it usually reproduces asexually.

**1.** During the first week of this term your instructor placed 10 duckweed (*Lemna*) plants in an illuminated aquarium. Each week since then, he or she has counted the number of plants in the aquarium. Those data are posted by the aquarium.

# Plagues

The plagues of flying insects mentioned in the Quran and Bible illustrate the impact of populations on the environment and other species. For example, a recent plague in Africa included hundreds of billions of locusts that darkened the sky and covered thousands of square kilometers of land. These locusts ate more than 150,000 kilograms of food—including cereal crops—per day, leaving almost a million people facing famine.

Sylvia Buchholz/REUTERS/Alamy Stock Photo

**Figure 22.4**   Plagues of locusts can devour crops, cause famines, and terrorize inhabitants.

**2.** From now until the end of the term, count the number of duckweed plants in the aquarium each week. Plot your data in figure 22.5*b*.

(a)

Courtesy Robert H. Mohlenbrock @ USDA-NRCS PLANTS Database, USDA NRCS, 1995. *Northeast Wetland Flora: Field Office Guide to Plant Species*. Northeast National Technical Center, Chester, PA

Robert H. Mohlenbrock/USDA NRCS

(graph with Y-axis labeled "Number of plants" and X-axis labeled "Time (days)")

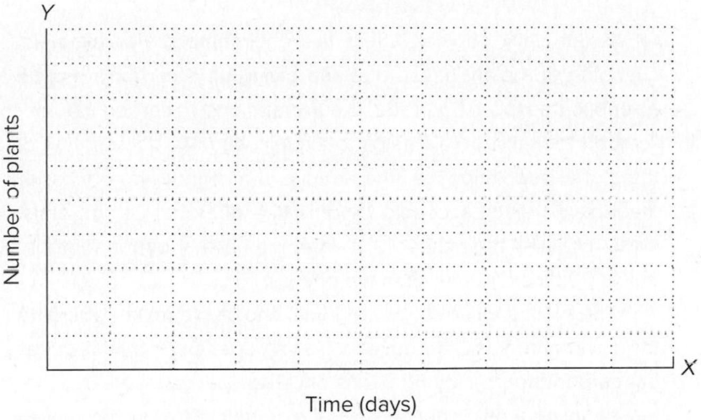

(b)

**Figure 22.5**  (*a*) Duckweed (*Lemna*) is a small flowering plant that grows in mats on the surface of stagnant ponds. (*b*) Population growth of duckweed (*Lemna*).

## Question 3

***a.*** What do you conclude about population growth of duckweed?

***b.*** What will eventually happen to the size of the population? Why?

## Growth of Human Populations

The rate of human population growth has diminished in recent years to about 1.1% per year, but our population

continues to increase by more than 80 million people per year. Consider these data:

| Year | Estimated Size of Human Population (millions) |
|------|------|
| 8000 B.C. | 5 |
| 4000 B.C. | 90 |
| A.D. 1 | 130 |
| 1650 | 550 |
| 1750 | 720 |
| 1800 | 910 |
| 1850 | 1100 |
| 1900 | 1600 |
| 1950 | 2400 |
| 1960 | 3000 |
| 1970 | 3700 |
| 1980 | 4600 |
| 1990 | 5300 |
| 2000 | 6200 |
| 2010 | 6900 |
| 2018 | 7500 |
| 2020 | 7810 |
| 2030 | 8500 (projected by the United Nations) |
| 2050 | 9700 (projected by the United Nations) |
| 2100 | 10,900 (projected by the United Nations) |

Plot these data in figure 22.6. Use a dashed line (----) to plot the projected estimates of population beyond 2020.

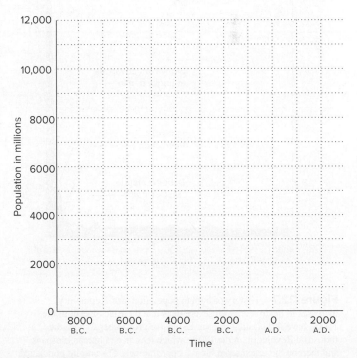

**Figure 22.6**  Growth of the human population.

## Question 4

***a.*** How does the shape of this graph (fig. 22.6) compare with those you made for bacteria (fig. 22.3) and duck-weed (fig. 22.5)?

***b.*** What do you conclude from this?

Another important feature of a population is its **doubling time.** In 1850, the doubling time for our population was 135 years. Today, the doubling time is about 61 years. This means that during the next 61 years we must double our resources if we are to maintain our current standard of living. *Improving* our standard of living will require that we more than double our resources.

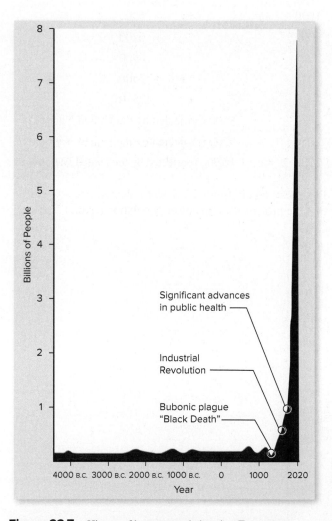

**Figure 22.7** History of human population size. Temporary increases in death rate, even severe ones like the Black Death of the 1400s, have little lasting impact. Explosive growth began with the Industrial Revolution in the 1700s, which lengthened human lifespans and thereby increased reproductive opportunities. The current population exceeds 7.8 billion, and at the current rate will double in 61 years.

## Question 5

***a.*** Population growth influences public health, economic stability, social structure, and the wellbeing of our environment. What is the importance of population growth to you?

***b.*** The doubling time for populations in developed countries far exceeds that of developing countries. What is the significance of this?

# Population Growth and Our Carbon Footprint

All of our daily choices affect the environment. For example, changing to a plant-based diet and giving up a car for a bicycle or public transportation reduce our emissions of carbon (i.e., our "carbon footprint"). This helps to explain why the poorest half of the global population (i.e., the people who cannot afford a car or meat-based diets) accounts for only 10% of global carbon emissions, and why the richest 10% of people have a carbon footprint that is 60 times greater than the poorest 10%.

The most consequential personal choice we make regarding the environment is reproduction. Having one fewer child reduces our carbon-footprint by 58.6 tons of carbon per year. The second-most consequential choice is living automobile-free, which saves 2.4 tons of carbon per year.

Monkeybusinessimages/Getty Images

**Figure 22.8** Having a child is a choice that has social and environmental consequences.

*Source:* Wynes, S. & Nicholas, K. A. (2017). The climate mitigation gap: Education and government recommendations miss the most effective individual actions. *Environmental Research Letters, 12*(7).

The human population has increased explosively during the past three centuries (fig. 22.7). Although the birth rate is now at about 18 births per 1000 people, the death rate has fallen to less than 8 per 1000 people per year. The difference between birth and death rates (10 per 1000) means that the human population is growing at a rate of about 1.1% per year. Here's what that means:

- Each hour the world's population grows by more than 8000 people.
- Each year there are more than 80 million more people on Earth. That annual increase in our population equals the combined population of Great Britain, Ireland, Iceland, Belgium, The Netherlands, Sweden, Norway, and Finland.
- In the 6 seconds it takes to read this sentence, 14 more people will be added to our population. Each of these people eats food; generates wastes; and, in his or her own way, affects our Earth.
- During the 20th century alone, the population of the world increased from 1.65 billion to 6 billion.
- In 1974, there were less than half as many people as there are now.

**Question 6**

*a.* What is the significance of this population growth?

*b.* Can this continue? Why or why not?

*c.* How many people is too many? Do you think the worldwide population has already reached its carrying capacity? Explain the basis for your answer.

*d.* What will happen when the worldwide population exceeds the earth's carrying capacity?

*e.* How has the growth of the human population affected the growth of other populations?

*f.* How does the growth of the human population affect ecosystems?

*g.* In some countries, the human population has shrunk in recent years. Is this good or bad? Why?

*h.* In early 2021, the U.S. population exceeded 330 million people and was growing at an annual rate of approximately 0.6% per year. At this rate of growth, how many people are added to the population in a year?

## INQUIRY-BASED LEARNING

### How does population growth respond to environmental conditions?

Observations: The environment is always changing. The growth and size of a population are strongly influenced by the environment. Population growth of microorganisms with short life cycles and fast growth rates is particularly sensitive to changes in abiotic factors.

Question: How does population growth respond to environmental stimuli?

a. Establish a working lab group and obtain Inquiry-Based Learning Worksheet 22 from your instructor.

b. Discuss with your group well-defined questions relevant to the preceding observation and question. Choose and record your group's best question for investigation.

c. Translate your question into a testable hypothesis and record it.

d. Review procedures 22.1 and 22.3, which use bacteria and duckweed to investigate population growth. Outline on Worksheet 22 your experimental design and supplies needed to test your hypothesis. Ask your instructor to review your proposed investigation.

e. Conduct your procedures, record your data, answer your question, and make relevant comments.

f. Discuss with your instructor any revisions to your questions, hypotheses, or procedures. Repeat your work as needed.

# Questions for Further Study and Inquiry

1.  How can the growth of a population be slowed by its own numbers?

2.  Some people are now realizing the ecological significance of population growth as they grapple with the ethics of bringing children onto a crowded planet. Although this exercise treated the problem only in biological terms, the reality of population growth is much more complex because it involves many political, social, and economic issues. What are some of these issues? How do they affect you now? How will they affect you later in life (e.g., when you want to retire)?

3.  Should we do anything to slow population growth? If so, what? If not, why?

4.  From a purely ecological standpoint, can the problem of world hunger ever be overcome by improved agriculture alone? What other components must a hunger-control policy include?

5.  How are issues such as deforestation, pollution, world hunger, and sustainability linked with population growth?

6.  The late Garrett Hardin (1915–2003), a famous biologist, wrote that "Freedom to breed will bring ruin to us all." Do you agree with him? Explain your answer.

7.  How is population growth important for industries such as fisheries? For conservation projects such as game reserves?

8.  In 2020, one million species faced extinction; this is more than ever before. How is this related to human population growth?

**DOING BIOLOGY YOURSELF**

Design a simple spreadsheet to facilitate calculation of population size for each generation based on different generation times or different numbers of offspring per individual.

**WRITING TO LEARN BIOLOGY**

Charles Darwin's ideas about natural selection as a driving force for evolution were strongly influenced by ideas in an essay entitled "Essay on Population" by Thomas Malthus. Go to the library and read about Malthus's ideas. How did they influence Darwin? How are they relevant to this exercise?

# Pollution
## The Effects of Chemical, Thermal, and Acidic Pollution

## Learning Objectives

By the end of this exercise you should be able to:

1. Describe how chemical pollution, thermal pollution, and acid rain affect the growth and reproduction of selected organisms.
2. Determine what factors can lead to excessive growth of algae.

Please visit connect.mheducation.com to review online resources tailored to this lab.

**S**eldom a day passes in which we don't hear a pollution-related story in the news. Oil spills, leaks of toxic chemicals, and noise pollution are a few of the pollution-related problems that affect our lives. The effects of pollution can be disastrous, as exemplified by oil spills along the coasts of Alaska and Spain; the leakage of 40 tons of toxic gas in 1984 in India (which killed thousands and injured hundreds of thousands of people), and the dumping of toxic wastes into our rivers, lakes, and oceans.

A **pollutant** is any physical or chemical agent that decreases the aesthetic value, economic productivity, or health of the biosphere. There are many types of pollutants, including noise, chemicals, radiation, and heat. Whether or not it makes headlines, pollution of any type affects fundamental aspects of life, namely our health and how organisms grow and behave. Because populations of organisms interact among themselves and with their environment, pollution always affects more than one organism. For example, consider the following food chain:

**algae → zooplankton → small fish → large fish → humans**

A pollutant that reduces a population at any step of this food chain will affect all other levels of the food chain because all steps of the chain are linked. Thus, water polluted with chemicals, such as herbicides that kill algae, will decrease populations of zooplankton that eat the algae; this ultimately affects populations of fish and other organisms that are part of the food chain.

In this exercise, you will study several types of pollution. Specifically, you will

- Simulate the effects of **acid precipitation** by examining seed germination and survival of organisms at acidic pH.

- Study chemical pollution by examining the effects of differing concentrations of nutrients and **pesticides** on growth of algae and shrimp.

- Study **thermal pollution** by examining how organisms survive at unnaturally high temperatures.

- Examine water polluted with an overabundance of algae.

- Use the Allium Test to assay the effects of a variety of pollutants on plant growth.

## SIMULATING THE EFFECTS OF ACID RAIN

Acid rain is a worldwide problem caused by atmospheric pollution (fig. 23.1a). Compounds such as nitrates and sulfates released into the atmosphere by automobiles and industries combine with water to form nitric and sulfuric acids that are deposited across the countryside in precipitation having a pH as low as 4.2. There, the acid decreases the pH of the soil, affects the availability of nutrients and metals, and usually diminishes plant growth. The effects of acid rain are often subtle but significant; for example, the cumulative pollution of decades of acid rain can destroy landmarks and reduce forests and crop yields (fig. 23.1b).

In this exercise, you will study the influence of acid rain on (1) seed germination and (2) growth and reproduction of brine shrimp. You will simulate acid rain with dilute solutions of sulfuric acid.

Be careful while you handle the cultures and do not get the sulfuric acid solutions on yourself. If you do, wash yourself immediately and thoroughly.

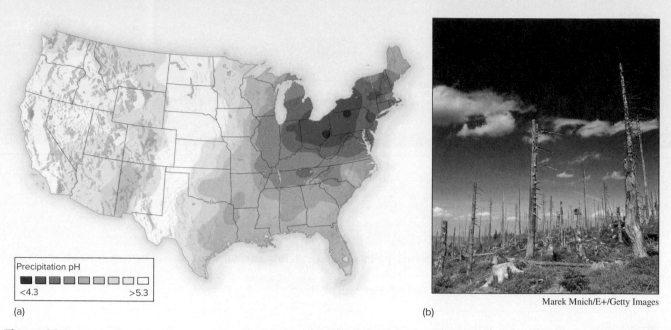

(a)

(b)

Marek Mnich/E+/Getty Images

**Figure 23.1** Acid rain and its effects. (*a*) The pH of snow and rain throughout the United States. (*b*) Acid rain often kills trees and damages landmarks such as sculptures and buildings.

 **SAFETY FIRST** Before coming to lab, you were asked to read this exercise so you would know what to do and be aware of safety issues. In the space below, briefly list the safety issues associated with today's procedures. If you have questions about these issues, contact your laboratory assistant before starting work.

### *Procedure 23.1* Observe seed germination

*1.* Examine the set of petri dishes labeled "Acid Rain and Seed Germination." Each of the dishes was inoculated several days ago with 50 seeds of corn (*Zea mays*). Dish 1 contains seeds soaked in a solution having a neutral pH (i.e., pH 7, which is near the pH at which corn seeds germinate in their typical environments), dish 2 contains seeds soaked in a solution having a pH of 4, and dish 3 contains seeds soaked in a solution having a pH of 2.

### Question 1

*a.* Examine dish 1 (pH 7). What percentage of the seeds germinated? A germinated seed is one in which the root or shoot has punctured the seed coat. Use the following formula to calculate the percentage germination:

$$\% \text{ Germination} = \frac{\text{number of germinated seeds}}{\text{total number of seeds}} \times 100$$

*b.* How do you think that lowering the pH will affect seed germination? Formulate a hypothesis for how lowering the pH will affect seed germination.

*2.* Compare the seeds in dish 1 with those in dishes 2 (pH 4) and 3 (pH 2). Calculate the percentage germination for each treatment. Record your results in table 23.1.

*3.* Do your data in table 23.1 support your hypothesis? Explain.

### Question 2

*a.* How did acidity affect seed germination? How accurate were your predictions in Question 1?

*b.* What does this tell you about the effect of acid rain on seed germination in natural environments?

*c.* How would the decreased growth of plants in response to acid rain affect animals in the same environment?

**Table 23.1**

**Germination of Seeds in Environments of Different pH**

| Treatment | Total Number of Seeds | Number Germinated | % Germination |
|---|---|---|---|
| pH = 7 | _____ | _____ | _____ |
| pH = 4 | _____ | _____ | _____ |
| pH = 2 | _____ | _____ | _____ |

**Table 23.2**

**Growth of Brine Shrimp in Environments of Different pH**

| | Counts of Number of Living Shrimp | | | | | | |
|---|---|---|---|---|---|---|---|
| Treatment | 1 | 2 | 3 | 4 | 5 | Total | Mean |
| pH = 7 | _____ | _____ | _____ | _____ | _____ | _____ | _____ |
| pH = 4 | _____ | _____ | _____ | _____ | _____ | _____ | _____ |
| pH = 2 | _____ | _____ | _____ | _____ | _____ | _____ | _____ |

## *Procedure 23.2* Observe growth of brine shrimp

Brine shrimp (*Artemia salina*) is a small crustacean that lives in salty aquatic environments such as the Great Salt Lake. Brine shrimp are often used as live food in aquariums.

1. Examine the three cultures labeled "Acid Rain and Brine Shrimp." Several days ago all of the cultures were inoculated with the same number of brine shrimp eggs. Culture 1 contains shrimp growing in a solution having a pH of 4, and culture 2 contains shrimp growing in a solution having a pH of 2. The pH of typical environments for brine shrimp ranges from 7 to 8.5.

2. Compare these cultures with that in dish 3 at neutral pH (pH = 7). Use your dissecting microscope to count the number of living shrimp in five randomly selected fields of view of each culture.

3. Record your results in table 23.2.

### Question 3

*a.* How does acidity affect hatching and growth of brine shrimp in your experiment? What does this tell you about the effect of acid rain on survival of brine shrimp?

*b.* Consider the food chain described in the introduction of this exercise. How would acid rain affect fish and humans in this food chain?

*c.* Comment on the validity of extrapolating the results of your experiments with seed germination and brine shrimp survival to the natural world.

## NUTRIENT ENRICHMENT: EUTROPHICATION

As naturally occurring nutrients are washed from the soil by rain, bodies of water such as lakes and ponds undergo a nutrient-enrichment process called **eutrophication** (fig. 23.2). Some instances of eutrophication occur naturally, whereas others are produced by human activities such as pollution. Most nutrients can be absorbed by an ecosystem without damaging the ecosystem's integrity. However, if the critical load of nutrients is exceeded, eutrophication occurs.

Nagel Photography/Shutterstock

**Figure 23.2** A eutrophic pond. The surface bloom of green algae indicates the abundance of nutrients in the water.

Table 23.3

**Effects of Nutrient Enrichment on Algal Cultures**

| Culture | Observations |
|---|---|
| 1 Eutrophication | |
| 2 Eutrophication, anaerobic | |
| 3 Control | |

Nitrogen and phosphorous are important nutrients that can limit plant growth. These nutrients are also major sources of anthropogenic eutrophication. These nutrients can enter aquatic systems from untreated sewage, agricultural runoff, excess fertilizer from lawns and gardens, and some detergents. In extreme situations, this excessive input of nutrients can cause rapid growth of algae and aquatic plants, thereby leading to the accelerated filling of lakes and ponds. In this exercise, you will study eutrophication by observing the growth rates of several organisms in both nutrient-enriched and unenriched media.

Several days ago algal cultures were started with similar amounts of algae. Culture 1 contains an excess of nitrogen, a condition similar to that of lakes enriched with raw sewage. Culture 2 contains large amounts of nitrogen and minimal oxygen to simulate anaerobic conditions that occur naturally when bacterial populations in water increase rapidly in response to added sewage. As the bacteria grow and reproduce, they rapidly deplete the oxygen supply in the water, thus producing anaerobic conditions. Culture 3 is an aerobically grown control containing normal amounts of nitrogen. This experiment relates the well-publicized and frequent occurrences of fish-kills in overly eutrophic lakes and ponds to nutrient stimulation of rapid bacterial growth that depletes the dissolved oxygen.

Fred Ward Productions

**Figure 23.3** Huge amounts of pesticides and herbicides, such as those being applied by this crop duster, are used in agriculture each year. One of the most important tasks facing agriculturists and other biologists is the development of integrated pest management and systems involving genetically engineered organisms that will avoid the need for such chemical excesses.

*c.* What does this tell you about the influence of excessive amounts of nutrients on algal growth?

*d.* How could you reverse the effects of eutrophication?

*e.* Can eutrophication ever be beneficial? If so, when?

### *Procedure 23.3* **Examine the effects of simulated eutrophication**

*1.* Examine the two algal cultures labeled "Eutrophication" and the culture labeled "Control."

*2.* Make at least three observations of each culture. Record your observations in table 23.3.

**Question 4**

*a.* Which culture has the most algal growth?

*b.* Which has the least?

### PESTICIDE POLLUTION

Pesticides are common pollutants, and the increasing demands for food by our expanding population have caused farmers to increase their reliance on pesticides to increase crop yields (fig. 23.3). These pesticides are often effective, but residues washed from the soil by rain contaminate lakes, ponds, and the underground water supply. These pollutants can harm organisms that drink and live in these waters.

## Table 23.4

**Effects of Pesticide on Survival of Brine Shrimp**

| Culture | Observations | | | |
|---|---|---|---|---|
| | Replicate 1 | Replicate 2 | Replicate 3 | Mean |
| Pesticide culture 1 | _____ | _____ | _____ | _____ |
| Pesticide culture 2 | _____ | _____ | _____ | _____ |
| Pesticide culture 3 | _____ | _____ | _____ | _____ |
| Control culture 4 | _____ | _____ | _____ | _____ |

## Table 23.5

**Effects of Temperature on Survival of Brine Shrimp**

| Culture | Observations | | | |
|---|---|---|---|---|
| | Replicate 1 | Replicate 2 | Replicate 3 | Mean |
| Culture 1 (room temperature) | _____ | _____ | _____ | _____ |
| Culture 2 (35°C) | _____ | _____ | _____ | _____ |

## Procedure 23.4 Examine the effect of a pesticide on the survival of brine shrimp

 Do not open the pesticide cultures or spill the liquids on yourself; the pesticides are toxic.

1. Examine the four cultures labeled "Pesticides." Several days ago these cultures were started with the same number of brine shrimp and kept at the same temperature.

2. Culture 1 contains the recommended dosage of the pesticide, culture 2 contains 1/10 the recommended dosage, and culture 3 contains 10-times the recommended dosage. Culture 4 is a control that lacks pesticide.

3. Examine three samples from each culture with your dissecting microscope. Record your observations of each culture in table 23.4.

### Question 5

*a.* Which culture contains the most living shrimp?

*b.* Which culture contains the fewest living shrimp?

*c.* What does this tell you about how pesticides affect the growth of aquatic organisms?

*d.* How might this pollution affect terrestrial organisms that depend on the water for drinking water?

## THERMAL POLLUTION

Excessive heat is a common pollutant. Many factories use water from lakes, reservoirs, or rivers to cool heat-generating equipment and release the hot water to reservoirs or ponds, where it raises the water temperature. Thermal pollution:

- Speeds biochemical reactions, thereby altering the growth of organisms and composition of communities.

- Speeds the evaporation of water, thereby concentrating other pollutants in the water.

- Decreases the oxygen supply in the water because warm water holds less oxygen than does cool water.

## Procedure 23.5 Examine the effect of temperature on the survival of brine shrimp

1. Use your dissecting microscope to examine the two cultures labeled "Thermal Pollution." Both cultures were started several days ago with the same number of brine shrimp.

2. Culture 1 was grown at room temperature (20–25°C), and culture 2 was grown at 35°C in oxygen-depleted water. Count the number of living brine shrimp in three fields of view. Record your observations of each culture in table 23.5.

### Question 6

*a.* Which culture contains more living shrimp?

**b.** Which culture contains fewer living shrimp?

**c.** What does this tell you about the influence of thermal pollution on growth of aquatic organisms?

**d.** How might thermal pollution of a lake or pond affect nearby terrestrial organisms?

## THE ALLIUM TEST

Many pollutants directly affect the growth and development of organisms. Thus, the most informative assay for these pollutants involves measuring growth of living organisms exposed to controlled treatments of suspected pollutants. The **Allium Test** provides this kind of assay. In the Allium Test, bulbs of the common onion, *Allium,* are subjected to solutions of water being investigated for pollutant toxicity. Roots develop quickly, and their number and length after five days estimate the effect of the potential pollutant on a common aspect of plant growth.

 Do not open the pesticide cultures or spill the liquids on yourself. Pesticides are toxic.

**Table 23.6**

**Growth of *Allium* Roots Exposed to Solutions Containing Different Pollutants**

| Solutions Assayed | Mean Number of Roots | | Mean Length of Roots | |
| | Treatment | Control | Treatment | Control |
| --- | --- | --- | --- | --- |
| _____ | _____ | _____ | _____ | _____ |
| _____ | _____ | _____ | _____ | _____ |
| _____ | _____ | _____ | _____ | _____ |
| _____ | _____ | _____ | _____ | _____ |
| _____ | _____ | _____ | _____ | _____ |
| _____ | _____ | _____ | _____ | _____ |
| _____ | _____ | _____ | _____ | _____ |

**Table 23.7**

**Root Growth of Replicate *Allium* Bulbs Exposed to a Specific Treatment Solution**

Treatment: _____

Control: _____

| Number of Roots | Length of Roots | Number of Roots | Length of Roots |
| --- | --- | --- | --- |
| _____ | _____ | _____ | _____ |
| _____ | _____ | _____ | _____ |
| _____ | _____ | _____ | _____ |
| _____ | _____ | _____ | _____ |
| _____ | _____ | _____ | _____ |
| _____ | _____ | _____ | _____ |
| _____ | _____ | _____ | _____ |
| _____ | _____ | _____ | _____ |
| _____ | _____ | _____ | _____ |
| _____ | _____ | _____ | _____ |

Mean = _____ Mean = _____ Mean = _____ Mean = _____

## Procedure 23.6 Use the Allium Test to assay a variety of pollutants

1. Examine the solutions provided by your instructor for your class to assay with the Allium Test. Record the solutions in table 23.6.

2. You may need to work in small groups to assay all of the solutions. Determine which solutions or treatments are your group's responsibility. Some of the treatments may involve different temperatures and some may involve solutions with unknown contents. Record in table 23.7 the treatment and control solutions tested by your group.

3. Obtain onion bulbs. You will need 5 to 10 bulbs for each treatment and 5 to 10 bulbs for controls for each treatment.

4. Trim the outer, loose layers of each bulb. Use a razor blade to trim exposed tissue from the root crown. Your instructor will demonstrate how to do this.

5. Obtain a beaker (or test tube) wide enough to support the onion but that will allow the root crown to protrude into the solution.

6. Fill 5 to 10 replicate beakers with the solution being tested.

7. Fill 5 to 10 beakers with the appropriate control solutions.

### Question 7
What is the appropriate control solution for your treatment? Explain your answer.

8. Put a trimmed onion bulb on your beaker so the root crown is completely submerged (fig. 23.4).

9. Incubate the treatments and controls in the dark at room temperature for three to five days.

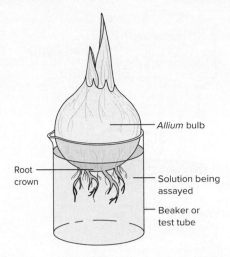

Root crown · Allium bulb · Solution being assayed · Beaker or test tube

**Figure 23.4** Allium Test setup. The number and length of the growing roots estimate the toxicity of the solution being assayed.

10. Check and supplement the solution levels in the beakers periodically if needed.

11. After four to seven days, determine the mean root length and number of roots for each bulb. Record the values in table 23.7. These are the results for the treatment solution that your group tested.

12. Record your mean values from table 23.7 in table 23.6.

13. In table 23.6 record the results for the solutions tested by the other groups.

14. Compare the number and length of roots on treated bulbs with those of controls.

### Question 8
a. Why did each group need to run controls, rather than one set of controls for everyone?

## INQUIRY-BASED LEARNING

### Using the Allium Test to Detect Variation in Water Quality

Observations: The presence and concentration of biologically active compounds can be measured directly or by their indirect effect on a quantifiable biological process, such as a bioassay. Root production by *Allium* is sensitive to a variety of compounds. The number of biologically active compounds that humans introduce each year into our environment is immense.

Question: How does root growth by *Allium* respond to environmental contaminants?

a. Establish a working lab group and obtain Inquiry-Based Learning Worksheet 23 from your instructor.

b. Discuss with your group and instructor a specific contaminant and question relevant to the preceding observation and question. Record it on Worksheet 23.

c. Translate your question into a testable hypothesis and record it.

d. Review the Allium Test procedure 23.6. Then outline on Worksheet 23 your experimental design and supplies needed to test your hypothesis. You may choose to use a bioassay other than the Allium Test. Ask your instructor to review your proposed investigation.

e. Conduct your procedures, record your data, answer your question, and make relevant comments.

f. Discuss with your instructor any revisions to your questions, hypotheses, or procedures. Repeat your work as needed.

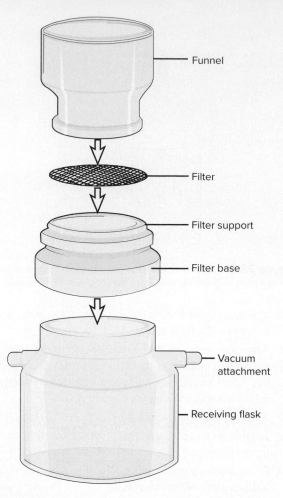

Funnel

Filter

Filter support

Filter base

Vacuum
attachment

Receiving flask

**Figure 23.5** Diagram of filtration apparatus used to collect organisms from a water sample.

*b.* The Allium Test is a common test in environmental science. What are some disadvantages to this test?

*c.* The Allium Test for thermal pollution may show that higher temperatures accelerate growth. Is this bad? Why or why not?

## ALGAL POLLUTION OF WATER SUPPLIES

Growing human populations have created great demands on groundwater supplies. To circumvent these demands, many areas have started using surface waters such as ponds and lakes to meet their water needs. Unlike groundwater, which is usually free of algae and cyanobacteria, surface water usually contains large amounts of algae and cyanobacteria that produce tastes and odors (*Anabaena, Nitella, Pandorina, Hydrodictyon*), clog pipes (*Synedra, Anacystis*), produce toxins (*Anabaena*), and form mats covering a lake or reservoir (*Oscillatoria, Spirogyra*).

Work in small groups to assess algal pollution in the water samples available in the lab.

### Procedure 23.7 Determine algal pollution of water samples

*1.* Dip a filter base and funnel (fig. 23.5) into 70% ethanol for 1 min. Shake off the excess alcohol and let the apparatus air dry.

*2.* Use sterilized forceps to place a filter having pores 0.45 μm in diameter on the filter base. Screw the funnel onto the support (fig. 23.5) so that the filter is held securely in place.

*3.* Add 200 mL of the water sample to the funnel. If the water sample is heavily polluted, decrease the sample to 100, 50, or 25 mL.

*4.* Use a hand vacuum pump or water aspirator (fig. 23.6) to evacuate the receiving flask. Algae will be trapped on the filter.

*5.* Remove and dry the filter in an oven at 45°C for 8 min (or overnight at room temperature).

*6.* Float the dried filter on approximately 5 mL of immersion oil in a small petri dish. The dry filter will become transparent in the oil. If the filter remains opaque after a few minutes, place the petri dish on a warm surface until the filter clears.

*7.* Remove the filter from the petri dish by dragging it across the side of the dish.

*8.* Cut the filter into fourths and place all of these pieces of filter on one clean microscope slide.

*9.* Use a dissecting microscope to examine the filter with reflected light at low total magnification (10–14×). You will see much algae and cyanobacteria on the filter.

*10.* Count the algae in each of 10 randomly selected fields of view and record the number in the following table:

**No. of Algae in Each Field of View**

| Field | 1 | 2 | 3 | 4 | 5 | 6 | 7 | 8 | 9 | 10 | Total |
|---|---|---|---|---|---|---|---|---|---|---|---|
| Number of Algae | __ | __ | __ | __ | __ | __ | __ | __ | __ | __ | ___ |

*11.* Use the following formula to calculate the total number of algae on your filter:

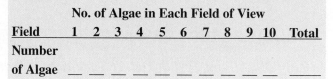

$$\text{Total no. of algae on filter} = \frac{1380 \text{ mm}^2}{\text{area of one field (mm}^2)} \times \frac{\text{total no. of algae counted}}{\text{no. of fields counted}}$$

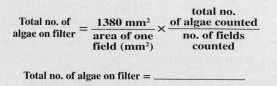

Total no. of algae on filter = _____

Calculate the area of the field of view with a stage micrometer or ruler (see Exercise 3). The value 1380 mm² is the area of the 47-mm filter that you used to collect the algae.

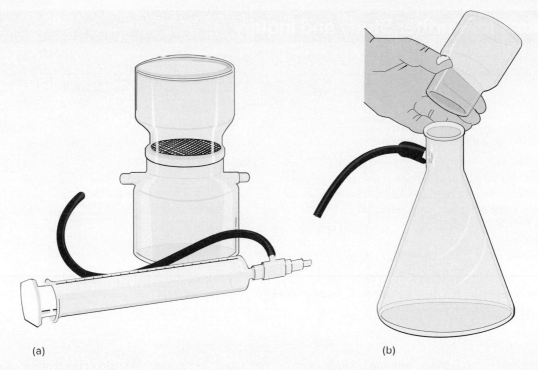

(a)                                                   (b)

**Figure 23.6**   Vacuum filtration using (*a*) a hand vacuum pump or (*b*) a water aspirator. Applying a vacuum speeds the filtration of samples.

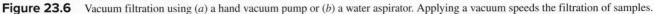

Area of field of view = _____

**12.** Use the following formula to calculate the number of algae per milliliter of sample:

$$\text{No. of algae mL}^{-1} = \frac{\text{no. of algae on filter}}{\text{volume of sample (mL)}}$$

No. of algae mL$^{-1}$ = _____

Algal populations exceeding 1000 organisms mL$^{-1}$ indicate that the water is over-enriched by sewage or other nutrients.

**Question 9**

*a.* How many algae per milliliter are in your water sample?

*b.* Is your sample polluted?

*c.* In what ways might these algae benefit other organisms in the water?

*d.* In what ways might they harm other organisms?

*e.* Suppose you are responsible for "cleaning up" a lake polluted with algae. What would you do, and why?

## BIOREMEDIATION

**Bioremediation** is the use of living organisms (usually microbes) to remove pollutants from an environment. For example, bacteria that metabolize hydrocarbons are often used to minimize the impact of oil spills. Similarly, several bacteria have been genetically engineered (Exercise 16) to metabolize ionic mercury and other dangerous chemicals from nuclear waste.

**Question 10**

Bacteria that remove dangerous chemicals have an obvious value for remedying pollution. Are there any drawbacks or limitations of bioremediation? If so, what are they?

## Questions for Further Study and Inquiry

*1.* What causes pollution?

*2.* Is all pollution bad? Why or why not?

*3.* What are some ways that you could "purify" water polluted with a pesticide?

*4.* What are the consequences of using water from ponds and rivers to cool industrial processes? What are the consequences of stopping the thermal pollution (forcing an industry to stop releasing the heated water)?

*5.* Polluting lakes with laundry water and sewage often produces an algal "bloom." Why don't populations of predators (i.e., zooplankton and fish) increase to offset this bloom and keep the algal population in check?

*6.* Diseases such as typhoid fever (caused by *Salmonella typhosa*) and dysentery are often associated with lakes polluted by sewage. Suppose that you live near a large lake in which sewage is dumped and that as a local health official you are in charge of reducing the incidence of typhoid in your area. How would you do this, assuming that you had the complete cooperation of city and other local officials? How would you do this if you couldn't prevent sewage from being dumped in the lake? What would be the consequences of your actions?

*7.* Define pollution. Can something be a pollutant in some situations but not in others? Explain your answer.

*8.* Does pollution always result from the demands of expanding populations? Why or why not?

 **DOING BIOLOGY YOURSELF**

Design your own assay similar in concept to the Allium Test. Use a living plant and test potential pollutants of your choice.

 **WRITING TO LEARN BIOLOGY**

Rachel Carson devoted much of her life to studying how organisms are affected by pollution. Her masterpiece, *Silent Spring*, raised public awareness of pollution and made the word *ecology* a household word. Go to the library and read an article about this remarkable woman. What was her message, and why is it still important?

# Survey of Prokaryotes
## Domains Archaea and Bacteria

---

## Learning Objectives

By the end of this exercise you should be able to:

1. Describe distinguishing features of Domains Archaea and Bacteria.
2. Describe the major differences between bacteria and cyanobacteria.
3. Identify representative examples of archaea, bacteria, and cyanobacteria.
4. Perform a Gram stain.

---

Please visit **connect.mheducation.com** to review online resources tailored to this lab.

---

**C**ellular organisms have evolved along two lines. Species whose cells lack membrane-bound organelles are **prokaryotes** (table. 24.1); bacteria are examples of prokaryotes. Species whose cells have membrane-bound organelles are **eukaryotes** and include plants, animals, fungi, and protists. About 5000 species of prokaryotes have been described, and many more await identification and description.

Prokaryotes were long thought to be a unified group commonly called bacteria. However, analyses as recently as 1996 of prokaryotic cell structure and genetics revealed two groups with surprisingly different DNA sequences. The sequences within both groups were strikingly different from each other and from those of eukaryotes. This new information has revealed three distinctive domains of organisms (fig. 24.1).

Species of **Domain Archaea** are prokaryotic and often inhabit, but are not restricted to, extreme and stressful environments. Species of **domain Bacteria** are also prokaryotic and the most abundant organisms on Earth (fig. 24.2). Species of **domain Eukarya** are all eukaryotic and include **kingdoms Fungi, Plantae, Animalia,** and the polyphyletic (multiple origins) **protists** (see Exercise 25 for the current status of protistan classification). Organisms in domain Eukarya are described in Exercises 25–31 and 36–40. This classification of living organisms into three domains is now widely accepted, but much phylogenetic information remains to be revealed. Discovering the relationships among life forms is an exciting and ongoing process.

## DOMAIN ARCHAEA

Domain Archaea likely includes the oldest forms of life on earth, and domains Bacteria and Eukarya probably diverged from Archaea independently. In domain Archaea, a single cell is called an **archaeon**. Archaeons are diverse prokaryotes that share ribosomal RNA sequences as well as several important biochemical characteristics that distinguish them from all other types of organisms. Archaeons have distinctive membranes, unusual cell walls, and unique metabolic cofactors.

Today's archaea are probably survivors of ancient lines that persist in habitats similar to those present when archaeons first evolved. These habitats are often extremely acidic, hot, or salty. This is why many archaea are called **extremophiles**. Many archaea can live in an anaerobic atmosphere rich in carbon dioxide and hydrogen as well as the more benign environments typical of bacteria and eukaryotes. Exercise 16 includes a procedure involving *Halobacterium salinarum,* a common archaea that inhabits salty environments.

## DOMAIN BACTERIA

A single cell of organisms in domain Bacteria is called a **bacterium** (pl. bacteria). Bacteria are distributed more widely than any other group of organisms, and individual cells are microscopic (1 µm or less in diameter, figs. 24.2 and 24.3). A single gram of soil may contain over a billion bacteria. Bacteria have cell walls, which give them three characteristic shapes (fig. 24.4).

- Bacillus (rod-shaped)
- Coccus (spherical)
- Spirillum (spiral)

Most bacteria are **heterotrophic,** meaning that they derive their energy from organic molecules made by other organisms.

## Table 24.1

**Prokaryotes Compared to Eukaryotes**

| Feature | Example |
|---|---|
| **Unicellularity.** All prokaryotes are basically single-celled. Even though some bacteria may adhere together or form filaments, their cytoplasm is not directly interconnected, and their activities are not as integrated and coordinated as cells of multicellular eukaryotes. | 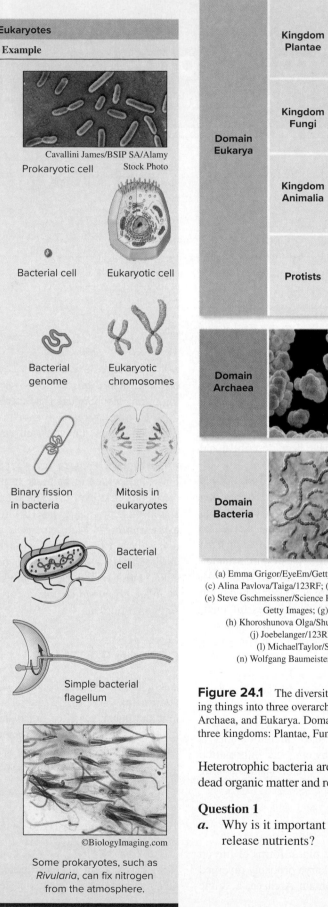 Cavallini James/BSIP SA/Alamy Stock Photo — Prokaryotic cell |
| **Cell Size.** Most prokaryotic cells are 1–10 μm in diameter, while most eukaryotic cells are over 10 times that size. | Bacterial cell — Eukaryotic cell |
| **Chromosomes.** Prokaryotic DNA exists as a single circle in the cytoplasm, while in eukaryotes, proteins are complexed with the DNA into multiple chromosomes. | Bacterial genome — Eukaryotic chromosomes |
| **Cell Division.** Prokaryotic cells divide by binary fission. The cells pinch in two. In eukaryotes, microtubules pull chromosomes to opposite poles during the cell division process, called mitosis. | Binary fission in bacteria — Mitosis in eukaryotes |
| **Internal Compartmentalization.** Unlike eukaryotic cells, bacterial cells contain little internal compartmentalization, no internal membrane systems such as Golgi or ER, and no cell nucleus. | Bacterial cell |
| **Flagella.** Prokaryotic flagella are simple, composed of a single fiber of protein that spins like a propeller. Flagella in eukaryotes are complex structures that whip back and forth, rather than rotating. | Simple bacterial flagellum |
| **Metabolic Diversity.** Prokaryotes possess many metabolic abilities not found among eukaryotes; some prokaryotes can perform several different kinds of anaerobic and aerobic photosynthesis, obtain their energy from oxidizing inorganic compounds, or fix atmospheric nitrogen. | ©BiologyImaging.com — Some prokaryotes, such as *Rivularia*, can fix nitrogen from the atmosphere. |

(a) Emma Grigor/EyeEm/Getty Images; (b) Don Farrall/Photodisc/Getty Images; (c) Alina Pavlova/Taiga/123RF; (d)Ro-ma Stock Photography/Corbis/Getty Images; (e) Steve Gschmeissner/Science Photo Library/Alamy Stock Photo; (f) Russell Illig/Getty Images; (g) Gary W. Carter/Corbis Super/Alamy Stock Photo; (h) Khoroshunova Olga/Shutterstock; (i) Alan and Sandy Carey/Getty Images; (j) Joebelanger/123RF; (k) Science History Images/Alamy Stock Photo; (l) MichaelTaylor/Shutterstock; (m) Power and Syred/Science Source; (n) Wolfgang Baumeister/Science Source; (o) Don Rubbelke/McGraw Hill; (p) Alfred Pasieka/Science Source

**Figure 24.1** The diversity of life. Biologists categorize all living things into three overarching groups called domains: Bacteria, Archaea, and Eukarya. Domain Eukarya is composed of protists and three kingdoms: Plantae, Fungi, and Animalia.

Heterotrophic bacteria are **decomposers** because they feed on dead organic matter and release nutrients locked in dead tissue.

## Question 1

*a.* Why is it important that decomposers such as bacteria release nutrients?

**Ribosome:**
site of protein synthesis

**Inclusion body:**
stored nutrients for
later use

**Mesosome:**
plasma membrane
that folds into the
cytoplasm and
increases surface area

**Fimbriae:**
hairlike bristles that
allow adhesion to
the surfaces

**Conjugation pilus:**
elongated, hollow
appendage used for
DNA transfer to other
bacterial cells

**Nucleoid:**
location of the bacterial
chromosome

**Plasma membrane:**
sheath around cytoplasm
that regulates entrance
and exit of molecules

**Cell wall:**
covering that supports,
shapes, and protects cell

**Glycocalyx:**
gel-like coating outside
cell wall; if compact, called
a capsule; if diffuse, called
a slime layer

**Flagellum:**
rotating filament present
in some bacteria that
pushes the cell forward

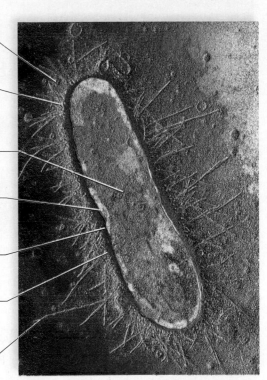

*Escherichia coli*          Sercomi/Science Source

**Figure 24.2**   The structure of a bacterial cell.

(a) Dr. Tony Brain/David Parker/Science
Photo Library/Science Source

(b) Dr. Tony Brain/David Parker/Science
Photo Library/Science Source

(c) Dr. Tony Brain/David Parker/Science
Photo Library/Science Source

(d) Dr. Tony Brain/David Parker/Science
Photo Library/Science Source

**Figure 24.3**   Four views of a contaminated pin, which would seem an unlikely site for bacteria to grow. (*a*) The tip of the pin, magnified
7×. When scanning electron micrographs are shown at increasing magnifications—(*b*) 35×, (*c*) 178×, and (*d*) 4375×—you see rod-shaped
bacteria growing there.

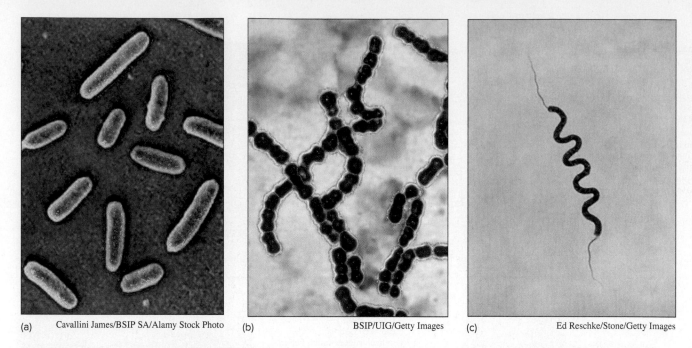

(a)     Cavallini James/BSIP SA/Alamy Stock Photo     (b)     BSIP/UIG/Getty Images     (c)     Ed Reschke/Stone/Getty Images

**Figure 24.4**    The three basic shapes of bacteria: (*a*) bacillus (*Pseudomonas*); (*b*) coccus (*Streptococcus*); and (*c*) spirillum (*Spirilla*), 400×.

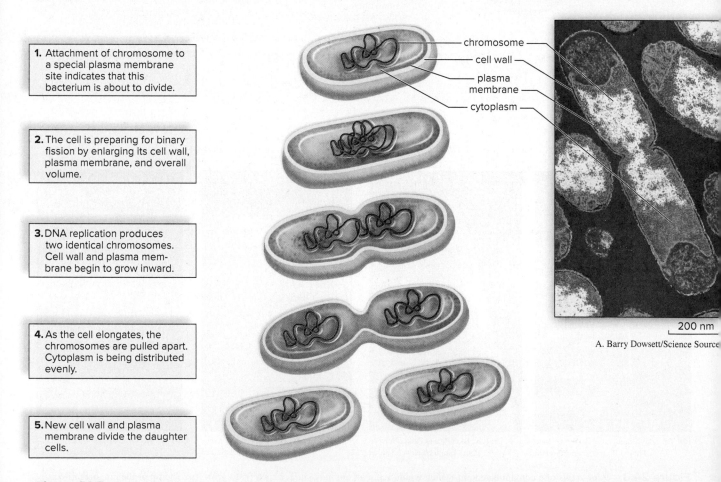

1. Attachment of chromosome to a special plasma membrane site indicates that this bacterium is about to divide.

2. The cell is preparing for binary fission by enlarging its cell wall, plasma membrane, and overall volume.

3. DNA replication produces two identical chromosomes. Cell wall and plasma membrane begin to grow inward.

4. As the cell elongates, the chromosomes are pulled apart. Cytoplasm is being distributed evenly.

5. New cell wall and plasma membrane divide the daughter cells.

chromosome
cell wall
plasma membrane
cytoplasm

200 nm

A. Barry Dowsett/Science Source

**Figure 24.5**    Binary Fission. First, DNA replicates, and as the cell lengthens, the two chromosomes separate, and the cells divide. The two resulting bacteria are identical. In step 3 a protein called FtsZ forms a ring at the site of cell separation. This protein is similar to that forming the mitotic spindle of eukaryotes.

## Table 24.2

**Important Bacterial Diseases That Affect Humans**

| Disease | Pathogen | Vector/Reservoir | Epidemiology |
|---|---|---|---|
| Anthrax | *Bacillus anthracis* | Animals, including processed skins | Bacterial infection that can be transmitted through contact or ingested. Rare except in sporadic outbreaks. May be fatal. |
| Botulism | *Clostridium botulinum* | Improperly prepared food | Contracted through ingestion or contact with wound. Produces acute toxic poison; can be fatal. |
| Chlamydia | *Chlamydia trachomatis* | Humans, Sexually Transmitted Infection | Urogenital infections with possible spread to eyes and respiratory tract. Occurs worldwide; increasingly common over past 30 years. |
| Cholera | *Vibrio cholerae* | Human feces, plankton | Causes severe diarrhea that can lead to death by dehydration; 50% peak mortality if the disease goes untreated. A major killer in times of crowding and poor sanitation; over 100,000 died in Rwanda in 1994 during a cholera outbreak. |
| Dental caries (tooth decay) | *Streptococcus* | Humans | A dense collection of this bacteria on the surface of teeth leads to secretion of acids that destroy minerals in tooth enamel—sugar alone will not cause cavities. |
| Gonorrhea | *Neisseria gonorrhoeae* | Humans only | Sexually Transmitted Infection on the increase worldwide. Usually not fatal. |
| Hansen's disease (leprosy) | *Mycobacterium leprae* | Humans, feral armadillos | Chronic infection of the skin; worldwide incidence about 10–12 million, especially in Southeast Asia. Spread through contact with infected individuals. |
| Lyme disease | *Borrelia bergdorferi* | Ticks, deer, small rodents | Spread through bite of infected tick. Lesion followed by malaise, fever, fatigue, pain, stiff neck, and headache. |
| Peptic ulcers | *Helicobacter pylori* | Humans | Infects the stomach, where it causes ulcers. About 40% of the world's population harbors *H. pylori*. Barry Marshall, who isolated the bacterium in 1982, drank a culture of *H. pylori*; he got an ulcer. |
| Plague | *Yersinia pestis* | Fleas of wild rodents: rats and squirrels | Killed ¼ of the population of Europe in the 14th century; endemic in wild rodent populations of the western United States today. |
| Pneumonia | *Streptococcus, Mycoplasma, Chlamydia* | Humans | Acute infection of the lungs, often fatal without treatment |
| Tuberculosis | *Mycobacterium tuberculosis* | Humans | An acute bacterial infection of the lungs, lymph, and meninges. Its incidence is on the rise, complicated by the development of new strains of the bacteria that are resistant to antibiotics. |
| Typhus | *Rickettsia typhi* | Lice, rat fleas, humans | Historically a major killer in times of crowding and poor sanitation; transmitted from human to human through the bite of infected lice and fleas. Typhus has a peak untreated mortality rate of 70%. |

**b.** What term best describes heterotrophic bacteria that feed on living tissue?

Bacteria that derive their energy from photosynthesis or the oxidation of inorganic molecules are **autotrophic.** However, photosynthesis in bacteria is often different from that in eukaryotes because molecular sulfur rather than oxygen is sometimes produced as a by-product.

A laboratory culture of bacteria usually consists of a tube of liquid nutrients (broth) containing growing bacteria or a tube or plate of solidified agar with bacteria growing on the surface.[1] The jellylike agar is melted, mixed with nutrients, and poured into tubes or plates to solidify. Many species of bacteria can be cultured in nutrient broth or on a layer of nutrient-rich agar. It may surprise you to know that most species of bacteria are *not* culturable in vitro. We just don't know enough about the nutrient and environmental requirements of these bacteria to grow them in the lab.

Bacteria reproduce asexually via **binary fission,** in which a cell's DNA replicates and the cell pinches in half without the nuclear and chromosomal events associated with mitosis (see Exercise 14) (fig. 24.5). Some bacteria have genetic recombination via **conjugation,** in which all or part of the genetic material of one bacterium is transferred to another bacterium and a new set of genes is assembled.

---

1   *Agar is a gelatinous polysaccharide used in culture media for microbiology labs. You'll learn more about agar and the red algae it comes from in Exercise 25.*

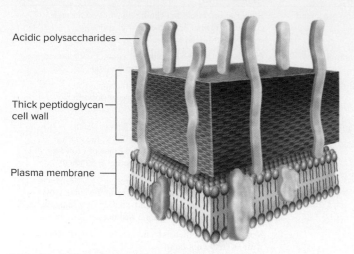

Acidic polysaccharides

Thick peptidoglycan cell wall

Plasma membrane

**(a) Gram-positive: thick peptidoglycan cell wall layer, no outer membrane**

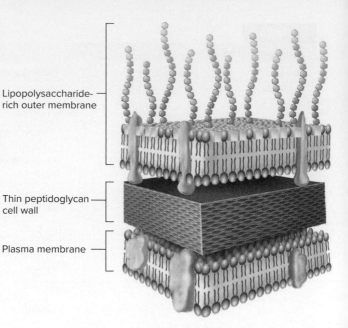

Lipopolysaccharide-rich outer membrane

Thin peptidoglycan cell wall

Plasma membrane

**(b) Gram-negative: thinner peptidoglycan cell wall layer, with outer membrane**

**Figure 24.6** Cell walls of Gram-positive and Gram-negative bacteria. (*a*) Cell walls of Gram-positive bacteria have a thick peptidoglycan layer and no outer membrane. Gram stains stick readily to this peptidoglycan layer. (*b*) Cell walls of Gram-negative bacteria have an outer membrane of lipopolysaccharides. Gram stains don't stick to this outer membrane.

Some bacteria are pathogenic (table 24.2); that is, they cause diseases such as pneumonia and tuberculosis. However, most bacteria are harmless to humans. Indeed, many beneficial bacteria live in and on your body. Nevertheless, you should handle all bacterial cultures with care. The preparation of wet mounts of bacterial cultures requires proper use of a transfer loop and sterilizing flame. Your instructor will demonstrate this aseptic technique.

**SAFETY FIRST** Before coming to lab, you were asked to read this exercise so you would know what to do and be aware of safety issues. In the space below, briefly list the safety issues associated with today's procedures. If you have questions about these issues, contact your laboratory assistant before starting work.

---

### *Procedure 24.1* Culture common bacteria

1. Obtain a sterile cotton swab and a closed petri dish containing sterile nutrient agar.

2. Open the packaged swab and drag the tip over a surface such as your teeth, face, or tabletop.

3. Open the petri dish and drag the exposed swab over the surface of the agar in the manner demonstrated by your instructor.

4. Close the lid and tape it shut. Label the dish with a wax pencil.

5. Turn the dish upside down and place it in the incubator or in a warm area.

---

6. After 24–48 h examine the agar for bacterial growth. Record your observations.

**Question 2**
What are the shape and size of each bacterial colony?

## Gram Stain

One of the most important techniques to classify bacteria is the **Gram stain,** based on the different structural and chemical compositions of bacterial cell walls (fig. 24.6). Gram staining is important because it often correlates with the sensitivity of a bacterium to antibiotics. **Gram-positive** bacteria (e.g., *Streptococcus, Micrococcus*) have a thick cell wall that retains a purple dye, whereas **Gram-negative** (e.g., *Escherichia coli, Serratia*) bacteria have a much thinner cell wall that does not retain the dye.

During the Gram stain technique, crystal violet and iodine are applied to stain all of the bacteria purple. Then alcohol is used to remove the stain from the surface of the Gram-negative cell walls that do not bind the stain. Finally, safranin is used to counterstain the Gram-negative cells with a red color contrasting with purple Gram-positive cells (fig. 24.7). In the following procedure, you will Gram-stain some of your bacteria to see the difference between Gram-negative and Gram-positive organisms.

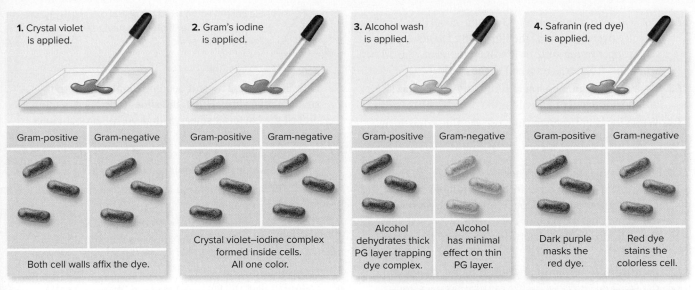

**1.** Crystal violet is applied.

Gram-positive | Gram-negative

Both cell walls affix the dye.

**2.** Gram's iodine is applied.

Gram-positive | Gram-negative

Crystal violet–iodine complex formed inside cells. All one color.

**3.** Alcohol wash is applied.

Gram-positive | Gram-negative

Alcohol dehydrates thick PG layer trapping dye complex. | Alcohol has minimal effect on thin PG layer.

**4.** Safranin (red dye) is applied.

Gram-positive | Gram-negative

Dark purple masks the red dye. | Red dye stains the colorless cell.

**(a)**

**Figure 24.7** The Gram stain procedure. (*a*) The thick peptidoglycan (PG) layer encasing Gram-positive bacteria traps crystal violet dye, so the bacteria appear purple in a Gram-stained smear (named after Hans Christian Gram, Danish bacteriologist, 1853–1938, who developed the technique). Because Gram-negative bacteria have an outer membrane covering their thin peptidoglycan layer, they do not retain the crystal violet dye. A red counterstain (usually a safranin dye) is applied to emphasize contrast with the purple Gram-positive cells. (*b*) A micrograph showing the results of a Gram stain with both Gram-positive and Gram-negative cells.

**(b)**

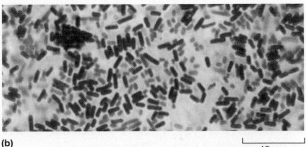

10 µm

De Agostini Picture Library/Getty Images

## Procedure 24.2  Observe stained bacteria with oil-immersion magnification

1. Obtain a microscope and a small bottle of immersion oil. Recall from Exercise 3 that the resolving power of a lens depends, among other things, on the amount of light that it gathers. More light improves resolution, and light is scattered when it passes through air. If a drop of immersion oil, a fluid with the same refractive index (ability to bend light) as glass, is placed between the objective lens and the specimen, then the lens can gather more light.

2. Examine the microscope, and verify with your instructor that the microscope is equipped with an oil-immersion objective. This objective can resolve micrometer-sized particles such as bacteria.

3. Rotate the low-power objective into observation position.

4. Obtain some prepared slides of stained bacteria from your instructor. These may be commercially prepared slides or slides with bacteria that your instructor has stained to help you practice with your microscope.

5. Place a slide on the stage with the specimen centered over the light path through the hole in the stage.

6. While watching from the side, slowly rotate the low-power objective as close as possible to the slide without the objective touching the slide. Adjust the diaphragm for medium light-intensity.

7. Look through the oculars and slowly adjust the coarse adjustment to increase the working distance. Stop when you see the color of the stained bacteria and are roughly focused on the smear of bacteria.

8. Improve the illumination and sharpen the image as much as possible with the fine-adjustment knob. At this low magnification you will only see small dots, at best.

9. Rotate a higher-power objective into position and refocus.

10. Rotate the nosepiece so that the alignment is halfway between the oil-immersion objective and the next lowest-power objective. There should not be an objective in correct position for observation. This position will allow you to place a drop of oil on the slide.

11. Put one drop of immersion oil on the coverslip directly over the spot of the light path. Do not touch the dropper to the slide or it will contaminate the oil when the dropper is returned to the bottle.

12. Rotate the oil-immersion lens directly into observation position and directly into the drop of oil.

13. While looking from the side, use the fine-adjustment knob to lower the objective until it *gently* touches the coverslip.

14. Look through the oculars and *slowly* rotate the fine-adjustment knob to increase the working distance. This rotation should be counterclockwise. Stop when the stained bacterial color appears. Slowly rotate the fine-adjustment knob back and forth until the bacteria are in focus.

15. Improve your resolution by adjusting the diaphragm.

16. Examine the sizes, shapes, and stains of the bacteria on the slide.

17. Repeat this entire procedure for each of the slides offered by your instructor.

18. When you finish your work, clean the oil from the slides and objectives with the lens paper provided.

## Procedure 24.3  Use known bacterial cultures to prepare and observe a Gram stain

1. Obtain a slide, coverslip, transfer loop, alcohol burner, and a culture of living bacteria.

2. Available cultures should include the following bacteria, among others:

   *Bacillus megaterium*—a large bacterium resistant to radiation, desiccation, and heat.

   *Rhodospirillum rubrum*—a photosynthetic purple bacterium.

   *Escherichia coli*—found in the human intestine; the most intensively studied of all bacteria.

   *Staphylococcus epidermidis*—a coccus found among the normal flora of skin.

3. Apply a loop of bacteria to a drop of water on a slide. If your cultures are in liquid broth, then add one drop of culture medium to your slide. Your instructor will demonstrate how to use sterile technique to open, sample, and close the culture of bacteria. Do not add a coverslip to the slide.

4. Heat the slide gently by holding it with a clothespin and passing it over the top of a flame three to four times. Drying time is critical to success. Check with your instructor to avoid heating the slide too little or too much. Be careful not to break the slide. If hot plates are available, hold the slide to the hot surface for 10 sec. This heat will adhere the bacteria to the slide.

 Be careful not to inhale or spill on your skin crystal violet or any other biological stain used in the following steps.

5. Examine figure 24.7 and become familiar with the staining procedure. When the slide has cooled, gently drench the bacterial smear with drops of crystal violet for 20 sec.

6. Rinse the slide for 2 sec with a gentle but steady stream of water from a squirt bottle.

7. Gently drench the bacterial smear with drops of iodine for 1 min.

8. Drop 95% alcohol (decolorizer) on the smear with an eyedropper until no purple shows in the alcohol coming off the slide. Quickly rinse the slide with water to remove the alcohol.

9. Gently drench the bacterial smear with drops of safranin for 20 sec.

10. Gently rinse the slide with water. Air dry the slide, or blot gently if necessary. Add a coverslip.

11. Observe the smear with your microscope using low power and then high power and/or the oil-immersion objective. You will not need a coverslip for heat-fixed slides with oil immersion.

12. Determine if the bacteria are Gram positive or Gram negative.

13. Repeat the Gram stain procedure using other known bacteria that you obtain from culture tubes.

14. Record your observations in table 24.3.

## Procedure 24.4  Use a Gram stain to observe living bacteria from your teeth

1. Obtain a slide, coverslip, transfer loop, and alcohol burner.

2. Use the wide end of a toothpick to scrape your teeth near the gum line.

3. Thoroughly mix what's on the tip of the toothpick in a small drop of water on a microscope slide.

4. Allow this bacterial smear to dry.

5. Repeat steps 4–13 of procedure 24.3.

**Question 3**

*a.* Which type of bacteria is most prevalent in the sample from your teeth?

# Deadly Food! Beware!

The bacterium *Clostridium botulinum* can grow in food products and produces a toxin called botulinum, the most toxic substance known. Microbiologists estimate that 1 gram of this toxin can kill 14 million adults! The good news is that *C. botulinum* requires anaerobic conditions for growth, which limits its prevalence. The bad news is that *C. botulinum* is extremely tolerant to stress; it can withstand boiling water (100°C) for short periods, but is killed at 120°C in 5 min. This tolerance makes *C. botulinum* a serious concern when people can vegetables. If home canning is not done properly, this bacterium will grow in the anaerobic conditions of the sealed container and be extremely poisonous. Several adults and infants die every year from botulism in the United States.

Tolerance to stress is enhanced in *C. botulinum* and many other bacteria by the formation of thick-walled **endospores** that surround their chromosome and a small portion of the surrounding cytoplasm. These highly resistant endospores (fig. 24.8) may later germinate and grow after decades or even centuries of inactivity. The endospores of *Clostridium botulinum* can germinate in poorly prepared canned goods, so never eat food from a swollen (gas-filled) can of food; you risk contracting botulism leading to nerve paralysis, severe vomiting, and death.

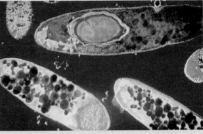

PTP/Phototake/Medical Images

**Figure 24.8** Endospores. The round orange circle in the upper cell is an endospore forming within a cell of *Clostridium botulinum*, the bacterium that causes the disease botulism. These resistant endospores enable the bacterium to survive in improperly sterilized canned and bottled foods.

| Table 24.3 | | | |
|---|---|---|---|
| **The Relative Size and Shape of Some Common Bacteria** | | | |
| **Bacterial Species** | **Gram Stain (+/−)** | **Relative Size** | **Shape** |
| | | | |
| | | | |
| | | | |
| | | | |
| | | | |
| | | | |

**b.** Is *Bacillus megaterium* Gram positive or Gram negative? How do you know?

## Bacterial Colony Morphology

A bacterial **colony** is a visible speck or patch of millions of bacterial cells that are typically the progeny of a single cell that reproduced on the agar's surface. A bacterial colony growing on the surface of nutrient agar often has distinctive characteristics depending on the species. Careful observation of the shape, color, size, texture, and margins of a colony is important for any bacterial identification. Figure 24.9 illustrates the most common features of bacterial colony morphology.

### Procedure 24.5 Evaluate the colony morphology of bacterial species

*1.* Obtain 48-h growth plates of 4–6 bacterial species provided by your instructor for your evaluation.

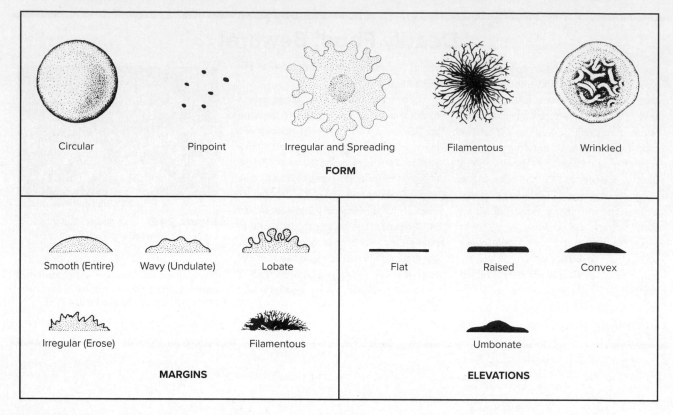

**Figure 24.9** Colony characteristics. Colonies are classified by form, margin, and elevation as well as color and diameter.

2. Familiarize yourself with the colony morphologies shown in figure 24.9.

3. Use a stereomicroscope or hand lens to evaluate a representative colony on each of the plates. Record your observations in table 24.4.

4. When you have completed your analysis, your instructor will provide the species names associated with each of the plates and the accepted colony descriptions for that species. Use this information to evaluate the accuracy of your observations.

**Table 24.4**

**An Evaluation of Bacterial Colony Morphology**

| Plate ID | Species | Colony Diameter (mm) | Color | Form | Elevation | Margin |
|----------|---------|----------------------|-------|------|-----------|--------|
| A | | | | | | |
| B | | | | | | |
| C | | | | | | |
| D | | | | | | |
| E | | | | | | |
| F | | | | | | |

## Nitrogen Fixation by Bacteria

Certain bacteria and cyanobacteria transform atmospheric nitrogen ($N_2$) into other nitrogenous compounds that can be used as nutrients by plants. This process is called **nitrogen fixation.** All organisms need nitrogen as a component of their nucleic acids, proteins, and amino acids. However, chemical reactions capable of breaking the strong triple bond between atoms of atmospheric nitrogen are limited to certain bacteria and cyanobacteria. This process uses an enzyme called nitrogenase along with ATP, energized electrons, and water to convert $N_2$ to ammonia ($NH_3$). Ammonia can be absorbed by plants and used to make proteins and other macromolecules.

*Rhizobium* is a bacterium that can fix nitrogen and can grow intimately with roots of some plants called legumes (e.g., clover, alfalfa, and soybeans). Such associations between *Rhizobium* and host roots form **nodules** on the roots (fig. 24.10). These resident nitrogen fixers provide ammonia to the plant while the plant provides sugars and other nutrients to the bacteria.

---

> ### *Procedure 24.6*  Observe root nodules
>
> 1. Observe the root systems on display and note the nodules.
> 2. Examine a prepared slide of a cross section of a nodule.

Nigel Cattlin/Alamy Stock Photo

**Figure 24.10**  Root nodules containing nitrogen-fixing bacteria.

## Question 4

*a.* Where are the bacteria? Are they between cells or inside cells?

*b.* Why is this relationship between a plant and bacterium called a mutualism?

*c.* How does *Rhizobium* benefit from this association?

*d.* How does the host plant benefit from the association?

## Bacterial Sensitivity to Inhibitors

Growth of some bacterial species is more sensitive to inhibitors such as antibiotics than is growth of other species. For example, an antibiotic may be more effective against *Staphylococcus* than against *Streptococcus*. This is important information to a physician who must select one of many available antibiotics to treat a bacterial infection.

To determine the effectiveness of an antibiotic, a medical laboratory may set up a **sensitivity plate.** A sensitivity plate is a petri dish of solid medium that has been uniformly inoculated on its entire surface with a known bacterium or an unknown sample from an infected patient. After inoculation, four to eight small paper disks—each soaked in a different antibiotic—are placed equidistant from each other on the culture surface. After 24 h, an effective antibiotic will produce a visible halo of clear surface around the disks where it inhibited growth of the bacteria—the larger the halo, the more effective the antibiotic (fig. 24.11). If the antibiotic was totally ineffective, the bacteria will grow to the edge of the paper disk.

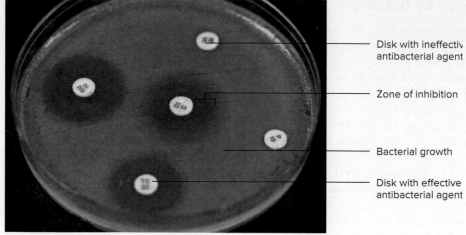

Disk with ineffectiv
antibacterial agent

Zone of inhibition

Bacterial growth

Disk with effective
antibacterial agent

Lisa Burgess/McGraw Hill

**Figure 24.11** A sensitivity plate is used to determine the effectiveness of one or more antibiotics, each of which is on the surface of a paper disk. Any disk containing an effective antibacterial agent inhibits nearby bacterial growth on the agar. If the agent is ineffective, then bacteria will grow up to the disk.

You've probably seen television commercials for products such as mouthwash or disinfectant that "kills germs on contact." Many of these products are developed and tested using sensitivity plates. The mouthwash is effective if no bacteria grow around a paper disk soaked in the mouthwash.

### Procedure 24.7   Examine sensitivity plates

1.  Obtain from your instructor one of each type of prepared sensitivity plate.

2.  Examine each plate, note the bacteria used to inoculate the plate, and note the types of disks distributed on the plate.

3.  Determine which disks inhibited bacterial growth strongly, weakly, or not at all.

4.  Record the bacterial species and your observations in table 24.5.

**Table 24.5**

**Inhibition of Four Bacterial Species by Various Growth Inhibitors**

| Antibiotic/Antiseptic | Plate 1 _____ | Plate 2 _____ | Plate 3 _____ | Plate 4 _____ |
|---|---|---|---|---|
| | | | | |
| | | | | |
| | | | | |
| | | | | |
| | | | | |
| | | | | |
| | | | | |
| | | | | |
| | | | | |
| | | | | |

## Question 5

Based on their appearance, which drugs or chemicals inhibit the growth of bacteria?

Sinclair Stamers/Science source

(a)

## Cyanobacteria (Blue-Green Algae)

Cyanobacteria are a major group of photosynthetic bacteria that grow in many environments. Most cyanobacteria are free-living, whereas others live symbiotically with plants and other organisms. Cyanobacteria are photosynthetic, and their pigments include **chlorophyll *a*** and the accessory pigments **phycocyanin** (blue) and **phycoerythrin** (red). Because of various proportions of these pigments, only about half of the cyanobacteria are blue-green in color; other cyanobacteria range in color from brown to olive green.

Cyanobacteria reproduce by fission and are often surrounded by a jellylike **sheath.** Because cyanobacteria are prokaryotes, they are not related to other algae, all eukaryotic. Cyanobacteria such as *Oscillatoria* often cause many of the disagreeable tastes, colors, and odors in water.

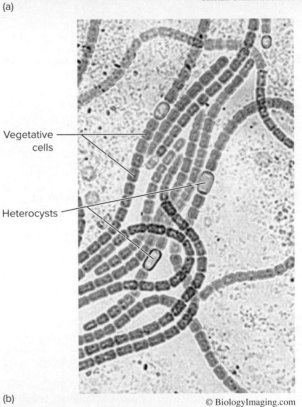

(b)                                    © BiologyImaging.com

### *Procedure 24.8*  Examine cyanobacteria

*1.* With your microscope, examine living material and prepared slides of *Oscillatoria* (fig. 24.12*a*). *Oscillatoria* grows as long chains of cells called **trichomes.**

## Question 6

Do all cells of a trichome of *Oscillatoria* appear similar?

*2.* Examine living material and a prepared slide of *Anabaena* or *Nostoc* (commonly called witch's butter or starjelly) (fig. 24.12*b*). Filaments of these cyanobacteria consist of small grapelike colonies. Trichomes of *Nostoc* consist of small vegetative cells and larger, thick-walled **heterocysts,** in which nitrogen fixation occurs.

*3.* Examine a wet mount of living *Gloeocapsa*, characterized by a thick, gelatinous sheath (fig. 24.12*c*).

*4.* Add a drop of dilute India ink to the slide of *Gloeocapsa* so that the sheath will stand out against the dark background. Often *Gloeocapsa* forms clusters of cells and therefore has a colonial body form. Locate one of these colonies.

*5.* Add a drop of methylene blue to a fresh slide of *Gloeocapsa* and determine if it enhances your observation more than does India ink.

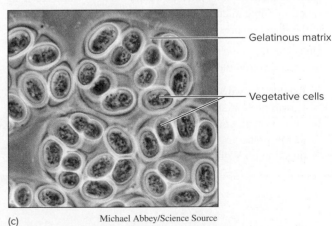

(c)                   Michael Abbey/Science Source

**Figure 24.12** Cyanobacteria. (*a*) *Oscillatoria* (200×); (*b*) *Nostoc* (400×); and (*c*) *Gloeocapsa* (400×).

## Question 7

*a.* Do adjacent cells of *Gleocapsa* share a common sheath?

*b.* What do you suppose is the function of the sheath?

*c.* Do clusters of *Gloeocapsa* represent multicellular organisms? Why or why not?

*d.* What is the best stain for *Gloeocapsa,* India ink or methylene blue?

6. Examine some living *Merismopedia,* which also form colonies. Sketch these cyanobacteria in the following space.

## Question 8

*a.* How is the shape of *Merismopedia* different from other cyanobacteria you studied in this exercise?

*b.* How would a colony attain this shape?

## INQUIRY-BASED LEARNING

*Do we reach every microenvironment with our household disinfectants?*

Observations: Bacteria grow virtually everywhere. Some common household products effectively inhibit bacterial growth on floors and tabletops but may leave behind resistant species. Some microenvironments, however, are not exposed to the selection pressure of disinfectants.

Question: Do some places in our surroundings harbor bacteria that resist household disinfectants especially well?

a. Establish a working lab group and obtain Inquiry-Based Learning Worksheet 24 from your instructor.
b. Discuss with your group a well-defined question relevant to the preceding observation and question. Record your question on Worksheet 24.

c. Translate your question into a testable hypothesis and record it.
d. Review the discussion of sensitivity plates and procedure 24.7. Outline on Worksheet 24 your experimental design and supplies needed to test your hypothesis. Ask your instructor to review your proposed investigation.
e. Conduct your procedures, record your data, answer your question, and make relevant comments.
f. Discuss with your instructor any revisions to your questions, hypotheses, or procedures. Repeat your work as needed.

1. Why do Gram-positive and Gram-negative bacteria stain differently?

2. What happens when milk is pasteurized?

3. What causes milk to sour?

4. What ecological roles are performed by cyanobacteria?

5. How do antibiotics kill bacteria? Why do they not affect viruses?

6. How could bacteria become resistant to an antibiotic?

7.  Antibiotic resistance is promoted by overprescription of antibiotics. How do we stop this trend? What problems are involved? Is antibiotic resistance inevitable?

## DOING BIOLOGY YOURSELF

Obtain two nutrient agar plates. Touch and drag the tip of your finger on the agar surface of one plate. Wash your hand and repeat the procedure on the other plate. Incubate the plates for 24–48 h. Then compare colony appearances and number. What can you conclude about the presence and diversity of bacteria on your plates? Is the appearance of a colony a good way to distinguish species of bacteria? Why or why not?

## DOING BIOLOGY YOURSELF

Doorknobs, sinks, tables, and so on are often laden with bacteria. Devise a protocol to swab areas of your workplace to determine which microenvironments harbor the most bacteria.

## WRITING TO LEARN BIOLOGY

There is a great diversity of roles in a typical ecosystem, some shared by a variety of organisms. What ecological roles are performed by cyanobacteria?

# Survey of Protists
## Algal Autotrophs

## Learning Objectives

By the end of this exercise you should be able to:
1. Discuss the distinguishing features of different groups of algae.
2. Appreciate the economic importance of algae.
3. Outline the events of "alternation of generations" in green algae.

Please visit **connect.mheducation.com** to review online resources tailored to this lab.

This exercise begins your study of **domain Eukarya. Eukaryotes** are organisms composed of cells having membrane-bound nuclei. This domain is commonly divided into three well-defined kingdoms—Fungi, Animalia, Plantae—along with a group of organisms with diverse origins called protists. In one sense, **protists** include all eukaryotes that lack the distinguishing characteristics of fungi, animals, or plants. Protists live in moist habitats and include simple eukaryotes such as amoebas, as well as multicellular organisms such as the brown alga, kelp.

**Fungi** have cell walls and are heterotrophic. **Heterotrophic** organisms feed on organic matter produced by other organisms because they cannot make their required organic compounds from inorganic substances; that is, heterotrophs require organic nutrients produced by other organisms.

**Animals** are ingestive-feeding heterotrophs that lack cell walls and can respond rapidly to external stimuli. Animals are multicellular.

**Plants** are multicellular, **autotrophic** organisms, meaning that they can synthesize all required organic compounds from inorganic substances using external energy, usually sunlight.

Protists share two common characteristics: Most protists are microscopic, and most are successful in moist habitats. Their high surface-to-volume ratio causes them to dry out quickly. Despite shared features, modern phylogenetic analyses of DNA sequences and cellular features reveal that protists are not a well-defined, monophyletic (single origin) group. Protists probably share common ancestry with fungi, plants, and animals. The diversity of protists clearly makes understanding their phylogeny and taxonomy a challenge.

Currently, we classify protists into several eukaryotic **supergroups** that each displays distinctive features that represent a separate evolutionary lineage. Notice that one supergroup, Archaeplastida, includes a lineage giving rise to land plants (table 25.1). Review information in your textbook to thoroughly examine the phylogenetic relationships among protists.

To recognize two major ecological roles of protists, this lab exercise examines autotrophic species commonly referred to as **algae**, Latin for "seaweeds." Exercise 26 examines heterotrophic species referred to as **protozoa** (fig. 25.1). Notice that some supergroups of protistans include both autotrophic species and heterotrophic species. Despite having photosynthesis in common, algae do not form a monophyletic group descended from a single common ancestor. They are polyphyletic (i.e., they have more than one common ancestor), and the details of their phylogenies are still being discovered. Diversity of protists is immense (fig. 25.2).

## INTRODUCTION TO ALGAE

One way to determine the importance of something is to remove it and see what happens. If we did that with algae, most people would be shocked by the result. Global oxygen production would immediately decline. A major food source, perhaps *the* major food source for the world's ecosystems,

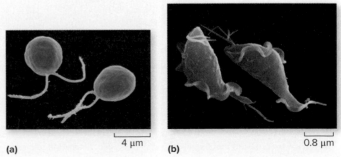
(a) Aaron J. Bell/Science Source; (b) David M. Phillips/Science Source

**Figure 25.1** Representative algal and protozoan protists. (*a*) *Chlamydomonas* sp., a common unicellular alga. (*b*) *Trichomonas vaginalis,* a protozoan causing trichomoniasis, a common sexually transmitted disease in humans.

## DOMAIN: Eukarya
## Protists

**CHARACTERISTICS**

Usually a complex single cell; photosynthesize, ingest, or absorb food; haploid life cycle

| Supergroup | Members | Distinguishing Features |
|---|---|---|
| Archaeplastida | Green algae, red algae, charophytes | Plastids; unicellular, colonial, and multicellular |
| Chromalveolata | Stramenopiles: brown algae, diatoms, golden brown algae, water molds | Most with plastids; unicellular and multicellular |
| | Alveolates: ciliates, apicomplexans, dinoflagellates | Alveoli support plasma membrane; unicellular |
| Excavata | Euglenoids, kinetoplastids, parabasalids, diplomonads | Feeding groove; unique flagella; unicellular |
| Amoebozoa | Amoeboids, plasmodial and cellular slime molds | Pseudopods; unicellular |
| Rhizaria | Foraminiferans, radiolarians | Thin pseudopods; some with tests; unicellular |
| Opisthokonta | Choanoflagellates, animals, nucleariids, fungi | Some with flagella; unicellular and colonial |

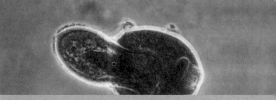

Assorted fossilized diatoms

*Onychodromus*, a giant ciliate ingesting one of its own kind

*Plasmodium*, a slime mold

*Blepharisma*, a ciliate with visible vacuoles

*Licmorpha*, a stalked diatom

*Acetabularia*, a single-celled green alga (chlorophyte)

*Nonionina*, a foraminiferan

*Ceratium*, an armored dinoflagellate

*Bossiella*, a coralline red alga

*Synura*, a colony-forming golden brown alga

*Amoeba proteus*, a protozoan

(a) Science History Images/Alamy Stock Photo; (b) Eric Grave/Science History Images/Alamy Stock Photo; (c) MichaelTaylor/Shutterstock; (d) Franz Neidl/Alamy Stock Photo; (e) Jeff Rotman/Alamy Stock Photo; (f) Borut Furlan/Getty Images; (g) Stephen Durr; (h) Roland Birke/Getty Images; (i) Scenics&Science/Alamy Stock Photo; (j) Eric Grave/Science History Images/Alamy Stock Photo; (k) Patrick W. Grace/Science Source

**Figure 25.2** Diversity of algal protists.

**Table 25.1**

**Eukaryotic Supergroups and Examples of Included Protists**

| Supergroup | Groups | Unifying Characteristics |
|---|---|---|
| EXCAVATA | euglenoids<br>flagellates | Unicellular flagellates, often with feeding groove; mitochondria highly modified in specialized parasites; secondary plastids (when present) derived from endosymbiotic green algae |
| ARCHAEPLASTIDA | red algae<br>green algae | Primary plastids having only two envelope membranes |
| CHROMALVEOLATA | dinoflagellates<br>ciliates<br>apicomplexans<br>diatoms<br>brown algae | Peripheral membrane sacs (alveoli); some dinoflagellates have secondary plastids derived from red algae, some have secondary plastids derived from green apicomplexan algae, and some have tertiary plastids derived from diatoms; some apicomplexans have secondary plastids derived from red or green algae. Straw-like flagellar hairs; fucoxanthin accessory pigment common in autotrophic forms |
| RHIZARIA | radiolarians<br>foraminiferans | Thin, cytoplasmic projections; secondary plastids (when present) derived from endosymbiotic green algae |
| AMOEBOZOA | slime molds<br>amoebas | Amoeboid movement by pseudopodia |
| OPISTHOKONTA | choanoflagellates | Swimming cells possess a single posterior flagellum. |

**Table 25.2**

**The Common Pigments in Addition to Chlorophyll *a*, Storage Products, and Cell Wall Components Characteristic of Common Algae**

| Groups | Predominant Organization | Pigment | Storage Product | Cell Wall |
|---|---|---|---|---|
| Green algae | Unicellular, filamentous, colonial | Chlorophyll *a,b* | Starch | Mainly cellulose |
| Brown algae | Filamentous, multicellular | Chlorophyll *a,c,* fucoxanthin | Laminarin, mannitol, lipids | Cellulose, alginates |
| Red algae | Multicellular | Chlorophyll *a,d,* phycobilins | Modified starch | Cellulose, agar, carrageenan |
| Diatoms | Unicellular | Chlorophyll *a,c,* fucoxanthin | Chrysolaminarin, lipids | Silica |
| Dinoflagellates | Unicellular | Chlorophyll *a,c* | Starch, lipids | Pectin, cellulose plates |
| Euglenoids | Unicellular | Chlorophyll *a,b* | Paramylon, lipids | Proteinaceous pellicle |

would be gone. Tens of thousands of irreplaceable algal species would be lost, along with their unique diversity of potentially useful chemicals, many with pharmaceutical value. The absence of algae would lead to rapid extinction of many hundreds of thousands of invertebrate animal species. Ecosystems would collapse. Fortunately we won't lose algae anytime soon; they have thrived for 1.5 billion years.

Algae are photosynthetic, eukaryotic organisms typically lacking multicellular sex organs. The major groups of algae are distinguished in part by their energy storage products, cell walls, and color, resulting from the type and abundance of colored **pigments** (substances that absorb light) in their plastids (table 25.2). Biologists often group algae by the colors of these pigments—for example, green algae, brown algae, and red algae.

Algae are also distinguished by their cellular organization (fig. 25.3). **Unicellular** algal species occur as single, unattached cells that may or may not be motile. **Filamentous**

algal species occur as chains of cells attached end to end. These filaments may be few to many cells long and may be unbranched or branched in various patterns. **Colonial** algae occur as groups of cells attached to each other in a nonfilamentous manner. For example, a colony may include several to many cells adhering to each other as a sphere, flat sheet, or other three-dimensional shape. Colonies are clonal—that is, their cells are identical. Multicellular organization is not typical of protists but describes algae of more complex design than simple colonies. Multicellular species have cells of different types and functions and show significant interdependence.

## GREEN ALGAE (SUPERGROUP ARCHAEPLASTIDA)

Green algae, collectively referred to as chlorophytes, are the most diverse and familiar algae in fresh water (fig. 25.4). However, a few genera live in salt water. Although

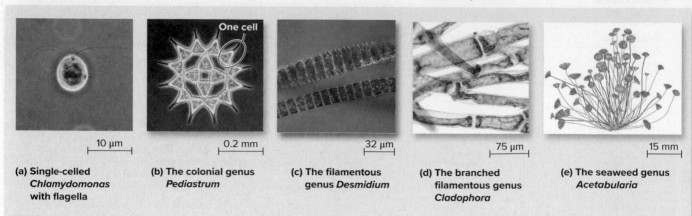

(a) DE AGOSTINI PICTURE LIBRARY; (b) Roland Birke/Phototake/Medical Images; (c) Marek Mis/Science Source; (d) Buccaneer/123RF; (e) Free_styler/Shutterstock

**(a) Single-celled *Chlamydomonas* with flagella**

**(b) The colonial genus *Pediastrum***

**(c) The filamentous genus *Desmidium***

**(d) The branched filamentous genus *Cladophora***

**(e) The seaweed genus *Acetabularia***

**Figure 25.3**   The diversity of algal body-types reflects their habitats. (*a*) The single-celled flagellate genus *Chlamydomonas* occurs in the phytoplankton of lakes. (*b*) The colonial genus *Pediastrum* consists of several cells arranged in a lacy star shape that helps to keep this alga afloat in water. (*c*) The filamentous genus *Desmidium* occurs as a twisted row of cells. (*d*) The branched filamentous genus *Cladophora* that grows attached to nearshore surfaces is large enough to see with the unaided eye. (*e*) The relatively large seaweed genus *Acetabularia* lives on rocks and coral rubble in shallow tropical oceans.

the common name chlorophyte means "green chlorophyll plants" (*chloro* + *phyta*), algae are not classified as plants. Green algae may be ancestral to land plants and share many characteristics with land plants, such as

- Chlorophyll *a,* which occurs in algae and green plants
- Chlorophyll *b,* which occurs in land plants and in the green and euglenoid algae
- Starch as the carbohydrate storage material
- Cell walls made of cellulose

Let's examine a few representatives of green algae.

## Unicellular Green Alga: *Chlamydomonas*

*Chlamydomonas* is a motile, unicellular alga found in soil, lakes, and ditches (fig. 25.4). It probably has the simplest structure and type of reproduction among green algae. The egg-shaped cells of *Chlamydomonas* contain a large chloroplast and a pyrenoid involved in the production and storage of starch.

### *Procedure 25.1*   Observe *Chlamydomonas*

1. Observe a drop of water containing living *Chlamydomonas* and note the movement of the cells. If necessary, review in Exercise 3 the proper use of a compound microscope and the associated videos at the lab manual's website.
2. If the movement is too fast, make a new preparation by placing one or two drops of methylcellulose on a slide and adding a drop of water containing *Chlamydomonas.*
3. Mix gently and add a coverslip.

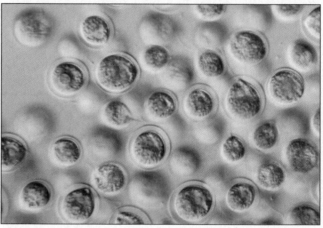

M. I. Walker/Science source

**Figure 25.4**   *Chlamydomonas* is a common alga that is rich in chlorophylls *a* and *b*. It is a single-celled green alga less than 100 µm long.

4. Note the **stigma,** which appears as a reddish, light-absorbing spot at the anterior end of the cell.
5. Examine prepared slides of *Chlamydomonas.*

### Question 1
*a.* Is the movement of *Chlamydomonas* smooth or does it appear jerky?

**b.** Can you see both flagella? (You may need to reduce the light intensity to see flagella.)

**c.** How does methylcellulose affect the movement of *Chlamydomonas*?

**d.** How does the stigma help *Chlamydomonas* survive?

## Asexual and Sexual Reproduction in *Chlamydomonas*

*Chlamydomonas* usually reproduces asexually via mitosis. Each mitosis produces two haploid daughter cells. As many as 16 daughter cells can form inside the cell wall of the original parent cell before being released.

Sexual reproduction in *Chlamydomonas* is a response to unfavorable environmental conditions. For sexual reproduction, vegetative cells of *Chlamydomonas* undergo mitosis to produce gametes. The gametes fuse to form a diploid **zygote,** the resting stage of the life cycle. In most species of *Chlamydomonas,* the gametes of two strains are **isogamous,** meaning they have an identical shape and appearance. For convenience, isogametes of *Chlamydomonas* are referred to as + or −. Gametes unite to form a diploid zygote. **Syngamy** is the pairing and fusion of morphologically similar haploid gametes to form a diploid zygote. Syngamy is similar to fertilization, but fertilization typically refers to fusion of unequal gametes such as egg and sperm cells. The zygote surrounds itself with a resistant surface and is called a **zygospore.** Under favorable conditions the zygote undergoes meiosis to produce haploid individuals called **spores.** Spores are reproductive cells capable of developing into an adult without fusing with another cell. Study the life cycle of *Chlamydomonas* shown in figure 25.5.

---

### Procedure 25.2  Observe syngamy

**1.** Place drops of + and − gametes of *Chlamydomonas* provided by your instructor next to each other on a microscope slide, being careful not to mix the two drops. Do not add a coverslip.

**2.** While you observe the drops through low power of the microscope, mix the two drops of isogametes.

**3.** Switch to high magnification and note the clumping gametes. Try to locate cells that have paired.

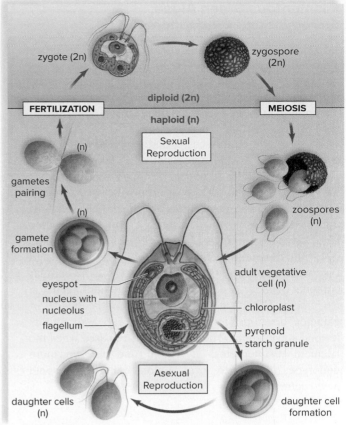

**Figure 25.5** *Chlamydomonas* life cycle. Individual cells of *Chlamydomonas* (a microscopic, biflagellated alga) are haploid and divide asexually, producing copies of themselves. At times, such haploid cells act as gametes, fusing to produce a zygote. The zygote develops a thick, resistant wall, becoming a zygospore (as shown in the lower right-hand side of the diagram). Meiosis of the zygospore then produces four haploid individuals. Only + and − strains can mate with one another, although both may also divide asexually and reproduce themselves.

### Question 2

**a.** Under what environmental conditions would a zygote not undergo meiosis immediately?

**b.** Are spores of *Chlamydomonas* haploid or diploid?

**c.** Which portions of the life cycle of *Chlamydomonas* are haploid?

**d.** Which are diploid?

## Filamentous Green Algae: *Spirogyra* and *Cladophora*

Two of the most common genera of filamentous green algae are *Spirogyra* and *Cladophora*. *Spirogyra* grows in running streams of cool fresh water and secretes mucilage that makes it feel slippery. *Cladophora* is also common in streams and has a much coarser appearance and texture.

*Spirogyra* reproduces sexually by a process called **conjugation** as well as asexually by **fragmentation**. During conjugation, filaments of opposite mating types lie side by side and form projections that grow toward each other. These projections touch and the separating wall dissolves, thus forming a **conjugation** tube (fig. 25.6, also see fig. 4.15). The cellular contents of the (−) strain then migrate through the conjugation tube and fuse with that of the nonmotile (−) strain. The cellular contents of these two strains function as nonflagellated isogametes. The zygote resulting from the fusion of gametes develops a thick, resistant cell wall and is termed a zygospore. The zygospore is released when the filament disintegrates, at which time the zygospore undergoes meiosis to form haploid cells that become new filaments.

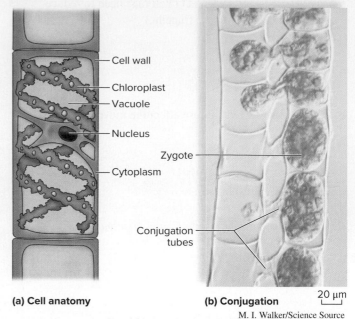

**(a) Cell anatomy**

Cell wall
Chloroplast
Vacuole
Nucleus
Cytoplasm
Zygote
Conjugation tubes

**(b) Conjugation**

20 µm

M. I. Walker/Science Source

**Figure 25.6** *Spirogyra* (watersilk). (*a*) Spiral, ribbonlike chloroplasts occur in each cell. (*b*) During conjugation, the cell contents of one filament (+strain) enter the cells of another filament (−strain) through a conjugation tube (200×).

### *Procedure 25.3* Examine *Spirogyra* and *Cladophora*

1. Obtain and examine a living culture of *Spirogyra*.

2. Prepare a wet mount with a small amount of living *Spirogyra*. Examine it with low and then high magnification. Notice the arrangement of the chloroplasts and whether filaments are branched.

3. Examine a prepared slide of *Spirogyra*.

4. In the following space, sketch a filament of *Spirogyra* and note the arrangement of its chloroplasts.

5. Examine a prepared slide of filaments of *Spirogyra* undergoing conjugation. Locate the conjugation tubes, gametes, and zygotes.

6. Prepare and examine a wet mount of living *Cladophora* (fig. 25.7). Then examine a prepared slide of *Cladophora*.

7. In the following space, draw a few cells of *Cladophora* showing their general shape and the filament's branching pattern.

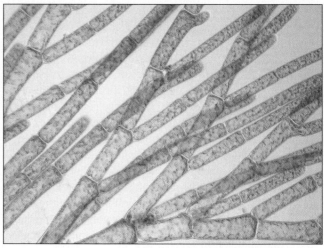

Blickwinkel/Alamy Stock Photo

**Figure 25.7** The green alga *Cladophora* forms branched filaments consisting of multinucleate cells (100×).

### Question 3

*a.* Are filaments of *Spirogyra* branched?

*b.* What is the arrangement of chloroplasts of *Spirogyra?*

*c.* Can you see any conjugation tubes? If you can't, examine the prepared slides that show these structures.

*d.* How do you think *Spirogyra* reproduces asexually?

**Question 4**

*a.* How is *Cladophora* morphologically similar to *Spirogyra?* How is it different?

*b.* What is the shape of its chloroplasts?

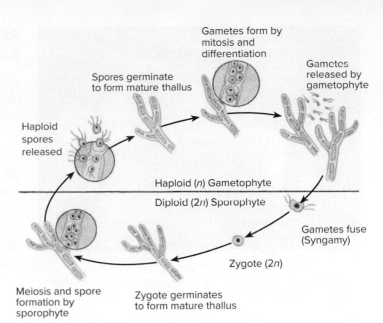

**Figure 25.8** Alternation of generations in *Cladophora.* During alternation of generations, the haploid (*n*) gametophytes alternate with diploid (2*n*) sporophytes. Gametophytes produce haploid gametes that fuse to form a diploid zygote, the first cell of the sporophyte generation. The zygote germinates and undergoes meiosis to form haploid spores, the first cells of the gametophyte generation. The gametophyte then uses mitosis to produce gametes, thereby completing the life cycle.

Mature *Cladophora* exists in diploid and haploid forms. The diploid stage of the life cycle produces spores and is called the **sporophyte.** The haploid stage of the life cycle produces gametes and is called the **gametophyte.** This phenomenon of alternating haploid and diploid stages of a life cycle is called **alternation of generations.** Alternation of generations is a reproductive cycle in which the haploid gametophyte produces gametes that fuse to form a zygote that germinates to produce a diploid sporophyte. Within the sporophyte, meiosis produces spores that germinate into gametophytes, thus completing the cycle (fig. 25.8). Alternation of generations occurs in many green algae, including *Cladophora,* and in all land plants.

You should become familiar with the concept of alternation of generations because it occurs frequently in the plant kingdom and we will refer to it repeatedly in this and future exercises. Refer to your textbook or instructor for more information.

## Colonial Green Alga: *Volvox*

*Volvox* consists of many *Chlamydomonas*-like cells bound in a common spherical matrix (fig. 25.9). Each cell in the sphere has two flagella extending outward from the surface of the colony. Synchronized beating of the flagella spins the colony through the water like a globe on its axis. *Volvox* is one of the most structurally advanced colonial forms of algae, so much so that some biologists consider *Volvox* to be multicellular. Some of the cells of a *Volvox* colony are functionally differentiated; a few specialized cells can produce new colonies, and eggs and sperm are formed by different cells in the colony.

*Volvox* reproduces by **oogamy.** Motile sperm swim to and fuse with the large nonmotile eggs to form a diploid zygote. The zygote enlarges and develops into a thick-walled zygospore released when the parent colony disintegrates. The zygospore then undergoes meiosis to produce haploid cells that subsequently undergo mitosis and become a new colony. During asexual reproduction, some cells of *Volvox* divide, bulge inward, and produce new colonies called **daughter colonies** that initially are held within the parent colony.

### *Procedure 25.4* Observe *Volvox*

*1.* Examine a prepared slide of *Volvox.*

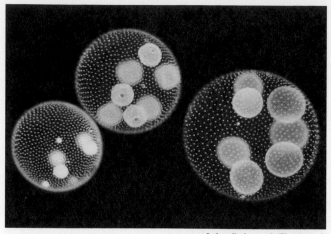

Lebendkulturen.de/Shutterstock

**Figure 25.9** Colonies of *Volvox*. Many parent colonies contain asexually produced daughter colonies (400×).

2. Obtain a living culture of *Volvox*. Place a drop of living *Volvox* on a depression slide.

3. Under low magnification, observe the large, hollow, spherical colonies for a few minutes to appreciate their elegance and beauty.

4. Search for flagella on the surface.

5. Complement your observations of this alga by re-examining prepared slides of *Volvox*.

**Question 5**

*a.* What is oogamy?

*b.* What are the tiny spheres inside the larger sphere of *Volvox*?

*c.* How do you suppose they get out?

*d.* How do you think the number of cells in a young *Volvox* colony compares to the number in a mature colony?

## RED ALGAE (SUPERGROUP ARCHAEPLASTIDA)

Red algae obtain their color from the presence of red **phycobilins** in their plastids. Red algae typically live in warm marine waters. The thallus of a red alga can be attached or free-floating, filamentous, or fleshy (figs. 25.10, 25.11).

*Procedure 25.5* **Examine *Polysiphonia*, *Porphyra*, and commercial products of red algae**

*1.* Obtain prepared slides of *Polysiphonia* and *Porphyra* and any living cultures that are available in lab.

*2.* Examine a prepared slide of *Polysiphonia*. Notice the thickness of the filaments compared with that of filamentous green algae.

*3.* Examine living *Polysiphonia*. This genus is highly branched and filamentous. As with other red algae, their life cycles can be quite complex. Gametophytes of these organisms are dioecious (i.e., they are either male or female).

*4.* Examine some prepared slides and living *Porphyra* if available. Compare the structure of *Polysiphonia* with that of *Porphyra*. "Blades" of *Porphyra* consist of two layers of cells separated by colloidal material.

*5.* Study the display of carrageenan, agar, and other products derived from red algae.

*6.* Pick up and feel a piece of agar, noting its texture. Agar is a gelatinous polysaccharide from red algae used as a solidifying agent in culture media for microbiology labs (fig. 25.11*b*). A 1% suspension of hot agar remains liquid until it cools to about body temperature.

A. Martin UW Photography/Getty Images        Premaphotos/Alamy Stock Photo

**Figure 25.10** Red algae come in many forms and sizes.

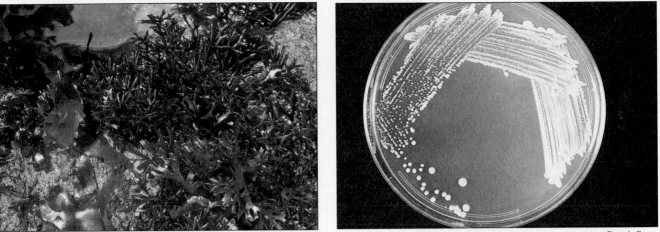

(a)          Heather Angel/Natural Visions      (b)          Dennis Strete

**Figure 25.11** A red alga and a common extract, agar. (*a*) Irish moss (*Chondrus crispus*) is a red alga that is commercially important as a source of carrageenan. Carrageenan is used as a stabilizer in paints and cosmetics, as well as in foods such as salad dressings and dairy products. (*b*) Microbiologists grow a variety of organisms on media solidified with agar (shown here) extracted from seaweeds such as *Gracilaria*. The bacteria shown here and the fungus shown in figure 27.10 are growing on agar. Agar is also used to make drug capsules, cosmetics, and gelatin desserts.

# BROWN ALGAE (SUPERGROUP CHROMALVEOLATA)

Brown algae are primarily marine and structurally complex; there are no unicellular or colonial brown algae. Brown algae usually grow in cool water and obtain their name from the presence of a brown pigment called **fucoxanthin.** Brown algae range in size from microscopic forms to kelps over 50 m long. Thalli of some brown algae are similar to those of land plants. Review table 25.2 for the characteristics of brown algae.

Among the larger brown algae is *Macrocystis,* a kelp reaching 100 m in length (fig. 25.12). The flat blades of this kelp float on the surface of the water, while the base is anchored far below the surface. Another ecologically important brown alga is sargasso weed (*Sargassum;* fig. 25.13), which forms huge floating masses that dominate the vast Sargasso Sea in the Atlantic Ocean northeast of the Caribbean. These mats are microhabitats for a variety of highly adapted and cryptically colored animals.

## *Fucus*

*Fucus* is another common genus of brown algae (fig. 25.14). *Fucus* (rockweed) typically attaches to rocks in the intertidal zone via a specialized structure called a holdfast. The outer surface of *Fucus* is covered by a gelatinous sheath. Tips of *Fucus* branches, called **conceptacles,** may be swollen and contain reproductive structures called **oogonia** (female) and **antheridia** (male), as shown in figure 25.14. Oogonia are multicellular sex organs that produce eggs. Antheridia are multicellular sex organs that produce sperm. Most protists do not have multicellular reproductive organs.

The life cycle of *Fucus* is similar to the common life cycle of animals. The mature thallus is diploid, and cells within reproductive structures undergo meiosis to produce gametes, thereby skipping the multicellular haploid stage common to many protists, plants, and fungi.

Karen Gowlett-Holmes/Getty Images

**Figure 25.12** Brown alga. The giant kelp, *Macrocystis pyrifera,* grows in relatively shallow water along coasts throughout the world and provides food and shelter for many different kinds of organisms.

Kingsley Stern

**Figure 25.13** *Sargassum,* a floating brown alga from which the Sargasso Sea got its name. *Sargassum* also lives in other oceans.

## Procedure 25.6 Examine *Fucus*

1. Refer to figure 25.14 as you examine *Fucus* in the lab. Use your dissecting microscope to examine a cross section of the flattened, dichotomously branched thallus of *Fucus.*

2. Notice the presence of swollen areas on the thallus of *Fucus.*

3. Work in a group to dissect one of these structures.

4. Examine prepared slides of antheridia and oogonia of *Fucus.* Sketch what you see.

**Question 6**

*a.* How does the structure of *Fucus* differ from that of the green algae that you examined earlier in this exercise?

*b.* Are all portions of the thallus photosynthetic? How can you tell?

*c.* Considering where *Fucus* lives, what do you think is the function of its gelatinous sheath?

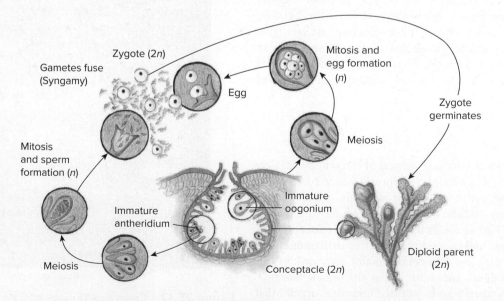

**Figure 25.14** Sexual reproduction in a monoecious species of *Fucus.* Some species of *Fucus* have separate male thalli and female thalli containing conceptacles with only antheridia, and only oogonia, respectively.

***d.*** Are the swollen structures solid masses or are they hollow?

**Question 7**

***a.*** Are the gametes of *Fucus* isogamous or oogamous?

***b.*** How does the structure of tissue surrounding the reproductive structures com.pare with that of green algae?

## Economic Importance of Brown Algae

In the Orient, some brown algae are used as food. One of these algae is *Laminaria,* a kelp marketed as "kombu." Brown algae are also important sources of **alginic acid,** a hydrophilic substance (i.e., it absorbs large quantities of water). Alginic acid is used as an emulsifier (an additive used to stabilize processed food and other products) in dripless paint, ice cream, pudding mixes, and cosmetics.

*Procedure 25.7* **Examine some commercial products of brown algae**

*1.* Taste a small piece of kelp packaged as a food product. How would you describe its taste and texture?

*2.* Observe the products on display and, in the case of foods, read their contents labels.

## DIATOMS (SUPERGROUP CHROMALVEOLATA)

Diatoms are unicellular algae containing chlorophylls *a* and *c* and xanthophyll pigments that give them their golden-brown color. Although diatoms are tiny, their great numbers, rapid rates of reproduction, and photosynthetic capacity make them vitally important as a primary link in the food chain of the oceans.

Diatoms have a hard cell wall made of silicon dioxide (glass) (fig. 25.15). These walls are arranged in overlapping halves, much like the halves of a petri dish. The glass walls of diatoms persist long after the remainder of the cell disintegrates (fig. 25.16) and may accumulate in layers of **diatomaceous earth** several hundred meters deep. This depth indicates how many diatoms have existed through the ages.

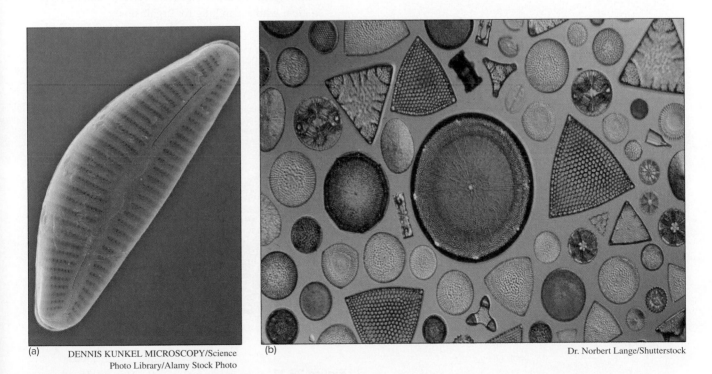

(a) DENNIS KUNKEL MICROSCOPY/Science Photo Library/Alamy Stock Photo

(b) Dr. Norbert Lange/Shutterstock

**Figure 25.15** Diatoms. (*a*) A pennate (bilaterally symmetrical) diatom. (*b*) Several different kinds of diatoms, including some centric (round) species (600×).

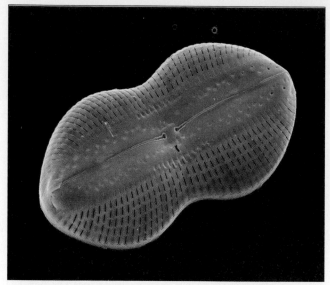

STEVE GSCHMEISSNER/Science Photo Library/Alamy Stock Photo

**Figure 25.16** A pennate diatom showing its ornate, silicon cell wall (1000×).

**Procedure 25.8** **Examine prepared slides of diatoms, living diatoms, and diatomaceous earth**

1. Examine a prepared slide of diatoms. Sketch some of the cells here. Some cells are long and thin, whereas others are disklike.

2. Prepare a wet-mount slide from a culture of living diatoms. Compare the shapes of the cells with those on the prepared slide.

3. Mount a small amount of diatomaceous earth in water on a microscope slide. Examine the diatomaceous earth with your microscope. Note the variety of shells, some broken and others intact. A mass of these shells is clean, insoluble, and porous.

**Question 8**
*a.* Can you see any pores in the walls of diatoms?

*b.* Are any of the diatoms moving?

*c.* If diatoms lack flagella, how do you explain their motility?

*d.* How would diatomaceous earth compare to sand as a material for swimming pool filters? Which would be better and why?

## DINOFLAGELLATES (SUPERGROUP CHROMALVEOLATA)

Dinoflagellates are all unicellular and characterized by the bizarre appearance of their cellulose plates and by the presence of two flagella located in perpendicular grooves (fig. 25.17). Blooms of a red-pigmented dinoflagellate called *Ptychodiscus brevis* produce a "red tide." Toxin production and oxygen depletion by these blooms of algae can kill massive numbers of fish. Dinoflagellates are important primary producers in oceans (second only to diatoms) and include many autotrophic and heterotrophic forms. Some dinoflagellates are bioluminescent, others live symbiotically with corals, and some are heterotrophic.

**Procedure 25.9** **Examine dinoflagellates**

1. Examine a prepared slide of *Peridinium* or *Ceratium*. Look for longitudinal and transverse flagella and flagellar grooves.

2. Prepare a wet-mount slide from a living culture of dinoflagellates. Dinoflagellates are quite small, so be patient while searching for organisms.

**Question 9**
How do the shapes of dinoflagellates compare with other unicellular algae that you have observed in this exercise?

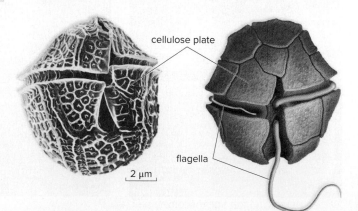

cellulose plate

flagella

2 μm

**Figure 25.17** Dinoflagellate *Gonyaulax*. Most dinoflagellates have rigid cellulose plates and pair of flagella in perpendicular grooves.

## EUGLENOIDS (SUPERGROUP EXCAVATA)

Euglenoids include mostly freshwater unicellular forms. Although plastids of euglenoids contain chlorophylls *a* and *b* (like the green algae), euglenoids are distinctive because they lack a cell wall, and instead have a pellicle made largely of protein. The protein makes the cell more flexible. Euglenoids are motile and have two flagella (fig. 25.18).

### Procedure 25.10   Observe *Euglena*

1. Observe living and prepared slides of *Euglena* available in the lab while referring to figure 25.18.
2. You may want to add a drop of methylcellulose in your preparation to slow the *Euglena*.
3. Note the colored **eyespot** near the base of the flagella.
4. Observe the movement and changing shapes of the organisms.

Chloroplasts of *Euglena* may contain a single pyrenoid, which appears as a clear, circular area within the plastid. *Euglena* is best known for its ability to be autotrophic, heterotrophic, and saprophytic. Its specific mode of nutrition is determined by current environmental conditions. This phenomenon illustrates why it is often impossible to distinguish plant from animal at the cellular level and why classification of protists seems so unwieldy. Our classification schemes for these and other organisms will improve as we learn more about them.

### Question 10
What is the function of the eyespot of *Euglena*?

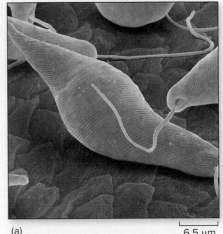

(a)        6.5 μm

Andrew Syred/Science Source

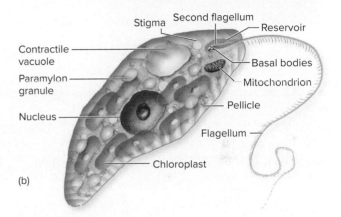

Stigma — Second flagellum — Reservoir
Contractile vacuole
Paramylon granule
Nucleus
Basal bodies
Mitochondrion
Pellicle
Flagellum
Chloroplast

(b)

**Figure 25.18** Euglenoids. (*a*) Micrograph of individuals of the genus *Euglena* (400×). (*b*) Diagram of *Euglena*. Paramylon granules store food reserves.

## INQUIRY-BASED LEARNING

### *How do algae respond to changing environmental stimuli?*

Observation: You learned in Exercise 23 how algae are affected by environmental changes involving nutrients, pollutants, and temperature changes. As a result, algal growth can be used to monitor environmental changes (e.g., eutrophication).

Question: How can algae be used to monitor environmental conditions?

a. Establish a working lab group and obtain Inquiry-Based Learning Worksheet 25 from your instructor.
b. Discuss with your group well-defined questions relevant to the preceding observation and question. Choose and record your group's best question for investigation.

c. Translate your question into a testable hypothesis and record it.
d. Outline on Worksheet 25 your experimental design and supplies needed to test your hypothesis. Ask your instructor to review your proposed investigation.
e. Conduct your procedures, record your data, answer your question, and make relevant comments.
f. Discuss with your instructor any revisions to your questions, hypotheses, or procedures. Repeat your work as needed.

## Questions for Further Study and Inquiry

1. What are examples of unicellular, filamentous, and colonial green algae?

2. How are green algae different from cyanobacteria?

3. What is meant by "alternation of generations"?

4. What is meant by the term "kelp"?

5. Are the stem, holdfast, and blade of brown algae the same as stems, roots, and leaves of land plants? Why or why not?

6. Brown algae contain chlorophyll. Why, then, do they appear brown and not green?

7. What are the main differences and similarities among the major groups of algae? How are these groups related to each other evolutionarily?

8. Describe three ways that algae affect your life.

9. Why are green algae considered to be ancestral to plants?

 **DOING BIOLOGY YOURSELF**

Would you expect environmental conditions to influence syngamy? Design two experiments to investigate the effects of two environmental conditions on the frequency of syngamy. How would you measure syngamy? Its frequency?

 **WRITING TO LEARN BIOLOGY**

Describe the plant-like and animal-like characteristics of *Euglena*. Which characteristics conclusively define a plant? Which ones define an animal?

# Survey of Protists
## Protozoan Heterotrophs

## Learning Objectives

By the end of this exercise you should be able to:
1. Describe the features characterizing protozoa and slime molds.
2. List examples, habitats, reproductive methods, and unique features of the protozoans.
3. Become familiar with representative protozoans and slime molds.

Please visit **connect.mheducation.com** to review online resources tailored to this lab.

Protozoans (*proto* = first, *zoan* = animal) are among the most versatile of all eukaryotes on earth. Protozoa, however, like algae, is a descriptive term rather than a taxonomic group. Protozoans are eukaryotes with an animal-like, heterotrophic ecology, which means they are active consumers and not photosynthetic. (Remember that heterotrophs get their nutritional energy from organic molecules made by other organisms.) Typically, protozoans have food vacuoles to enclose food particles for digestion, and contractile vacuoles to expel excess water. Their single cells employ a variety of features for motility, and they live everywhere from a drop of pond water to the intestines of termites. Review table 25.1 and the supergroups of protists that include common protozoans.

## AMOEBAS (SUPERGROUP AMOEBOZOA)

Amoebas occur throughout the world in marine, freshwater, and terrestrial environments. Their unifying characteristic is the presence of **pseudopods,** which are movable extensions of cytoplasm used for locomotion and gathering food. Amoebas are likely polyphyletic, lack flagella, and most reproduce asexually.

### Amoeba

*Amoeba* is a genus among many organisms commonly called amoebas that has a structure and physiology typical of most amoeboid genera (fig. 26.1). *Amoeba* are **phagocytic,** meaning they engulf food particles and form a **food vacuole** surrounded by a membrane. They then secrete enzymes into the food vacuole for **intracellular digestion.** A **contractile**

**vacuole** maintains the cell's water balance by accumulating and expelling excess water. Other common amoebas include *Difflugia,* which makes a protective case of sand grains called a **test.** Test is a general term referring to a secreted or partially secreted covering, much like a shell (fig. 26.2). Examine a prepared slide of *Difflugia.* Also examine the amoeba *Entamoeba histolytica,* a parasite that causes dysentery in humans.

### Procedure 26.1    Observe *Amoeba* movement and structure

1. Review microscope use and procedures in Exercise 3, as well as the associated videos at the lab manual's website. Use a dissecting microscope (see Exercise 3) to examine a culture of living *Amoeba.* Locate individuals on the bottom of the culture.

2. Prepare a wet mount of living *Amoeba* by using an eyedropper to remove a few drops from the bottom of the culture of organisms.

3. Put the drops in a depression slide if one is available or use a standard slide.

4. Cover the preparation with a coverslip and examine it under low power (10×) (see Exercise 3). Soon the *Amoeba* should move by extending their pseudopods.

5. If nutrient broth is available, add a drop to the preparation and observe the *Amoeba*'s response.

6. Examine a prepared slide of stained *Amoeba,* and locate the structures shown in figure 26.1.

7. Examine any other live amoebas and prepared slides that are available. Draw the basic structure of these organisms.

(a)

M I (Spike) Walker/Alamy Stock Photo

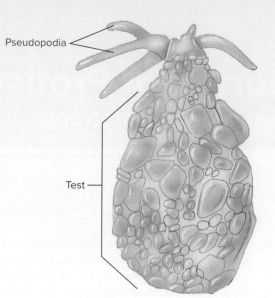

**Figure 26.2** *Difflugia oblongata*, a common freshwater amoeba with a sand-grain case. The case consists of cemented mineral particles collected by the amoeba (150×).

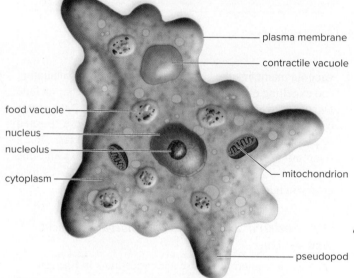

plasma membrane

contractile vacuole

food vacuole

nucleus

nucleolus

cytoplasm

mitochondrion

pseudopod

(b)

**Figure 26.1** Amoebas. (*a*) Light micrograph of *Amoeba proteus* showing blunt pseudopodia (160×). (*b*) Anatomy of *Amoeba*.

## Question 1

***a.*** Can you detect moving cytoplasm in the extending pseudopods of *Amoeba?*

***b.*** What do you suppose the living *Amoeba* is moving toward or away from?

***c.*** How does the *Amoeba* respond to nutrient broth?

***d.*** About how long would it take an *Amoeba* to move across the field of view on low power?

***e.*** Why is a contractile vacuole of a protozoan often more difficult to see than a food vacuole?

***f.*** Why would excess water tend to accumulate in *Amoeba?*

## SLIME MOLDS (SUPERGROUP AMOEBOZOA)

Bizarre plasmodial slime molds such as *Physarum* stream along the damp forest floor in a mass of brightly colored protoplasm called a **plasmodium** in which individual cells are indistinguishable. Plasmodia are coenocytic (multinucleate) because their nuclei are not separated by cell walls, and the entire plasmodium resembles a moving mass of slime. The plasmodium of a slime mold should not be confused with the sporozoan genus with the same name.

Often plasmodia of slime molds live beneath detached bark on decomposing tree trunks. Slime mold plasmodia occasionally occur on lawns or on mulch beneath shrubs. The moving protoplasm of a plasmodium conspicuously pulsates back and forth as it engulfs and digests bacteria, yeasts, and other organic particles.

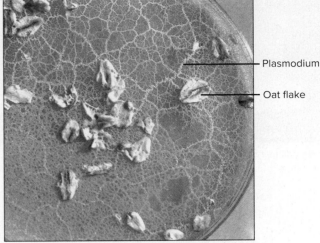

Plasmodium

Oat flake

©BiologyImaging.com

**Figure 26.3** Slime mold (*Physarum*). Slime molds share characteristics with fungi but also with protistans. Their life cycle includes a plasmodial stage, which is a thin, multinucleate mass of streaming protoplasm with amoeboid or flagellated cells.

*c.* What is a possible function of cytoplasmic movement in *Physarum?*

### Procedure 26.2   Examine a culture of *Physarum*

1. Obtain a dissecting microscope and a petri plate containing a culture of *Physarum* growing on oatmeal flakes.

2. Examine the yellowish trails of the colony under the highest magnification of the microscope. Look for cytoplasmic movement.

3. Look for signs of the organism beginning to condense into darkly tipped sporangia. Sporangia are easily visible without a microscope.

The *Physarum* on demonstration is in the vegetative plasmodial stage and feeds on the oatmeal (fig. 26.3). However, if environmental conditions become less than optimal (e.g., if food or moisture decreases), the plasmodium may dry into a hard resistant structure called a **sclerotium** and remain dormant until conditions improve. Or, if light is available, the diploid plasmodium will move to the illuminated area and coalesce. The condensed structure will grow sporangia, and meiosis will produce spores for dispersal (fig. 26.4). Light is associated with an open environment that allows successful reproductive dispersal; dispersal under or within a tree trunk would be ineffective. Haploid spores produced by meiosis in the sporangia germinate as amoeboid or flagellated organisms. These haploid stages may later fuse as gametes and grow into a new plasmodium.

### Question 2
*a.* Is cytoplasmic movement of *Physarum* apparent?

*b.* Is the movement in a particular direction?

## FORAMINIFERANS (SUPERGROUP RHIZARIA)

These marine organisms are called "shelled amoebas" because they surround themselves with a secreted test and have long, thin, rather stiff pseudopods protruding from their tests (fig. 26.5). The test is made of calcium carbonate and is perforated with pores. The marine fossil record is replete with old and well-preserved tests of forams and is often used by oil companies to locate oil-bearing strata.

(a) Science Photo Library/Alamy Stock Photo

Plasmodium, *Physarum*

(b) DP Wildlife Fungi/Alamy Stock Photo

Sporangia, *Hemitrichia*

⊢ 1 mm ⊣

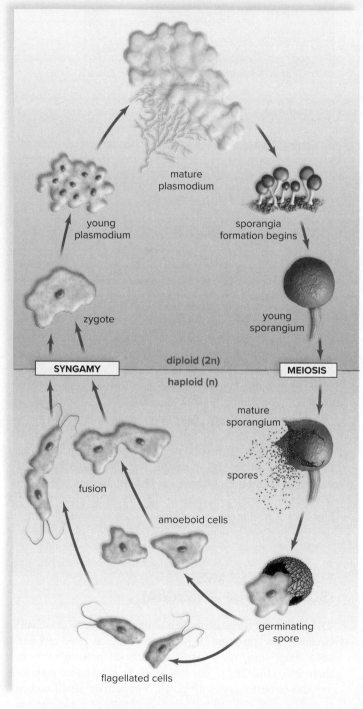

mature plasmodium

young plasmodium

sporangia formation begins

zygote

young sporangium

**SYNGAMY**

diploid (2n)

haploid (n)

**MEIOSIS**

mature sporangium

spores

fusion

amoeboid cells

germinating spore

flagellated cells

### Procedure 26.3 Examine foram tests

1. Obtain a prepared slide of foram tests.
2. Search with low magnification the edges of the cover slips for the foraminiferan tests. They are relatively heavy and shift to the side easily.

### Question 3

How could fossilized forams in different geological layers of rock or sediment indicate the probability of finding oil?

## FLAGELLATES (SUPERGROUP EXCAVATA)

Flagellates have at least one **flagellum** and are likely the most primitive protozoans. Flagellates are parasitic as well as free-living heterotrophs.

### *Trypanosoma*

Among flagellates, trypanosomes are pathogenic and cause African sleeping sickness and Chagas' disease. Charles Darwin may have died from Chagas' disease, for during his later years he suffered from general fatigue, irregular fever, and heart damage. All of these are symptoms of Chagas' disease transmitted by the bite of an assassin bug, which resembles a common stinkbug. Trypanosomes are common in the tropics and spread by infection from biting insects such as mosquitoes, sand flies, and tsetse flies (fig. 26.6*b*).

**Figure 26.4** Life cycle of a plasmodial slime mold. When food or moisture is scarce, a diploid, mature plasmodium stops moving and forms sporangia. Haploid spores form by meiosis. The spores wait until conditions are favorable to germinate. Spores can give rise to flagellated or amoeboid gametes; the two forms convert from one to the other readily. Fusion of the gametes (syngamy) forms the diploid zygote, which gives rise to the mobile, feeding plasmodium by mitosis.

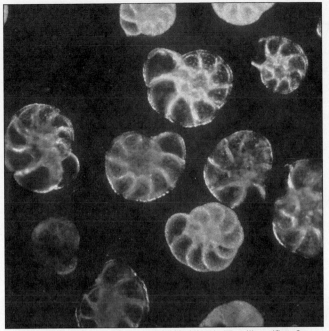

(a) Alfred Pasieka/Photolibrary/Getty Images

(b)                                    8.3 μm
                        Claude Carre/Science Source

**Figure 26.5** Foraminifera. (*a*) The calcereous tests, or shells, of some representative foraminiferans (40×). (*b*) Podia, which are thin cytoplasmic projections, extend through pores in the calcareous test of this living foram (40×).

**Question 4**

*a.* How large is a trypanosome relative to *Amoeba?*

*b.* What alga does a trypanosome superficially resemble?

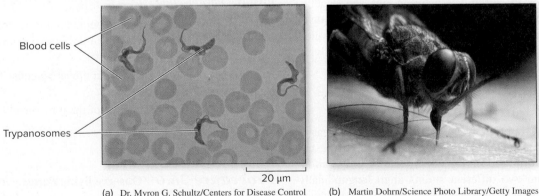

(a)  Dr. Myron G. Schultz/Centers for Disease Control        (b)  Martin Dohrn/Science Photo Library/Getty Images
                    and Prevention

20 μm

**Figure 26.6** A parasite and its vector. (*a*) *Trypanosoma* among red blood cells. The nuclei (dark-staining bodies), anterior flagella, and undulating, changeable shape of the trypanosomes are visible in this photomicrograph (400×). (*b*) The tsetse fly, shown here sucking blood from a human arm, can carry and transmit trypanosomes.

# African Sleeping Sickness

Human African trypanosomiasis, also known as sleeping sickness, is a vector-borne parasitic disease. Parasites of genus *Trypanosoma* are transmitted to humans by the bite of tsetse flies (*Glossina* spp.; fig. 26.6b), which acquire their infection from humans or other infected animals (World Health Organization Fact Sheet No. 259). The most vulnerable people are in rural populations dependent on agriculture, fishing, animal husbandry, or hunting. *Trypanosoma brucei gambiense (T.b.g)* accounts for 95% of reported cases of sleeping sickness (fig. 26.7). After continued control efforts, the number of cases reported in 2009 dropped below 10,000 for the first time in 50 years. This trend continued in 2010 with 7139 new cases reported.

For unknown reasons, many regions of 36 sub-Saharan affected countries have tsetse flies, but no sleeping sickness. A person can be infected for months or even years without developing the major signs and symptoms of the disease, which include fever, headaches, joint pains, and itching. Advanced stages disrupt the sleep cycle and disrupt coordination.

### Figure 26.7

*Trypanosoma brucei.* (*a*) Micrograph of *Trypanosoma brucei*, a causal agent of African sleeping sickness, among red blood cells. (*b*) This drawing shows the general structure of *Trypanosoma brucei*.

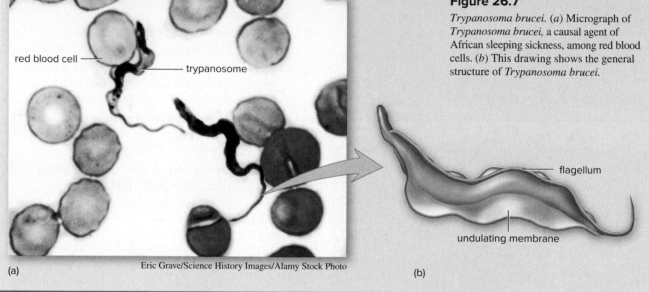

red blood cell — trypanosome

Eric Grave/Science History Images/Alamy Stock Photo

(a)

flagellum

undulating membrane

(b)

## CILIATES (SUPERGROUP CHROMALVEOLATA)

More than 8000 species of ciliates have been described, all having characteristically large numbers of **cilia.** Review in your textbook the difference between cilia and flagella. Many ciliates also have two types of nuclei: **micronuclei** and **macronuclei** (fig. 26.8). Macronuclei, which develop from micronuclei, control cellular function, and they divide when ciliates reproduce asexually by mitosis.

### *Paramecium*

This free-living, freshwater genus is widely studied and easily observed. *Paramecium,* like most ciliates, undergoes a sexual process called **conjugation** (fig. 26.9b). During conjugation, individuals from two different strains align longitudinally and exchange nuclear material (fig. 26.9b). This exchange seems to stimulate metabolism of the individuals and is usually followed by frequent mitosis. Asexual reproduction is more common than conjugation and includes mitosis of the micronucleus and transverse fission (fig. 26.9a) of the macronucleus and cell body.

### *Procedure 26.5*  Examine conjugating and dividing *Paramecium*

1. Obtain a prepared slide of conjugating *Paramecium* and one of transversely dividing (binary fission) *Paramecium.*
2. Notice the alignment of the conjugating cells. Their nuclei are in close proximity.
3. Notice the plane of division of the transversely dividing cells.

### *Procedure 26.6*  Observe living *Paramecium*

1. Prepare a wet mount from a culture of living organisms.
2. Add a drop of methylcellulose to your wet mount to slow the *Paramecium* and make it easier to examine.

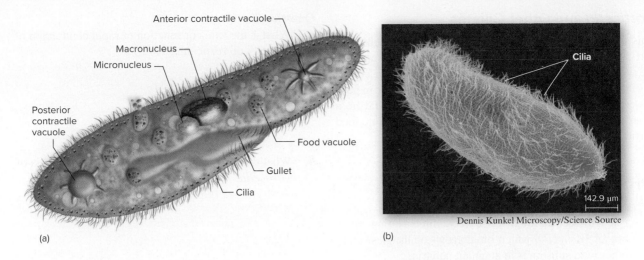
(a)

(b)

Dennis Kunkel Microscopy/Science Source

**Figure 26.8** *Paramecium.* (*a*) The main features of this familiar ciliate include cilia, two nuclei, and numerous specialized organelles. (*b*) Scanning electron micrograph of *Paramecium* showing numerous cilia on the cell surface.

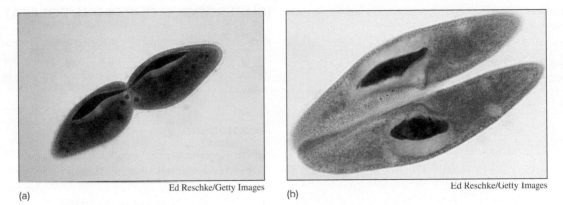
(a)                    Ed Reschke/Getty Images

(b)                    Ed Reschke/Getty Images

**Figure 26.9** Reproduction among paramecia. (*a*) A mature *Paramecium* divides asexually by transverse fission (100×). (*b*) During conjugation, individuals exchange genetic material. Conjugation is a sexual process.

***3.*** Describe aspects of their movement in comparison to *Amoeba.*

***4.*** Identify as many structures as possible in figure 26.8.

**Question 5**

***a.*** Are cilia visible on living or prepared *Paramecium?*

***b.*** Does *Paramecium* rotate as it moves?

***c.*** How does movement of *Paramecium* compare with that of *Amoeba?* With a flagellated alga?

**Question 6**

***a.*** Why is the division of *Paramecium* cells called "transverse" fission?

***b.*** Why is transverse fission not a sexual process?

***c.*** What are the advantages and disadvantages of conjugation in *Paramecium?*

## *Vorticella* and Other Ciliates

This freshwater ciliate is sessile (i.e., attached to a substrate) and has two notable features: (1) a contractile stalk that attaches the organism to the substrate and (2) a cell body with a corona of cilia. To feed, *Vorticella* extends its contractile stalk to push the cell body as far as possible from the substrate and from other individuals. Then it rapidly beats its cilia to capture food particles. This is a type of filter feeding (fig. 26.10).

### Procedure 26.7   Examine *Vorticella* and other ciliates

1. Use a dissecting microscope to examine a living colony of *Vorticella,* which often grows on the glass and other hard substrates in stagnant aquaria.

2. Tap the sides of the dish and observe the contraction of each stalk.

3. When the stalks are extended, notice the whirling action of the cilia at the open end of each bell-shaped individual.

4. Examine a prepared and stained slide of *Vorticella* and draw its general shape.

5. Examine other cultures or prepared slides of ciliates available in the lab. Draw the general shape of these organisms and describe the movement of living specimens.

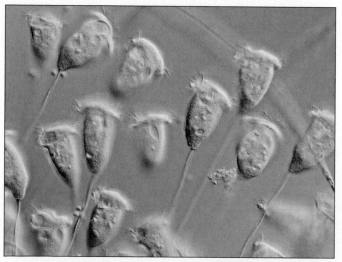

blickwinkel/Alamy Stock Photo

**Figure 26.10**   *Vorticella* are heterotrophic, feed largely on bacteria, and have retractable stalks (80×).

## Question 7

*a.* What is the value or function of rapid contraction of the stalk of *Vorticella?*

*b.* What is the probable function of the moving cilia of *Vorticella?*

## APICOMPLEXANS (SUPERGROUP CHROMALVEOLATA)

Apicomplexans are typically nonmotile parasites of animals. Frequently, these parasites have complex life histories and life stages with various morphologies occurring in multiple hosts.

### *Plasmodium*

This pathogen is the best-known apicomplexan and has been the most common killer of humans in history. *Plasmodium* cause malaria, and mosquitoes of the genus *Anopheles* transmit *Plasmodium* from human to human (fig. 26.11). These malarial parasites infect and rupture red blood cells, causing cycles of fever and chills. Sexual reproductive stages occur in the mosquito host, and asexual stages occur in the human host.

### Procedure 26.8   Examine a blood smear from a victim of malaria

1. Obtain a prepared slide with *Plasmodium.*

2. Locate the infected blood cells. The organisms are inside the infected blood cells, not in the surrounding plasma as are trypanosomes.

3. Locate and compare infected and uninfected blood cells. The infected cells will exhibit only faint signs of the circular ring stage of the parasite. Sketch what you see (fig. 26.12).

4. Review in your textbook the life cycle of *Plasmodium* (fig. 26.11).

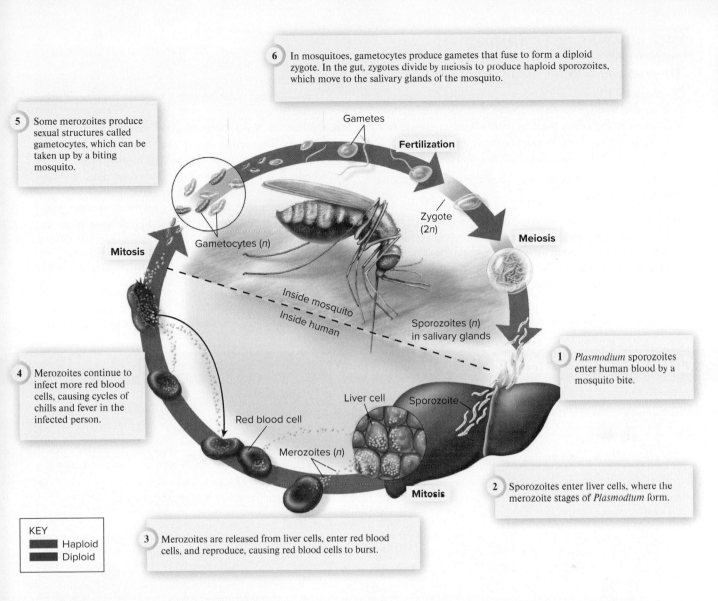

**6** In mosquitoes, gametocytes produce gametes that fuse to form a diploid zygote. In the gut, zygotes divide by meiosis to produce haploid sporozoites, which move to the salivary glands of the mosquito.

**5** Some merozoites produce sexual structures called gametocytes, which can be taken up by a biting mosquito.

Gametes

**Fertilization**

**Mitosis**

Gametocytes (*n*)

Zygote (2*n*)

**Meiosis**

Inside mosquito

Inside human

Sporozoites (*n*) in salivary glands

**4** Merozoites continue to infect more red blood cells, causing cycles of chills and fever in the infected person.

**1** *Plasmodium* sporozoites enter human blood by a mosquito bite.

Liver cell

Sporozoite

Red blood cell

Merozoites (*n*)

**Mitosis**

**2** Sporozoites enter liver cells, where the merozoite stages of *Plasmodium* form.

KEY
Haploid
Diploid

**3** Merozoites are released from liver cells, enter red blood cells, and reproduce, causing red blood cells to burst.

**Figure 26.11** Diagram of the life cycle of *Plasmodium falciparum,* the agent of malaria. This life cycle requires two alternate hosts, humans and *Anopheles* mosquitoes.

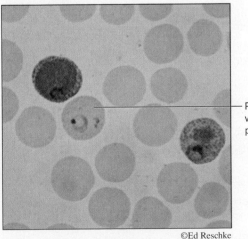

Red blood cell with malarial parasite

©Ed Reschke

**Figure 26.12** Malarial parasite, *Plasmodium*, within a red blood cell (500×).

## INQUIRY-BASED LEARNING

### Are protozoa sensitive to dissolved nutrients?

Observations: Protozoa are particularly sensitive to changes in their aquatic world, including changes in the amount and type of nutrients in the water. Nutrient broth elicits a response from protozoans seeking food. Nutrients include a variety of organic compounds, including fats, carbohydrates, and proteins.

Question: What kinds of nutrients elicit the strongest response from protozoa?

**a.** Establish a working lab group and obtain Inquiry-Based Learning Worksheet 26 from your instructor.

**b.** Discuss with your group ways to measure protozoan response to introduced solutes. Pose a well-defined question relevant to the preceding observation and question. Record your question on Worksheet 26.

**c.** Translate your question into a testable hypothesis and record it.

**d.** Outline on Worksheet 26 your experimental design and supplies needed to test your hypothesis. Ask your instructor to review your proposed investigation.

**e.** Conduct your procedures, record your data, answer your question, and make relevant comments.

**f.** Discuss with your instructor any revisions to your questions, hypotheses, or procedures. Repeat your work as needed.

## Questions for Further Study and Inquiry

*1.* What requirements might make culturing parasitic zoomastigotes difficult in the lab?

*2.* Why do some scientists call conjugation "sexual reproduction" and others do not?

*3.* Is the cell the fundamental unit of life in plasmodial slime molds? Or is the "whole organism," regardless of cellular composition, the fundamental unit? Explain your answer.

*4.* What functions do cilia, flagella, and pseudopods have in common?

*5.* What factors may account for the ubiquitous occurrence and great structural diversity of unicellular organisms?

*6.* In what sense are protists "primitive" and in what sense are they "advanced"?

*7.* Why are unicellular organisms that reproduce by mitosis considered immortal?

# Survey of the Kingdom Fungi
## Molds, Sac Fungi, Mushrooms, and Lichens

## Learning Objectives

By the end of this exercise you should be able to:
1. Describe the characteristic features of the kingdom Fungi.
2. Discuss variation in structures and sequence of events of sexual and asexual reproduction for the major phyla of the kingdom Fungi.

Please visit connect.mheducation.com to review online resources tailored to this lab.

Fungi are as cute as a colorful mushroom and as ugly as the fuzzy slime on a spoiled pork chop. We slice some species to put on our salads, while others would kill us within minutes of eating them. Some species produce antibiotics vital to medicine, while others cause fatal disease. Fungi decompose dead organisms and recycle vital nutrients, while just as quickly they can attack our crops and our refrigerated foods. They do all this, yet they are among the organisms least understood by introductory biology students.

Fungi are basically filamentous strands of cells that secrete enzymes and feed on the organic material on which they are growing. That organic material may be humus in the soil where mushrooms grow or a stale loaf of bread where mold thrives. It may be the skin between your toes inhabited by athlete's foot fungus or a decaying animal on the forest floor being decomposed by fungi digesting the animal's dead tissue. Fungi not only cause disease; they are also important decomposers that recycle nutrients from dead organisms.

The basic structure of a fungus is the **hypha** (pl., *hyphae*)—a slender filament of cytoplasm and nuclei enclosed by a cell wall (fig. 27.1). A mass of these hyphae makes up an individual organism and is collectively called a **mycelium.** A mycelium can permeate soil, water, or living tissue; fungi certainly seem to grow everywhere. In all cases, the hyphae of a fungus secrete enzymes for **extracellular digestion** of the organic substrate. Then the mycelium and its hyphae absorb the digested nutrients. For this reason, fungi are called **absorptive heterotrophs.** Heterotrophs obtain their energy from organic molecules made by other organisms.

Fungi feed on many types of substrates. Most fungi obtain food from dead organic matter and are called **saprophytes.** Other fungi feed on living organisms and are **parasites.** Many of the parasitic fungi have modified hyphae called **haustoria,** which are thin extensions of the hyphae that penetrate living cells and absorb nutrients.

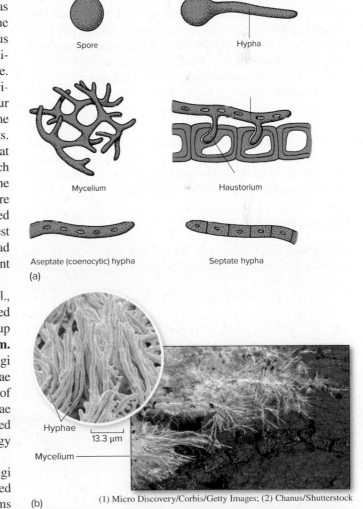

(a)

Aseptate (coenocytic) hypha    Septate hypha

Spore    Hypha    Mycelium    Haustorium

Hyphae
13.3 μm
Mycelium

(b)

(1) Micro Discovery/Corbis/Getty Images; (2) Chanus/Shutterstock

**Figure 27.1** Fungal mycelium. (*a*) Fundamental elements of fungal structure. (*b*) This mycelium, composed of hyphae, is growing through leaves on a forest floor in Maryland.

Hyphae of some species of fungi have crosswalls called **septa** that separate cytoplasm and nuclei into cells. Hyphae of other species have incomplete or no septa (i.e., are aseptate) and therefore are **coenocytic** (multinucleate). Notably, the cell walls of fungi are usually not cellulose, but instead are made of **chitin,** the same polysaccharides that comprise the exoskeleton of insects and crustaceans.

## REPRODUCTION IN KINGDOM FUNGI

Reproduction in fungi can be sexual or asexual. Subtle variation in the patterns and morphology of sexual reproduction distinguish one phylum from another.

### Asexual Reproduction

Asexual reproduction is usually fast and efficient. A single organism can do it. However, it does not include genetic recombination, because meiosis and fertilization are not part of asexual reproduction. Fungi commonly reproduce asexually by mitotic production of haploid vegetative cells called **spores** produced in **sporangia, conidiophores,** and other related structures. Spores are microscopic and surrounded by a covering well suited for the rigors of distribution into the environment. *Pilobolus,* an interesting fungus, points its sporangia toward the sun. This orientation of an organism to light is called **phototaxis.** *Pilobolus* ejects its entire sporangium as far as two meters to distribute its spores.

**Budding** and **fragmentation** are two other methods of asexual reproduction. Budding, which is mitosis followed by an uneven distribution of cytoplasm, is common in yeasts. After budding, an outgrowth of the original cell detaches and matures into a new organism. Fragmentation is the breaking of an organism into one or more pieces, each of which can develop into a new individual.

### Sexual Reproduction

The sexual life history of fungi and other eukaryotes includes the familiar events of vegetative growth, genetic recombination, meiosis, and nuclear fusion (karyogamy). However, the timing of these events is unique in fungi. It's unique because the mature vegetative form of most fungi is haploid. Fungi reproduce sexually when hyphae of two genetically different individuals of the same species encounter each other. In many fungal life cycles, haploid cells ($n$) from two mating strains will fuse their cytoplasms, **plasmogamy,** and become dikaryotic ($n + n$) with two nuclei per cell. Later the nuclei will fuse, via a process called **karyogamy,** to become diploid ($2n$) zygotes. Karyogamy is equivalent to fertilization.

A generalized life cycle of fungi (fig. 27.2) illustrates four important features of fungi:

- Nuclei of a fungal mycelium are haploid during most of the life cycle.

- Gametes are produced by mitosis and differentiation of haploid cells rather than directly from meiosis of diploid cells.

- Meiosis quickly follows formation of the zygote, the only diploid stage.

- Haploid cells produced by meiosis are not gametes; rather, they are spores that grow into a mature haploid organism. Recall that asexual reproduction produces spores by mitosis. In both cases, haploid spores grow into mature mycelia.

None of these features of the sexual cycle are unique to fungi, but together they describe the typical fungal life cycle.

### Classification of Fungi

The great diversity of fungi results from modifications of hyphae into varied and specialized reproductive structures often unique to a phylum, genus, or species. The five most diverse phyla of fungi are **Chytridiomycota, Zygomycota, Ascomycota, Basidiomycota,** and **Glomeromycota.**

As you examine members of these phyla, carefully note variations on the fundamental structure of vegetative mycelia

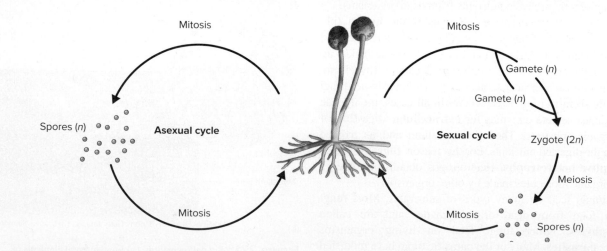

**Figure 27.2** Generalized life cycle of a fungus.

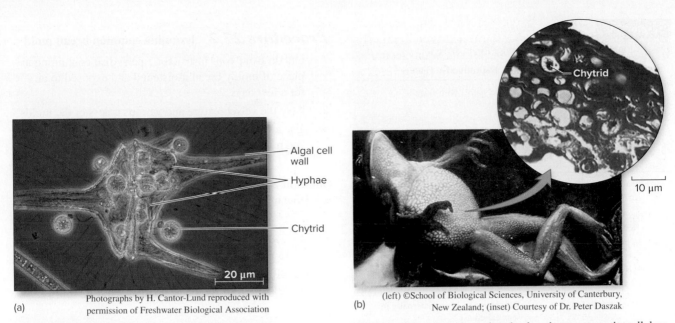

(a)

Photographs by H. Cantor-Lund reproduced with
permission of Freshwater Biological Association

(b)

(left) ©School of Biological Sciences, University of Canterbury,
New Zealand; (inset) Courtesy of Dr. Peter Daszak

**Figure 27.3** Chytridiomycota. (*a*) Chytrids growing on a freshwater alga. The colorless chytrids produce hyphae that penetrate the cellulose cell walls of the dinoflagellate *Ceratium hirundinella,* absorbing organic materials from the alga. Chytrids use these materials to produce spherical flagellate spores that swim away to attack other algal cells. (*b*) Frog killed by chytridiomycosis. Lesions formed by the chytrid can be seen on the abdomen of this frog.

and specialized structures associated with sexual and asexual reproduction. The names of phyla of fungi are derived from sexual reproductive structures rather than asexual structures (table 27.1). However, each phylum has various modifications for both sexual and asexual reproduction. Our research and understanding of less diverse phyla of fungi (that are not covered in this exercise) continue to develop.

## PHYLUM CHYTRIDIOMYCOTA (CHYTRIDS)

Molecular genetic evidence indicates that chytrids may be the most ancient fungi. They are typically aquatic saprobes or parasites on plants, animals, and protists (fig. 27.3). They have flagella, which is unusual for fungi. They also have absorptive nutrition and chitinous cell walls and share proteins and nucleic acids common to other fungi. Their distinctive reproductive feature is motile spores with flagella. In some areas, infections by chytrids have significantly reduced the population of many amphibians (fig. 27.3*b*).

 **SAFETY FIRST** Before coming to lab, you were asked to read this exercise so you would know what to do and be aware of safety issues. In the space below, briefly list the safety issues associated with today's procedures. If you have questions about these issues, contact your laboratory assistant before starting work.

*Procedure 27.1* **Examine a representative of phylum Chytridiomycota**

*1.* Obtain a culture of *Allomyces* or *Chytridium,* two common genera of chytrids.

*2.* Prepare a wet mount and use your microscope to examine chytrid morphology. If necessary, review in Exercise 3 the proper use of a compound microscope.

*3.* Be prepared to compare chytrid morphology with the morphology common to the other major fungal groups.

**Question 1**

*a.* Are hyphae apparent?

*b.* Are the cells motile?

Table 27.1

**Reproductive Feature of the Major Phyla of Fungi**

| Phylum | Key Reproductive Feature |
|---|---|
| Chytridiomycota (chytrids) | Motile spores with flagella |
| Zygomycota (zygote fungi) | Resistant zygosporangium as sexual stage |
| Ascomycota (sac fungi) | Sexual spores borne internally in sacs called asci |
| Basidiomycota (club fungi) | Sexual spores borne externally on club-shaped structures called basidia |
| Glomeromycota (plant symbionts) | Reproduction typically asexual. Aseptate hyphae. |

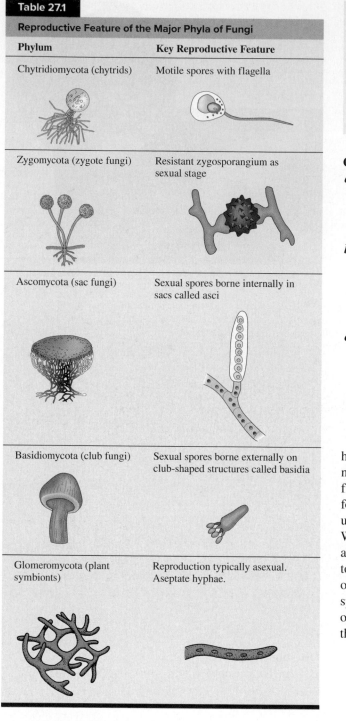

## PHYLUM ZYGOMYCOTA (BREAD MOLDS)

Zygomycetes (750 species), which include common bread molds, derive their name from resting sexual structures called **zygosporangia** that characterize the group. Most zygomycetes are saprophytic and their vegetative hyphae lack septa (i.e., they are aseptate).

## Procedure 27.2   Examine common bread mold

1. Obtain from your instructor a petri dish containing a piece of moldy bread moistened and exposed to air for a few days.
2. Note the velvet texture and various colors of the molds.
3. Use a dissecting microscope to examine the mycelia and notice that they grow as a tangled mass of hyphae.

### Question 2

*a.* How many species of mold are on the bread?

*b.* Is pigment distributed uniformly in each mycelium? If not, where is the pigment concentrated in each mold?

*c.* What is the adaptive significance of spores forming on ends of upright filaments rather than closer to the protective substrate?

A common genus of bread mold is *Rhizopus*. Its hyphae are modified into **rhizoids** (holdfasts), **stolons** (connecting hyphae), **sporangiophores** (supportive stem-like filaments), and **sporangia** (sites of asexual haploid spore formation), as shown in figure 27.4. Sporangiophores are upright hyphal filaments supporting sporangia (fig. 27.5). Within a sporangium, haploid nuclei become asexual spores and are separated by cell walls. These spores are released to the environment when the mature sporangium breaks open. *Rhizopus* appears dark from the thousands of black sporangia of a growing mycelium; hence, a common name of *Rhizopus* is black bread mold. This asexual formation of thousands of haploid spores is an efficient way to reproduce.

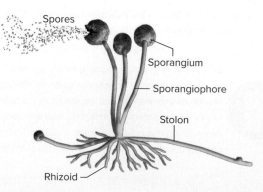

**Figure 27.4**   Vegetative and asexual reproductive structures of *Rhizopus,* a bread mold.

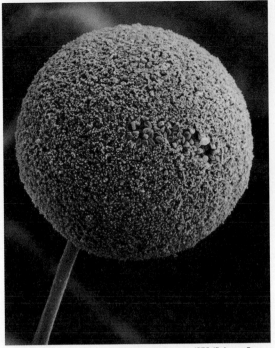

Dr. Jeremy Burgess/SPL/Science Source

**Figure 27.5** A sporangium of *Rhizopus stolonifer* (500×) releases thousands of spores when it matures and breaks open.

Sexual reproduction is a more involved interaction requiring contact between two genetically different strains, i.e. a + strain and a – strain). Because *Rhizopus* is isogamous (meaning that gametes from both strains look alike), the strains are not called male and female, but instead are called + and –. In procedure 27.3 you will observe contact between + and – strains. On one side of the plate a + strain of *Rhizopus* was introduced, and on the other side a – strain was introduced. These strains grew toward each other and formed reproductive structures called **gametangia** where they came into contact.

## Sexual Reproduction in *Rhizopus*

1. Sexual reproduction begins when hyphae of each strain touch each other (fig. 27.6).

2. Where the hyphae touch, **gametangia** form and appear as swellings. Within the gametangia many nuclei differentiate to serve as gametes from each strain.

3. The wall between the gametangia breaks down, and cells fuse (plasmogamy). Fusion of nuclei (karyogamy) follows and occurs in the zygosporangium.

4. A typically massive and elaborate **zygosporangium** differentiates around the zygotes. Except for the zygotes, all other nuclei are haploid.

5. Soon after the zygosporangium forms, the zygotes undergo meiosis. One or more of the resulting haploid spores soon germinate.

6. The hyphae of the germinating cells break out of the zygosporangium, produce a sporangiophore via mitosis, and form spores asexually.

7. The spores are released and produce a new generation of mycelia.

## Procedure 27.3 Examine *Rhizopus*

1. Obtain a culture of asexually reproducing *Rhizopus* from your instructor. This culture is growing in a sealed petri plate containing nutrient-fortified agar. Do not remove the top of the dish because you may contaminate the culture and unnecessarily release spores into the room.

2. Your instructor has prepared a demonstration slide of *Rhizopus* stained with lactophenol cotton blue for you to examine. Examine this slide and sketch what you see.

3. Your instructor has also prepared some cultures of *Rhizopus* (or *Mucor*, or *Phycomyces*) for observing the structures for sexual reproduction. Obtain one of these plates. Locate the midline where the different strains have come in contact.

4. Locate sexual gametangia and zygosporangia along the midline of this living culture. In addition, dark, asexual sporangia will also be apparent throughout the culture.

5. Examine a prepared slide of *Rhizopus* with developed zygosporangia (figs. 27.6 and 27.7). Then observe the zygosporangia where the strains have touched. Be careful not to confuse zygosporangia with dark sporangia on top of sporangiophores.

## Question 3

*a.* In what structure is the dark pigment of *Rhizopus* concentrated?

*b.* Is *Rhizopus* reproducing sexually as well as asexually in the same petri dish? How can you tell?

(a)

667 µm

Carolina Biological Supply Company/Phototake/Medical Images

Zygosporangium

Richard H. Gross/Biological Photography

**Figure 27.7** Sexual conjugation in *Rhizopus* (200×).

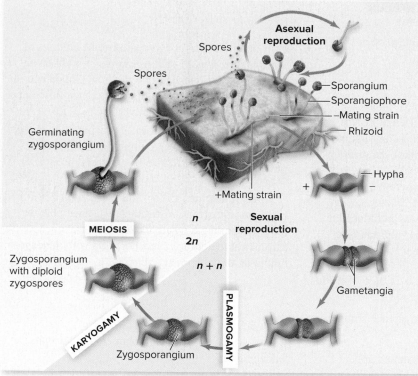

(b)

**Figure 27.6** *Rhizopus* (*a*) and its life cycle (*b*). Hyphae grow and feed on the surface of the bread or other material and produce clumps of erect, sporangium-bearing stalks. If both + and − strains are present in a colony, they may grow together, and their nuclei may fuse to form diploid (2*n*) zygotes. These zygotes, which are the only diploid cells of the life cycle, form within a thick, dark structure called a zygosporangium. Meiosis occurs in the zygosporangium, and vegetative, haploid hyphae grow from the resulting haploid (*n*) cells.

# PHYLUM ASCOMYCOTA (SAC FUNGI)

Phylum Ascomycota (30,000 species) includes yeasts, some molds, morels, and truffles (fig. 27.8). Its name is derived from a microscopic, sac-shaped, sexual reproductive structure called an **ascus.**

Ascomycetes reproduce asexually by forming spores called **conidia.** Modified hyphae called **conidiophores** partition nuclei in longitudinal chains of beadlike conidia (fig. 27.9). Each conidium contains one or more nuclei.

(a)                    ©BiologyImaging.com

(b)        ©BiologyImaging.com

**Figure 27.8**    Representatives of Ascomycota. All visible structures of fleshy fungi, such as the ones shown here, arise from an extensive network of filaments (hyphae) that penetrate and interweave with the substrate on which they grow. (*a*) *Helvella esculenta* is a poisonous ascomycete known as "false morel." Many people have died after mistaking this mushroom for an edible species. (*b*) A cup fungus in the rain forest of the Amazon Basin.

Conidia form on the surface of conidiophores (in contrast to spores that form within sporangia in *Rhizopus*). When mature, conidia are released in large numbers and germinate to produce new organisms. *Aspergillus* and *Penicillium* are common examples of fungi that form conidia. For some ascomycetes, sexual reproduction has never been observed. However, most are likely ascomycetes that have lost their ability to sexually reproduce.

Many ascomycetes are economically important. For example, species of *Penicillium* are used to produce antibiotics. *Penicillium roquefortii*, which is abundant in caves near Roquefort-sur-Soulzon, France, gives a unique flavor to Roquefort cheese. *Aspergillus oryzae* is used to brew Japanese saki and to enrich food for livestock.

### Procedure 27.4    Examine fungi with conidia

1. Obtain a culture plate of living *Aspergillus*, *Penicillium*, or *Neurospora*. Notice the soft texture of the colonies.
2. Use a dissecting microscope to examine the colonies' hyphae and their reproductive conidia. Note the rounded tufts of these reproductive cells.
3. Conidia are quite small. You will examine them more closely in procedure 27.5.

### Procedure 27.5    Examine *Penicillium*

1. Examine a prepared slide of *Penicillium* and a pure, living culture provided by your instructor (fig. 27.10).
2. Notice the formation of conidia.

### Question 4
What is the relative size of *Penicillium* hyphae compared with *Rhizopus* hyphae?

Yeasts are common unicellular ascomycetes and include about 40 genera. Most of their reproduction is asexual by cell fission or budding (i.e., the formation of a smaller cell from a larger one). However, sexual reproduction can occur. Occasionally, two sexually reproducing yeast cells will fuse to form one cell with two nuclei, a process

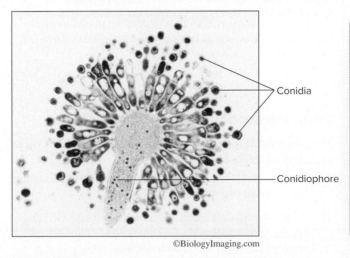

Conidia

Conidiophore

©BiologyImaging.com

**Figure 27.9**    Thin section through a conidiophore of *Aspergillis* sp.

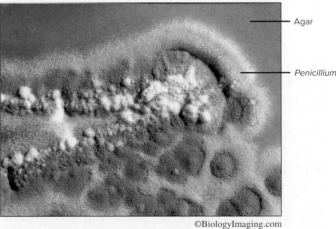

Agar

*Penicillium*

©BiologyImaging.com

**Figure 27.10**    *Penicillium*, an ascomycete, growing on agar.

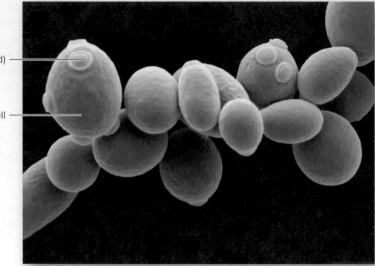

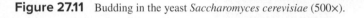
Daughter cell (bud)

Mother cell

Mediscan/Alamy Stock Photo

**Figure 27.11**   Budding in the yeast *Saccharomyces cerevisiae* (500×).

called **plasmogamy.** This cell functions as an ascus in which fertilization is followed immediately by meiosis. The resulting haploid ascospores function directly as new yeast cells. Yeasts do not form conidia.

The yeast used to produce wine and beer is usually a strain of *Saccharomyces cerevisiae*. The yeasts and other fungi growing naturally on grapes used for making wine may impart a unique flavor to a wine more than does the specific variety of grapes.

### *Procedure 27.6*   Examine *Saccharomyces*, a yeast, and *Peziza*, a cup fungus

1. Obtain and examine a stock culture of *Saccharomyces* (fig. 27.11).

2. Prepare a wet mount of the yeast from a culture dispensed by your instructor. Only a small amount of yeast is needed to make a good slide.

3. Review the description of sexual reproduction in cup fungi.

4. Examine a prepared slide of a cross section through the ascocarp of *Peziza* (fig. 27.12). Locate the asci.

**Question 5**

*a.* Do you see chains of yeast cells produced by budding?

*b.* How is the structure of yeast hyphae different from that of molds?

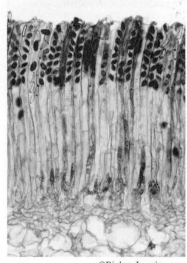

©BiologyImaging.com

**Figure 27.12**   Asci from the lining of a cup of *Peziza*, a cup fungus. The diploid ($2n$) zygote in each ascus divides meiotically to produce four haploid ($n$) nuclei. Each of these nuclei then divides mitotically. These meiotic and mitotic divisions produce a column of eight ascospores in each ascus (400×).

### Sexual Reproduction in Ascomycota

Most ascomycetes such as *Peziza* follow a sexual life cycle including ascus production.

1. Sexual reproduction begins with contact of monokaryotic hyphae from two mating strains (fig. 27.13).

2. Where the hyphae touch, large multinucleate swellings appear (antheridia and ascogonia) and eventually fuse. Haploid nuclei of the two strains intermingle in the swelling (ascogonium).

©BiologyImaging.com

(b)

(a) Robert Marien/Corbis/Getty Images

**Figure 27.13** The sexual life cycle of an ascomycete. Representative ascomycetes include (*a*) morels and (*b*) cup fungi, both of which undergo (*c*) a sexual life cycle including formation of characteristic asci.

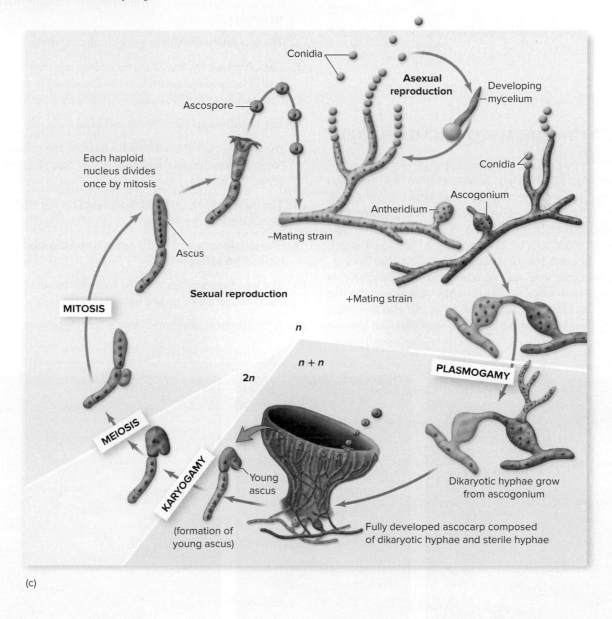

(c)

3. A **dikaryotic** mycelium grows from this swelling. Each dikaryotic cell has one nucleus from each parent; the nuclei do not fuse immediately.

4. Tightly bundled dikaryotic hyphae grow and mingle with **monokaryotic** hyphae from each parent to form a cup-shaped **ascocarp.**

5. Dikaryotic cells lining the inside of the ascocarp form sac-shaped **asci** (sing., ascus).

6. The nuclei fuse (karyogamy) in each ascus to form a zygote.

7. After fusion, meiosis produces four haploid **ascospores.**

8. Subsequent mitosis produces eight ascospores within each mature ascus.

9. The asci on the surface of the ascocarp rupture and release ascospores into the environment.

10. Each ascospore can produce a new mycelium.

**Question 6**
What is the difference between dikaryotic and diploid cells?

## PHYLUM BASIDIOMYCOTA (CLUB FUNGI)

Basidiomycetes (25,000 species) are probably the most familiar fungi (fig. 27.14). They include mushrooms, puffballs, shelf fungi, and economically important plant pathogens such as rusts and smuts. *Agaricus campestris* is a common field mushroom, and its close relative *A. bisporus* is cultivated for more than 60,000 tons of food per year in the United States. However, just one bite of *Amanita phaloides,* the "destroying angel" mushroom, may be fatal. Phylum Basidiomycota derives its name from the characteristic sexual reproductive structure, the basidium. Some club fungi can also reproduce asexually by conidia formation, as do sac fungi. But here we will describe only sexual reproduction involving their familiar fruiting body, a mushroom.

### Sexual Reproduction in Basidiomycota

1. Haploid hyphae from different mating strains permeate the substrate (fig. 27.15).

2. Septa form between the nuclei in the hyphae and form **monokaryotic primary mycelia.** A monokaryotic mycelium has one nucleus in each cell.

3. Cells of the primary mycelia of different mating strains touch, fuse (plasmogamy), and produce a **dikaryotic secondary mycelium.**

4. The secondary mycelium grows in the substrate.

5. The dikaryotic hyphae of the secondary mycelium eventually coalesce and protrude above the substrate as a tight bundle of hyphae called a basidiocarp (mushroom).

6. The **basidiocarp** forms a cap and gills.

7. Dikaryotic, club-shaped **basidia** form on the surface of the gills. Basidia are the sites of sexual reproduction, especially meiosis.

8. The two nuclei in each basidium fuse (karyogamy) to form a diploid zygote.

9. The zygote soon undergoes meiosis and produces haploid **basidiospores.**

10. The basidiospores are released from the basidia lining the gills and are dispersed by wind.

11. A basidiospore germinates into a new mycelium.

(a) ©BiologyImaging.com  (b) siloto/Shutterstock  (c) ©BiologyImaging.com

**Figure 27.14**  Representative basidiomycetes. (*a*) Fly amanita (*Amanita muscaria*). Many species of *Amanita* are poisonous. (*b*) A common stinkhorn fungus (*Phallus impudicus*). (*c*) Earthstar (*Geaster*).

(a)

©BiologyImaging.com

**Figure 27.15** Mushrooms (*a*) and their life cycle (*b*). Coalescing mycelia from compatible strains produce dikaryotic hyphae, which coalesce and form the familiar fruiting bodies of mushrooms. On the gills of the mushroom cap the nuclei in dikaryotic cells fuse, undergo meiosis, and produce basidiospores that germinate into new mycelia. The basidium is the site of karyogamy, which is the fusion of nuclei to form a zygote.

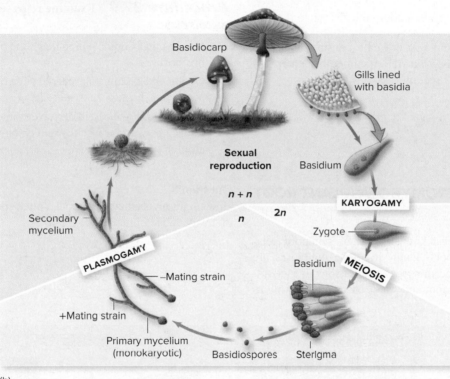

(b)

---

### Procedure 27.7 Examine some common mushrooms and their relatives

1. Examine a specimen of an earthstar (*Geaster*). Earthstars are oddly structured basidiomycetes with an array of support structures shaped much like a star (fig. 27.14*c*).

2. Examine some mushrooms. Mushrooms are familiar examples of the aboveground portions of extensive mycelia permeating the soil. Note the mushroom's stalk and umbrellalike **cap** (also called the **pileus**).

3. Find the **gills** on the undersurface of the cap. Gills are lined with microscopic, club-shaped cells called **basidia** where sexual reproduction occurs. Phylum

Basidiomycota is sometimes called the "club fungi" and derives its name from these characteristic basidia.

### Procedure 27.8 Examine *Coprinus,* a common mushroom

1. Examine a prepared slide of gills from the cap of *Coprinus,* a common mushroom (fig. 27.16).

2. Note the dark basidiospores in rows along the surface of the gills. Interestingly, the gills form perpendicular to the ground and allow spores to free-fall and disperse. Research suggests that gravity influences the orientation of gills.

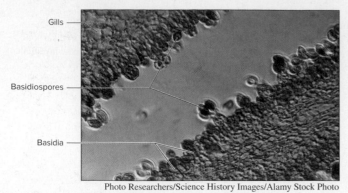

Gills

Basidiospores

Basidia

Photo Researchers/Science History Images/Alamy Stock Photo

**Figure 27.16** Gills of the mushroom *Coprinus,* a basidiomycete (400×).

## Question 7

How many spores would you estimate are present on the gills of a single cap of *Coprinus?* Remember that a prepared slide shows only a cross section.

## PHYLUM GLOMEROMYCOTA (PLANT ROOT SYMBIONTS)

Glomeromycetes are best known for their symbiotic relationship with plant roots. Fungi in these symbioses are called **mycorrhizae,** and they act as extensions of root systems to greatly aid nutrient and water absorption as well as maintain beneficial soil consistency. Only in recent years have we realized their (1) immense biomass, (2) role in the global carbon budgets, and (3) value in facilitating nitrogen and nutrients absorption for at least 90% of all plant species. These plant species provide their symbionts with as much as 20% of the plant's fixed carbon as carbohydrates in return for the benefits of mycorrhizae.

**Arbuscular** mycorrhizae penetrate plant root cell walls and closely contact root cells' plasma membranes. All arbuscular mycorrhizae are glomeromycetes (fig. 27.17). **Ectomycorrhizae** are typically basidiomycetes and extend between host root cells but don't penetrate the cell wall.

### *Procedure 27.9*  **Examine representatives of mycorrhizae.**

1. Obtain and examine a prepared slide of arbuscular mycorrhizae.

2. Obtain and examine a prepared slide of ectomycorrhizae.

3. Try to find in your field of view examples of the difference between arbuscular and ectomycorrhizal penetration of the root tissue.

## Question 8

How might mycorrhizae promote a plant species occupation of a wide range of habitats?

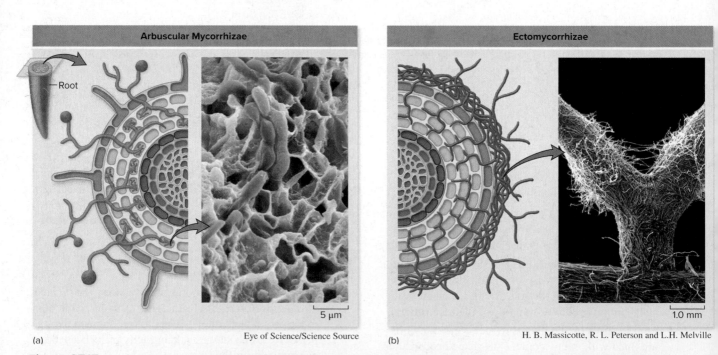

Arbuscular Mycorrhizae

Root

5 μm

Eye of Science/Science Source

(a)

Ectomycorrhizae

1.0 mm

H. B. Massicotte, R. L. Peterson and L.H. Melville

(b)

**Figure 27.17** Mycorrhizae. (*a*) In arbuscular mycorrhizae, fungal hyphae closely contact the cell's plasma membrane. (*b*) Ectomycorrhizae shown here on a tree root grow around and extend between the root cells but never penetrate the cell wall.

## Question 9

To understand a forest's carbon budget, should we consider rising heat's effect on plant physiology? The effect of excess heat on fungal growth? The effect on a forest's photosynthetic rate? What might happen?

## LICHENS

Lichens (25,000 species) are common, brightly colored organisms found on most hard substrates from the tropics to the artic. Trees, rocks, and firm soil provide all the support these slow-growing organisms need. A lichen includes an ascomycete (rarely other fungi) living symbiotically with a photosynthetic alga (a protist) or cyanobacterium. **Symbiosis** means living in a close and sometimes dependent association. About 26 genera of algae occur in different species of lichens. However, the green algae *Trebouxia* and *Trentepohlia* and the cyanobacterium *Nostoc* are in 90% of lichen species.

Lichens reproduce asexually by releasing fragments of tissue or specialized, stress-resistant packets of fungal and algal cells. Each of the two components (fungus and alga) may reproduce sexually by mechanisms characteristic of their phylum, and the new organisms may continue the lichen association.

The durable construction of fungi, linked with photosynthetic algae, enables lichens to proliferate in the harshest terrestrial habitats. Lichens have three basic growth forms: **crustose, foliose,** and **fruticose.** The thallus of crustose lichens grows close to the surface of a hard substrate such as rock or bark. The lichen is flat and two-dimensional. Foliose lichens adhere to their substrate, but some of the thallus peels and folds away from the substrate in small sheets.

Fruticose lichens are three-dimensional and often grow away from the substrate with erect stalks. The tips of the stalks are often sites of ascus formation by the sexually reproducing ascomycete symbiont.

Lichens are extremely sensitive to air pollution. This is probably because they are adapted to efficiently absorb nutrients and minerals from the air. This makes lichens particularly susceptible to airborne toxins.

### Procedure 27.10 Examine lichens

1. Examine a prepared slide of a cross section of a lichen thallus and note the intimate contact between the symbionts (fig. 27.18).
2. Examine the dried lichens on display and note the three basic growth forms: crustose, fruticose, and foliose (fig. 27.18).

## Question 10

*a.* What are the advantages of having an alga and a fungus in a lichen? What could each organism contribute to the partnership?

*b.* Would you expect lichens to grow best in rural or urban environments? Why?

## INQUIRY-BASED LEARNING

### *How effective are antimicrobial properties of common fungi?*

Observation: Many fungi produce antibiotics that hinder or stop the growth of microbes. These antibiotics are adaptive because they reduce competition and protect the fungus from predators.

Question: Do common fungi such as bread mold produce antimicrobial compounds?

a. Establish a working lab group and obtain Inquiry-Based Learning Worksheet 27 from your instructor.
b. Discuss with your group well-defined questions relevant to the preceding observation and question. Choose and record your group's best question for investigation.

c. Translate your question into a testable hypothesis and record it.
d. Outline on Worksheet 27 your experimental design and supplies needed to test your hypothesis. Ask your instructor to review your proposed investigation.
e. Conduct your procedures, record your data, answer your question, and make relevant comments.
f. Discuss with your instructor any revisions to your questions, hypotheses, or procedures. Repeat your work as needed.

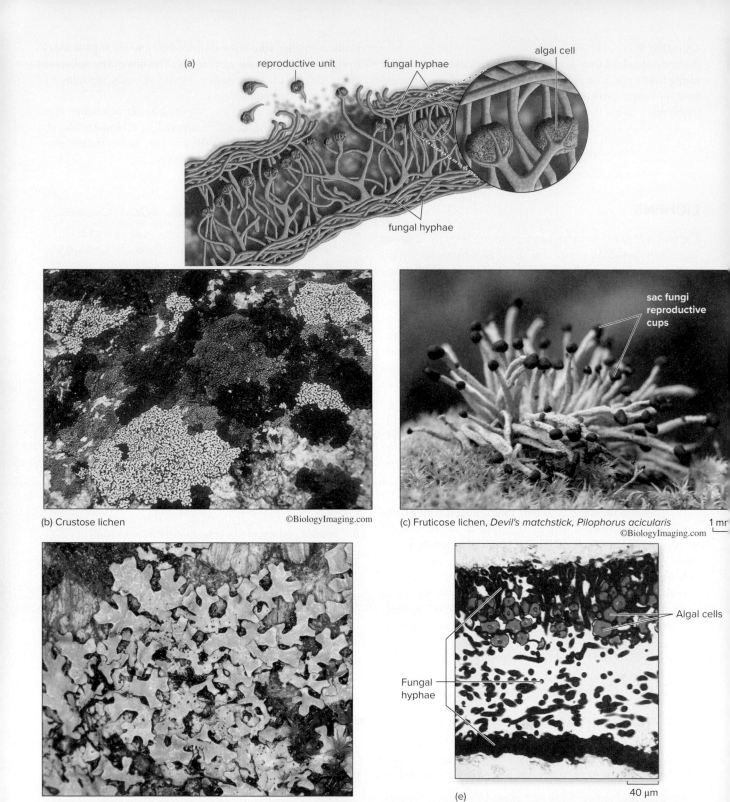

(a) reproductive unit    fungal hyphae    algal cell

fungal hyphae

(b) Crustose lichen    ©BiologyImaging.com

(c) Fruticose lichen, *Devil's matchstick, Pilophorus acicularis*    1 mr

sac fungi reproductive cups

©BiologyImaging.com

(d) Foliose lichen    ©BiologyImaging.com

Fungal hyphae

Algal cells

(e)    40 µm

Ed Reschke

**Figure 27.18** Lichen morphology. (*a*) A section of a compact crustose lichen shows the placement of the algal cells and the fungal hyphae, which encircle and penetrate the alga. (*b*) Thin sheets of crustose lichens adhere tightly to hard surfaces. (*c*) Fruticose lichens are shrublike. (*d*) Foliose lichens are leaflike. (*e*) This cross section of a lichen thallus shows the distribution of symbiotic algal cells and fungal hyphae. Algal cells are most dense at the upper surface to efficiently collect light, conduct photosynthesis, and provide carbohydrates to the hyphae.

1. Mushrooms often sprout from soil in rows or circles commonly called "fairy rings." How would you explain the shapes of these formations?

2. What advantages does asexual reproduction have over sexual reproduction?

3. Does dominance of the haploid condition in a fungal life cycle offer an adaptive advantage? Why or why not?

4. Compare and contrast the structure of a fungal mycelium with the structure of a filamentous alga.

5. What is the advantage of maintaining a dikaryotic condition rather than immediate nuclear fusion?

6. In fungi, the only distinction between a spore and a gamete is function. Explain.

7. Describe three ways that fungi affect your life.

8. What are the major differences in the five phyla of fungi?

9. Draw and label a life cycle for one fungal phylum.

10. What products or activities by fungi benefit other organisms? What products or activities of fungi are harmful to other organisms?

 **DOING BIOLOGY YOURSELF**
Wine-making is a multimillion-dollar industry, but the biology of wine-making is simple. Make your own special brand of wine by following the instructions in Exercise 12. Good luck!

**WRITING TO LEARN BIOLOGY**
Compare and contrast the fundamental life cycle of a fungus with that of plants and animals. When does meiosis occur in the sequence of events? Which stages are haploid and which are diploid?

# Survey of the Plant Kingdom
## Liverworts, Mosses, and Hornworts of Phyla Hepatophyta, Bryophyta, and Anthocerophyta

## Learning Objectives

By the end of this exercise you should be able to:
1.   Describe the life histories and related reproductive structures of bryophytes.
2.   Describe the distinguishing features of liverworts, mosses, and hornworts.
3.   Describe some of the key adaptations that allow liverworts, mosses, and hornworts to live on land.
4.   Describe the role of bryophytes in the environment.

Please visit **connect.mheducation.com** to review online resources tailored to this lab.

**P**lants evolved from a freshwater green algal species more than 550 million years ago and today comprise a remarkably diverse group of multicellular organisms. With few exceptions, plants are autotrophic, contain chlorophyll *a*, and have cell walls containing cellulose. Life cycles of all members of the plant kingdom are variations on alternation of generations. Be sure to review in your textbook this generalized life cycle and be familiar with the major stages presented in figure 28.1. In this and upcoming exercises, you'll learn how various modifications of this generalized life cycle characterize all of the phyla that comprise the plant kingdom.

You will study the following major groups of plants:

| Phylum | Example |
|--------|---------|
| Hepatophyta | liverworts |
| Bryophyta | mosses |
| Anthocerophyta | hornworts |
| Pterophyta | ferns |
| Lycophyta | club mosses |
| Cycadophyta | cycads |
| Ginkgophyta | *Ginkgo* |
| Coniferophyta | conifers |
| Gnetophyta | gnetophytes |
| Anthophyta | flowering plants |

These groups of plants are distinguished by morphology, life cycle, and the presence or absence of vascular tissues (fig. 28.2). Pay special attention to variations in each of these characteristics as you survey the plant kingdom in upcoming weeks.

Bryophytes include liverworts, mosses, and hornworts and are the most primitive group of terrestrial plants.

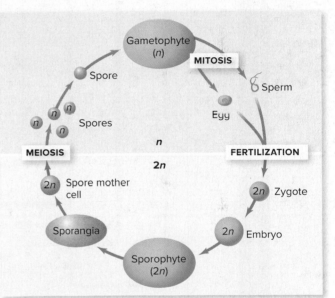

**Figure 28.1** Alternation of generations. In a generalized plant life cycle, gametophytes, which are haploid (*n*), alternate with sporophytes, which are diploid (*2n*). Antheridia (male) and archegonia (female) are the sex organs produced by the gametophyte; they produce sperm and eggs, respectively. The sperm and egg fuse during fertilization to produce the first diploid cell of the sporophyte generation, the zygote. Meiosis occurs within sporangia, which are the spore-producing organs of the sporophyte. The resulting spores are haploid and are the first cells of the gametophyte generation.

Bryophytes are green, have rootlike structures called **rhizoids,** and may have stem and leaflike parts. Bryophytes do not generally possess specialized vascular tissues, which transport materials between roots and shoots. This lack of developed vascular tissues in bryophytes typically limits their distribution to moist habitats because their rhizoids neither penetrate the soil very far nor absorb many nutrients. Also, the lack

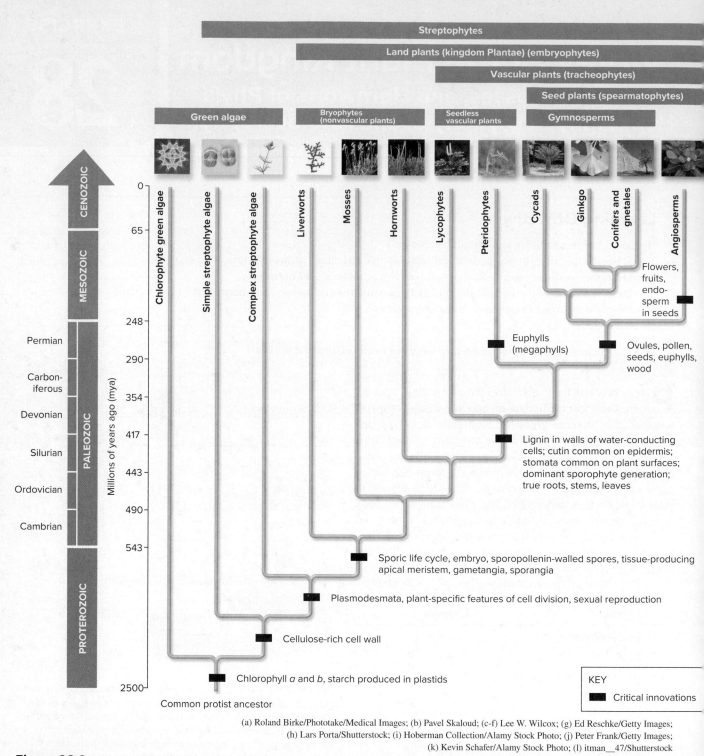

**Figure 28.2** The evolutionary history of plants. The evolution of plants has produced several important adaptations, including protected multicellular embryos that enabled plants to successfully invade a variety of terrestrial habitats. Land plants gradually evolved several adaptations that enabled them to thrive in terrestrial habitats. Streptophyte algae refers to those algal species with specific features in common with land plants.

(a) Roland Birke/Phototake/Medical Images; (b) Pavel Skaloud; (c-f) Lee W. Wilcox; (g) Ed Reschke/Getty Images; (h) Lars Porta/Shutterstock; (i) Hoberman Collection/Alamy Stock Photo; (j) Peter Frank/Getty Images; (k) Kevin Schafer/Alamy Stock Photo; (l) itman__47/Shutterstock

of vascular tissues necessitates that their photosynthetic and nonphotosynthetic tissues be close together. Because vascular tissues, along with supporting tissues, are often absent, bryophytes are relatively small and inconspicuous. Despite their diminutive size, however, bryophytes occur throughout the world in habitats ranging from the tropics to Antarctica.

There are approximately 24,000 species of bryophytes, more than any other group of plants except the flowering plants. Bryophytes fix $CO_2$, degrade rocks to soil, stabilize soil, and reduce erosion. Humans have used bryophytes in a number of ways, including as a fuel, in the production of Scotch whiskey, and as packing materials. In many terrestrial ecosystems, bryophytes reduce the leaching of nitrogen and other nutrients from the soil.

The plant body of bryophytes is called a **thallus** (pl., *thalli*). Liverwort thalli are flattened dorsoventrally (from back and front plane, rather than from side to side plane) and are bilaterally symmetrical (i.e., have two equal halves). For comparison, moss thalli are erect and radially symmetrical (circular). Hornwort thalli are similar to those of liverworts.

The life cycle of bryophytes is characterized by a distinct alternation of generations that includes both gametophyte and sporophyte phases, *but the gametophyte is the dominant phase*

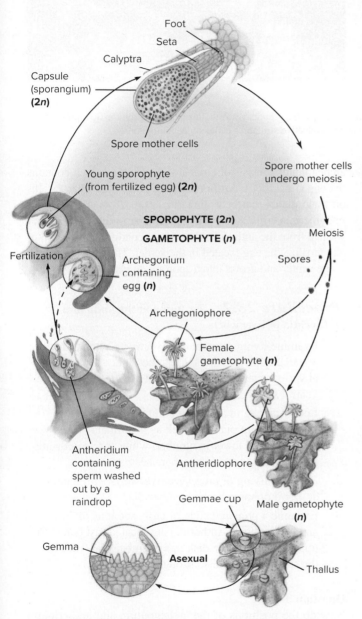

**Figure 28.3** Life cycle of *Marchantia,* a liverwort. During sexual reproduction, spores produced in the capsule germinate to form independent male and female gametophytes. The archegoniophore produces archegonia, each of which contains an egg; the antheridiophore produces antheridia, each of which produces many sperm. After fertilization, the sporophyte develops within the archegonium and produces a capsule with spores. *Marchantia* reproduces asexually by fragmentation and gemmae.

(fig. 28.3) because it is larger and lives longer. Bryophytes have multicellular sex organs in which gamete-producing cells are enclosed in a jacket of sterile cells. **Antheridia** are male sex organs that produce swimming, biflagellate sperm. Bryophytes require free water—usually provided as rain or dew—for sexual reproduction because their sperm must swim to eggs. These sperm fertilize eggs produced in **archegonia,** the female sex organs. The fertilized egg is called a **zygote.** This zygote divides and matures in the **archegonium** to produce the **sporophyte,** which remains attached to and nutritionally dependent on the **gametophyte.** The mature sporophyte produces haploid spores (via meiosis), each of which can develop into a gametophyte.

Some bryophytes have a sperm-delivery system that improves their rate of sexual reproduction. For example, fertile moss plants produce scented compounds that attract tiny insects such as mites and springtails; sperm can attach to these insects as they wade through dew on these plants. When the insects are lured to another moss plant by its scent, the insects deliver the hitchhiking sperm and, in the process, increase the rate of fertilization.

Before continuing this exercise, review figure 28.1; be sure you understand the ploidy (i.e., haploid versus diploid) of the gametophyte and sporophyte generations of the life cycle.

## PHYLUM HEPATOPHYTA: LIVERWORTS

Liverworts are the earliest land plants. Although many liverworts are "leafy," we will restrict our observations to a thallus-type liverwort, *Marchantia.* The gametophytic thallus of this liverwort grows as a large, flat, photosynthetic structure on the surface of the ground (fig. 28.4).

### Liverwort Gametophyte

***Procedure 28.1*** **Examine the thallus of** *Marchantia*

1. Observe some living *Marchantia* and note the Y-shaped (dichotomous) growth. Rhizoids extend downward from the lower (ventral) surface of the thallus.

**Question 1**
What are the functions of rhizoids?

2. View the upper (dorsal) surface of the thallus with a dissecting microscope and note the pores in the center of the diamond-shaped areas. Obtain a prepared slide of a thallus of *Marchantia* and locate a pore in cross section. These pores in the dorsal surface of the thallus overlie air chambers containing **chlorenchyma** (chloroplast-containing) cells.

(a)                                                Ed Reschke/Stone/Getty Images          (b)                                          Ed Reschke/Getty Images

**Figure 28.4** *Marchantia.* The flat, leafy thallus of this liverwort grows close to the ground. (*a*) A thallus bearing upright male reproductive structures called antheridiophores (3×). (*b*) A thallus bearing upright female reproductive structures called archegoniophores (4×).

**Question 2**
What is the function of these pores?

> *3.* Gently return the living materials to their containers.

## Asexual Reproduction in Liverworts

Liverworts can reproduce asexually via fragmentation. In this process, the older, central portions of the thallus die, leaving the growing tips isolated to form individual plants.

Structures called **gemmae cups** occur on the dorsal (upper) surface of some thalli near the midrib (fig. 28.5). Gemmae cups represent another means of asexual reproduction by liverworts. Inside the gemmae cups are lens-shaped outgrowths called **gemmae** (sing., *gemma*), which are splashed out of the cup by falling drops of rain. If a gemma lands in an adequate environment, it can produce a new gametophyte plant.

Examine a prepared slide of gemmae cups. Also examine available live or preserved material. In the following space, diagram and label what you see, and compare it to figure 28.5.

## Sexual Reproduction in Liverworts

Many species of *Marchantia* are **dioecious,** meaning that they have separate male and female plants. Gametes from each plant are produced in specialized sex organs borne on upright stalks (fig. 28.4). **Archegoniophores** are specialized stalks on female plants that bear archegonia. Each flask-shaped archegonium consists of a **neck** and a **venter,** which contains the **egg** (fig. 28.6*a*). **Antheridiophores** are specialized stalks on male plants that bear antheridia (fig. 28.4). Sperm form in antheridia (fig. 28.6*b*). Flagellated sperm are released and washed from the antheridia during wet conditions and eventually fertilize the egg located in the venter. The zygote remains in the venter and grows into a sporophyte plant.

> ***Procedure 28.2*** **Examine archegonia and antheridia of liverworts**
>
> *1.* Examine living or prepared liverworts having mature archegoniophores that bear archegonia. Archegonia at various stages of development are located on the ventral surface.
> *2.* Locate an egg in an archegonium. Notice a pattern of evolution in plants as well as in animals—eggs are larger and fewer in number, while sperm cells are smaller but greater in number.
> *3.* Examine living or preserved liverworts with mature antheridiophores bearing antheridia.
> *4.* Examine a prepared slide of cross sections of an antheridiophore. Antheridia are located just below the upper surface of the disk in a chamber that leads to the surface of the disk through a pore.

**Question 3**
How do the positions of the archegonium and antheridium relate to their reproductive function?

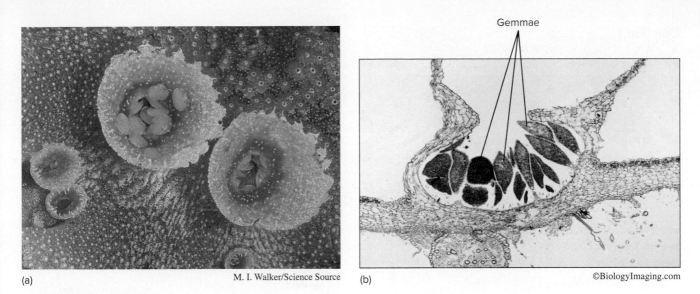

(a) M. I. Walker/Science Source    (b) ©BiologyImaging.com

**Figure 28.5** (*a*) Gemmae cups ("splash cups") containing gemmae on the gametophytes of a liverwort. Gemmae are splashed out of the cups by raindrops and can then grow into new gametophytes, each identical to the parent plant that produced it by mitosis (5×). (*b*) Longitudinal section of a gemmae cup (10×).

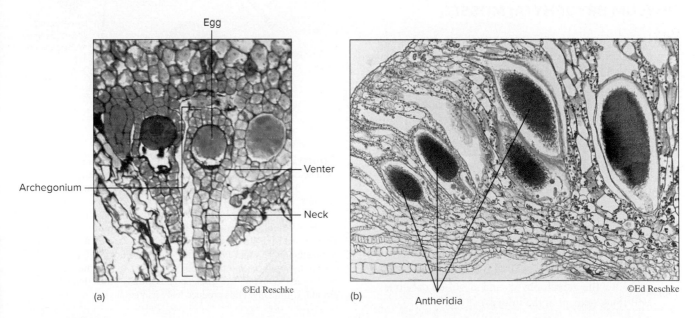

(a) ©Ed Reschke    (b) ©Ed Reschke

**Figure 28.6** (*a*) Section through an archegonium of *Marchantia*. A single egg differentiates within the archegonium (320×). (*b*) Section through an antheridiophore, showing individual antheridia (90×).

5. Return materials and slides to their containers.

## Liverwort Sporophyte

### *Procedure 28.3* Examine sporophytes of liverworts

*1.* Examine a prepared slide of a sporophyte of *Marchantia*. The nonphotosynthetic sporophyte is connected to the gametophyte by a structure called the **foot. Spores** are produced by meiosis in a **capsule** located on a **seta** (stalk) that extends downward from the foot (fig. 28.3).

*2.* Locate elongate cells called **elaters** among the spores. Elaters help disperse spores by twisting. In humid conditions the elaters coil, but when it is dry the elaters expand, pushing the spores apart and rupturing the spore case to release the spores.

*3.* Gamete release, fertilization, spore release, and germination are most efficient in individually specific environmental conditions.

## Question 4

*a.* What is the function of the foot?

*b.* Are the spores haploid or diploid?

*c.* What is the functional significance of the response of elaters to moisture?

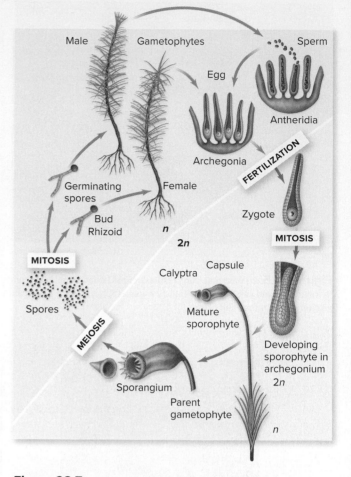

**Figure 28.7** Moss life cycle. Haploid *(n)* sperm are released from antheridia on the male gametophytes. The sperm then swim through water to the archegonia and down their necks to fertilize the eggs. The resulting diploid *(2n)* zygote develops into a diploid sporophyte. The sporophyte grows out of the archegonium and differentiates into a slender seta with a swollen capsule at its apex. The capsule is covered with a cap, or calyptra, formed from the archegonium. The sporophyte grows on the gametophyte and eventually produces spores by meiosis. The spores germinate, giving rise to gametophytes. The gametophytes intially grow along the ground. Ultimately, buds produce leafy gametophytes.

## PHYLUM BRYOPHYTA: MOSSES

Mosses are often more visible than liverworts because of their greater numbers and more widespread distribution and because gametophyte plants of mosses are leafy and usually stand upright. Mosses also withstand desiccation better than do liverworts. Consequently, mosses often grow in a greater diversity of habitats than do liverworts. The moss gametophyte is radially symmetrical and is the most conspicuous phase of the moss life cycle (fig. 28.7).

## Moss Gametophyte

### Procedure 28.4 Examine mosses

1. Observe the living moss on display called *Polytrichum* (fig. 28.8). The "leafy" green portions of the moss are the gametophytes and are often only one cell thick (except at the midrib).
2. Make a wet mount of a single leaflet and examine it with low magnification.

## Question 5

*a.* How many cells thick is the leaflet?

*b.* Is there a midrib vein?

*c.* Are pores visible on the leaf surface?

*d.* How does the symmetry of a moss gametophyte compare with that of a liverwort gametophyte?

*e.* In what ways are mosses morphologically similar to liverworts? In what ways are they different?

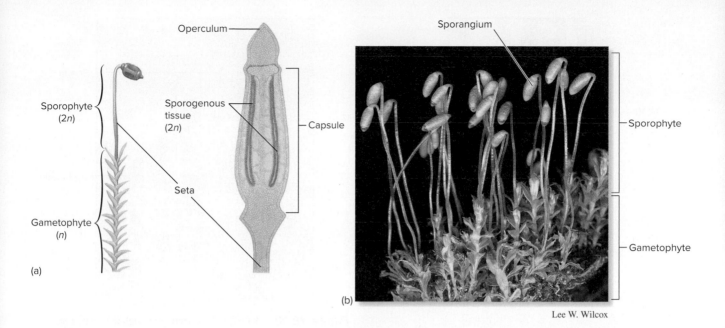

**Figure 28.8** The structure of a moss. (*a*) Diagram of the parts of a mature mass sporophyte. (*b*) A hair cap moss, *Mnium*. The leaves belong to the gametophyte. Each of the stalks, with the capsule at its summit, is a sporophyte. Although moss sporophytes may be green and perform a limited amount of photosynthesis when they are immature, they are soon completely dependent, in a nutritional sense, on the gametophyte.

Moss gametophytes have specialized cells that aid in the absorption and retention of water. Mats of moss act, in effect, like sponges. The following procedure demonstrates the water-absorbing potential of mosses.

### Procedure 28.5 Water absorption by moss

1. Weigh 3 g of *Sphagnum,* a peat moss, and 3 g of paper towel.
2. Add the moss and towel to separate beakers each containing 100 mL of water.
3. After several minutes, remove the materials from the beaker.
4. Measure the amount of water left in each beaker by pouring the water into a 100-mL graduated cylinder. Remember that 1 mL of water weighs 1 g.
5. Record your data.

**Question 6**
*a.* How many times its own weight did the moss absorb?

*b.* How does this compare with the paper towel?

*c.* Why is *Sphagnum* often used to ship items that must be kept moist?

6. Return all materials to their containers and clean your work area.

## Asexual Reproduction in Mosses

Unlike liverworts, mosses lack structures such as gemmae for asexual reproduction. Mosses reproduce asexually by fragmentation.

## Sexual Reproduction in Mosses

Most mosses, like liverworts, are dioecious (i.e., male and female reproductive structures are on separate individuals). Archegonia or antheridia are borne either on tips of the erect gametophyte stalks or as lateral branches on the stalks. The apex of stalks of the female plant (the plant that bears archegonia) appears as a cluster of leaves, with the archegonia buried inside.

### Procedure 28.6 Examine archegonia and antheridia of mosses

1. Examine living or preserved mosses having mature archegonia.

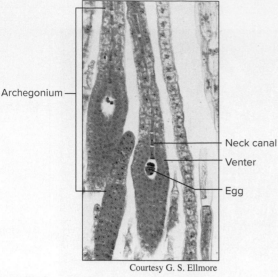

Courtesy G. S. Ellmore

**Figure 28.9** A longitudinal section through the tip of a female gametophyte of a moss (45×).

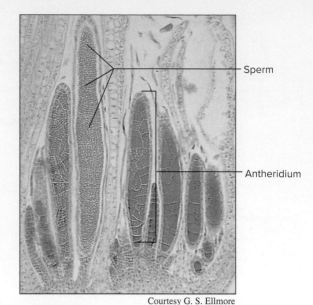

Courtesy G. S. Ellmore

**Figure 28.10** A longitudinal section through the tip of a male gametophyte of a moss (45×).

*2.* Examine a prepared slide of moss archegonia (fig. 28.9). Note the canal that leads through the neck and terminates in the venter of the archegonium. When the archegonium matures, cells lining the neck disintegrate and form a canal leading to the egg. Sperm, following a chemical attractant released by the archegonium, swim through this canal to reach the egg.

**Question 7**
Where is the egg located in the archegonium?

*3.* Examine living or preserved mosses having mature antheridia. The male plant (i.e., the plant that bears antheridia) has a platelike structure on the tip with the "leaves" expanding outward to form a rosette. This structure is sometimes called a "moss flower" because of its appearance or a "splash cup" because of its function (i.e., the dispersal of sperm by falling raindrops).

*4.* Examine a prepared slide of moss antheridia, which appear as elongate, saclike structures (fig. 28.10).

*5.* Locate the outer sterile jacket and the inner mass of cells destined to become sperm.

**Question 8**
Are sperm haploid or diploid?

## Moss Sporophyte

Moss sporophytes consist of **capsules** located atop stalks, called **setae,** that extend upward from the moss gametophyte (fig. 28.8*a*). A sporophyte is attached to the gametophyte by a structure called a **foot.**

**Question 9**
Is the sporophyte more prominent in mosses or liverworts?

The capsule atop the seta is covered by the hoodlike **calyptra,** which is a thin hood of tissue produced by the archegonium that covers the developing sporohyte as it matures. The calyptra falls off when the capsule matures. Inside the capsule are numerous haploid spores formed by meiosis.

**Question 10**
What is the adaptive significance of the seta of the sporophyte growing well above the mat of the gametophytes?

If enough living moss is available in the lab, remove the calyptra from a sporophyte capsule. On the tip of the capsule is a lidlike structure called the **operculum.** Remove the operculum and notice the hairlike teeth lining the opening of the lidlike structure. These teeth help control the release of spores from the capsule. In wet weather these teeth bend inward and prevent release of the spores. In dry weather they bend outward, facilitating distribution of spores by the wind.

Crush the capsule in water on a microscope slide and note the large number of spores released.

**Question 11**

*a.* What process produces spores?

*b.* Is the capsule haploid or diploid?

Moss spores germinate and form a photosynthetic **protonema,** which resembles a branching, filamentous alga. Leafy moss plants arise from "buds" located along the protonema.

**Question 12**

Can you think of any evolutionary implications of the similarity between a moss protonema and a filamentous green alga?

## PHYLUM ANTHOCEROPHYTA: HORNWORTS

The hornworts are the smallest group of bryophytes; there are only about 100 species in six genera. Hornworts have several features that distinguish them from most other bryophytes. The sporophyte is shaped like a long, tapered horn that protrudes from a flattened thallus. Also, archegonia are not discrete organs. Rather, they are embedded in the thallus and are in contact with surrounding vegetative cells.

sporophyte

gametophyte

**Figure 28.11** *Anthoceros,* a hornwort. The "horns" of a hornwort are sporophytes that grow up into the air, whereas the gametophytes grow flat along the ground. At maturity, the tips of hornwort sporophytes open and release spores.

The most familiar hornwort is *Anthoceros,* a temperate genus (fig. 28.11). Examine some living and preserved *Anthoceros.* Locate the gametophyte thallus and the rhizoids extending from its lower surface. Also locate the hornlike sporophytes extending from the upper surface. Spores are produced in the horn of the sporophyte. If prepared slides are available, locate spores in various stages of development within the sporophyte.

## ORGANIZING WHAT YOU'VE LEARNED

Go to page 362, which follows Exercise 31. Complete the portions of the table that refer to the plants you studied in today's lab.

## INQUIRY-BASED LEARNING

### *What are the roles of bryophytes in the environment?*

Observation: Although they are usually overlooked, bryophytes often grow in places that other plants cannot grow and are important parts of many ecosystems. Find a place where bryophytes are growing on campus. Consider what these plants are doing in that ecosystem and why they are not in other ecosystems on campus.

Question: What conditions are best suited for the growth of bryophytes?

**a.** Establish a working lab group and obtain Inquiry-Based Learning Worksheet 28 from your instructor.

**b.** Discuss with your group well-defined questions relevant to the preceding observation and question. Choose and record your group's best question for investigation.

**c.** Translate your question into a testable hypothesis and record it.

**d.** Outline on Worksheet 28 your experimental design and supplies needed to test your hypothesis. Ask your instructor to review your proposed investigation.

**e.** Conduct your procedures, record your data, answer your question, and make relevant comments.

**f.** Discuss with your instructor any revisions to your questions, hypotheses, or procedures. Repeat your work as needed.

1. Compare and contrast the complexity of bryophytes and algae regarding their morphology, habitat, asexual reproduction, and sexual reproduction.

2. What event begins the sporophyte phase of the life cycle? Where does this event occur in liverworts and mosses?

3. What event begins the gametophyte phase of the life cycle? Where does this event occur in liverworts and mosses?

4. What features distinguish a moss from a liverwort?

5. Diagram the life cycle of a liverwort, indicating which stages are sporophytic and which are gametophytic.

6. Diagram the life cycle of a moss, indicating which stages are sporophytic and which are gametophytic. Also indicate where meiosis and syngamy occur.

7. How did liverworts get their name?

8. What ecological roles do liverworts, mosses, and hornworts play in the environment?

9. Is the sporophyte of mosses ever independent of the gametophyte? Explain.

10. Why do you think that bryophytes are sometimes referred to as the amphibians of the plant kingdom?

11. What limits the height of mosses?

## DOING BIOLOGY YOURSELF

Liverworts and mosses are surprisingly diverse. Make a simple collection of living liverworts and mosses from two or three sites. Bring the collection to class and compare yours with those of other students. Use reference books in your lab or library to identify the plants that you've collected.

## WRITING TO LEARN BIOLOGY

Because water is required for the swimming sperm to reach the archegonium, would you say that this means that bryophytes are not truly land plants? Why or why not?

# Survey of the Plant Kingdom
## Seedless Vascular Plants of Phyla Pterophyta and Lycophyta

## Learning Objectives

By the end of this exercise you should be able to:
1. Discuss similarities and differences between ferns and other plants you have studied in the lab.
2. Describe the life cycles of ferns and their allies.
3. Describe the distinguishing features of true ferns, club mosses, whisk ferns, and horsetails.
4. Name some roles of ferns and fern allies.

Please visit **connect.mheducation.com** to review online resources tailored to this lab.

Seedless vascular plants include two phyla of nonflowering plants having a vascular system of fluid-conducting xylem and phloem: phylum Pterophyta (true ferns, whisk ferns, and horsetails) and phylum Lycophyta (club mosses) (table 29.1). The life cycle of seedless vascular plants includes both gametophyte and sporophyte phases, but in

### Table 29.1

**The Phyla of Extant Seedless Vascular Plants**

| Phylum | Examples | | Key Characteristics | Approximate Number of Living Species |
|---|---|---|---|---|
| Pterophyta | Ferns | | Primarily homosporous (a few heterosporous) vascular plants. Sperm motile. External water necessary for fertilization. Leaves are megaphylls that uncoil as they mature. Sporophytes and virtually all gametophytes photosynthetic. About 365 genera. | 11,000 |
| | Horsetails | | Homosporous vascular plants. Sperm motile. External water necessary for fertilization. Stems ribbed, jointed, either photosynthetic or nonphotosynthetic. Leaves scalelike, in whorls, nonphotosynthetic at maturity. One genus (*Equisetum*). | 15 |
| | Whisk ferns | | Homosporous vascular plants. Sperm motile. External water necessary for fertilization. No differentiation between root and shoot. No leaves; one of the two genera has scalelike enations and the other leaflike appendages. | 6 |
| Lycophyta | Club mosses | | Homosporous or heterosporous vascular plants. Sperm motile. External water necessary for fertilization. Leaves are microphylls. About 12–13 genera. | 1275 |

contrast to bryophytes (Exercise 28), in this group the sporophyte is the dominant phase (fig. 29.1). The sporophyte is the conspicuous adult stage in the life cycle of seedless vascular plants.

All ferns and fern allies possess **sporophylls** (*sporo* = spore forming, *-phyll* = leaf). Sporophylls are leaflike structures of the sporophyte generation that bear spores. They may be large **megaphylls** (*mega* = large, *-phyll* = leaf) with several to many veins (as in the megaphylls of true ferns), or they may be smaller **microphylls** (*micro* = small, *-phyll* = leaf) with one vein (as in whisk ferns, scouring rush, and club mosses). **Sporangia,** which form on sporophylls, are where spores are produced by meiosis (review fig. 28.1 and compare to fig. 29.1). Sporangia occur somewhere on all

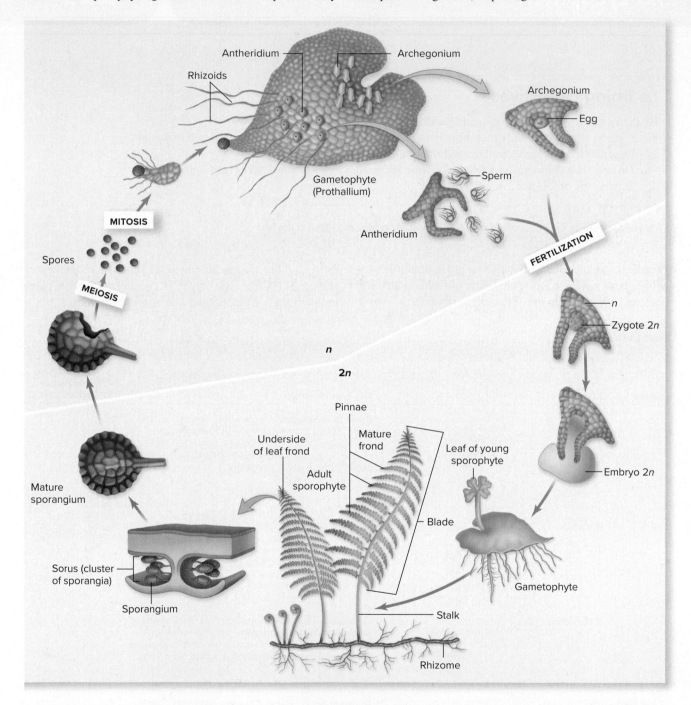

**Figure 29.1** Fern life cycle. The haploid gametophyte (i.e., the prothallium) grows in moist places. Rhizoids are anchoring structures that project from the lower surface of a prothallium. Eggs and sperm develop in archegonia and antheridia, respectively, on gametophytes' lower surface. Sperm released from antheridia swim to archegonia and fertilize the single egg. The zygote—the first cell of the diploid sporophyte generation—starts to grow within the archegonium and eventually becomes much larger than the gametophyte. Most ferns have horizontal stems, called rhizomes, that grow below the ground. On the sporophyte's leaves (fronds) are clusters of sporangia within which meiosis occurs and spores are formed. When released, the spores germinate and become new gametophytes called prothallia.

plants. In ferns, the sporangia are on the backs of leaves; this is why the leaves are called sporophylls (fig. 29.2). Like bryophytes, ferns require water for fertilization.

**Question 1**

*a.* Like bryophytes, ferns require water for fertilization. Why? Explain your answer.

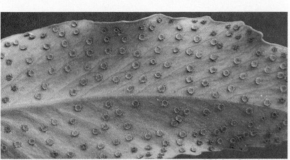
(a)                                    ©BiologyImaging.com

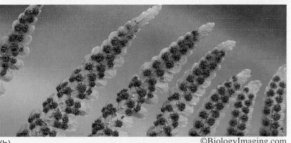
(b)                                    ©BiologyImaging.com

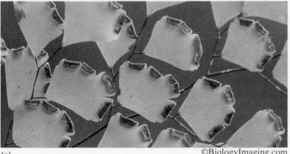
(c)                                    ©BiologyImaging.com

**Figure 29.2** Fern sporangia. Most ferns have sporangia aggregated into clusters, called sori, on the undersides of their leaves. (*a*) In some ferns, such as the Japanese holly fern (*Crytomium falcatum*), each sorus is covered by a flap of leaf tissue called an indusium (also see fig. 29.3). (*b*) Other ferns bear uncovered sori, as shown here in the autumn fern (*Dryopteris erythrosora*). (*c*) In still other ferns, as in the maidenhair fern (*Adiantum tenerum*), sori are enfolded by the edge of the frond itself.

*b.* What are the primary functions (e.g., support, transport, photosynthesis, reproduction) of the major organs shown in fig. 29.1?

## PHYLUM PTEROPHYTA (FERNS)

True ferns (phylum Pterophyta) inhabit almost all types of environments and possess characteristics of the more advanced seed plants as well as the less advanced bryophytes. Ferns have well-developed vascular tissue. Unlike bryophytes, ferns have an independent sporophyte (fig. 29.1) and stomata, which are pores that open and close on leaves and, in doing so, regulate gas exchange.

The diversity of ferns is striking; they range from majestic tree ferns to bizarre staghorn ferns. Tree ferns reach heights of up to 16 m. Along with other plants, these ferns once formed forests that were transformed into coal deposits. Today, humans use ferns as decorations and to grow rice. Review the fern life cycle shown in figure 29.1.

**Question 2**

*a.* Which parts of the life cycle are haploid?

*b.* Which are diploid?

*c.* What is the role of the gametophyte in ferns?

*d.* How is this different from that of the gametophyte in bryophytes?

*e.* In ferns, how does the size of the gametophyte compare with that of the sporophyte?

*f.* How does this compare with bryophytes, and why is this important?

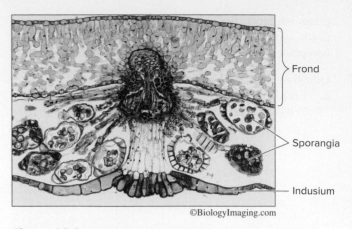

©BiologyImaging.com

**Figure 29.3** Fern sorus (20×). In most ferns, sori are on the undersides of leaves.

Fern sporophytes grow indefinitely via underground stems called **rhizomes** (fig. 29.1). Examine the fern rhizomes on display. Also examine the different ferns available in the lab and note the different shapes of the leaflike fronds. Identify the **stalk, blade,** and **pinnae.**

**Question 3**

*a.* How many veins are present in each frond?

*b.* What tissues comprise a vein of vascular tissue?

*c.* What is the function of the stalk? The blade? The pinnae?

*d.* Rhizomes are involved in the asexual reproduction of ferns. How could this happen?

Groups of sporangia called **sori** (singular = **sorus**) form on the underside of fern fronds (fig. 29.2). Sporangia may be protected by a shield-shaped **indusium** (fig. 29.3), a specialized outgrowth of a frond. Meiosis in the sporangium produces haploid spores, the first stage of the gametophyte.

### *Procedure 29.1*   **Examine sori**

*1.* Scrape a sorus into a drop of water on a microscope slide and use the low power on your microscope to observe the sorus.

**Question 4**
Are any spores in the sporangium?

 Be careful when handling acetone; it is a strong solvent. Wear safety glasses.

*2.* Place a few drops of acetone on the sporangium while you observe it with a dissecting microscope.
*3.* Watch the sporangium for a few minutes, adding acetone as needed.
*4.* Describe what you see.

**Question 5**
*a.* Did the application of acetone cause the spores of the fern to disperse?

*b.* How is the mechanism for spore dispersal in ferns similar to that of bryophytes?

*5.* Examine prepared slides of fern sori, referring to figure 29.3. Diagram and label each structure that you see, listing its function.

## Fern Reproduction

When conditions are favorable, fern spores germinate and form a threadlike **protonema.** Subsequent cellular divisions produce an independent, heart-shaped **prothallium** ("valentine plant").

### Question 6
***a.*** Is the prothallium haploid or diploid?

***b.*** Is the prothallium sporophyte or gametophyte?

Rhizoids and male and female reproductive structures occur on the underside of the prothallium. However, a prothallium rarely fertilizes itself because the antheridia and archegonia mature at different times. Globe-shaped antheridia form first, followed by archegonia. Archegonia are vase-shaped and are located near the cleft of the heart-shaped prothallium. After producing sperm, antheridia drop off, leaving sperm to swim to the archegonia of other prothallia.

### *Procedure 29.2* **Observe archegonia and antheridia**

***1.*** Observe archegonia and antheridia (see fig. 29.1) on prepared slides.
***2.*** Observe archegonia and antheridia on a living prothallium.

### Question 7
***a.*** What is the adaptive significance of having these structures on the lower surface of the prothallium rather than on the upper surface?

***b.*** What is the adaptive significance of having sperm and egg produced at different times?

The zygote develops in the archegonium and is nutritionally dependent on the gametophyte for a short time (fig. 29.4). Soon thereafter, the sporophyte becomes leaflike and crushes the prothallium.

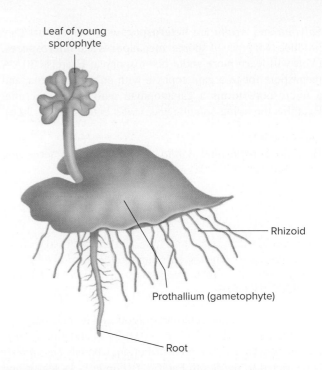

Figure 29.4  Young fern sporophyte growing out of its gametophyte parent (10X mag.). Shortly after this stage, the gametophyte dies and begins to decompose.

Fronds of the growing sporophyte break through the soil in a coiled position called a **fiddlehead** (fig. 29.5). The fiddlehead then unrolls to display the frond, a single leaf. Fiddleheads are considered a culinary delicacy in some parts of the world.

Most terrestrial ferns are **homosporous;** this means that they produce one kind of spore that develops into a single kind of gametophyte that produces both antheridia and archegonia (see fig. 29.1). Conversely, aquatic ferns such as

©BiologyImaging.com

**Figure 29.5**  A fiddlehead of a tropical tree fern.

*Salvinia* and *Azolla* are **heterosporous,** meaning that they produce two types of spores: **megaspores** and **microspores.** (You will learn more about heterospory in Exercise 30.) A megaspore forms a gametophyte with only archegonia, and a microspore forms a gametophyte with only antheridia. Examine the living *Salvinia* and *Azolla* in the lab (fig. 29.6).

**Question 8**
How do *Salvinia* and *Azolla* differ from other ferns you examined earlier?

## WHISK FERNS

Whisk ferns are close relatives of ferns and include only two extant representatives: *Psilotum* (fig. 29.7) and *Tmesipteris*. *Psilotum* has a widespread distribution, whereas *Tmesipteris* is restricted to the South Pacific. *Psilotum* lacks leaves and roots and is homosporous.

### *Procedure 29.3*   Examine *Psilotum*

*1.* Examine a prepared slide of a *Psilotum* sporangium (fig. 29.8).

*2.* If available, examine living *Psilotum* plants in the lab.

**Question 9**
*a.* How would you describe the branching pattern of *Psilotum?*

*b.* Are any roots present?

*c.* Are any leaves present?

*d.* Where are the sporangia?

*e.* Where does photosynthesis occur in *Psilotum?*

(a)      Source: Larry Allain/National Wetlands Research Center/USGS

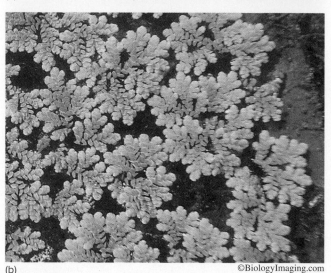

(b)                                    ©BiologyImaging.com

**Figure 29.6**   (*a*) *Salvinia* and (*b*) *Azolla* are ferns that grow in aquatic habitats.

Segundo Pérez/Alamy Stock Photo

**Figure 29.7**   Whisk ferns (*Psilotum* sp.) are so called because their branching pattern gives the impression of a whisk broom. The stems bear lobed sporangia (fig. 29.8).

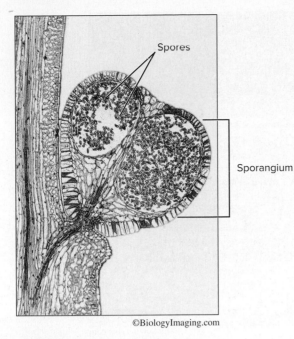

©BiologyImaging.com

**Figure 29.8** *Psilotum* sporangium (18×).

## HORSETAILS

*Equisetum* (also called scouring rush) is the only extant genus of horsetails, another small but distinctive group of ferns (fig. 29.9). *Equisetum* is an example of a plant whose vegetative structure identifies the plant better than does its reproductive structure: *Equisetum* is distinguished by its jointed and ribbed stem. Examine the living *Equisetum* plants available in lab.

Stuart Wilson/Science Source

**Figure 29.9** Scouring rush, *Equisetum telmateia,* of the phylum Pterophyta. This species forms two types of erect stems, a green, photosynthetic type and a brownish type terminating in spore-producing cones. The spores produced by meiosis in the cones give rise to a single kind of tiny, green, nutritionally independent gametophyte.

**Question 10**

*a.* Where are the leaves?

*b.* What part of the plant is photosynthetic?

*c.* Which part of the life cycle of *Equisetum* is dominant, the sporophyte or gametophyte?

Feel the ribbed stem of an *Equisetum* plant. Its rough texture results from siliceous deposits in its epidermal cells. During frontier times, *Equisetum* was used to clean pots and pans, sand wooden floors, and scour plowshares, thus accounting for its common name of "scouring rush."

Strobili of *Equisetum* occur at the tips of reproductive stems. Within a strobilus, sporangia form atop umbrellalike modified branches called **sporangiophores.** Elaters in sporangia of *Equisetum* help disperse spores. Examine prepared slides of *Equisetum* strobili (fig. 29.10). Diagram, label, and state the function of each major structure composing the strobilus.

**Question 11**

How do elaters aid in the dispersal of spores?

## PHYLUM LYCOPHYTA CLUB MOSSES

During the Devonian and Carboniferous periods (300–400 million years ago), club mosses, whisk ferns, and scouring rush were among the dominant plants on earth. Indeed, most of our coal deposits consist largely of these plants. However, the "giant" species of these phyla are now extinct, and modern representatives are relatively small compared to their giant ancestors.

Club mosses (phylum Lycophyta) possess true roots, stems, and leaves. Most asexual reproduction by club mosses occurs via rhizomes. If one is available, locate the rhizome on a *Lycopodium*, a club moss. Study the aboveground portion of the *Lycopodium* plant (fig. 29.11). *Lycopodium* is evergreen, as are some ferns, most gymnosperms, and some angiosperms.

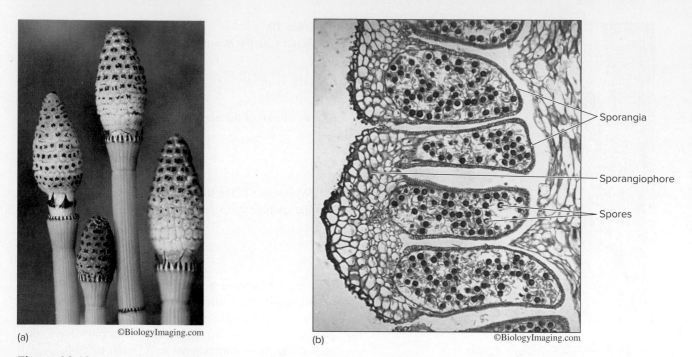

(a) ©BiologyImaging.com

(b) ©BiologyImaging.com

**Figure 29.10** *Equisetum.* (*a*) Strobili occur at the tips of reproductive stems. (*b*) Cross-section of a strobilus showing spores within sporangia at the tips of sporangiophores (20×).

(a) ©Ed Reschke

(b)

**Figure 29.11** *Lycopodium,* a lycophyte. (*a*) The sporophyte develops from an underground stem called a rhizome. (*b*) The spore-producing structures occur in club-like structures called strobili, or cones, which are aggregates of closely packed sporangium-bearing branches or leaves.

## Question 12

**a.** How could a rhizome be involved in asexual reproduction?

**b.** How is a rhizome different from a rhizoid?

**c.** Does the rhizome have leaves?

**d.** What are the shape and size of the leaves?

**e.** What is the significance of the form of the leaves?

**f.** Is a midvein visible?

**g.** What does the term "evergreen" mean?

**h.** Is being evergreen a good characteristic for classifying plants? Why or why not?

Sporangia of *Lycopodium* occur on small modified leaves called **sporophylls** clustered in **strobili** (cones) that form at the tips of branches. Species with these cones probably shared common ancestry with the familiar cone-bearing gymnosperms such as pine (*Pinus*) (see Exercise 30).

### *Procedure 29.4* Examine club mosses

**1.** Examine strobili on a living *Lycopodium* plant. Also examine prepared slides of strobili of *Lycopodium* (fig. 29.12).

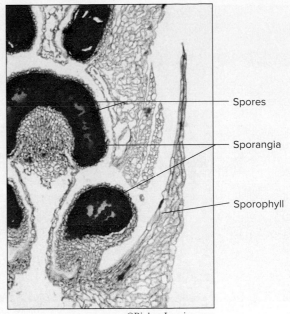

©BiologyImaging.com

**Figure 29.12** Sporophyll and sporangia of *Lycopodium*, a club moss (15×).

**2.** Diagram, label, and state the function of each major feature of the strobilus.

**3.** Examine prepared slides of spores of *Lycopodium* as well as those available on plants growing in the lab. Sketch what you see.

## Question 13

**a.** How many sporangia occur on each sporophyll?

**b.** Can you see why spores of *Lycopodium* are sometimes called "vegetable sulfur"?

**c.** Why are the spores a good dry lubricant?

©BiologyImaging.com

**Figure 29.13**   *Selaginella,* a club moss.

blickwinkel/Alamy Stock Photo

**Figure 29.14**   Quillworts (*Isoetes*) are so-named because of their narrow, quill-like leaves. Most quillworts live in aquatic environments.

*d.*   Can you see why these plants are sometimes referred to as "resurrection plants"?

*e.*   How does the formation of strobili in *Equisetum* compare with that in *Lycopodium* and *Selaginella?*

*d.*   Which is the dominant part of the *Lycopodium* life cycle, the sporophyte or gametophyte?

*4.*   Examine living *Selaginella* plants (fig. 29.13). Many species of *Selaginella* produce two types of strobili, typically red and yellow. If strobili are present, examine spores derived from these cones.

*5.*   Examine hydrated and dehydrated resurrection plants (*Selaginella lepidophylla*).

**Question 14**

*a.*   Are spores of *Selaginella* similar in size?

*b.*   What is this condition called?

*c.*   What is the functional significance of the difference in the appearance of dehydrated and rehydrated *Selaginella?*

*Isoetes,* or quillwort, is an aquatic or semi-aquatic lycopod that usually grows in clear ponds and slow-moving streams. At the branching points along the stem of *Isoetes,* you'll see an unusual runnerlike organ. These proplike axes are called **rhizophores** and have structural features that are intermediate between stems and roots.

**Procedure 29.5**   **Examine *Isoetes***

*1.*   If available, examine living *Isoetes* (quillwort) plants (fig. 29.14).

*2.*   Compare and contrast the following features of *Isoetes* with *Lycopodium* and *Selaginella*:

- Shape of aerial portion of plant
- Branching patterns
- Shape, size, and arrangement of leaves

## ORGANIZING WHAT YOU'VE LEARNED

Go to page 362, which follows Exercise 31. Complete the portions of the table that refer to the phyla you studied in today's lab. Also, to review the structures and characteristics of seedless vascular plants, complete table 29.2 on the next page.

## Table 29.2

**Summary of the Structures Common to Seedless Vascular Plants**

| Plant Structure | Sporophyte or Gametophyte | Function |
|---|---|---|
| Prothallium | | |
| Pinna | | |
| Spore | | |
| Frond | | |
| Annulus | | |
| Sporangium | | |
| Antheridium | | |
| Archegonium | | |
| Microspore | | |
| Megaspore | | |
| Microphyll | | |
| Megaphyll | | |

# INQUIRY-BASED LEARNING

## *What happens during the "resurrection" of a "resurrection plant"?*

Observation: The "resurrection" of a "resurrection plant" (*Selaginella*) involves the rapid absorption of water. This, in turn, triggers other changes in the plant that "revive" the plant.

Question: What changes accompany the "resurrection" of a "resurrection plant"?

a. Establish a working lab group and obtain Inquiry-Based Learning Worksheet 29 from your instructor.

b. Discuss with your group well-defined questions relevant to the preceding observation and question. Choose and record your group's best question for investigation.

c. Translate your question into a testable hypothesis and record it.

d. Outline on Worksheet 29 your experimental design and supplies needed to test your hypothesis. Ask your instructor to review your proposed investigation.

e. Conduct your procedures, record your data, answer your question, and make relevant comments.

f. Discuss with your instructor any revisions to your questions, hypotheses, or procedures. Repeat your work as needed.

1. How are ferns and fern allies similar to and different from bryophytes? What are the implications of these similarities and differences?

2. What structures and features do ferns possess that bryophytes do not that may have contributed to their success in a broader range of environments?

3. What are the advantages of vascular tissues in land plants?

4. What are the distinguishing features of club mosses, whisk ferns, and horsetails? How are these plants different from ferns?

## WRITING TO LEARN BIOLOGY

What problems did plants face as they moved onto land? What adaptations of mosses, liverworts, ferns, and other seedless plants are relevant to the transition?

# Survey of the Plant Kingdom
## Gymnosperms of Phyla Cycadophyta, Ginkgophyta, Coniferophyta, and Gnetophyta

## Learning Objectives

By the end of this exercise you should be able to:

1.   Describe the distinguishing features of the gymnosperms.
2.   Understand the life cycle of pine, a representative gymnosperm.
3.   Understand the evolutionary significance of pollen and seeds.
4.   Identify the parts and understand the function of a cone.
5.   Identify the parts and understand the function of a seed.

Please visit **connect.mheducation.com** to review online resources tailored to this lab.

Gymnosperms were the first group of plants to (1) protect their developing embryos in nutrient-containing seeds and (2) make the evolutionary transition from swimming sperm to pollen-enclosed sperm. These adaptations helped make the gymnosperms a remarkably successful group of plants.

Gymnosperms produce exposed seeds borne on scale-like structures called **cones** (strobili). Like ferns, gymnosperms have a well-developed alternation of generations. Unlike most ferns, however, gymnosperms are **heterosporous;** that is, they produce two types of spores (fig. 30.1). **Microspores** occur in male cones and form male gametophytes. **Megaspores**

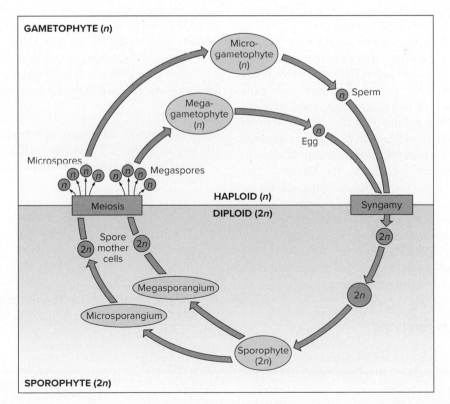

**Figure 30.1**   Diagram of the life cycle of a heterosporous vascular plant.

## Table 30.1

**The Five Phyla of Extant Gymnosperms**

| Phylum | Examples | | Key Characteristics | Approximate Number of Living Species |
|---|---|---|---|---|
| Cycadophyta | Cycads | | Heterosporous vascular seed plants. Sperm flagellated and motile but confined within a pollen tube that grows to the vicinity of the egg. Palmlike plants with pinnate leaves. Secondary growth slow compared with that of the conifers. Ten genera. | 320 |
| Ginkgophyta | *Ginkgo* | | Heterosporous vascular seed plants. Sperm flagellated and motile but conducted to the vicinity of the egg by a pollen tube. Deciduous tree with fan-shaped leaves that have evenly forking veins. Seeds resemble a small plum with fleshy, ill-scented outer covering. One species: *Ginkgo biloba* | 1 |
| Coniferophyta | Conifers (including pines, spruces, firs, yews redwoods, and others) | | Heterosporous seed plants. Sperm not motile; conducted to egg by a pollen tube. Leaves mostly needlelike or scalelike. Vascular. Trees, shrubs. About 50 genera. | 630 |
| Gnetophyta | Gnetophytes | | Heterosporous vascular seed plants. Sperm not motile; conducted to egg by a pollen tube. The only gymnosperms with vessels. Trees, shrubs, vines. Three very diverse genera (*Ephedra, Gnetum, Welwitschia*). | 65 |

occur in female cones and form female gametophytes. Gametophytes of gymnosperms are microscopic and completely dependent on the large, free-living sporophyte.

In gymnosperms, pollen grains are the male gametophyte. Pollination is the transfer of pollen from male cones (where the pollen is produced) to female cones, which house the eggs. In gymnosperms, pollen is carried from male cones to female cones by wind. Gymnosperms were the first plants to evolve that did not need free water to transfer sperm to egg, and were therefore able to colonize terrestrial habitats. Gymnosperms were the dominant land plants during the Age of Dinosaurs (i.e., during the Mesozoic Era, 225–65 million years ago). Today, we use gymnosperms for products ranging from lumber (e.g., pine, fir, spruce, cedar) to soaps, varnish, nail polish, gum, food, and perfume.

Gymnosperms include four phyla: Cycadophyta, Ginkgophyta, Coniferophyta, and Gnetophyta (table 30.1). The last of these, the Gnetophyta, is discussed only briefly here because the phylum consists of a few rare genera.

## PHYLUM CYCADOPHYTA

The Cycadophyta (cycads) once flourished, but today the phylum consists of only about 10 genera and 320 species. Cycads resemble palms because they have unbranched trunks and large, closely packed leaves that are evergreen and tough (fig. 30.2). Unlike conifers, cycads have pollen and seed cones on separate plants. The cones, which are surrounded by leaves at the top of the stem, can be more than a meter long and weigh as much as 35 kg.

Sperm of cycads are flagellated. Examine a branch of a cycad such as *Zamia* bearing developing seeds. The seeds are fleshy and exposed to the environment.

### Question 1

*a.* Based on your knowledge of reproduction in bryophytes (Exercise 28) and ferns (Exercise 29), is the presence of flagellated sperm in cycads surprising? Why or why not?

*b.* Is the possession of flagellated sperm a primitive or advanced characteristic in the plant kingdom? Why?

(a)

©BiologyImaging.com

(b)

©BiologyImaging.com

**Figure 30.2**  Cycads. (*a*) A male cycad (*Cycas revoluta*) with a strobilus. (*b*) A female cycad with strobili. *Zamia* sp. is the only genus of cycad native to the United States. The starchy roots and stems (mostly underground) of cycads were used by Native Americans for food.

David Stephenson/Lexington Herald-Leader/Tribune News Service/Getty Images

**Figure 30.3**  Maidenhair tree, *Ginkgo biloba,* is the only living representative of the phylum Ginkgophyta, a group of plants abundant 200 million years ago. Among living seed plants, only the cycads and *Ginkgo* have swimming sperm. This photograph shows *Ginkgo* leaves and fleshy seeds.

## PHYLUM GINKGOPHYTA

The Ginkgophyta consists of one species, *Ginkgo biloba* (maidenhair tree), a large dioecious tree that does not bear cones. *Ginkgo* are hardy plants in urban environments and tolerate insects, fungi, and pollutants. Males are usually planted because females produce fleshy, smelly, and messy fruit that superficially resemble cherries. Leaves of *Ginkgo* have a unique shape (fig. 30.3). *Ginkgo* has not been found in the wild and would probably be extinct but for its cultivation in ancient Chinese gardens.

### Question 2
What does dioecious mean?

## PHYLUM CONIFEROPHYTA

The Coniferophyta are a large group of cone-bearing plants that includes the 5000-year-old bristlecone pines, the earth's oldest living plants. The cones they bear are reproductive structures of the sporophyte generation that consist of several scalelike sporophylls arranged about a central axis (fig. 30.4). Sporophylls, also present in ferns and their allies, are modified leaves specialized for reproduction. Sporophylls bear spores. In conifers, sporophylls of male cones are called **microsporophylls** (see fig. 30.8). On the surface of each microsporophyll is a layer of cells called a **microsporangium** that produces spores. Male cones live only a few weeks; after these cones release their pollen, they fall off the tree. Sporophylls of female cones are **megasporophylls;** each of these "scales" of the female cone

(a)                                                      ©BiologyImaging.com          (b)                                                    ©BiologyImaging.com

**Figure 30.4**   Pinecones. (*a*) First-year ovulate (seed) cones open for pollination. (*b*) Second-year ovulate pinecones at time of fertilization.

bears two spore-producing **megasporangia** on its upper surface. Microsporangia and megasporangia are patches of cells near the central axis of the sporophylls composing the cones on the respective sporophylls. Male cones are small and similar in all conifers (fig. 30.5). However, female cones are variable; they may be small (1–3 cm in *Podocarpus*) and fleshy (a *Juniperus* "berry") or large and woody (6–40 cm in *Pinus*).

In this exercise you'll study pine (*Pinus*), a representative and widely distributed conifer. Pine has considerable economic value because it is used to produce lumber, wood pulp, pine tar, resin, and turpentine. Other examples of conifers include spruce, cedar, and fir.

## The Pine Sporophyte

The life cycle of *Pinus* is typical of conifers (fig. 30.6). You are already familiar with the sporophyte of pine—it is the tree.

### Procedure 30.1   Examine pine twigs and leaves

1.   Examine pine twigs having leaves (needles) and a terminal bud. Notice that the leaves are borne on short branches only a few millimeters long. The length and number of leaves distinguish many of the species of *Pinus*.
2.   Examine a prepared slide of a cross section of a pine leaf; locate the structures labeled in figure 30.7.

**Question 3**
*a.*   How are needles (i.e., leaves) arranged?

©BiologyImaging.com

**Figure 30.5**   Pollen-bearing cones of *Pinus*. These cones are usually found on the lower branches of pine trees. The sporophylls of the cones have opened to release pollen.

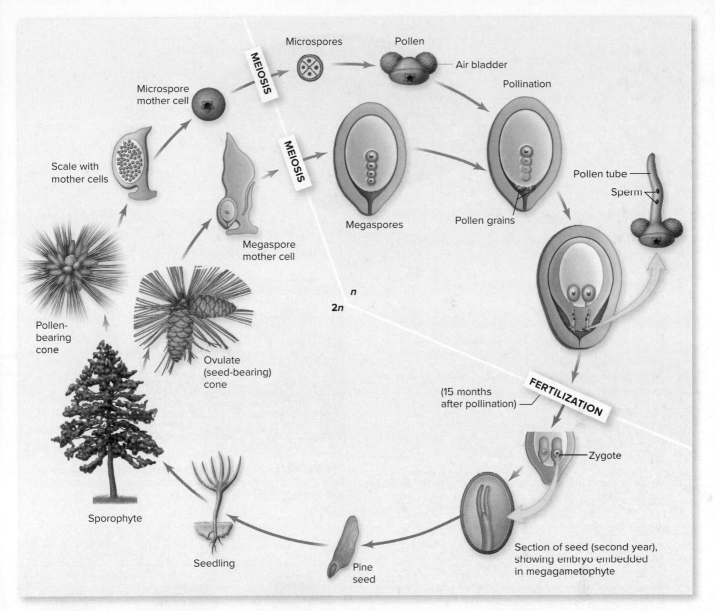

**Figure 30.6** Pine life cycle. In seed plants, the gametophyte generation is greatly reduced. A germinating pollen grain is the mature microgametophyte of a pine. Pine microsporangia are borne in pairs on the scales of the delicate pollen-bearing cones. Megagametophytes, in contrast, develop within the ovule. The familiar seed-bearing cones of pines are much heavier than the pollen-bearing cones. Two ovules, and ultimately two seeds, are borne on the upper surface of each scale of a cone. In the spring, when the seed-bearing cones are small and young, their scales are slightly separated. Drops of sticky fluid, to which the airborne pollen grains adhere, form between these scales. Pollination occurs more than a year before the ovule produces a mature female gametophyte. These pollen grains germinate, and slender pollen tubes grow slowly toward the egg. When a pollen tube grows to the vicinity of the megagametophyte, sperm are released, fertilizing the egg and producing a zygote there. The development of the zygote into an embryo occurs within the ovule, which matures into a seed. Seeds mature up to six months after fertilization. Eventually, the seed falls from the cone and germinates, the embryo resuming growth and becoming a new pine tree.

*b.* How many leaves are in a bundle?

*c.* Pine "needles" are modified leaves. How are pine needles different from the leaves of broad-leaved trees such as maples and oaks?

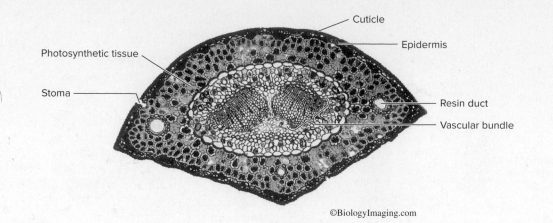

Cuticle

Epidermis

Photosynthetic tissue

Stoma

Resin duct

Vascular bundle

©BiologyImaging.com

**Figure 30.7**   Cross section of a pine leaf (needle) (40×).

*d.* Why are pines called evergreens?

*e.* What other plants have you studied in the lab that are evergreen?

*f.* What is the function of each of the structures labeled in figure 30.7?

*g.* How do the structural features of pine leaves adapt the tree for life in dry environments?

Male cones usually form on the lower branches of pine trees, and female cones usually form on the upper branches. Examine pine branches with staminate (male) and ovulate (female) cones. Examine a prepared slide of a young staminate cone and note the pine pollen in various stages of development. Each scale (microsporophyll) of the male cone bears a microsporangium, which in turn produces diploid **microspore mother cells** (fig. 30.8). Microspore mother cells undergo meiosis to produce **microspores** that develop via mitotic divisions into microgametophytes called pollen grains (fig. 30.9). Each **pollen grain** consists of four nuclei and a pair of bladderlike wings. The gametophytes of gymnosperms are reduced to only a few cells.

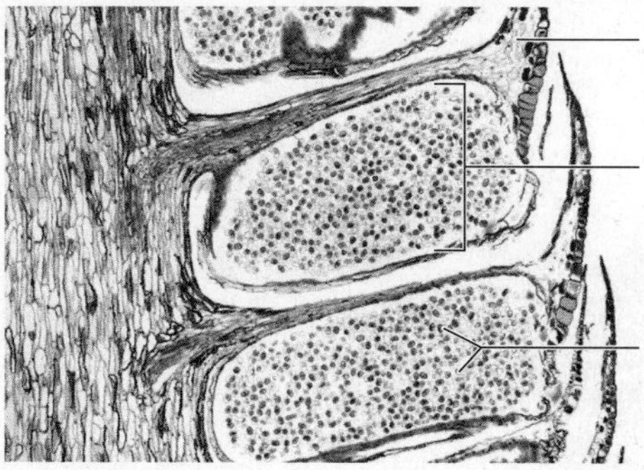

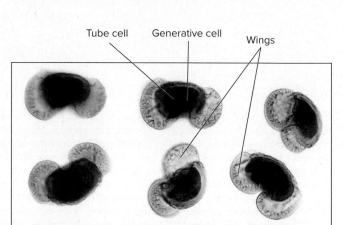

Microsporophyll

Microsporangium

Microgametophytes from microspores

©BiologyImaging.com

**Figure 30.8**   A single microsporophyll (with microsporangium) from a male cone. Within the microsporangium, microspore mother cells form microspores via meiosis (10×).

Tube cell   Generative cell   Wings

Dr. Keith Wheeler/Science Source

**Figure 30.9**   Pollen grains of *Pinus,* each with enclosed male gametophyte. Each gametophyte includes a small generative cell and a larger tube cell. When the pollen grain germinates, the pollen tube will emerge between the two bladder-shaped wings (20×).

> *Procedure 30.2*   **Examine staminate pinecones and pollen**
>
> *1.* If male cones are available, prepare a wet mount of some pine pollen; notice their characteristic shape.

2. Remove a scale from a mature staminate cone and tease open the microsporangium.

3. Prepare a wet mount of the microsporangium and its contents, and examine the contents with your microscope.

4. In the space below, draw the shape of a pine pollen grain. If male cones are not available, examine a prepared slide of pine pollen grains; sketch the shape of the pollen.

**Question 4**

*a.* Are all of the cones the same size?

*b.* Recall that in ferns the antheridia and archegonia on a prothallium mature at different times to avoid self-fertilization. How might the different locations of male and female cones on a pine tree help to do the same?

*c.* What is the probable function of the wings of a pine pollen grain?

Each ovuliferous scale of the female cone bears two megasporangia, each of which produces a diploid **megaspore mother cell.** Each megaspore mother cell undergoes meiosis to produce a **megaspore** that develops via mitotic divisions into a **megagametophyte.** The tissue of the megasporangium immediately surrounding the megagametophyte is the **nucellus** (nutritive tissue) surrounded by **integuments** (which will form the seed coat). A megagametophyte and its surrounding tissues constitute an ovule and contain at least one archegonium with an egg cell.

**Procedure 30.3  Examine ovulate pinecones and ovules**

1. Examine young living or preserved ovulate cones. These cones will develop and enlarge considerably before they are mature.

2. Examine a prepared slide of a young ovulate cone ready for pollination.

3. Examine a prepared slide of an ovulate cone that has been sectioned through an ovule (fig. 30.10). An ovule develops into a seed.

4. Examine a mature ovulate cone and notice its spirally arranged **ovuliferous scales.** These scales are analogous to microsporophylls of staminate cones. At the base of each scale you'll find two naked seeds. Notice that the seeds are exposed to the environment.

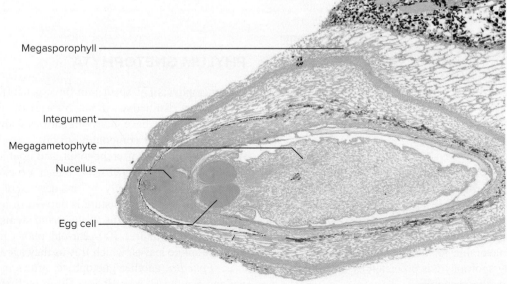

Megasporophyll

Integument

Megagametophyte

Nucellus

Egg cell

©BiologyImaging.com

**Figure 30.10**  Pine ovule section before fertilization. The egg will be fertilized later to form a new sporophyte inside the seed. The nucellus is a nutritive tissue, and integuments form the seed coat (15×). When this specimen was prepared, the soft megagametophyte shrank away from the firmer exterior layers.

## Question 5

*a.* On which surface of the scale are the seeds located?

*b.* How large is a staminate cone compared to a newly pollinated ovulate cone? Compared to a mature ovulate cone?

## Pollination and Seed Formation in Pine

**Pollination** is the transfer of pollen to a receptive surface. Pollen grains released in late spring are carried by wind to ovulate cones, in which the grains sift through the ovuliferous scales and stick in a drop of resin at the micropylar end of an ovule. Pollination occurs more than a year before the ovule produces a mature female gametophyte. The pollen grain then germinates and grows a tube into the archegonium, where it releases its two nonmotile sperm nuclei. One of these sperms disintegrates; the other fuses with the egg to form a zygote. The zygote, while in the ovule, develops into the embryo of a new sporophyte. The ovule is now a seed and consists of an embryo, a seed coat (integuments of the megasporangia), and a food supply (tissue of the megagametophyte). Seeds mature up to six months after fertilization.

> ### Procedure 30.4    Examine a pine seed
>
> 1. Examine a prepared slide of a pine seed. Locate the embryo, seed coat, and food supply. Seeds are released when the cone dries and the scales separate. This usually occurs 13–15 months after pollination.
> 2. Examine some mature pine seeds, noting the winglike extensions of the seed coat.

### Seed

The evolution of seeds is one of the most significant events in the history of the plant kingdom. Indeed, the evolution of seeds is one of the factors responsible for the dominance of seed plants in today's environment because a seed permits a small but multicellular sporophyte to remain dormant until conditions are favorable for continued growth. While dormant, the young sporophyte is protected by a seed coat and is surrounded by a food supply.

Refer to the pine life cycle shown in figure 30.6 to answer Question 6. If plant material and time are available, examine other gymnosperms in the lab.

## Question 6

*a.* Where are spores located?

*b.* What is the male gametophyte?

*c.* What is the female gametophyte?

*d.* What is an ovule?

*e.* What is an integument?

*f.* How are other gymnosperms similar to pine?

*g.* How are they different?

*h.* What evolutionary advantages might arise from not needing free water for fertilization?

## PHYLUM GNETOPHYTA

The gnetophytes (71 species in three genera) include some of the most distinctive (if not bizarre) of all seed plants (fig. 30.11). They have many similarities with angiosperms, such as flowerlike compound strobili, vessels in the secondary xylem, loss of archegonia, and double fertilization. The slow-growing *Welwitschia* plants are extraordinary in appearance and live only in the deserts of southwestern Africa. Most of their moisture is derived from fog that rolls in from the ocean at night. *Welwitschia* stems are short and broad (1 m). Mature, 100-year-old plants have only two wide, straplike leaves, which fray as they age.

*Ephedra*, another gnetophyte, grows in North America and Southeast Asia. It is a shrub with small, scalelike leaves and conelike reproductive structures. The stimulant ephedrine, which is used to prevent low blood-pressure, is extracted from *Ephedra*.

(a) ©Robert & Linda Mitchell

(b) LOETSCHER CHLAUS/Alamy Stock Photo

(c) ©BiologyImaging.com

**Figure 30.11** Gnetophytes. (*a*) Leaves and immature seeds of *Gnetum*. *Gnetum* grows as a shrub or woody vine in tropical or subtropical forests. (*b*) *Welwitschia* plants in the Namib Desert of southwest Africa. (*c*) *Ephedra* is the only living genus of Gnetophyta found in the United States. Most photosynthesis in *Ephedra* occurs in its photosynthetic stems, not its small, scaleline leaves. Dietary supplements containing ephedrine, which is extracted from *Ephedra*, are illegal, except those used in traditional Chinese remedies.

## ORGANIZING WHAT YOU'VE LEARNED

Go to page 362, which follows Exercise 31. Complete the portions of the table that refer to the phyla you studied in today's lab.

## INQUIRY-BASED LEARNING

### *How much pollen is released by a male pinecone?*

Observation: Male pinecones produce and release large amounts of pollen. In some people this pollen contributes to hay fever and other sinus-related problems.

Question: How much pollen is released by a male pinecone?

**a.** Establish a working lab group and obtain Inquiry-Based Learning Worksheet 30 from your instructor.

**b.** Discuss with your group well-defined questions relevant to the preceding observation and question. Choose and record your group's best question for investigation.

**c.** Translate your question into a testable hypothesis and record it.

**d.** Outline on Worksheet 30 your experimental design and supplies needed to test your hypothesis. Ask your instructor to review your proposed investigation.

**e.** Conduct your procedures, record your data, answer your question, and make relevant comments.

**f.** Discuss with your instructor any revisions to your questions, hypotheses, or procedures. Repeat your work as needed.

## Questions for Further Study and Inquiry

1.  What is the difference between pollination and fertilization?

2.  What does the term *gymnosperm* mean? Is this term an appropriate description? Explain your answer.

3.  How is alternation of generations different in ferns and pines?

4.  How are the environmental agents for uniting sperm and egg different in pines and bryophytes?

5.  Compare and contrast alternation of generations in mosses, ferns, and pines.

6.  What is the evolutionary significance of pollen and seeds?

7.  What are the distinguishing features of each group of plants that you studied in this exercise?

**WRITING TO LEARN BIOLOGY**
Describe the ecological advantages that the production of seeds has over the reproductive process in ferns.

# Survey of the Plant Kingdom
## Angiosperms

## Learning Objectives

By the end of this exercise you should be able to:
1. Relate the life cycle of angiosperms to the other phyla of the plant kingdom.
2. Discuss the events associated with development of microspores, megaspores, gametophytes, gametes, seeds, and fruit.
3. List and discuss the reasons why angiosperms are considered to be the most advanced land plants.
4. Understand the structure and function of flowers.

Please visit connect.mheducation.com to review online resources tailored to this lab.

Flowering plants (phylum Anthophyta—also referred to as angiosperms)—are the most abundant, diverse, and widespread of all land plants. They owe their success to several factors, including their structural diversity, efficient vascular systems, and mutualisms with fungi and insects (fig. 31.1). Angiosperms' flowers attract insects and other pollinators, and thereby make reproduction more efficient. Fruits, which contain seeds and nutritive tissue, help disperse angiosperms to new environments.

BiologyImaging.com

**Figure 31.1** This bee is pollinating a rock rose in central Texas. The evolution of flowers and the success of flowering plants correlate with the development of mutualisms with insects and other pollinators.

## PHYLUM ANTHOPHYTA

Angiosperms range in size from 1 mm (*Wolffia*) to over 100 m tall (*Eucalyptus*). As in gymnosperms, the sporophyte of angiosperms is large and heterosporous. The microgametophyte (pollen) and megagametophyte (embryo sac) are completely dependent on the sporophyte. Angiosperms are important to humans because our world economy is overwhelmingly based on them. We eat and use **vegetative structures** (roots, stems, leaves) as well as **reproductive structures** (flowers, fruits, seeds) of angiosperms.

There are more than 250,000 living species of angiosperms. Until the late 1990s, most botanists grouped angiosperms into two groups, depending on the number of **cotyledons** (seed leaves) in the embryo: species having one cotyledon were called **monocots,** and those with two cotyledons were called **dicots.** However, genetic analyses revealed that species traditionally called dicots evolved from more than one common ancestor. As a result, the vast majority of species once classified as dicots are now known as **eudicots.** (The other former dicots comprise several small lineages that will not be considered in this exercise.) These studies also showed that all monocots evolved from a common ancestor; as a result, their classification as monocots remains unchanged (fig. 31.2). The "typical" features of these groups of plants are described in table 31.1.

Having studied gymnosperms, you probably concluded that vegetative adaptations of angiosperms and gymnosperms are similar. This is true. However, in today's exercise you will study two unique adaptations of angiosperm reproduction: flowers and fruit.

Figure 31.2 In monocots such as this lily (*Lilium* sp.), flower parts usually occur in multiples of three. A lily has six stamens, three petals, three petal-like sepals, and a three-chambered ovary.

## STRUCTURE AND FUNCTION OF FLOWERS

Luckily for florists, angiosperms include a seemingly infinite variety of flowers, ranging from the microscopic flowers of

<table>
<tr><td colspan="1">Table 31.1</td></tr>
</table>

**Characteristics and Examples of the Two Major Classes of Angiosperms**

**Monocots (~70,000 species)**

1. One cotyledon per embryo
2. Flower parts in sets of three
3. Parallel venation in leaves
4. Multiple rings of vascular bundles in stem
5. Lack a true vascular cambium (lateral meristem)
6. Pollen grains have one pore.

*Examples:* Orchids, barley, corn, lily, wheat, daffodils, onion, bamboo, banana

**Eudicots (~170,000 species)**

1. Two cotyledons per embryo
2. Flower parts in sets of four or five
3. Reticulate (i.e., netted) venation in leaves
4. One ring of vascular bundles or cylinder of vascular tissue in stem
5. Have a true vascular cambium (lateral meristem)
6. Pollen grains have three pores.

*Examples:* Zucchini, roses, oaks poppy, peas, beans, mint, daisies, tomato

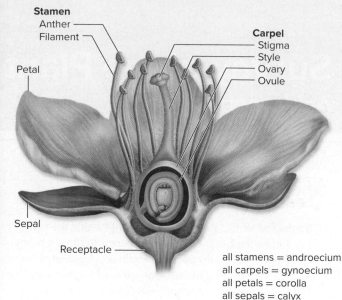

all stamens = androecium
all carpels = gynoecium
all petals = corolla
all sepals = calyx

Figure 31.3 The parts of a flower. This is a generalized flower with four primary parts: sepals, petals, stamens, and carpels. The carpels and stamens are the fertile parts of a flower. The roles of carpels and stamens in the life cycle of angiosperms are shown in figure 31.4.

*Lemna* (duckweed) to the rare, monstrous blossoms of *Rafflesia,* which can be up to 1 m across. In today's exercise, we'll consider only the "typical" flower depicted in figure 31.3.

Examine the flower model(s) available in the lab and the living flowers on display. Identify the following parts:

- **Peduncle**—flower stalk.
- **Receptacle**—the part of the flower stalk that bears the floral organs; located at the base of the flower; usually not large or noticeable.
- **Sepals**—the lowermost or outermost whorls of structures, which are usually leaflike and protect the developing flower; the sepals collectively constitute the **calyx.**
- **Petals**—whorls of structures located inside and usually above the sepals; may be large and pigmented (in insect-pollinated flowers) or inconspicuous (in wind-pollinated plants); the petals collectively constitute the **corolla.**
- **Androecium**—the male portion of the plant that rises above and inside the petals; consists of **stamens,** each of which consists of a **filament** atop which is located an **anther;** inside the anthers are **pollen grains,** which are the microgametophytes and contain the male gametes.
- **Gynoecium**—the female portion of the plant that rises above and inside the androecium; consists of one or more **carpels,** each made up of an **ovary, style,** and **stigma;** the ovary contains **ovules** that contain the megagametophyte; the

megagametophyte is called the embryo sac and contains female gametes.

The sepals and petals are usually the most conspicuous parts of a flower, and a variety of flower types are distinguished by their sepal and petal characteristics. In radially symmetrical flowers such as tulips, the different whorls of the flower consist of similarly shaped parts that radiate from the center of the flower and are equidistant from each other. In other flowers such as snapdragons, one or more parts of at least one whorl are different from other parts of the same whorl. These flowers are usually bilaterally symmetrical.

### Procedure 31.1  Examine flowers and their parts

1. Obtain a flower provided in the laboratory.
2. Remove the parts of the flower, beginning at the lower, outside whorls of sepals and petals.
3. Locate the petals and sepals and determine their arrangement and point of attachment.

#### Question 1
How would you describe this flower?

4. Remove the petals and sepals.
5. Locate and remove a stamen and place it on a slide. Examine the stamen with low magnification.
6. Examine a prepared slide of pollen grains. Use high magnification to locate the generative and tube nucleus of a pollen grain. The **generative nucleus** is usually small, spindle-shaped, and off center (you cannot see it easily). The **tube nucleus** is larger and centered. You'll learn more about pollen nuclei later in this exercise.
7. Locate the gynoecium of the flower, and make longitudinal and transverse sections. The gynoecium consists of one or more fused carpels, each with an interior cavity called a locule containing ovules.

#### Question 2
How many carpels (locules) are apparent?

8. Examine the other types of flowers available in the lab. Repeat procedure 31.1 to guide your examination. Preserved specimens of *Lilium* (lily) and *Ranunculus* (buttercup) may also be provided for you to study and dissect.

## Alternation of Generations in Flowering Plants

The life cycle of flowering plants involves the alternation of a multicellular haploid stage with a multicellular diploid stage, as is typical for all plants (fig. 31.4, also see fig. 30.1). The diploid **sporophyte** produces haploid spores by meiosis. Each haploid spore develops into the **gametophyte** by mitosis and cellular differentiation. In angiosperms the sporophyte is the large, mature organism with flowers that you easily recognize. The gametophyte within the flower is reduced to a pollen grain (that contains a sperm nucleus) or an embryo sac (that produces an egg) within an ovule. Be sure to review your textbook's description of alternation of generations.

Production of spores in the sporophyte by meiosis is part of a larger process called **sporogenesis.** Flowering plants produce two types of spores: **microspores** and **megaspores.** Production of gametes by the gametophytes is called **gametogenesis.**

## Microsporogenesis, Production of Pollen, and Microgametogenesis

Many of the terms associated with flowers are similar to those you've learned in earlier exercises, except that with angiosperms those terms have a prefix. For example, sporogenesis (i.e., the production of spores) in angiosperms is called microsporogenesis. **Microsporogenesis** is the production of microspores within **microsporangia** of a flower's anthers via meiosis of **microspore mother cells** (microsporocytes) (fig. 31.5). These microspores grow and mature into microgametophytes, which are also known as pollen grains (fig. 31.6). The haploid nuclei in a mature pollen grain include a **tube nucleus** (or vegetative nucleus) and a **generative nucleus.** During pollination, pollen grains are transferred to the stigma, where they germinate and grow a tube through the style to the ovary. The tube nucleus controls the growth of the pollen tube, and the generative nucleus replicates to produce two **sperm nuclei.**

Pollen of some plants cause allergies in many people. However, studies of pollen are important to science beyond the treatment of allergies. For example, geologists examine pollen brought up in sediment cores during oil drilling. Dark brown to black pollen often indicate temperatures too high for oil deposits and suggest that a well will likely produce natural gas. Orange pollen suggest the less intense heat associated with high-quality oil production. In addition, examination of fossil pollen tells us about the diversity of ancient flora and climatic change through the ages, and helps us locate ancient seas and their shorelines where pollen is known to accumulate.

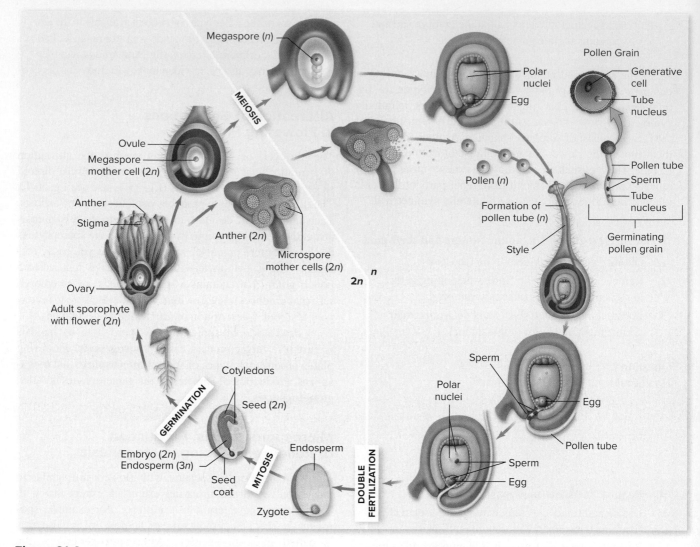

**Figure 31.4** Angiosperm life cycle. Eggs form within the embryo sac inside the ovules, which, in turn, are enclosed in the carpels. The pollen grains, meanwhile, form within the sporangia of the anthers and are shed. Fertilization is a double process. A sperm and an egg fuse to produce a zygote; at the same time, another sperm fuses with the polar nuclei to produce the endosperm. The endosperm is the tissue, unique to angiosperms, that nourishes the embryo and young plant.

## Procedure 31.2 Examine stages of microsporogenesis and microgametogenesis in *Lilium*

*1.* Examine fresh or preserved specimens of dehiscent (split-open) and predehiscent anthers (fig. 31.7).

### Question 3

*a.* What differences can you detect between the structures of dehiscent and predehiscent anthers?

*b.* Which stage is the most mature?

*2.* Examine a prepared cross section of a young anther. Note the immature sporangia tissue that will form microsporocytes.

*3.* Examine a prepared cross section of a lily anther showing microsporocytes in early prophase I (fig. 31.8*a*).

*4.* Examine a prepared slide showing stages of meiosis II (fig. 31.8*b*).

*5.* Examine a prepared slide showing pollen tetrads of microspores produced by meiosis (fig. 31.8*c*).

*6.* Examine a prepared slide showing mature pollen with two or more nuclei (fig. 31.8*d*).

*7.* Examine living or prepared pollen from various plants if available. Note any differences among pollen grains and differences between pollen of monocots and eudicots.

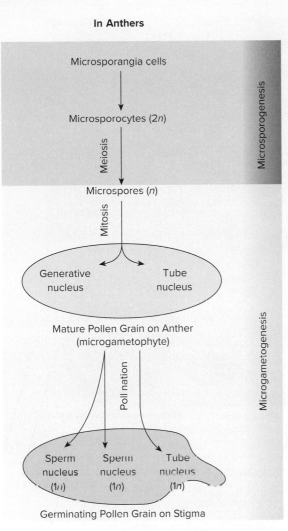

**In Anthers**

Microsporangia cells

↓

Microsporocytes (2n)

| Meiosis

↓

Microspores (n)

| Mitosis

↓

Generative nucleus    Tube nucleus

Mature Pollen Grain on Anther (microgametophyte)

Pollination

Sperm nucleus (1n)    Sperm nucleus (1n)    Tube nucleus (1n)

Germinating Pollen Grain on Stigma

*Microsporogenesis*

*Microgametogenesis*

**Figure 31.5**  Microsporogenesis and microgametogenesis occur in the anthers of flowers. Mature pollen grains (see fig. 31.6) are microgametophytes. Micrographs of microsporogenesis and microgametogenesis are shown in figure 31.8.

---

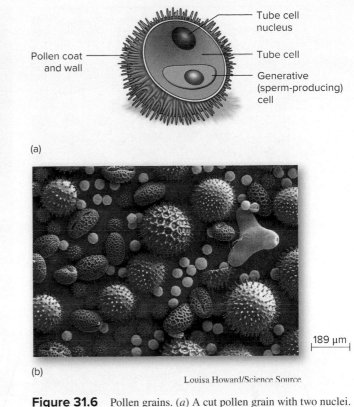

(a)

Tube cell nucleus

Tube cell

Generative (sperm-producing) cell

Pollen coat and wall

(b)

189 μm

Louisa Howard/Science Source

**Figure 31.6**  Pollen grains. (*a*) A cut pollen grain with two nuclei. (*b*) Scanning electron micrograph of pollen grains of different species (1200×).

**Question 4**

***a.***  How many pollen grains germinated?

***b.***  Can you see vegetative and generative nuclei in the pollen tubes?

---

**Procedure 31.3**  **Examine germinated pollen grains**

Examine a prepared slide of germinated pollen grains.

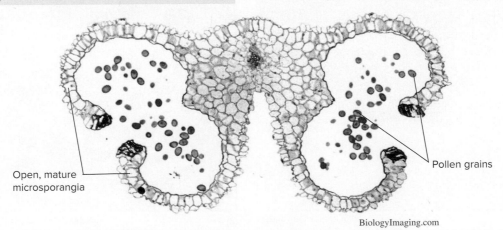

Open, mature microsporangia

Pollen grains

BiologyImaging.com

**Figure 31.7**  *Lilium* anther (40×). Mature pollen grains are the product of microsporogenesis and microgametogenesis within the microsporangia of flower anthers. A higher-magnification view of developing pollen grains is shown in figure 31.8.

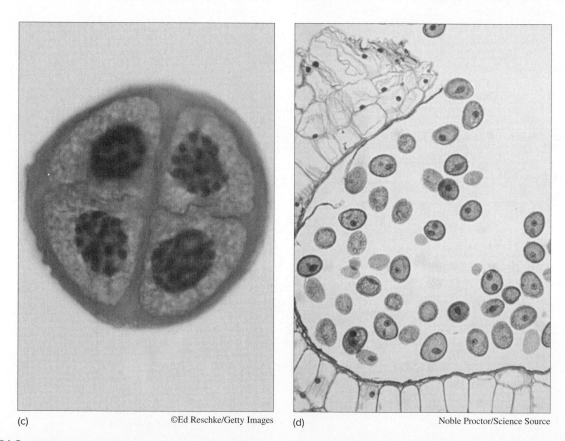

(a)      M. I. Walker/Science Source

(b)      Biophoto Associates/Science Source

(c)      ©Ed Reschke/Getty Images

(d)      Noble Proctor/Science Source

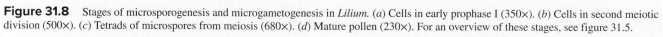

**Figure 31.8** Stages of microsporogenesis and microgametogenesis in *Lilium*. (*a*) Cells in early prophase I (350×). (*b*) Cells in second meiotic division (500×). (*c*) Tetrads of microspores from meiosis (680×). (*d*) Mature pollen (230×). For an overview of these stages, see figure 31.5.

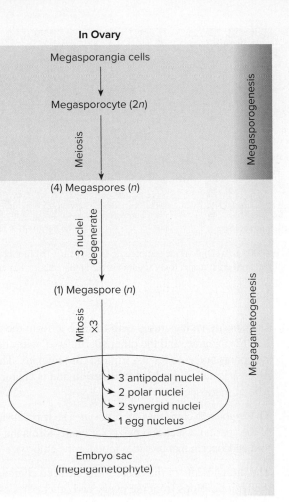

**In Ovary**

Megasporangia cells

↓

Megasporocyte (2*n*)

Megasporogenesis

Meiosis

↓

(4) Megaspores (*n*)

3 nuclei degenerate

↓

(1) Megaspore (*n*)

Megagametogenesis

Mitosis ×3

↓

→ 3 antipodal nuclei
→ 2 polar nuclei
→ 2 synergid nuclei
→ 1 egg nucleus

Embryo sac
(megagametophyte)

**Figure 31.9** Megasporogenesis and megagametogenesis occur in the ovaries of flowers. The embryo sac is the mature megagametophyte. The locations of these processes are shown in figures 31.10 and 31.11.

## Megasporogenesis, Production of Ovules, and Megagametogenesis

**Megasporogenesis** is the production of megaspores (fig. 31.9); it occurs in the sporangia of the flower ovary by meiosis of **megaspore mother cells** (megasporocytes). These megaspores undergo **megagametogenesis;** that is, they develop into **megagametophytes.** The megagametophyte and its surrounding tissues are called an **ovule.** Ovules usually have two coverings called **integuments.** The entire haploid structure is called the **embryo sac** and consists of only six to ten nuclei, one of which is an egg. A seven- or eight-celled embryo sac is most common (fig. 31.10).

---

### *Procedure 31.4* Examine stages of megasporogenesis and megagametogenesis in *Lilium*

*1.* Examine a cross section of a *Lilium* ovary and locate the six megasporangia. Within each of these megasporangia a megasporocyte will form and develop. The stages for development of a megasporocyte are shown in figures 31.9, 31.10, and 31.11.

*2.* Examine a prepared slide of a cross section of a *Lilium* ovary showing a diploid megasporocyte within the sporangium before meiosis (fig. 31.11*a*).

*3.* Examine a prepared slide showing the four-nucleate embryo sac after meiosis (fig. 31.11*b*). Meiosis produced four haploid megaspores in the embryo sac. In most angiosperms, three of the four nuclei degenerate and the single remaining nucleus passes through two mitotic divisions before the next stage. *Lilium* is atypical because all four products of meiosis contribute nuclei to the embryo sac, but only one develops into the egg cell.

*4.* Examine a prepared slide showing the eight-nucleate embryo sac (fig. 31.11*c*). Because of the large size of the embryo sac, it is seldom possible to observe all eight nuclei in the same section.

*5.* At the end of the megagametophyte toward the micropyle (the small opening where the pollen tube enters the ovule), locate the eight nuclei including the egg and two synergid nuclei associated with fertilization

---

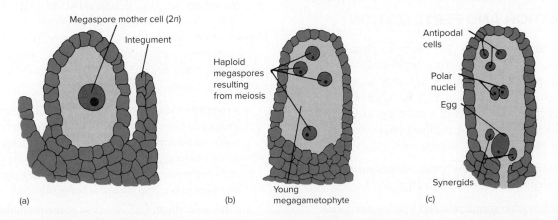

Megaspore mother cell (2*n*)

Integument

Haploid megaspores resulting from meiosis

Young megagametophyte

Antipodal cells

Polar nuclei

Egg

Synergids

(a)          (b)          (c)

**Figure 31.10** Megasporogenesis and megagametogenesis in *Lilium.* (*a*) The megaspore mother cell (megasporocyte) is diploid and undergoes meiosis. (*b*) Four haploid megaspores result from meiosis. (*c*) The mature megagametophyte contains eight nuclei, one of which is the egg.

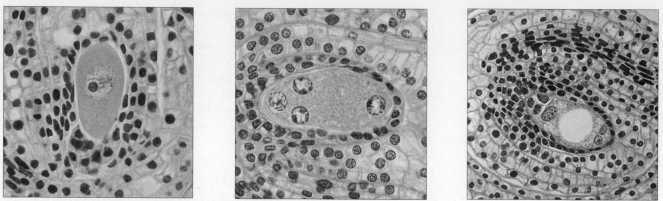

(a) Biophoto Associates/Science Source  (b) Biophoto Associates/Science Source  (c) Ed Reschke/Getty Images

**Figure 31.11**    Stages of megasporogenesis and megagametogenesis in *Lilium* (200×). (*a*) Diploid megasporocyte in lily ovary. (*b*) Embryo sac with four nuclei produced by meiosis. (*c*) Embryo sac (mature megagametophyte) with eight nuclei (only six are visible in this image). For an overview of these stages, see figure 31.9.

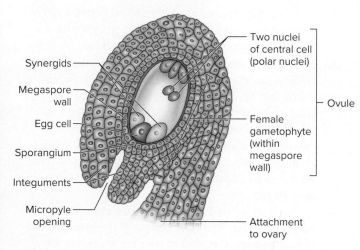

**Figure 31.12**    Mature female gametophyte within an ovule.

(figures 31.11c and 31.12). At the opposite end locate three antipodal cells that usually do not participate in reproduction. In the center are two polar nuclei that migrated from each pole of the megagametophyte.

## POLLINATION AND FERTILIZATION

Biologists have long been interested in plant pollination, for obvious economic reasons. Angiosperm reproduction depends on pollination. However, you may have a more personal interest in pollen if you suffer from hay fever (fig. 31.6).

After the development of a microgametophyte (pollen grain) with sperm and a megagametophyte (ovule) with an egg, sexual reproduction in angiosperms occurs as follows:

1. Pollination occurs when pollen is transported to the surface of the flower's stigma (fig. 31.13).

2. The pollen grain germinates, and a pollen tube grows through the stigma and style to the ovary. Its growth is governed by the style and the tube nucleus of the pollen grain.

3. One sperm nucleus fuses with the egg to form the diploid ($2n$) zygote, and the other sperm fuses with the two polar nuclei to form a triploid ($3n$) nucleus. This process is called **double fertilization** and is characteristic of angiosperms.

4. The zygote develops into the embryo. The triple fusion of the sperm nucleus and two polar nuclei forms the triploid endosperm that provides food for the embryo.

5. The integuments of the ovule form the seed coat, and the fruit develops from the ovary and other parts of the flower.

## SEED AND EMBRYO DEVELOPMENT

A **seed** is a mature ovule that includes a seed coat, a food supply, and an embryo. Seeds range in size from tiny (e.g., a broomrape seed weighs only a few micrograms) to massive (e.g., an avocado seed can weigh more than 50 g). Embryology of the mature zygote occurs within the seed and before germination. This embryology and its controlling factors are complex, but various stages of development are easily observed.

The stages of embryo development in the seed of *Capsella* (a eudicot) are shown in figure 31.14. The developing embryo grows, absorbs endosperm, and stores those nutrients in "seed leaves" called cotyledons. Development includes the following stages:

- **Proembryo stage.** During development, the **zygote** divides to form a mass of cells called the **embryo.** Initially the embryo consists of a **basal cell, suspensor,** and a two-celled proembryo. The suspensor is the column of cells that pushes the embryo into the **endosperm.** The endosperm is extensive but is being digested.

- **Globular stage.** Cell division of the proembryo soon leads to the globular stage that is radially symmetrical and has little internal cellular organization.

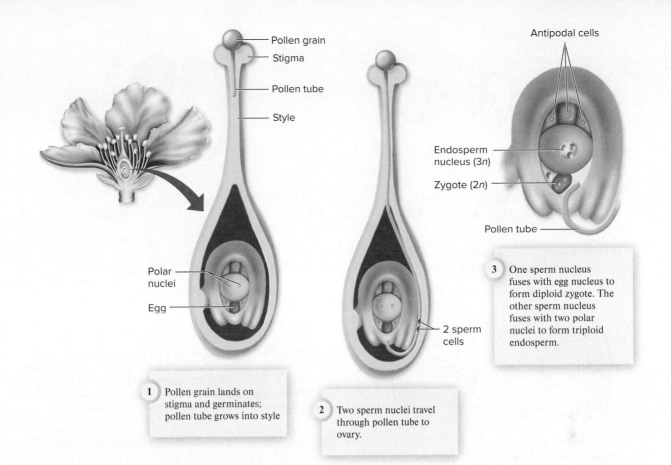

**Figure 31.13** Pollen germination and double fertilization. Pollen sticks to a stigma on a flower of the same species. A pollen tube grows toward the ovule and transports two sperm nuclei. One sperm nucleus fertilizes the egg to form a zygote, and the other fertilizes the polar nuclei to yield the endosperm.

- **Heart-shaped stage.** Differential division of the globular stage produces bilateral symmetry and two **cotyledons** forming the heart-shaped embryo. The enlarging cotyledons store digested food from the endosperm. Tissue differentiation begins, and root and shoot meristems soon appear.

- **Torpedo stage.** The cotyledons and root axis soon elongate to produce an elongate torpedo-stage embryo. Procambial tissue appears and will later develop into vascular tissue.

- **Mature embryo.** The mature embryo has large, bent cotyledons on each side of the **stem apical meristem.** The **radicle,** later to form the root, is differentiated toward the suspensor. The radicle has a **root apical meristem** and **root cap.** The **hypocotyl** is the region between the apical meristem and the radicle. The endosperm is depleted, and food is stored in the cotyledons. The **epicotyl** is the region between attachment of the cotyledons and the stem apical meristem; it has not elongated in the mature embryo.

### Procedure 31.5 Examine development of a *Capsella* embryo

1. Obtain and examine prepared slides showing the various stages of embryo development.

2. The cross section most likely passes through an entire fruit of *Capsella* and shows a number of sectioned, developing seeds each in a slightly different stage. Some seeds were sectioned off-center and the stage may not be obvious. Nevertheless, each slide should include an example of at least one stage.

3. Locate as many of the following stages of seed development as possible on your slide: globular, heart, torpedo, and mature embryos. These stages are continuous, so each slide may have multiple stages and intermediates between stages.

4. You may need to examine several slides. Find examples of as many stages as you can.

5. Compare your observations with figure 31.14.

### Question 5
*a.* Why is the endosperm being digested?

*b.* Is *Capsella* a monocot or a eudicot? How can you tell?

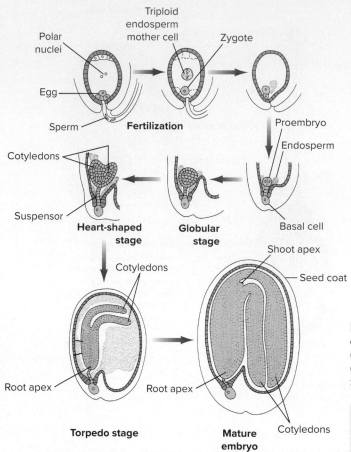

**Figure 31.14** Stages of development in a *Capsella* embryo. This development transforms a zygote into a mature embryo.

# SEED STRUCTURE

 Do not eat any seeds or fruits used in this laboratory.

## Procedure 31.6 Observe parts of a bean seed

*1.* Obtain some beans that have been soaked in water for 24 h.

*2.* Peel off the seed coat and separate the two cotyledons. Between the cotyledons you'll see the young root and shoot.

*3.* Examine the opened seed and compare its structure with that shown in figure 31.15. Look for these features:

• **Micropyle**—a small opening on the surface of the seed through which the pollen tube grew.

• **Hilum**—an adjacent, elliptical area at which the ovule was attached to the ovary.

• **Cotyledon**—food for the embryo.

• **Embryo with young root and shoot**—develops into the new sporophyte.

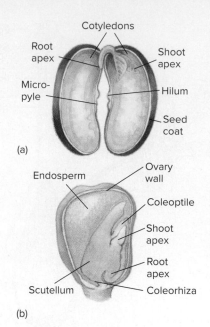

**Figure 31.15** Seed structure of a garden bean (eudicot) and corn (monocot). (*a*) The two cotyledons in each seed of garden bean (*Phaseolus vulgaris*) absorb the endosperm before germination. (*b*) Corn (*Zea mays*) has seeds in kernels (grains); the single cotyledon is an endosperm-absorbing structure called a scutellum.

*4.* Add a drop of iodine to the cut surface and observe the staining pattern. (Recall from Exercise 6 that iodine stains starch black.) Show this pattern on figure 31.15.

*5.* Repeat steps 2–4 with soaked peas.

**Question 6**
How are seeds of peas and beans similar? How are they different?

## Procedure 31.7 Examine a corn grain with embryo

*1.* Examine a prepared slide of a corn grain (fig. 31.16). Identify the following features:

• **Endosperm**

• **Scutellum** (cotyledon; this structure helps absorb the endosperm)

• **Coleoptile** (sheath enclosing shoot apical meristem and leaf primordia of grass embryos)

• **Root**

• **Root cap**

• **Coleorhiza** (sheath enclosing embryonic root of grass embryo)

• **Shoot apical meristem**

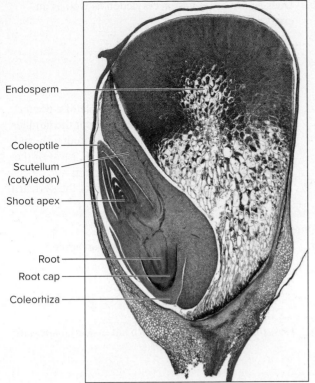

Endosperm

Coleoptile

Scutellum
(cotyledon)

Shoot apex

Root

Root cap

Coleorhiza

Dr. Keith Wheeler/Science Source

**Figure 31.16** The structure of a corn grain, longitudinal section. The embryo is nourished by the starchy endosperm (7×).

2. Use a razor blade to longitudinally split a water-soaked corn grain.
3. Add a drop of iodine to the cut surface and observe the staining pattern. Indicate this pattern on figure 31.16.

**Question 7**

*a.* What does this staining pattern tell you about the content of the endosperm and embryo?

*b.* Do mature seeds of monocots or eudicots store most of their food in cotyledons? How can you tell?

## FRUIT

A **fruit** is a mature, ripened ovary plus any associated tissues. Therefore, a fruit contains seeds. Are you surprised that tomatoes, beans, and okra are fruit? A mature fruit is often larger than the ovary at the time of pollination and fertilization, which indicates that a great deal of development occurs while the seeds are maturing.

Most fruits are either dry or fleshy. Dry fruits crack or split at maturity and release their seeds. Sometimes the dry wall surrounds the seed until it germinates. The fruit wall is usually tough and hard and is sometimes referred to as stony. The seeds of fleshy fruits remain in the tissue until germination.

A typical fruit has an outer wall called a **pericarp** composed of an **exocarp, mesocarp,** and **endocarp.** Within the pericarp are seeds, various partitions, and placental tissues. The fruit often includes the receptacle of the flower. The structure of a fruit depends on the number of carpels in each flower and whether or not the carpels are fused.

*Procedure 31.8* **Observe the diversity of fruits**

*1.* Use the following descriptions to study and classify fruits available in the lab. This list is not complete but describes a few common types of fruit.

*Procedure 31.9* **Observe the structure of a bean, sunflower, corn grain, apple, and tomato**

*1.* Examine the pod of a string bean. This single carpel has two seams that can open to release seeds. Remove the seeds; then locate and remove the seed coat. Split the seed and locate the embryo.

**Question 8**

*a.* Does the pod appear to be a single carpel with one cavity containing seeds?

*b.* Is the micropyle near the attachment of the seed to the pod?

*2.* Crack the outer coat of a sunflower fruit and remove the seed. Locate the embryo and determine whether sunflower is a monocot or eudicot.

**Question 9**

*a.* Before today would you have referred to the uncracked sunflower achene as a fruit? Why or why not?

*b.* Is sunflower a monocot or eudicot?

## Dichotomous Key to Major Types of Fruit

**I.** Fleshy fruits

  **A.** Simple fruits develop from a single ovary.

    **1.** Flesh mostly of ovary tissue

      **a.** Endocarp hard and stony; ovary superior and single-seeded (cherry, olive, peach, plum, coconut): **drupe**

      **b.** Endocarp fleshy or slimy; ovary usually many-seeded (tomato, grape, green pepper): **berry**

    **2.** Flesh mostly of receptacle tissue (apple, pear, quince): **pome**

  **B.** Complex fruits develop from more than one ovary.

    **1.** Fruit from many carpels on a single flower (raspberry, blackberry): **aggregate fruit**

    **2.** Fruit from carpels of many flowers fused together (pineapple, mulberry): **multiple fruit**

**II.** Dry fruits

  **A.** Fruits that split open at maturity (usually more than one seed)

    **1.** Split occurs along two seams in the ovary. Seeds borne on one of the halves of the split ovary (pea and bean pods, peanuts): **legume**

    **2.** Seeds released through pores or multiple seams (poppies, irises, lilies): **capsule**

  **B.** Fruits that do not split open at maturity (usually one seed)

    **1.** Pericarp hard and thick, with a cup at its base (acorn, chestnut, hickory): **nut**

    **2.** Pericarp thin and winged (maple, ash, elm): **samara**

    **3.** Pericarp thin and not winged

      (sunflower, buttercup): **achene**

      (cereal grains): **caryopsis**

**3.** Examine an ear of corn from which the husks have been removed without disturbing the "silk." The strands of silk are styles of the gynoecium.

### Question 10

**a.** If a corn grain is actually a fruit, where is the pericarp?

**4.** Examine an apple, which is an example of a pome. Section the apple transversely and count the number of carpels. Then section an apple longitudinally and compare its structure with figure 31.18. Most of the outer flesh of the apple is derived from tissue other than the ovary. Locate the outer limit of the pericarp and the limit of the endocarp.

### Question 11

**a.** How many carpels are fused to form an apple?

**b.** How might the fleshy pericarp aid in seed dispersal?

**5.** Examine a tomato and section it transversely. Compare its structure with that shown in figure 31.18. The jelly-like material is the placenta giving rise to the ovules.

### Question 12

How many carpels are fused to form a tomato?

### Our Uses of Angiosperms

We have countless uses for angiosperms, including wood (e.g., oak, maple), food (e.g., wheat, rice, corn, apples, tomatoes), textiles (e.g., jute), shade, decorations, insect repellants, spices, and landscaping. Modern agriculture is based almost entirely on angiosperms.

We also used angiosperms as sources of numerous drugs. For example, Taxol (a drug used to treat cancer) and morphine (a painkiller) are derived from flowering plants, as are illicit drugs such as cocaine, opium, and heroin. So, too, is caffeine, a drug ingested by more than 100 million Americans every day.

## ORGANIZING WHAT YOU'VE LEARNED

Go to page 362, which follows this exercise. Complete the portions of the table that refer to the phyla you studied in today's lab.

Coffee, which is grown in more than 50 countries worldwide, is the world's second-most traded commodity (petroleum is first). Coffee is a $60 billion industry. Brazil is the world's leading producer of coffee; it produces 30% of the world's crop. Worldwide, nearly 7 million tons of green coffee beans are produced per year. About 80% of this coffee comes from *Coffea arabica*, and the remaining 20% from *Coffea canephora*. It takes three to five years for a coffee plant to reach maturity and produce its scented white flowers and cherry-like fruit (figure 31.17). The red/purple coffee fruits are called "cherries." Except for peaberry coffee, whose fruits each contains one smaller, rounder seed, each coffee fruit contains two seeds which are referred to as "beans" (despite the fact that they are not true beans).

In the United States, coffee is the second-most popular drink (water is first). Coffee drinkers drink an average of 3.1 9-ounce cups of coffee per day. Fifty-four percent of Americans at least 18 years old drink coffee every morning, and another 30% drink coffee occasionally. About 30% of people in the United States who drink coffee every day drink specialty drinks such as lattes and cappuccinos; only about 35% of coffee drinkers like their coffee black.

The average coffee-drinker spends $165 per year on coffee. Coffee drinkers who get their coffee from a coffeehouse stand in line there about 45 hours per year.

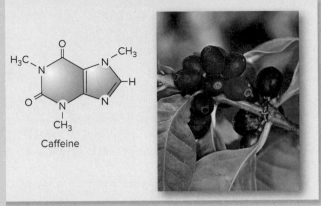

Caffeine

GFC Collection/Alamy Stock Photo

**Figure 31.17** Coffee fruit (i.e., "beans") contain caffeine, which, in nature, is a natural pesticide that paralyzes and kills certain insects that attack coffee plants. In humans, caffeine is a stimulant that temporarily wards off drowsiness and restores alertness.

## INQUIRY-BASED LEARNING

### *Grocery store botany: What plant organs are sold at your local grocery store?*

Observation: The produce sections of grocery stores and markets contain a vast array of plant organs, each having a characteristic shape, taste, and texture. These plants have been the subject of intensive artificial selection over many generations, and today we enjoy the many varied products of that selection.

Question: What plant organ (e.g., modified root, stem, leaf, flower) are you purchasing when you buy items for a salad? Assume that your salad includes beets, radishes, sweet potatoes, celery, carrots, broccoli, asparagus, eggplant, lettuce, and onion.

a. Establish a working lab group and obtain Inquiry-Based Learning Worksheet 31 from your instructor.

b. Discuss with your group well-defined questions relevant to the preceding observation and question. Choose and record your group's best question for investigation.

c. Translate your question into a testable hypothesis and record it.

d. Outline on Worksheet 31 your experimental design and supplies needed to test your hypothesis. Ask your instructor to review your proposed investigation.

e. Conduct your procedures, record your data, answer your question, and make relevant comments.

f. Discuss with your instructor any revisions to your questions, hypotheses, or procedures. Repeat your work as needed.

## Drupe

**pericarp**
- exocarp (skin)
- mesocarp (flesh)
- endocarp (pit contains seed)

Peter Fakler/Alamy Stock Photo

(a) A drupe is a fleshy fruit with a pit containing a single seed produced from a simple ovary.

## True Berry

- chamber of ovary has many seeds
- exocarp

D. Hurst/Alamy Stock Photo

(b) A berry is a fleshy fruit having seeds and pulp produced from a compound ovary.

## Legume

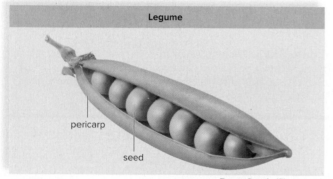

- pericarp
- seed

Danny Smythe/Shutterstock

(c) A legume is a dry dehiscent fruit produced from a simple ovary.

## Samara

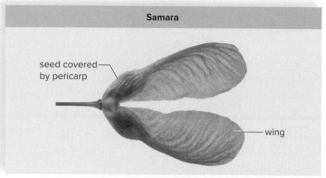

- seed covered by pericarp
- wing

Robert Llewellyn/Corbis/Getty Images

(d) A samara is a dry indehiscent fruit produced from a simple ovary.

## Aggregate Fruit

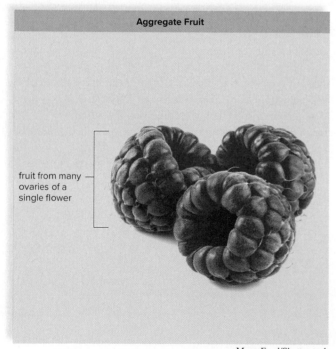

- fruit from many ovaries of a single flower

MarcoFood/Shutterstock

(e) An aggregate fruit contains many fleshy fruits produced from simple ovaries of the same flower.

## Multiple Fruit

- one fruit
- fruits from ovaries of many flowers

Ingram Publishing/Alamy Stock Photo

(f) A multiple fruit contains many fused fruits produced from simple ovaries of individual flowers.

**Figure 31.18**  Diagrams of fruit and their originating flowers.

1. What is meant by "double fertilization"?

2. What features of seeds and fruits have enabled angiosperms to become so widespread?

3. What are the similarities and differences between cones and flowers?

4. What is the difference between a fruit and a vegetable?

5. In what ways does human welfare depend on seed plants?

6. How can you distinguish a monocot from a eudicot? If a plant has six stamens and leaves with parallel veins, is the plant a monocot or a eudicot?

7. How are insects such as bees important for the reproduction of angiosperms?

8. What advantages do flowers give angiosperms over gymnosperms?

9. How is microsporogenesis different from megasporogenesis?

10. What are the major parts (and their functions) of your favorite flower?

## WRITING TO LEARN BIOLOGY

What are the functions of a flower? Describe how a flower of your choice is adapted to each function.

After each of your studies of plants (Exercises 28–31), complete the relevant parts of this table to help you summarize what you've learned.

| Exercise Number | Phylum | Common Name | Representatives | Dominant Generation | How Propagated |
|---|---|---|---|---|---|
| 28 | Hepatophyta | | | | |
| 28 | Bryophyta | | | | |
| 28 | Anthocerophyta | | | | |
| 29 | Pterophyta | | | | |
| 29 | Lycophyta | | | | |
| 30 | Cycadophyta | | | | |
| 30 | Ginkgophyta | | | | |
| 30 | Coniferophyta | | | | |
| 30 | Gnetophyta | | | | |
| 31 | Anthophyta | | | | |

# Plant Anatomy
## Vegetative Structure of Vascular Plants

## Learning Objectives

By the end of this exercise you should be able to:
1. Describe the functions of roots, stems, and leaves.
2. Distinguish between primary and secondary growth.
3. Describe the functional significance of the internal and external structure of roots, stems, and leaves.
4. Explain what causes growth rings in wood.

Please visit **connect.mheducation.com** to review online resources tailored to this lab.

The structure of plants varies greatly among species; compare, for example, an oak tree with a cactus. However, these structural differences are typically quantitative rather than qualitative; that is, the differences among roots, stems, and leaves result not from tissues unique to one species or another, but rather from different arrangements and proportions of the same tissues. These differences among plants represent a variety of adaptations that result in the universal evolutionary outcomes of survival and reproduction. This exercise will concentrate on the structure of roots, stems, and leaves of flowering plants.

## ROOTS

During seed germination, a **radicle** or young **primary root** emerges from the seed and grows down. The primary root soon produces numerous **secondary roots** and forms a root system that absorbs water and minerals, anchors the plant, and stores food. Root systems have different morphologies. For example, a **taproot system** has a large main root and smaller secondary roots branching from it (e.g., carrot). In a **fibrous root system,** the primary and secondary roots are similar in size (fig. 32.1). Examine the displays of taproot and fibrous root systems available in the lab.

    **Primary growth** of roots and all primary tissues is formed by **apical meristems.** A meristem is a localized area of cellular division. Apical meristems occur at the tips of roots and stems. Primary growth (i.e., growth in length) produces herbaceous (nonwoody) tissue. **Secondary growth** refers to growth in girth resulting from nonapical meristems, some of which are discussed later in this exercise.

## Question 1

*a.* How do taproot systems and fibrous root systems help plants survive and reproduce?

*b.* Would one type of root system provide more adaptive advantages in a particular environment such as a rain forest? A desert? Explain your answer.

## The Root Apex

Examine the root tips of two-day-old seedlings of radish (*Raphanus*) and corn (*Zea*) with your dissecting microscope. Refer to figure 32.2 and identify the **root cap, root apical meristem, zone of elongation,** and **zone of maturation.**

    The cone of loosely arranged cells at the root cap perceives gravity and protects the root apical meristem. The root cap protects the root by secreting mucilage and sloughing cells as the root grows through the soil (fig. 32.3). The root apical meristem is behind the root cap and produces all of the new cells for primary growth. These cells elongate in the zone of elongation, which is 1–4 mm behind the root tip. This elongation produces primary growth.

(a)

Peter Anderson/Getty Images

(b)

redmal/E+/Getty Images

**Figure 32.1** Two common types of root systems of vascular plants. (*a*) The taproot system consists of a prominent, thickened taproot. (*b*) The fibrous root system consists of many similarly sized roots. Fibrous root systems form extensive networks in the soil and minimize soil erosion.

### Question 2

*a.* In the following space, sketch the root tips that you examined. In which area of a root tip are cells largest? In which area are they smallest?

*b.* Aside from their size, do all cells in the root tip appear similar? Why is this significant?

Re-examine the two-day-old radish seedling with your dissecting microscope. Note the many **root hairs** in the zone of maturation. Root hairs are outgrowths of epidermal cells and are short-lived. Root hairs increase the surface area of the root.

### Question 3

*a.* Why do you think root hairs occur only in the zone of maturation? Why aren't they in the zone of elongation?

*b.* What is the function of root hairs?

## Primary Tissues of the Root

The root apical meristem produces cells that differentiate into primary tissues of the root. The outer layer of cells is the **epidermis.** Just inside the epidermis is the **cortex,** whose cells contain numerous **amyloplasts,** which are starch-containing plastids. The inner layer of the cortex is the **endodermis,** which affects water flow and regulates the flow of nutrients to the vascular tissue in the center of the root. Immediately inside the endodermis is the **pericycle,** which can become meristematic and produce **secondary roots** (figs. 32.4 and 32.5).

### *Procedure 32.1* Examine primary tissues of a root

*1.* Examine a prepared slide of a cross section of a buttercup (*Ranunculus,* a eudicot) root (figs. 32.4 and 32.5). Sketch what you see. Label and state the function of each tissue that is present. (Note: Your

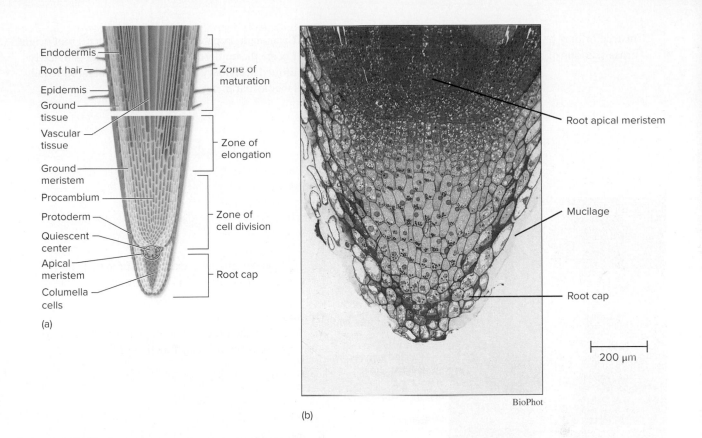

(a)

(b)

BioPhot

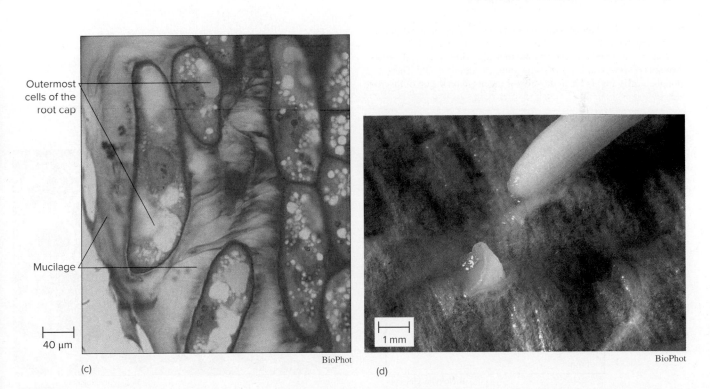

(c)

(d)

BioPhot

BioPhot

**Figure 32.2** (*a*) Root tip showing root cap, root apical meristem, zones of structure and function. (*b*) The root apical meristem is covered by a thimble-shaped root cap that protects the meristem as the root grows through the soil (60×). (*c*) Mucilage produced by the root cap lubricates the root as it grows through the soil (see fig. 32.3) (420×). (*d*) In plants such as corn, the root cap can easily be removed from the rest of the root (6×).

instructor may have you examine a slide comparing cross sections of roots of monocots and eudicots.)

2. Examine a prepared slide labeled "lateral root origin." Locate the epidermis, cortex, pericycle, and newly formed secondary root. Sketch the lateral root and label its parts.

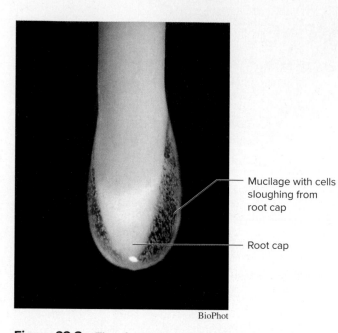

Mucilage with cells sloughing from root cap

Root cap

BioPhot

**Figure 32.3** Tips of roots secrete large amounts of mucilage, a lubricant that helps the root force its way through the soil. The mucilage is secreted primarily by the root cap. Movement of the root through soil is also aided by the sloughing of root-cap cells. These sloughed cells are visible in the drop of mucilage on the tip of this root.

3. If time permits, also examine a prepared slide of a cross section of a corn (*Zea*, a monocot) root.

**Question 4**

*a.* Based on the presence of amyloplasts, what do you suppose is the primary function of the cortex?

*b.* Do secondary roots arise inside the primary root or on its surface?

*c.* How does the structure of a monocot root differ from that of a eudicot?

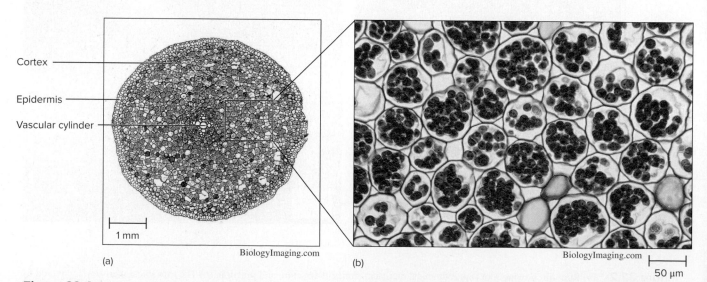

Cortex

Epidermis

Vascular cylinder

1 mm

BiologyImaging.com

(a)

(b)

BiologyImaging.com

50 μm

**Figure 32.4** Transverse sections of a root of a buttercup (*Ranunculus*), a eudicot. (*a*) Overall view of mature root. The vascular cylinder includes tissues specialized for long-distance transport of water and solutes, whereas the epidermis forms a protective outer layer of the root (16×). (*b*) Detail of cortex (250×). Each parenchyma cell in the cortex contains many amyloplasts, which store starch. In this image, the starch in the amyloplasts has been stained with iodine (Exercise 6).

In the center of the buttercup root is the **vascular** (fluid-conducting) **cylinder** composed of **xylem** and **phloem** (fig. 32.5). Xylem transports water and minerals; phloem transports water and most organic compounds in the plant, including carbohydrates. Water-conducting cells in the xylem of angiosperms are called **tracheids** and **vessel elements** and are dead and hollow at maturity. Tracheids are long, spindle-shaped cells with thin areas or apertures called **pits** where the cell walls of adjacent cells overlap (fig. 32.6). Water moves through these pits from one cell to the next. Vessels are stacks of cylindrical cells with thin or completely open end-walls. Water moves through vessel elements in straight, open tubes. These tubes are usually stained red or blue in slide preparations of buttercup roots.

Conducting cells in phloem are called **sieve cells** and **sieve tube members** and are alive at maturity. Phloem cells are small, thin-walled, and arranged in bundles that alternate with the poles of xylem (fig. 32.5). Sieve tube members are usually stained green.

## Procedure 32.2   Examine carrot root

1.  Prepare two thin cross sections of a carrot root.
2.  Stain one slice with iodine (a stain for starch) and examine it with your microscope.
3.  Stain the other section of carrot root with phloroglucinol. Phloroglucinol stains **lignin,** a molecule that strengthens xylary cell walls.

 Phloroglucinol contains hydrochloric acid. Do not spill any on yourself or your belongings!

### Question 5
*a.*   Where is starch located in a carrot root?

*b.*   What can you conclude from this observation?

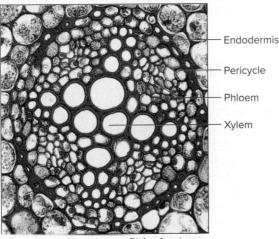

Endodermis

Pericycle

Phloem

Xylem

BiologyImaging.com

**Figure 32.5**   A cross section through the center of a root of a buttercup (*Ranunculus*), a eudicot (125×). The phloem and xylem are vascular tissues of the vascular cylinder shown in figure 32.4*a*.

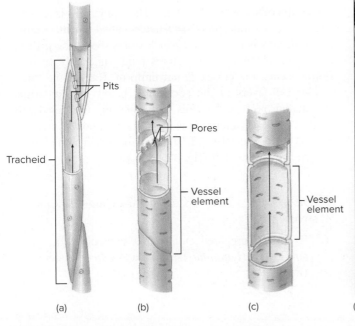

Pits

Tracheid

Pores

Vessel element

Vessel element

(a)       (b)       (c)

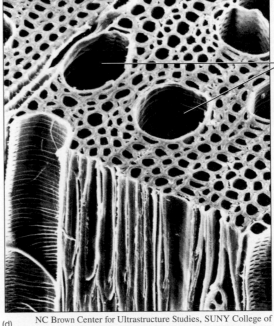

Vessel elements

(d)   NC Brown Center for Ultrastructure Studies, SUNY College of Environmental Science and Forestry, Syracuse, NY

**Figure 32.6**   Comparison of vessel elements and tracheids. (*a*) In tracheids, water passes from cell to cell through pits, which are thin areas where cell walls overlap. (*b, c*) In vessel elements, water moves through pores, which may be simple or interrupted by bars. (*d*) The large, hollow tubes shown in this scanning electron micrograph of the wood of a red maple (*Acer rubrum,* a eudicot) are vessel elements (350×).

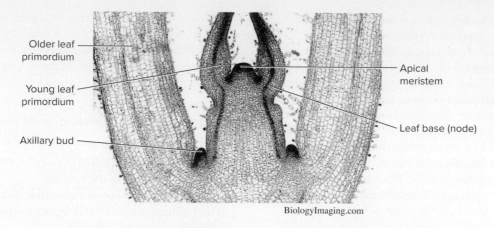

Older leaf primordium

Young leaf primordium

Axillary bud

Apical meristem

Leaf base (node)

BiologyImaging.com

**Figure 32.7**   Shoot tip of *Coleus,* a eudicot. The apical meristem is the site of rapid cell division and primary growth. Young leaves are produced at the tip of the shoot. Shoot tips, unlike root tips, are not covered by a cap (20×).

## STEMS

Stems are often conspicuous organs whose functions include support and the transport of water and solutes. Some stems (e.g., cacti) also photosynthesize and store food.

### The Shoot Apex

Examine a living *Coleus* plant and note the arrangement of leaves on the stem. Keeping this observation in mind, examine a prepared slide of a longitudinal section of a shoot tip of *Coleus* (fig. 32.7). The dome-shaped **shoot apical meristem** is not covered by a cap as was the root. The shoot apical meristem produces young leaves (**leaf primordia**) that attach to the stem at a **node.** An **axillary bud** between the young leaf and the stem forms a branch or flower.

### Question 6
*a.*   How does the absence of a cap at a shoot apex differ from the apical meristem of a root?

*b.*   Are all cells in the shoot apex the same size? Why is this significant?

### External Features of a Mature Woody Stem

Examine the features of a dormant twig (fig. 32.8). A **terminal bud** containing the apical meristem is at the stem tip and is surrounded by **bud scales. Leaf scars** from shed leaves occur at regularly spaced **nodes** along the length of the stem. The portions of stem between the nodes are called **internodes. Vascular bundle scars** may be visible within the leaf scars. Axillary buds protrude from the stem just distal to each leaf scar.

Search for clusters of **bud scale scars.** The distance between clusters or from a cluster to the terminal bud indicates the length of yearly growth.

### Primary Tissues of Stems

An **epidermis** covers the stem. The epidermis is coated with a waxy, waterproof substance called **cutin.** Below the epidermis is the **cortex,** which stores food. The **pith** in the center of the stem also stores food (fig. 32.9). In sunflower stems, the cortex is not uniform. Rather, the three to four cell-layers of the cortex just below the epidermis are smaller, rectangular cells with unevenly thickened cell walls. These are **collenchyma** cells; they support elongating regions of the plant (fig. 32.10). In stems, xylem and phloem are arranged in bundles (fig. 32.10).

### *Procedure 32.3*   Examine the primary tissues of stems

*1.*   Examine a prepared slide of a cross section of a sunflower (*Helianthus*) or alfalfa (*Medicago*) stem (fig. 32.9).

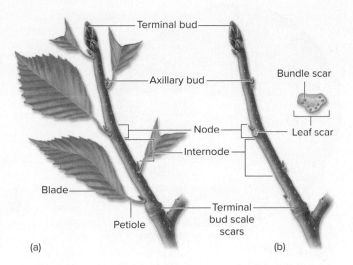

Figure 32.8  A woody twig. (*a*) In summer. (*b*) The twig in its dormant winter condition.

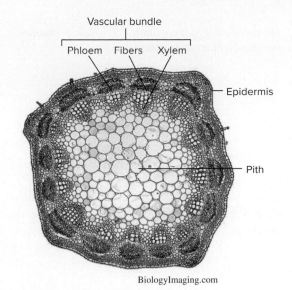

BiologyImaging.com

Figure 32.9  Cross-sectional view of a stem of alfalfa (*Medicago* sp.), a eudicot (10×). Note the ring of vascular bundles surrounding the pith. A high-magnification view of a vascular bundle from this stem is shown in figure 32.10.

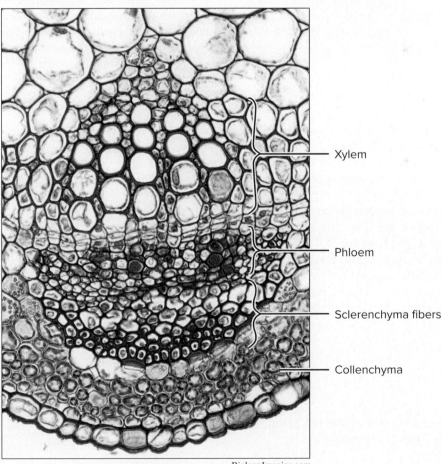

BiologyImaging.com

Figure 32.10  Cross section of a vascular bundle of a sunflower stem (485×). The sclerenchyma fibers support the stem and protect the vascular tissue. The walls of fibers are much thicker than those of adjacent cells.

> **2.** On the slide you are examining, locate a vascular bundle. Sketch what you see.

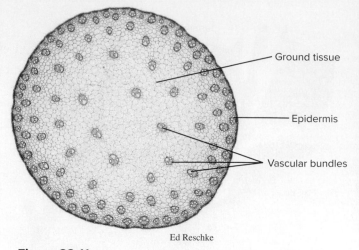

Ed Reschke

**Figure 32.11** Cross section of a stem of corn (*Zea mays*), a monocot (5×). Unlike in eudicots such as alfalfa (fig. 32.9), bundles of vascular tissue in monocots occur throughout the ground tissue. The stem is surrounded by an epidermis.

### Question 7

*a.* What is the significance of a coating of cutin on the epidermis?

*b.* How does the arrangement of xylem and phloem in stems differ from that in roots?

The darkly stained, thick-walled cells just outside the phloem in figure 32.10 are **sclerenchyma fibers,** which function in support. Sclerenchyma fibers, which can be more than a meter long in hemp and jute, are used to make linen, rope, and burlap.

The ring of vascular bundles in sunflower stems is typical of **eudicots,** which are flowering plants with two **cotyledons** (seed leaves; see Exercise 31). Examine a prepared slide of a cross section of a corn stem (fig. 32.11). Corn is a **monocot** (a flowering plant with only one cotyledon). In the following space, sketch a cross section of a sunflower stem and a corn stem. Note the distribution of the vascular bundles.

### Question 8

How does the arrangement of vascular bundles differ in stems of monocots as compared to eudicots?

## Secondary Growth of Stems

Between the xylem and phloem and each vascular bundle in eudicot stems is a meristematic tissue called **vascular cambium.** The vascular cambium is a secondary meristem that produces secondary growth (i.e., growth in girth). The vascular cambium is cylindrical and produces secondary xylem to its inside and secondary phloem to its outside.

### Question 9

How is secondary growth different from primary growth?

> *Procedure 32.4* **Examine secondary growth in woody stems**
>
> *1.* Examine cross sections of 1-, 2-, and 3-year-old basswood (*Tilia*) stems. The vascular cambium is a narrow band of cells between the xylem and phloem (fig. 32.12).
>
> *2.* Compare the structures of the three stems. The secondary xylem of older stems consists of concentric annual rings made of alternating layers of large and small cells. The large cells are formed in the spring, and the smaller cells are produced in summer.
>
> *3.* In the following space draw cross sections of 1-, 2-, and 3-year-old stems.

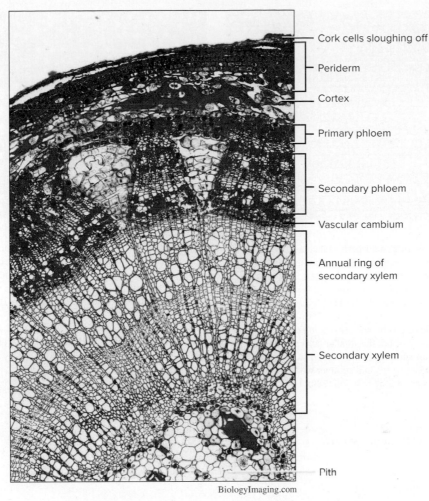

Cork cells sloughing off

Periderm

Cortex

Primary phloem

Secondary phloem

Vascular cambium

Annual ring of secondary xylem

Secondary xylem

Pith

BiologyImaging.com

**Figure 32.12** A cross section of a portion of a young linden (*Tilia*) stem showing secondary growth (30×). The vascular cambium produces secondary xylem (wood) to the inside and secondary phloem to the outside. Note the annual rings in the secondary xylem. A close-up of a growth ring from pine is shown in figure 32.13.

### Question 10

*a.* How do you account for this seasonal production of different-sized cells?

*b.* What is the common name for secondary xylem?

*c.* What is "grain" in wood?

*d.* Aside from conducting water and minerals, what is another important function of secondary xylem?

**4.** Examine a prepared slide of a cross section of secondary xylem of pine (*Pinus*), a gymnosperm (fig. 32.13). In cone-bearing plants such as pine, the conducting cells of xylem are all tracheids. The absence of vessel elements gives wood of these plants a relatively uniform appearance.

**5.** Now examine a prepared slide of a cross section of secondary xylem of an angiosperm (fig. 32.14). Sketch what you see.

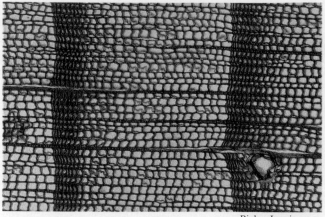

**Figure 32.13** The wood of gymnosperms (such as this pine) consists almost exclusively of tracheids. These water-conducting cells are relatively small and help to support the plant. Although water moves slower through tracheids than through vessel elements, tracheids are less likely to be disabled by air bubbles that form in response to freezing and wind-induced bending of branches. The larger cells on the right side of this photo form during the wet days of spring and are called spring wood; the smaller cells at the left form during the drier days of summer and are called summer wood. The change in density between spring and summer wood produces a growth ring, which appears as "grain" in wood (280×).

## Question 11

***a.*** What are the large cells in the wood?

***b.*** What is their function?

***c.*** Which type of wood do you think transports more water per unit area, a gymnosperm or an angiosperm? Why?

## Bark

Bark includes all tissues outside of the vascular cambium, including the secondary phloem (fig. 32.12). When viewed in cross section, secondary phloem consists of pyramidal masses of thick- and thin-walled cells. The thin-walled cells are the conducting cells.

The increase in stem circumference resulting from activity of the vascular cambium eventually ruptures the epidermis. The ruptured epidermis is replaced by a tissue called the **periderm** that, like the epidermis, functions to minimize water loss. Periderm consists of cork cells produced by another secondary meristem called the **cork cambium.**

Locate the cork cambium in cross sections of 1-, 2-, and 3-year-old basswood stems. The cork cambium is a

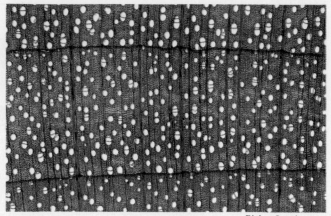

**Figure 32.14** Unlike the xylem of gymnosperms, which contains only tracheids, the xylem of angiosperms (such as this oak) also contains vessel elements. These vessel elements are much wider than tracheids and appear in this micrograph as large circles. Vessel elements are an adaptation for increased rates of water flow in angiosperms (see fig. 32.6c) (40×).

band of thin-walled cells located beneath the epidermis. The cork cambium produces cork cells to the outside and cork parenchyma to the inside. Cork cells stain red because of the presence of suberin, a water-impermeable lipid.

## Question 12

Is the amount of cork similar in 1-, 2-, and 3-year-old stems? If not, how does it differ? Why is this important?

As the stem diameter continues to increase, the original periderm ruptures and new periderms form in the underlying tissues. Tissues outside the new periderm die and form encrusting layers of bark.

Gas exchange through peridermal tissues occurs through structures called **lenticels** (fig. 32.15). Examine a prepared slide of a lenticel and locate lenticels on a mature woody stem.

## Question 13

How does a lenticel differ from the remainder of the periderm?

## LEAVES

With few exceptions, most photosynthesis occurs in leaves, although some may occur in green stems. Leaves typically consist of a **blade** and a **petiole.** The petiole attaches the leaf blade to the stem. **Simple leaves** have one blade connected to the petiole, whereas **compound leaves** have several

![A lenticel cross section and simple/compound leaf diagrams]

**Figure 32.15** A lenticel in a cross section of part of a young stem of elderberry (*Sambucus*). Gas exchange across the cork layer of the stem occurs through lenticels (80×).

BiologyImaging.com

Lenticel

Cork cambium

Cork

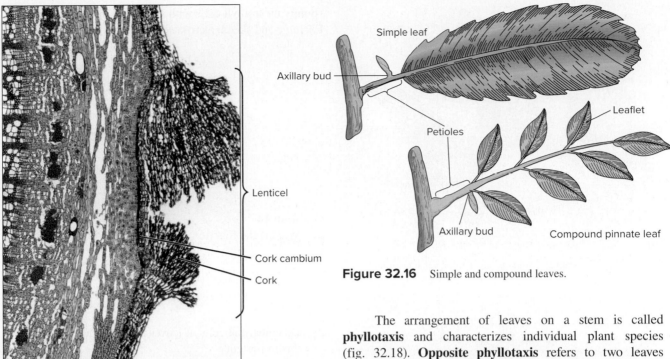

Simple leaf

Axillary bud

Petioles

Leaflet

Axillary bud

Compound pinnate leaf

**Figure 32.16** Simple and compound leaves.

leaflets sharing one petiole (fig. 32.16). **Palmate** leaflets of a compound leaf arise from a central area, as your fingers arise from your palm. **Pinnate** leaflets arise in rows along a central midline. Examine the simple and compound leaves on display, and sketch a representative of each type of leaf.

Leaves are also classified according to their **venation** (i.e., arrangement of veins) (fig. 32.17). **Parallel veins** extend the entire length of the leaf with little or no crosslinking. **Pinnately veined** leaves have one major vein (i.e., a midrib) from which other veins branch. **Palmately veined** leaves have several veins each having branches. Veins of vascular tissue in leaves are continuous with vascular bundles in stems.

Examine the leaves on display in the lab and determine their venation. Sketch a few of these leaves to show their venation. List the names of common leaves on demonstration and indicate whether each is simple, pinnately compound, or palmately compound. Also indicate whether venation in each leaf is parallel, pinnate, or palmate.

The arrangement of leaves on a stem is called **phyllotaxis** and characterizes individual plant species (fig. 32.18). **Opposite phyllotaxis** refers to two leaves per node located on opposite sides of the stem. **Alternate phyllotaxis** refers to one leaf per node, with leaves appearing first on one side of the stem and then on another. **Whorled phyllotaxis** refers to more than two leaves per node. Examine the plants on display in the lab to determine their phyllotaxis.

### What Good Is Bark?

You've doubtless seen chips of bark used as garden mulch that is spread around plants. Because bark cells (secondary phloem and periderm) were once alive, they contain many nutrients that are released into the soil as the bark decays. However, sawdust (wood) has few nutrients in it and may result in nitrogen deficiency due to rapid decomposition. Consequently, sawdust is a poor mulch for plants.

### Tree Girdling

You've probably heard of "tree girdling," which is the stripping of a ring of bark from a branch. (In the wild, porcupines also girdle trees.) Girdling removes secondary phloem from a tree, thereby leaving no pathway for photosynthate to move from the leaves to the roots. As a result, the shoot accumulates sugars and grows rapidly. Moreover, the roots cannot grow or provide water and nutrients to the shoot. The following spring it is impossible to send sugars from the roots to the girdled branch to renew growth. Thus, the branch or tree dies the year after it is girdled.

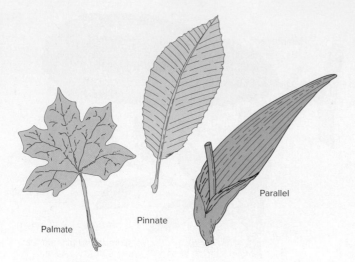

**Figure 32.17** Palmate, pinnate, and parallel venation of leaves.

spongy mesophyll cells with numerous intercellular spaces. Examine and sketch a cross section of a corn leaf.

**Question 14**

*a.* What is the function of stomata?

*b.* Do epidermal cells of leaves have a cuticle? Why is this important?

*c.* What is the significance of chloroplasts being concentrated near the upper surface of the *Ligustrum* leaf?

*d.* What are the functions of air spaces near the lower surface of the leaf?

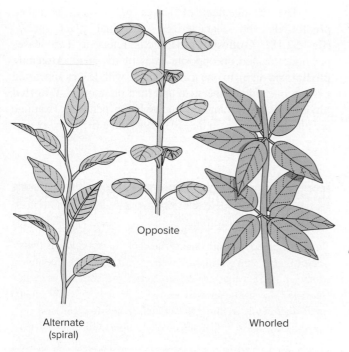

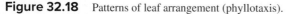

**Figure 32.18** Patterns of leaf arrangement (phyllotaxis).

*e.* How is the internal anatomy of a corn leaf different from that of *Ligustrum*?

## Internal Anatomy of a Leaf

Examine a cross section of a leaf of *Ligustrum* (privet) (figs. 32.19 and 13.9). The leaf is only 10–15 cells thick—pretty thin for a solar collector! The epidermis contains pores called **stomata,** each surrounded by two guard cells (you will study stomata again in the next exercise). Just below the upper epidermis are closely packed cells called **palisade mesophyll** cells; these cells contain about 50 chloroplasts per cell. Below the palisade layers are

*f.* How is it similar?

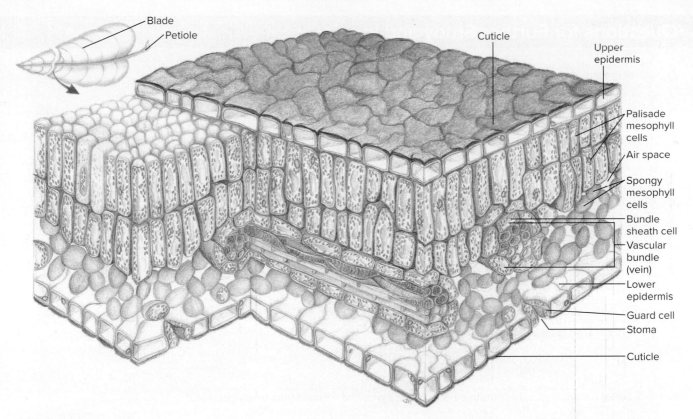

**Figure 32.19** *Ligustrum* leaf, cross section. Most photosynthesis occurs in the densely packed palisade mesophyll cells, just beneath the upper epidermis of the leaf. Gas exchange occurs through stomata, which are usually most abundant on the lower side of the leaf. Water loss is minimized by the waxy cuticle that covers the leaf.

**g.** Based on the arrangement of vascular tissue, how could you distinguish the upper versus lower surfaces of a leaf?

**h.** Examine prepared slides of leaves of corn (*Zea mays,* a monocot) and pine (*Pinus,* a gymnosperm). What differences are there in the structures of these leaves? How do these structural differences correlate with functional differences?

## INQUIRY-BASED LEARNING

### *How do plants sense and respond to light and gravity?*

Observation: The ability of plant roots to grow downward and shoots to grow upward is adaptive because it increases the plants' chances of encountering water (by the roots) and light (by the shoot).

Question: When the roots of a corn (*Zea mays*) seedling grow downward, is this a negative response to light or a positive response to gravity?

**a.** Establish a working lab group and obtain Inquiry-Based Learning Worksheet 32 from your instructor.

**b.** Discuss with your group well-defined questions relevant to the preceding observation and question. Choose and record your group's best question for investigation.

**c.** Translate your question into a testable hypothesis and record it.

**d.** Outline on Worksheet 32 your experimental design and supplies needed to test your hypothesis. Ask your instructor to review your proposed investigation.

**e.** Conduct your procedures, record your data, answer your question, and make relevant comments.

**f.** Discuss with your instructor any revisions to your questions, hypotheses, or procedures. Repeat your work as needed.

1.  What is the function of xylem? Phloem? Vascular cambium? Epidermis?

2.  How are the structure and function of stomata similar to lenticels? How are they different?

3.  How is the internal anatomy of a stem different from that of a root?

4.  How is primary growth different from secondary growth?

5.  How is a leaf structurally adapted for its function?

6.  Why does a stem typically contain more sclerenchyma and collenchyma than does a leaf?

7.  An old friend tells you that 30 years ago she nailed a sign into a tree trunk at a height of 1 meter. She now says the sign is 25 meters up in the tree. Should you believe her? Why or why not?

8.  Compare and contrast (*a*) monocot and dicot roots and (*b*) monocot and dicot stems.

9.  What are three examples of plants growing on your campus that have (a) simple leaves and (b) compound leaves?

**DOING BIOLOGY YOURSELF**
Choose a couple of defined environments nearby, such as a vacant field or riverside. Survey the variety of leaf morphologies of the dominant plants. What can you conclude about the predominance of monocots versus eudicots in each environment?

**WRITING TO LEARN BIOLOGY**
Why is leaf abscission especially important for temperate plants?

**WRITING TO LEARN BIOLOGY**
Would you expect to find annual rings in wood of a tropical dicot tree? Why or why not?

# Plant Physiology
## Transpiration

## Learning Objectives

By the end of this exercise you should be able to:
1. Describe the structure and function of stomata.
2. Describe how environmental conditions such as wind and light influence stomatal opening.
3. Measure the transpiration rate of a plant.

Please visit **connect.mheducation.com** to review online resources tailored to this lab.

To survive and reproduce, land plants must cope with many environmental challenges. Among the biggest of these problems are obtaining water and avoiding desiccation (water loss). Solutions to these and other challenges require expenditure of energy and an efficient structure and physiology. Structural and functional solutions to problems such as maintaining water balance are called adaptations, and the study of these adaptations is among the most interesting aspects of biology (fig. 33.1).

In this exercise, you will study the "water physiology" of plants—how plants are adapted to control their water content and evaporative loss. Before starting today's exercise, review in your textbook the structure and function of xylem and stomata, and review how water moves through plants (fig. 33.2).

The movement of water through plants and evaporation from aerial plant parts is called **transpiration.** Most transpiration occurs through stomata of leaves. However, transpiration also occurs in flowers. Demonstrate this with the following procedure.

Magda Urbanova/Shutterstock

**Figure 33.1** The droplets of water at the edges of these leaves of a strawberry plant are formed by guttation. Guttation results from root pressure (i.e., a positive pressure in the xylem) common in small plants growing in moist soil on cool, damp mornings. Although most water movement in plants results from evaporation of water from leaves, guttation may be an important way of moving water in short, herbaceous plants.

## Procedure 33.1 Visualize transpiration in flowers

1. Obtain a small beaker containing a solution of food color from your lab instructor at the beginning of your laboratory period.

2. Working in groups of three or four, cut the stem of a white flower such as periwinkle (*Catharanthus*) or carnation (*Dianthus*) so that the cut is about 0.5 cm from the base of the flower. Do not cut all the way through the stem.

3. Remove the remainder of the stem about 2.0 cm below the partial cut.

4. Float the flower in the food color so that the partial cut is submerged.

5. Set the beaker containing the flower and dye in an illuminated area, and examine the flower after 2–3 h.

**Question 1**
*a.* In which part of the flower is the dye located?

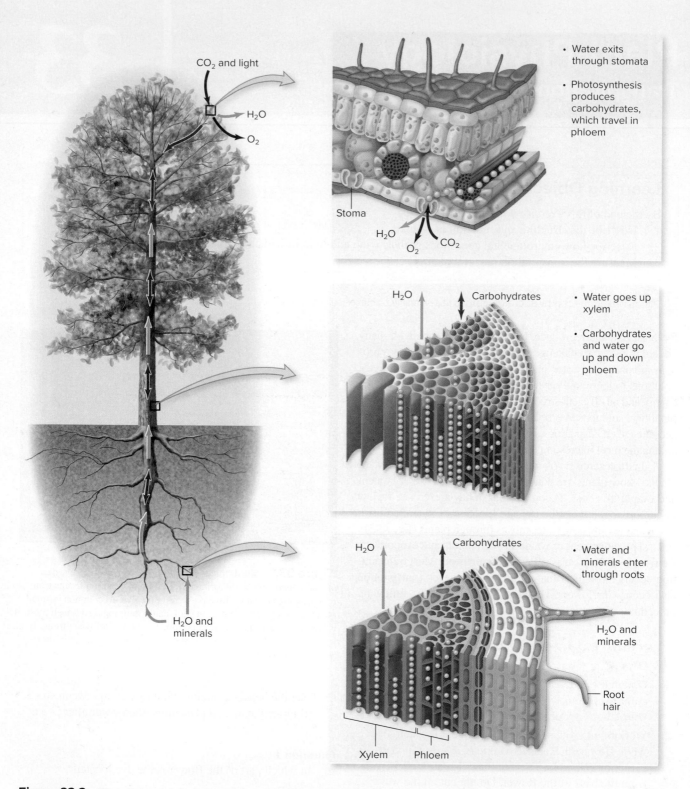

**Figure 33.2** Flow of materials in a plant. Water with dissolved minerals enters the plant through root hairs and moves upward through the xylem of the vascular system. Most of this water is eventually lost through the leaves during transpiration. Dissolved minerals are used by plant cells for metabolic reactions. Carbohydrates made during photosynthesis move throughout the plant in the phloem. These carbohydrates are either stored or used directly as an energy source for metabolism and growth.

**b.** What do you conclude from this observation?

## PLANT–WATER RELATIONS

Leaves of most plants have epidermal pores called **stomata** (sing., *stoma*), formed by the separation of a pair of guard cells (fig. 33.3). Together, the guard cells and the pore between them form a **stoma** (plural, *stomata*).

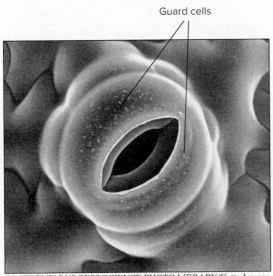

Guard cells

DR JEREMY BURGESS/SCIENCE PHOTO LIBRARY/Getty Images

**Figure 33.3**  A scanning electron micrograph of a stoma in a leaf of a tobacco plant. Guard cells change shape and, in doing so, open and close a pore. In dry conditions, stomata close and minimize water loss (100×). Water vapor passes out through the stomata, and carbon dioxide enters by the same portals. The mechanism of stomatal opening and closing is described in figure 33.4.

In Exercise 13, you learned that $CO_2$ is essential for photosynthesis. Carbon dioxide in the air reaches the photosynthetic cells of the leaf via stomata surrounded by guard cells. If enough water is available, guard cells absorb water and become turgid. When guard cells are turgid, they expand and bend to form a stomatal pore through which $CO_2$ enters and water vapor exits a leaf. When water is scarce, guard cells lose their turgidity and shrink. As a result, guard cells touch each other and the stomatal pore closes (fig. 33.4). Thus, the water status of a plant determines whether stomata are open or closed and indirectly determines whether gases move in and out of the leaf.

## STOMATAL STRUCTURE

Guard cells in dicots are bean-shaped and attached to each other at their ends. Microfibrils form transverse bands around the cell so that when water enters the cells they cannot expand around the middle. Instead, they lengthen and bow apart, thereby opening the stomatal pore. Use procedure 33.2 to examine living guard cells and how they respond to changing environmental conditions.

### Procedure 33.2  Stomatal structure

**1.** Remove part of a leaf from a well-watered *Kalanchöe* or *Zebrina* plant that has been illuminated for 4–5 h. Bend the leaf until it snaps; then pull the surface portion of the leaf away from the rest of the leaf in a slow downward motion (your instructor will demonstrate this technique). A thin sheet of transparent epidermis will tear away from the leaf.

**2.** Quickly place the piece of epidermis in a few drops of distilled water on a microscope slide (do not let the tissue desiccate) and place a coverslip on the tissue. Be careful not to trap any air bubbles beneath the tissue or coverslip.

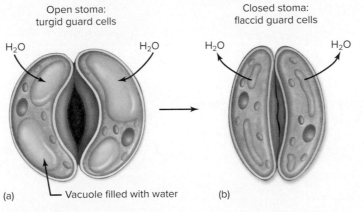

Open stoma: turgid guard cells

Closed stoma: flaccid guard cells

$H_2O$   $H_2O$   $H_2O$   $H_2O$

(a)   Vacuole filled with water   (b)

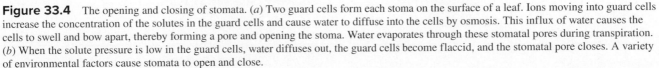

**Figure 33.4**  The opening and closing of stomata. (*a*) Two guard cells form each stoma on the surface of a leaf. Ions moving into guard cells increase the concentration of the solutes in the guard cells and cause water to diffuse into the cells by osmosis. This influx of water causes the cells to swell and bow apart, thereby forming a pore and opening the stoma. Water evaporates through these stomatal pores during transpiration. (*b*) When the solute pressure is low in the guard cells, water diffuses out, the guard cells become flaccid, and the stomatal pore closes. A variety of environmental factors cause stomata to open and close.

3. Examine the tissue with your microscope. Diagram the epidermis and its stomata.

4. Place a few drops of 10% NaCl at one edge of the coverslip covering the tissue.
5. Wick this solution under the coverslip by touching the solution at the opposite edge of the coverslip with a piece of paper towel.
6. Reexamine the stomata. Sketch what you see.

7. Observe the demonstration of an epidermal peel of *Zea mays* (corn), a monocot. Diagram cells of the epidermis of corn.

b. Are the densities of stomata similar on the upper and lower surfaces of *Kalanchöe* and *Zebrina?*

c. When the guard cells of *Kalanchöe* and *Zebrina* are turgid, they form a stomatal pore. How?

d. Are stomatal densities similar in *Zea mays* and *Kalanchöe* and *Zebrina?*

e. Are stomata of the two plants arranged in a pattern, or do they occur randomly?

**Question 2**
a. Approximately how many stomata are in a mm² of the lower surface area of *Kalanchöe* and *Zebrina?*

f. How is the structure of individual stomata different in *Zea mays* than in *Kalanchöe* and *Zebrina?*

## Question 3

*a.* Did the NaCl solution cause stomata to close? Why or why not?

*b.* What is the advantage of closed stomata when water is in short supply?

*c.* What are the disadvantages?

### Procedure 33.3  Stomatal density in other plants

Unlike *Kalanchöe* and *Zebrina*, not all plants have an epidermis that can be removed easily from their leaves. We can examine the stomata of these plants by creating a mold of the epidermis made with nail polish.

Before starting this procedure, examine leaves on the plants available in lab. For each plant, which surface of the leaf—that is, the upper or lower surface—do you think will have the most stomata? Which do you think will have the fewest? Why? Write your hypotheses here:

Test your hypotheses by making molds of the surfaces of the leaves, and then calculating the stomatal density of each surface. Here's how:

*1.* Obtain two leaves from a plant available in lab.

*2.* Paint a coating of clear nail polish onto the upper and lower sides of the leaves. The polish should cover 1–2 cm². 

Handle the nail polish carefully; it can be toxic if ingested or inhaled.

*3.* Wait for the polish to dry completely. This will take 15–20 min. The dried polish is now a mold of the leaf surface. Commercially available "liquid bandage" also works for this procedure.

*4.* When the polish has completely dried (i.e., it is no longer sticky), apply a piece of clear cellophane tape to the patch of nail polish. (Do not use opaque tapes such as Scotch matte-finish "Magic" tape; instead, use clear carton-sealing tape.)

*5.* Lay the leaves flat on the table and use your thumb to gently press the hardened polish onto the leaf.

*6.* Peel the nail polish from the leaf by gently pulling on a corner of the tape. The patch of nail polish is the impression of the leaf surface that you will examine.

*7.* Tape your mold of the leaf surface to a clean microscope slide (with the sticky side of the tape down). If the mold doesn't come off the leaf cleanly, use forceps to remove the polish.

*8.* Examine the nail-polish mold with the high-power objective of your light microscope. Search for areas that are clean and intact and in which you can see stomata. Write the name of the plant and the leaf surface (i.e., upper or lower) you are examining on the slide.

*9.* Count the number of stomata in at least three different fields of view; then convert those numbers into stomatal densities (i.e., stomata mm⁻²; see Exercise 3). Record your results here:

Name of plant _____

**Upper surface**

Number of stomata
  per field of view    _____ _____ _____

Density (stomata mm$^{-2}$)  _____ _____ _____

**Lower surface**

Number of stomata
  per field of view    _____ _____ _____

Density (stomata mm$^{-2}$)  _____ _____ _____

## Question 4

What is the average density of stomata on the upper versus lower surface of the leaf? Why is this significant?

*10.* If other plants are available in lab (e.g., monocots versus eudicots, aquatic plants versus desert plants), formulate hypotheses about the abundance of stomata on the upper and lower surfaces of their leaves. Write your hypotheses here:

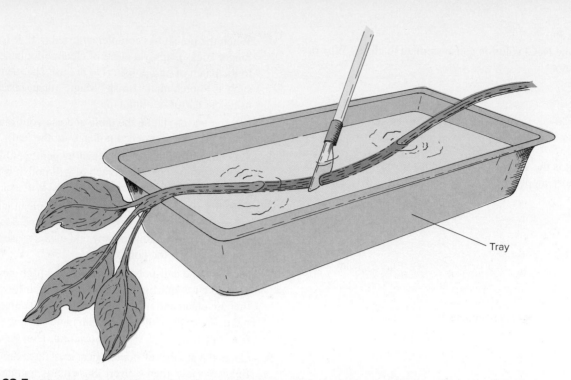

**Figure 33.5** Method of cutting submerged stems. Cutting the stem underwater avoids introducing air into the vascular tissue.

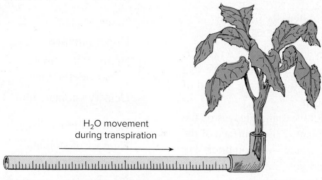

H₂O movement
during transpiration

Pipet with graduations

**Figure 33.6** Experimental setup for measuring transpiration. As water is transpired from the leaves, the water moves through the pipet.

| Table 33.1 | |
| --- | --- |
| **Experimental Conditions for Measuring Transpiration** | |
| **Condition** | **Transpiration Rate (mL H₂O h⁻¹)** |
| Control (standard illumination) | _____ |
| Darkness or reduced light | _____ |
| Light; mild breeze (created by a small fan) | _____ |
| Dark; mild breeze | _____ |
| Light; leaves coated with petroleum jelly | _____ |
| Light; all leaves removed | _____ |

**11.** Test your hypotheses by making molds of the leaves' surfaces and calculating the density of stomata. Record your results here:

**Question 5**

What do you conclude about the density of stomata in the different types of plants you examined? For example, how did the density of stomata vary on the upper versus lower epidermis of the plants you examined?

# TRANSPIRATION

Transpiration is the loss of water from plants; it is strongly influenced by environmental conditions such as the levels of $CO_2$, light, wind, and humidity. Use procedure 33.4 to quantify the effects of light and wind on transpiration.

## Procedure 33.4  Quantify transpiration

1. Obtain a sunflower (*Helianthus*) plant from your lab instructor. If sunflower plants are not available, use the other stems available in the lab.
2. Submerge the stem in water and obliquely cut the stem with a sharp scalpel (fig. 33.5).
3. While it is still submerged in water, attach the severed stem to a water-filled tube connected to a 1.0-mL pipet as shown in figure 33.6. Keep the pipet in a horizontal position. Be sure that no air bubbles are in the system and that no water is on the leaves.
4. Place the stem in an illuminated area. Transpiration will move water through the pipet regardless of whether the stem is vertical or horizontal.
5. Allow about 10 min for the rate of water movement in the pipet to stabilize. Note the position of the meniscus in the pipet.
6. Allow transpiration to occur for 15 min.
7. Determine and record the volume of water (mL) transpired by observing the distance that the meniscus moved.
8. Multiply this value by 4 to convert to mL per hour. Record this rate of transpiration in table 33.1 as the control value.
9. Continue to determine the transpiration rate under the conditions in the order listed in table 33.1.

## Question 6

State your conclusion about the influence of light, dark, breeze, clogged stomata, and removed leaves on water movement in plants.

Light:

Dark:

Breeze:

Clogged stomata:

Removed leaves:

# INQUIRY-BASED LEARNING

## How do transpiration rates vary in different species of plants?

Observation: In this lab you measured the rates of transpiration through individual parts of plants. All plants transpire, and their rates of transpiration are influenced by the environments in which they grow.

Question: Do transpiration rates vary among different species of plants?

a. Establish a working lab group and obtain Inquiry-Based Learning Worksheet 33 from your instructor.
b. Discuss with your group well-defined questions relevant to the preceding observation and question. Choose and record your group's best question for investigation.
c. Translate your question into a testable hypothesis and record it.
d. Outline on Worksheet 33 your experimental design and supplies needed to test your hypothesis. Ask your instructor to review your proposed investigation.
e. Conduct your procedures, record your data, answer your question, and make relevant comments.
f. Discuss with your instructor any revisions to your questions, hypotheses, or procedures. Repeat your work as needed.

# Questions for Further Study and Inquiry

1.  What structural features of plants minimize water loss?

2.  Why was it important in today's exercise to sever the stems while they were submerged in water?

3.  What adaptations help plants conserve water?

## DOING BIOLOGY YOURSELF

Assume that a maple tree in your front yard has 100,000 leaves, each having an average area of 35 cm$^2$. If the transpiration rate of each leaf is 0.05 mL H$_2$O h$^{-1}$ cm$^{-2}$ leaf area, how many liters of water move through the plant in a day?

## WRITING TO LEARN BIOLOGY

Describe any adaptations that would improve water economy in plants. In which environments might you expect to find the adaptations you described?

# Plant Physiology
## Tropisms, Nutrition, and Growth Regulators

<hr/>

## Learning Objectives

By the end of this exercise you should be able to:
1. Define the terms phototropism and gravitropism.
2. Describe symptoms of mineral deficiency in plants.
3. Explain how the quality and quantity of light affect seed germination.
4. Explain the modes of action of auxin and gibberellic acid.

Please visit **connect.mheducation.com** to review online resources tailored to this lab.

**P**lants respond to a variety of environmental stimuli such as light and gravity, and require several nutrients such as calcium and potassium. Plant growth and development are controlled by internal chemical signals called **growth regulators.** Plants have a small number of growth regulators that influence many processes and traits that we see. This is in contrast to animals, which have many more hormones and chemical regulators. Auxin and gibberellic acid are examples of growth regulators in plants.

In this exercise, you will study a variety of common physiological responses of plants. Throughout this exercise, remember that each of these responses is an adaptation for the survival and reproduction of plants in varied environments.

## PLANT TROPISMS

A **tropism** is a movement in response to an external stimulus. The direction of movement is determined by the direction of the most intense stimulus. Two tropisms you will study in today's exercise are **phototropism,** which is directed growth in response to light (fig. 34.1), and **gravitropism,** which is directed growth in response to gravity (fig. 34.2).

### Phototropism

> *Procedure 34.1*   **Observe phototropism**
>
> *1.* Obtain from your instructor six 10- to 14-day-old radish (*Raphanus*) seedlings. These seeds have been grown in diffuse, overhead light.
> *2.* At the beginning of the lab period, place your seedlings approximately 25 cm from a 100-Watt light so that light strikes the shoots at a right angle.

> *3.* Use a protractor to measure curvature of the seedlings every 30 min for 2 h.
> *4.* Record your results in table 34.1.

**Question 1**
*a.* In which direction did the seedlings curve?

*b.* Is this curvature positive or negative phototropism?

*c.* What is the adaptive significance of phototropism?

*d.* Would you expect roots to react similarly to light? Why or why not?

Gaertner/Alamy Stock Photo

**Figure 34.1** Sunflowers (*Helianthus annus*) get their common name because the flowers of some varieties track the sun across the sky, much like radiotelescopes track satellites. This "solar tracking" by sunflowers and other plants (e.g., cotton, alfalfa, beans) helps the plants regulate the amount of light that they absorb. Desert plants such as the "compass plant" use phototropism to orient their leaves parallel to the sun's rays and minimize the amount of light that they absorb; this prevents overheating and desiccation. Other plants orient their leaves perpendicular to the sun's rays to maximize the amount of light absorbed for photosynthesis.

Martin Shields/Alamy Stock Photo

**Figure 34.2** Negative gravitropism by a stem. This plant was placed on its side in the dark 24 h before this picture was taken. The stems have curved away from the pull of gravity.

**Table 34.1**

| Observations of Phototropism | |
| --- | --- |
| Time (h) | Mean Curvature (degrees) |
| 0.5 | _____ |
| 1.0 | _____ |
| 1.5 | _____ |
| 2.0 | _____ |

## Gravitropism

Plants may perceive gravity by the movement of starch-laden amyloplasts within cells (fig. 34.3). Growing corn roots show their response to gravity by curving down (fig. 34.4).

Yesterday, your instructor placed *Zea mays* (corn) seedlings having roots approximately 1 cm long in a glass beaker. Seedlings labeled "H" had their root oriented

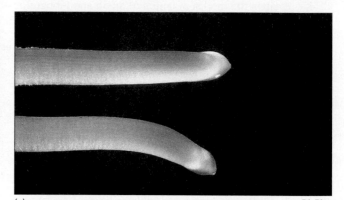

(a)           BioPhot

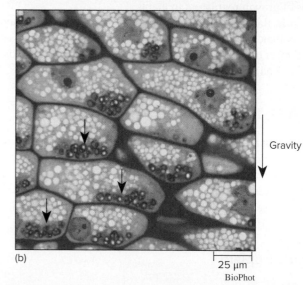

Gravity

(b)         25 μm
BioPhot

**Figure 34.3** Perception of gravity in root caps. (*a*) Cells in the center of a root cap contain numerous starch-laden amyloplasts located in the lower part of the cells (5×). (*b*) Amyloplasts (arrows) sediment to the bottom of the cell when roots are oriented horizontally (280×). This sedimentation of amyloplasts has long been thought to be the basis for how roots perceive gravity (upper photo) (see fig. 34.4). To see a higher-magnification view of the root tip and root cap, see figures 32.2*b* and 32.3.

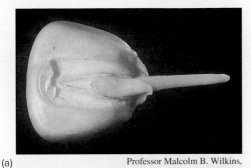

(a)

Professor Malcolm B. Wilkins,
Botany Dept., Glasgow University

(b)

Professor Malcolm B. Wilkins,
Botany Dept., Glasgow University

**Figure 34.4** Gravitropism by horizontally oriented roots of corn. Downward curvature (*a*) begins within 30 min and (*b*) is completed within a few hours. Curvature results from faster elongation of the upper side of the root than of the lower side.

horizontally, whereas those labeled "D" had their root pointing down. The terminal 3–4 mm was removed from the roots of seedlings labeled with an asterisk (*), whereas the other seedlings were not cut.

### Procedure 34.2   Observe root gravitropism

1. Obtain the containers of corn seedlings oriented horizontally and vertically.
2. Examine the experimental setup and the direction of root growth.
3. Record your observations in table 34.2.

**Question 2**

*a.* Which roots grew down?

*b.* Which one(s) didn't?

*c.* What do you conclude from these observations?

### Table 34.2

#### Observations of Gravitropism in Roots

| Treatment or Orientation | Direction of Growth |
| --- | --- |
| Intact roots oriented vertically (V) | _____ |
| Intact roots oriented horizontally (H) | _____ |
| Detipped roots oriented vertically (V*) | _____ |
| Detipped roots oriented horizontally (H*) | _____ |

*d.* Where in roots does the differential growth occur that produces gravitropism?

*e.* Are roots positively or negatively gravitropic?

*f.* What is the adaptive significance of gravitropism?

### Procedure 34.3   Examine stem gravitropism

1. Obtain three containers of tomato plants (*Lycopersicon*) or sunflower (*Helianthus*) plants.
2. Turn the potted plants horizontally.
3. Measure the distance from the stem tip to the table's surface. Record your measurements in table 34.3.
4. Every 30 min remeasure the distance between the stem tips and table, and record your results.
5. Some of the plants available in the lab were turned horizontally and inverted yesterday; others were left upright.
6. Examine stem curvature after 24 h and record your observation in table 34.3.
7. Make a simple sketch of the seedling stem curvatures after 24 h.

**Question 3**

*a.* How are the stems oriented after 24 h?

*b.* What is the adaptive advantage of a gravitropic response?

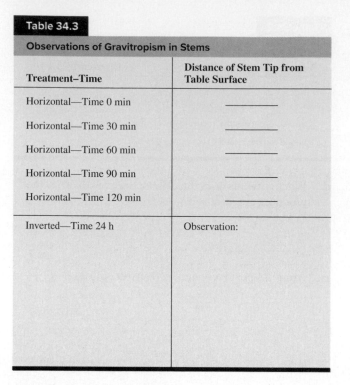

**Table 34.3**

| Observations of Gravitropism in Stems | |
| --- | --- |
| Treatment–Time | Distance of Stem Tip from Table Surface |
| Horizontal—Time 0 min | _____ |
| Horizontal—Time 30 min | _____ |
| Horizontal—Time 60 min | _____ |
| Horizontal—Time 90 min | _____ |
| Horizontal—Time 120 min | _____ |
| Inverted—Time 24 h | Observation: |

## SEED GERMINATION

Several environmental factors affect seed germination. For example, seeds of many plants germinate only in response to certain types of light. The pigment that absorbs light affecting seed germination (and several other developmental responses) is **phytochrome.** Phytochrome alternates between two forms, depending on the light it has absorbed. Phytochrome is activated by red light (660 nm) and inactivated by far-red light (730 nm) and/or darkness (fig. 34.5). The activation and deactivation reactions are reversible, and the ultimate physiological effect induced by the phytochrome depends on which wavelengths were absorbed last. More than 50 different developmental processes are affected by phytochrome absorption of red and far-red light.

Grand Rapids lettuce seeds (*Lactuca sativa*) are excellent models for examining the effects of light on germination because they are sensitive to light and germinate quickly. The seeds are dormant when they are first shed, and germination is poor even with adequate oxygen and heat. Dormancy is broken by light absorption by phytochrome. The phytochrome usually exists in the red-absorbing form. Activation (absorption of red light) causes increased water absorption by the radicle cells, and the radicle grows and elongates. Germination begins.

### *Procedure 34.4*   Observe germination of lettuce seeds

1. Obtain about 300 Grand Rapids lettuce seeds from your laboratory instructor.
2. Working in small groups, place 50 seeds in each of six petri dishes containing water-soaked filter paper.

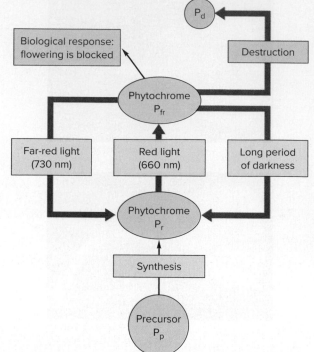

**Figure 34.5**   How phytochrome works. Phytochrome is synthesized in the $P_r$ form from amino acids, designated $P_p$ for phytochrome precursor. When exposed to red light, $P_r$ changes to $P_{fr}$, which is the active form that elicits a response in plants. $P_{fr}$ is converted to $P_r$ when exposed to far-red light, and it also converts to $P_r$ or is destroyed in darkness. The destruction product is designated $P_d$.

3. Cover the petri dishes with lids and expose the seeds to the following treatments:

   *Treatment 1:* Continuous darkness (wrap the dish in metal foil)

   *Treatment 2:* Continuous light

   *Treatment 3:* Red light for 10 min, then darkness

   *Treatment 4:* Far-red light for 10 min, then darkness

   *Treatment 5:* Red light for 10 min, followed by far-red light for 10 min, then darkness

   *Treatment 6:* Red light for 10 min, far-red light for 10 min, red light for 10 min, then darkness

4. During your next laboratory period determine the percentage of germination resulting from each treatment.

5. Record your results in table 34.4.

**Question 4**

What do you conclude about the influence of light on germination of lettuce seeds?

## Table 34.4

### Germination of Lettuce Seeds under Six Different Light Treatments

| Treatment | Percentage Germination |
|---|---|
| 1. Dark | _____ |
| 2. Constant room light | _____ |
| 3. Red light, then dark | _____ |
| 4. Far-red light, then dark | _____ |
| 5. Red light, followed by far-red light, then dark | _____ |
| 6. Red light, far-red light, red light, then dark | _____ |

## Table 34.5

### Germination of Onion Seeds in Light and Dark

| Treatment | Percentage Germination |
|---|---|
| Room light | _____ |
| Dark | _____ |

Germinating seeds of different plants often respond differently to light. Germination of onion seeds, for example, is affected by light versus dark. Determine their sensitivity with procedure 34.5.

### *Procedure 34.5* Observe germination of onion seeds

1. Divide 50 onion seeds by placing 25 in each of two petri dishes containing water-soaked filter paper.
2. Put lids on the petri dishes and label the lids "light" and "dark." Place one petri dish in room light and the other dish in darkness.
3. Examine the seeds during your next lab period and record the percentage germination in table 34.5.

### Question 5

*a.* Does light promote or inhibit germination of onion seeds?


*b.* Is this response different from that of lettuce?


*c.* How could seed germination be influenced by depth of planting? How might phytochrome be involved in such a response?


## PLANT GROWTH IS AFFECTED BY LIGHT

Many seeds germinate in the dark and push their stems above the soil to receive light. If the seedlings do not receive light, they grow abnormally; the leaves grow longer and the plant appears pale and spindly. Such a plant is **etiolated.** Chlorophyll does not develop until the plant is exposed to light. Activation of phytochrome helps promote normal growth. Use procedure 34.6 to observe etiolation.

### *Procedure 34.6* Examine etiolation

1. Obtain two groups of bean (*Phaseolus*) seedlings. They were planted 10–14 days ago, with one group being grown in the dark and one group in light.
2. Complete the first two lines of table 34.6 with your observations.
3. After completing and recording your observations, reverse the two treatments and examine the plants during your next lab period.
4. Complete the last two lines of table 34.6 with your final observations.

### Question 6

*a.* How did the seedlings manage to grow at all while in the dark?


*b.* Did etiolation include stem elongation?


*c.* What is the advantage of rapid stem elongation in a seed germinating in the dark?


## PLANT NUTRITION

Growth of green plants requires suitable temperature and adequate amounts of carbon dioxide, oxygen, water, light, and a group of essential elements. Carbon, hydrogen, and

Table 34.6

**Plant Growth under Different Light Treatments**

| Treatment | Leaf Size | Stem Diameter | Height | Color of Shoots | Stem Strength |
|---|---|---|---|---|---|
| Light | | | | | |
| Dark | | | | | |
| Light then dark | | | | | |
| Dark then light | | | | | |

**Table 34.7**

**Hoagland's Solution, a Common Nutrient Medium for Healthy Plant Growth**

| Macronutrients | Grams/Liter | Micronutrients | Grams/Liter |
|---|---|---|---|
| $Ca(NO_3)_2 \cdot 4H_2O$ | 1.18 | $H_3BO_3$ | 0.60 |
| $KNO_3$ | 0.51 | $MnCl_2 \cdot 4H_2O$ | 0.40 |
| $MgSO_4 \cdot 7H_2O$ | 0.49 | $ZnSO_4$ | 0.05 |
| $KH_2PO_4$ | 0.14 | $CuSO_4 \cdot 5H_2O$ | 0.05 |
| | | $H_2Mo_4 \cdot 4H_2O$ | 0.02 |
| | | Ferric tartrate | 0.50 |

oxygen compose 98% of the fresh weight of plants. The remaining 2% is composed of 13 other elements classified either as macronutrients or micronutrients:

- **Macronutrients** are nutrients needed in relatively large amounts (100–2000 ppm).[1] Nitrogen (N), phosphorus (P), calcium (Ca), potassium (K), magnesium (Mg), and sulfur (S) are macronutrients.

- **Micronutrients** are nutrients needed in relatively small amounts (<100 ppm). Iron (Fe), chlorine (Cl), copper (Cu), manganese (Mn), zinc (Zn), molybdenum (Mo), and boron (B) are examples of micronutrients.

Although plants need more macronutrients than micronutrients, all are essential for plant growth. The absence of any essential micronutrient or macronutrient will ultimately kill a plant. Before your lab period, seeds were germinated in distilled water and washed sand. Then the young seedlings were transferred to a variety of growth media. Control plants were watered with a solution containing all essential nutrients. Hoagland's Solution is a common mixture of nutrients

used to provide for healthy growth (table 34.7). Your control plants were watered with a solution similar to Hoagland's. Other groups of seedlings were watered with solutions deficient in an essential element. Follow procedure 34.7 to observe the effects of nutrient deficiency.

### *Procedure 34.7*    **Examine plants with symptoms of nutrient deficiency**

1. Examine the plants on demonstration that have been grown for their first month in nutrient solutions lacking Ca, N, P, Mg, K, S, and Fe.

2. Record the symptoms for deficiency of each nutrient in table 34.8. Do not refer to table 34.9 until you have completed your observations.

3. After examining all the seedlings, refer to table 34.9 and determine which deficiencies are most obvious and easily diagnosed.

1   *One part per million (ppm) equals one second in 277 h (11.5 days) and one drop in 50 L (13.2 gallons).*

## Table 34.8

### Symptoms of Nutrient Deficiency

| Nutrient Solution | Symptom(s) |
|---|---|
| Complete (containing all required minerals) | |
| Complete solution minus Ca | |
| Complete solution minus N | |
| Complete solution minus P | |
| Complete solution minus Mg | |
| Complete solution minus K | |
| Complete solution minus S | |
| Complete solution minus Fe | |

## Table 34.9

### Functions and Deficiency Symptoms of Some Major Elements

| Element | Major Functions | Deficiency Symptoms |
|---|---|---|
| Calcium (Ca) | Component in pectin compounds of middle lamella. Present in organic acids bound to proteins. Plays a role in nitrogen metabolism and membrane integrity. | Deficiency may cause ion uptake imbalance, particularly with magnesium. Young leaves are affected first. Tips and margins of leaves become light green and later necrotic. Tips of leaves become limp. Terminal bud often dies. |
| Iron (Fe) | Electrons transported in cytochromes. | Effects localized on new leaves. Leaves chlorotic. Veins remain green. |
| Magnesium (Mg) | Constituent of chlorophyll. Important cofactor for enzymes in respiration and in phosphate metabolism. | Older leaves become chlorotic between the veins at the tips and margins. Usually not characterized by necrotic spots. Root system frequently overdeveloped. Leaf margins may cup upward. |
| Nitrogen (N) | Major component of amides, amino acids, and proteins. Present in membranes, organelles, and the cell wall. Balance of carbohydrates and nitrogenous substances necessary. | Leaves often more erect. Stem and leaves stunted with excess root development. Foliage, especially older leaves, chlorotic. Unable to flower. |
| Phosphorus (P) | Constituent of phospholipids, nucleic acids, and nucleoproteins. Important in respiration and energy transfer. | Small plants, narrow leaves, root system larger but fewer laterals. Accumulation of sugars in older leaves promotes the synthesis of purple anthocyanin pigments. Stiff but weak stems and leaves. Older leaves yellowed. Other leaves dark green. |
| Potassium (K) | Not known to be structurally part of organic compounds. Role is likely catalytic and regulatory. Needed to activate several enzyme systems. | Internodes short; stems weak. Localized chlorotic or molting of older leaves, particularly at the tips and margins. Later stages may have necrotic mottling. Leaf margins frequently curled under. |
| Sulfur (S) | Component of proteins. Component of iron-sulfur proteins of the electron transport system. | Younger leaves light green. Veins lighter than intervein area. |

## Question 7

How would you "cure" a plant suffering from these deficiency symptoms?

## Commercial Fertilizers

Commercial fertilizer contains three primary nutrients: nitrogen, phosphorus, and potassium. It may also contain other macronutrients and/or micronutrients. Examine the bag of fertilizer in the lab—its contents are described by three numbers on the bag's label. For example, a bag labeled "10–12–8" is 10% nitrogen (as ammonium or nitrate), 12% phosphorus (as phosphoric acid), and 8% potassium (as mineral potash).

# PLANT GROWTH REGULATORS

Plant growth and development are controlled by internal chemical signals called **growth regulators.** In this exercise you will study some common physiological responses of plants to growth regulators such as auxin and gibberellic acid (fig. 34.6). Remember that each of the plant responses is part of an overall design for survival and reproduction in varied environments.

## Indoleacetic Acid (IAA)

Indoleacetic acid (IAA), also known as **auxin,** is the best-known growth regulator in plants. It is produced by shoot tips and dramatically affects cell growth (fig. 34.7). Auxin also inhibits development of axillary buds. That is, IAA promotes apical dominance.

FUNCTIONS OF SOME OF THE MAJOR PLANT HORMONES

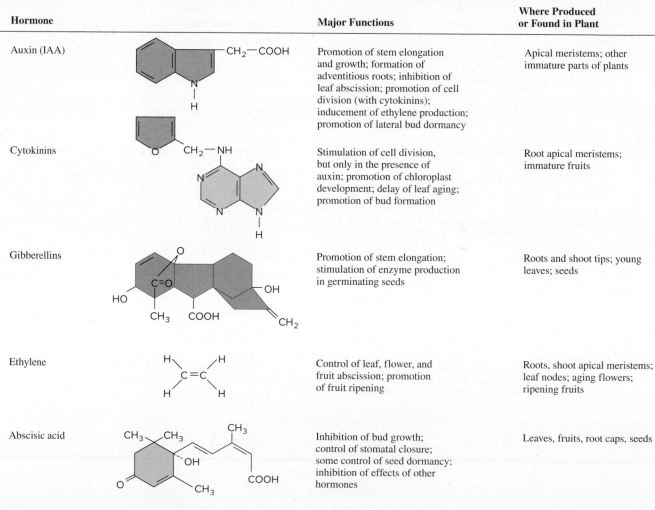

| Hormone | | Major Functions | Where Produced or Found in Plant |
|---|---|---|---|
| Auxin (IAA) | | Promotion of stem elongation and growth; formation of adventitious roots; inhibition of leaf abscission; promotion of cell division (with cytokinins); inducement of ethylene production; promotion of lateral bud dormancy | Apical meristems; other immature parts of plants |
| Cytokinins | | Stimulation of cell division, but only in the presence of auxin; promotion of chloroplast development; delay of leaf aging; promotion of bud formation | Root apical meristems; immature fruits |
| Gibberellins | | Promotion of stem elongation; stimulation of enzyme production in germinating seeds | Roots and shoot tips; young leaves; seeds |
| Ethylene | | Control of leaf, flower, and fruit abscission; promotion of fruit ripening | Roots, shoot apical meristems; leaf nodes; aging flowers; ripening fruits |
| Abscisic acid | | Inhibition of bud growth; control of stomatal closure; some control of seed dormancy; inhibition of effects of other hormones | Leaves, fruits, root caps, seeds |

**Figure 34.6**    Functions of some of the major plant hormones.

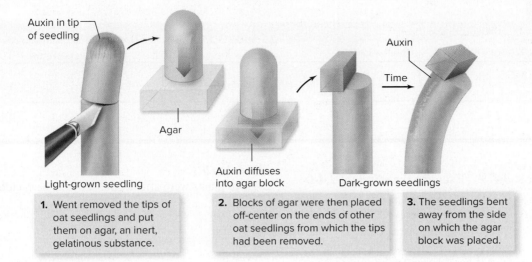

Auxin in tip of seedling

Agar

Auxin diffuses into agar block

Auxin

Time

Light-grown seedling

Dark-grown seedlings

1. Went removed the tips of oat seedlings and put them on agar, an inert, gelatinous substance.

2. Blocks of agar were then placed off-center on the ends of other oat seedlings from which the tips had been removed.

3. The seedlings bent away from the side on which the agar block was placed.

**Figure 34.7** Went's demonstration of how auxin affects plant growth. Frits Went, a Dutch plant physiologist, discovered how auxin controls plant growth. Went concluded that auxin promoted cell elongation and that it accumulated on the side of a grass seedling bent away from the light.

### Table 34.10

**Effects of Auxin on *Coleus* Plants**

| Treatment | Observations |
|---|---|
| Control (no treatment) | |
| Shoot tip removed | |
| Shoot tip removed and the cut surface coated with 1% IAA in lanolin | |
| Shoot tip removed and the cut surface coated with lanolin alone | |

### Procedure 34.8  Examine the effects of auxin

1. Examine the *Zea* or *Coleus* plants subjected to the treatments described in table 34.10.
2. Record your observations in table 34.10.

**Question 8**

*a.* What is the effect of replacing the shoot tip with IAA?

*b.* Why was it necessary to apply only lanolin to one plant?

## 2,4-Dichlorophenoxyacetic Acid

2,4-dichlorophenoxyacetic acid (2,4-D) is a synthetic auxin used frequently in agriculture as a herbicide. 2,4-D is one of the most widely used herbicides in the world; it kills weeds in grass lawns by selectively eliminating broad-leaved dicots. 2,4-D is useful because it is synthetic and there are no enzymes in plants to detoxify it or break it down.

**Table 34.11**

**Effects of 2,4-D on Monocots and Dicots**

| Type of Plant Treated with 2,4-D | Observations |
|---|---|
| Monocots | |
| Dicots | |

## Procedure 34.9 Examine the effects of 2,4-D

Be careful when handling 2,4-D; it can be especially irritating to your eyes.

1. Examine plants treated 2 weeks ago with 2,4-D. The pots originally contained a mixture of monocots (e.g., corn, oats) and dicots (e.g., peas, beans).
2. Record your observations in table 34.11.

**Question 9**

Which plants were affected most by the herbicide?

## Gibberellic Acid (GA)

Dwarfism often results from a plant's inability to synthesize active forms of **gibberellic acid (GA),** another plant growth regulator. Gibberellic acid mobilizes enzymes during seed germination, promotes stem elongation, and is used commercially to increase the production of grapes (fig. 34.8).

(a)     Omikron/Science Source

(b)

Amnon Lichter, The Volcani Center

**Figure 34.8** Gibberellins affect several aspects of plant growth and development. (*a*) Gibberellic acid stimulates cell elongation in cabbage. Cabbage (*Brassica oleracea*), a biennial native to the seacoast of Europe, will "bolt" when the heads are treated with gibberellin. (*b*) Applying gibberellic acid increases the space between grapes. As a result, larger grapes (right) develop because more space between grapes allows them to grow larger. This is important because it increases yields (and income) for farmers.

**Table 34.12**

**Effects of Gibberellic Acid on Corn Plants**

| Plant and Treatment | Observations |
|---|---|
| Normal, untreated | |
| Dwarf, untreated | |
| Normal, treated with GA | |
| Dwarf, treated with GA | |

## *Procedure 34.10*  Examine effects of gibberellic acid

1. Examine four trays of bean or corn plants treated in the following ways:

   *Tray 1:* Normal plants—untreated

   *Tray 2:* Dwarf plants—untreated

   *Tray 3:* Normal plants—treated with two or three drops of GA on alternate days for two weeks

   *Tray 4:* Dwarf plants— treated with two or three drops of GA on alternate days for two weeks

2. Record your observations in table 34.12.

**Question 10**

*a.* What is the effect of GA on dwarfism?

*b.* Did applying GA to normal plants have any effect? If so, what?

## INQUIRY-BASED LEARNING

### *How does light influence seed germination?*

Observation: Seed germination is cued to a variety of environmental stimuli. Some of these stimuli are seasonal.

Question: Is seed germination of tropical species of plants more sensitive to light than seed germination of temperate species?

a. Establish a working lab group and obtain Inquiry-Based Learning Worksheet 34 from your instructor.

b. Discuss with your group well-defined questions relevant to the preceding observation and question. Choose and record your group's best question for investigation.

c. Translate your question into a testable hypothesis and record it.

d. Outline on Worksheet 34 your experimental design and supplies needed to test your hypothesis. Ask your instructor to review your proposed investigation.

e. Conduct your procedures, record your data, answer your question, and make relevant comments.

f. Discuss with your instructor any revisions to your questions, hypotheses, or procedures. Repeat your work as needed.

1. How are phototropism and gravitropism similar? How are they different?

2. What is the adaptive significance of seed germination being influenced by environmental conditions such as light and water?

3. What comprises commercial "plant food"?

4. What are some commercial uses for plant growth-regulating substances?

5. Some seeds must pass through the digestive tract of an animal before they can germinate. What is the adaptive significance of this feature?

6. How are growth regulators in plants similar to hormones in animals? How are they different?

**DOING BIOLOGY YOURSELF**

Design an experiment to determine whether seed germination by tropical species of plants is more sensitive to light than that by temperate species.

**WRITING TO LEARN BIOLOGY**

Describe the characteristics of plants you would choose to culture were you to travel in space. Base your choices on nutrient requirements, phototropic sensitivity, and gravitropic sensitivity.

# Bioassay
## Measuring Physiologically Active Substances

## Learning Objectives

By the end of this exercise you should be able to:
1. Explain the concept of a bioassay.
2. Construct a dose-response curve.
3. Use a bioassay to estimate the concentration of a physiologically active substance.

Please visit **connect.mheducation.com** to review online resources tailored to this lab.

**B**ioassay is a method used by biologists to determine the presence or effect of a physiologically active substance by testing its effect on a living organism and comparing this activity with that of a standard. This technique, which is usually combined with other chemical analyses, typically measures a substance's biological activity rather than its concentration—the substance's concentration is later determined by comparing the magnitude of this biological activity to the activities elicited by a series of known concentrations of the substance. The unknown substance can be any type of chemical, such as a nutrient, herbicide, pesticide, vitamin, or pollutant.

Bioassays are important and useful because complex biological systems seldom allow the direct measurements used in chemistry labs. Interfering compounds are often present, thus making controls especially important.

Sometimes a bioassay is used to estimate the presence and relative concentration of a substance; other times a bioassay is used to measure the effect of an unknown substance or mixture *relative* to that of a particular concentration of a known substance. For example, if a leaf extract inhibits seed germination, then the "full-strength" extract could be described in terms of its having an effect *equivalent to* that of a specific concentration of a known germination-inhibitor. Finally, a bioassay could also be useful in cases where an effect cannot be quantified, but only described qualitatively. For example, a bioassay could be used to monitor environmental pollutants by determining whether soil is contaminated with substances that affect plant growth and development.

In the following procedures you will bioassay for the level of activity of three plant growth regulators: auxin, 2,4-D, and gibberellic acid. Each of these growth regulators has important uses. For example, synthetic auxins such as 2,4-D are used extensively as herbicides because they are inexpensive, relatively nontoxic to humans, and selectively kill (by causing excessive growth) many broadleaf weeds. These compounds

are also used to produce roots on cuttings, inhibit the preharvest dropping of fruit, produce seedless cucumbers, and inhibit the sprouting of lateral buds ("eyes") on potatoes. Similarly, gibberellic acid is used by some sugarcane growers to improve yields, by California growers of "Thompson seedless" grapes to produce larger grapes, and by brewers to stimulate the germination of barley seeds for the production of alcoholic beverages. These germinated barley seedlings are called malt; their germination involves the conversion of starch to sugars, which are later converted to ethanol during anaerobic fermentation.

In this lab, you will assay (i.e., measure) the effect of auxin on the curvature of split stems, the effect of 2,4-D on root elongation, and the effect of gibberellic acid on enlargement of radish cotyledons. All of these bioassays will be done at a constant pH. This will stabilize the reactivities and potency of the chemicals.

## BIOASSAY FOR AUXIN

You will measure how much a stem curves in response to known concentrations of auxin (see fig. 34.6). Using these raw data you will construct a standard curve (graph) showing the relationship between known concentrations of auxin and the degree of curvature of a stem. You will then bioassay a solution of auxin of unknown concentration by measuring its effect on stem curvature. Using the standard curve you can graphically determine the unknown concentration of auxin based on the amount of curvature induced by the unknown solution.

**SAFETY FIRST** Before coming to lab, you were asked to read this exercise so you would know what to do and be aware of safety issues. In the space below, briefly list the safety issues associated with today's procedures. If you have questions about these issues, contact your laboratory assistant before starting work.

## Procedure 35.1 Bioassay auxin

1. Working in a group of three to five students, obtain a tray of dark-grown pea seedlings 8 to 10 days old.

2. Locate the third internode from the top of the seedlings. Cut 35 uniform sections from the seedlings (one per seedling) each 3 cm long. The cuts should be made just below the internode.

3. Use a razor blade to split 3/4 of the length of each stem segment.

4. Soak the split stems in distilled water for 20–30 min to remove some of the endogenous, or naturally occurring, auxin.

5. Label each of seven petri dishes to receive auxin at one of the concentrations listed in table 35.1. Add to each petri dish 20 mL of the appropriate solution listed in table 35.1. Each solution has been adjusted to pH 5.9 with 10 mM phosphate buffer.

6. Place five split stems in each of the seven petri dishes.

7. Incubate the soaking sections at room temperature in the dark. (Light redistributes auxin.)

8. Examine the stems after 24 h.

9. Use a protractor to measure the curvature for each stem for each treatment and record your data in table 35.1. Calculate and record the mean curvature for each treatment.

10. Plot the mean curvature (degrees) versus the $\log_{10}$ of the known concentrations of auxin on one of the sheets of graph paper at the end of this exercise. The resulting graph is known as a **dose-response curve.** Because the graph is logarithmic, it will include much error, but it will nevertheless show you the general pattern of how auxin influences curvature of the stems.

11. Determine the unknown concentration of auxin by first locating the mean curvature value of the unknown treatment on the y-axis of your standard curve. Draw a straight line from this point parallel to the x-axis until the line intersects the standard curve. Draw a line from this intersection straight down until it intersects the x-axis. This point on the x-axis marks the concentration of the unknown solution.

12. Record the concentration of the unknown solution in table 35.1.

### Question 1

**a.** What are your conclusions?

**b.** According to your data, what is the approximate concentration of auxin in the unknown solution?

**c.** How does auxin affect curvature of stems?

**d.** What is the relationship between auxin concentration and stem curvature?

## BIOASSAY FOR 2,4-D

You will measure the effect of 2,4-D on elongation of cucumber roots. You'll then bioassay a solution of 2,4-D of unknown concentration by comparing its effect on root elongation with that induced by known concentrations of 2,4-D.

 Be careful when handling 2,4-D; it can be especially irritating to your eyes.

### Table 35.1

**Concentrations of Auxin Used to Bioassay a Solution of Unknown Concentration**

| Auxin Concentration | $\mathrm{Log_{10}}$ Auxin Concentration | Curvature (degrees) | | | | | Mean |
|---|---|---|---|---|---|---|---|
| Buffer alone | _____ | ____ | ____ | ____ | ____ | ____ | ____ |
| 0.0001 mM auxin | −4 | ____ | ____ | ____ | ____ | ____ | ____ |
| 0.001 mM auxin | −3 | ____ | ____ | ____ | ____ | ____ | ____ |
| 0.01 mM auxin | −2 | ____ | ____ | ____ | ____ | ____ | ____ |
| 0.1 mM auxin | −1 | ____ | ____ | ____ | ____ | ____ | ____ |
| 10 mM auxin | 1 | ____ | ____ | ____ | ____ | ____ | ____ |
| Unknown | _____ | ____ | ____ | ____ | ____ | ____ | ____ |

## Table 35.2

**Concentrations of 2,4-D Used to Bioassay a Solution of Unknown Concentration**

| Concentration of 2,4-D | Log$_{10}$ Concentration of 2,4-D | Mean Root Length (mm) |
|---|---|---|
| Buffer alone | ———— | ———— |
| 0.001 mg/L | −3 | ———— |
| 0.01 mg/L | −2 | ———— |
| 0.1 mg/L | −1 | ———— |
| 1 mg/L | 0 | ———— |
| 10 mg/L | 1 | ———— |
| Unknown | ———— | ———— |

## Table 35.3

**Concentrations of Gibberellic Acid Used to Bioassay a Solution of Unknown Concentration**

| Concentration of Gibberellic Acid | Log$_{10}$ Concentration of Gibberellic Acid | Mean Weight (mg) |
|---|---|---|
| Buffer alone | ———— | ———— |
| 0.001 mg/L | −3 | ———— |
| 0.01 mg/L | −2 | ———— |
| 0.1 mg/L | −1 | ———— |
| 1 mg/L | 0 | ———— |
| 10 mg/L | 1 | ———— |
| Unknown | ———— | ———— |

## Procedure 35.2  Bioassay 2,4-D

1. Working in a group of three to five students, label each of seven petri dishes to receive 2,4-D at one of the concentrations listed in table 35.2.
2. Place several pieces of filter paper into each petri dish. Using sterile paper will minimize growth of bacteria and fungi in the dish.
3. Add to each petri dish 10 mL of the appropriate solution listed in table 35.2. Each solution has been adjusted to pH 5.9 with 10 mM phosphate buffer.
4. Place 20 cucumber seeds in each petri dish and store each dish in the dark.
5. After 5 days, measure the length of the primary root of each seedling. Calculate and record the mean length for roots in each treatment in table 35.2.
6. Plot the mean root length versus the log$_{10}$ of the known concentrations of 2,4-D on one of the sheets of graph paper at the end of this exercise.
7. Determine the unknown concentration of 2,4-D in a manner similar to that described in step 11 of procedure 35.1 for auxin bioassay. Record the concentration of the unknown solution in table 35.2.

### Question 2

*a.* What are your conclusions?

*b.* According to your data, what is the approximate concentration of 2,4-D in the unknown solution?

# BIOASSAY FOR GIBBERELLIC ACID

You will measure the effect of gibberellic acid on enlargement of radish cotyledons (see figs. 34.6 and 34.8). You'll then bioassay a solution of gibberellic acid of unknown concentration by comparing its effect on cotyledon enlargement with that induced by known concentrations of gibberellic acid.

## Procedure 35.3  Bioassay gibberellic acid

1. Working in a group of three to five students, obtain a flat of dark-grown radish seedlings that are approximately 1.5 days old.
2. Label each of seven petri dishes to receive gibberellic acid at one of the concentrations listed in table 35.3.
3. Add to each petri dish 5 mL of the appropriate solution listed in table 35.3. Each solution has been adjusted to pH 5.9 with 10 mM phosphate buffer.
4. Excise the smaller cotyledon and all of the hypocotyl (the portion of seedling between the cotyledon and root) from 105 seedlings.
5. Place pads of sterile filter paper in each of seven petri dishes. Moisten the filter paper with distilled water.
6. Place 15 excised cotyledons on the moistened filter paper in each petri dish. Use cotyledons having similar sizes in all treatments.
7. Place the dishes in a humid environment (e.g., inside a plastic bag) under a constant fluorescent light.
8. After 3 days, weigh the cotyledons to the nearest milligram and calculate the mean weight of a cotyledon in each of the treatments. Record these mean weights in table 35.3.
9. Plot the mean weight of the cotyledons versus the log$_{10}$ of the known concentrations of gibberellic acid on one of the sheets of graph paper at the end of this exercise.
10. Determine the unknown concentration of gibberellic acid in a manner similar to that described in step 11 of procedure 35.1 for auxin bioassay. Record the concentration of the unknown solution in table 35.3.

**Question 3**

*a.* What are your conclusions?

*b.* According to your data, what is the approximate concentration of gibberellic acid in the unknown solution?

## INQUIRY-BASED LEARNING

*How do plant growth regulators influence plant tissues and organs?*

Observation: Plant growth regulators such as gibberellic acid and auxin have a wide variety of effects on plant growth and development. In some instances, these effects are not the same on roots and shoots of plants.

Question: How does gibberellin affect the growth of plant stems?

**a.** Establish a working lab group and obtain Inquiry-Based Learning Worksheet 35 from your instructor.

**b.** Discuss with your group well-defined questions relevant to the preceding observation and question. Choose and record your group's best question for investigation.

c. Translate your question into a testable hypothesis and record it.

d. Outline on Worksheet 35 your experimental design and supplies needed to test your hypothesis. Ask your instructor to review your proposed investigation.

e. Conduct your procedures, record your data, answer your question, and make relevant comments.

f. Discuss with your instructor any revisions to your questions, hypotheses, or procedures. Repeat your work as needed.

## Questions for Further Study and Inquiry

*1.* How could you use a bioassay to study a suspected pollutant in soil? In water? In air?

*2.* Bioassays are often the first step involved in studying the effects of a potential drug. Why?

*3.* What are the strengths of a bioassay? What are the limitations?

*4.* What are the problems associated with extrapolating from a logarithmic curve?

### DOING BIOLOGY YOURSELF

Establishing a dose-response curve for an unknown or unfamiliar substance, or for an extract that may contain more than one substance, is often a goal of a bioassay. Design an experiment to bioassay an extract of leaf litter for the inhibition or promotion of seed germination.

### WRITING TO LEARN BIOLOGY

Briefly describe the value of bioassays as opposed to direct chemical measurement. What assumptions must be made to validate that a bioassay for a chemical is accurate?

# Survey of the Animal Kingdom

## Phyla Porifera and Cnidaria

Please visit connect.mheducation.com to review online resources tailored to this lab.

**A**nimals are uniformly characterized as **eukaryotic, multicellular,** and **ingestive-feeding heterotrophs.** Heterotrophs derive their energy from organic molecules made by other organisms. To many people, the most ancient animals, such as **phylum Porifera** (sponges) and **phylum Cnidaria** (sea jellies and corals), may appear primitive and unsophisticated. However, as we discuss in this and upcoming exercises the remarkable diversity of animals, remember that "primitive" refers to characteristics derived early in the phylogenetic history of an organism's evolution. Primitive does not imply simple, less important, or poorly functioning charactersitics. Instead, it refers to characteristics that have stood the test of time. The bodies and body cavities of sponges and jellyfish are structurally uncomplicated and lack the complex behavior and sensory capabilities of most higher animals. But don't let that simplicity fool you. What sponges and sea jellies (sometimes called jellyfish) lack in complexity is more than compensated for by their extraordinarily elegant design. After all, their overall body-plans (table 36.1) have persisted in changing environments for many millions of years; their "simple" morphology accomplishes the primary functions of food getting, reproduction, and adaptive responses to their environment as does the more-complex morphology of other more familiar animals.

## PHYLUM PORIFERA

Sponges are the simplest of the major animal phyla and comprise 5150 species (fig. 36.1). Most sponges live in the ocean, but a few encrust rocks and wood in freshwater. Sponges lack tissues and organs and are typically **asymmetrical** assemblages of cells. Bodies of asymmetrical organisms have no symmetry or pattern such as left and right halves or anterior and posterior regions. Sponge cells are so loosely assembled and self-sufficient that if a sponge is forced through a mesh, the disassociated cells will survive. Even more remarkably, the disassociated cells of some species can reassemble as a functioning organism.

### Grantia

Examine a preserved specimen of *Grantia*, one of the simplest sponges (fig. 36.2). At first glance, sponges such as *Grantia* appear plantlike because they are **sessile** (i.e., attached to a substrate). Some sponges even appear green because symbiotic algae live in their bodies. However, sponges are **filter-feeding** heterotrophs and have no photosynthetic pigments. Notice that *Grantia* is a tubular, open-ended chamber surrounded by a thin, porous, folded wall of cells. Phylum Porifera gets its name from the many pores in the chamber walls.

## Sponge Reproduction

Sponges reproduce asexually and sexually. Asexual reproduction includes budding and the release of stress-resistant aggregates of amoebocytes called **gemmules.** In favorable conditions, amoebocytes in a gemmule can grow into a mature organism. During sexual reproduction, choanocytes and amoebocytes differentiate into gametes. Eggs remain in the gelatinous, interstitial (between cells) **mesenchyme,** but sperm are released into the water and are captured by choanocytes or amoebocytes of other sponges. The captured sperm are transported to eggs and fertilization occurs. After a brief development, the embryo is expelled from the sponge. Most species of sponges are hermaphroditic (i.e., produce both male and female gametes), but they usually produce eggs and sperm at different times.

## PHYLUM CNIDARIA (COELENTERATA)

Phylum Cnidaria (also called coelenterates) includes **class Hydrozoa** (hydras), **class Scyphozoa** (sea jellies), **class Cubozoa,** and **class Anthozoa** (anemones and corals). Almost all cnidarians are marine carnivores. Their bodies are **radially symmetrical** and more complex than sponges (fig. 36.7). Radial symmetry describes a body plan with repetitive body areas arranged in a circle around a central point like the pieces of a pie. Sensory organs are exposed to the environment in all directions around the perimeter. This is evolutionarily adaptive for a slow or nonmotile organism.

The body wall of cnidarians has two cellular layers, an **ectoderm** on the outside and an **endoderm** (sometimes called the gastroderm) lining the interior gastrovascular cavity (fig. 36.8). A gelatinous **mesoglea** separates the two body layers. Cells of cnidarians are organized into true tissues (nervous, muscular, and reproductive), but organs are organized loosely, at best.

Cnidarians have two basic body-plans: **polyps** and **medusae** (fig. 36.8). Polyps are cylindrical animals with a mouth surrounded by tentacles atop the cylinder (i.e., the end facing away from the substrate). Polyps are usually attached to the substrate and may be solitary or colonial. In contrast to polyps, medusae are usually free-floating and umbrella-shaped. Their mouths point downward and are surrounded by hanging tentacles. The classes of cnidarians are distinguished primarily by the relative dominance of the polyp stage or the medusa stage in the life cycle. Some cnidarians occur only as polyps, others only as medusae, and still others alternate between these two forms. This alternation is a form of **polymorphism,** which means "many forms."

The life cycle of many cnidarians is characterized by alternation between polyp and medusa (i.e., polymorphism; fig. 36.8). During the life cycle, medusae produce and release eggs and sperm into water for fertilization, although some species retain their eggs. After fertilization the zygote develops into a swimming mass of ciliated cells called a **planula larva.** A planula eventually settles to the

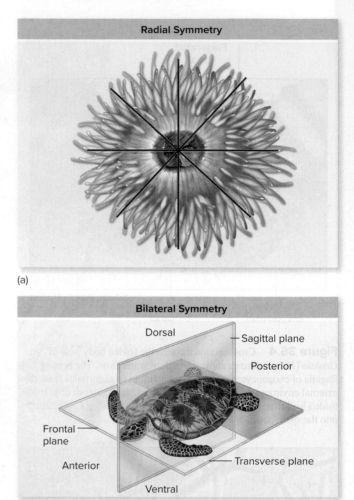

**Figure 36.7** A comparison of radial and bilateral symmetry. (*a*) Radially symmetrical animals, such as this sea anemone, can be bisected into equal halves in any two-dimensional plane. (*b*) Bilaterally symmetrical animals, such as this turtle, can only be bisected into equal halves in one plane (the sagittal plane).

substrate and develops into a polyp. The polyp may reproduce asexually by budding other polyps or may continue the sexual cycle by budding immature medusae called **ephyrae.** An ephyra develops into a mature medusa.

All cnidarians are carnivores that capture their prey (small fishes and crustaceans) with tentacles that ring their mouth. Captured prey are pushed through the mouth into the **gastrovascular cavity** (GVC), where a combination of intracellular and **extracellular digestion** occurs followed by phagocytosis of small food particles and intracellular digestion (fig. 36.9). These tentacles are armed with stinging cells called **cnidocytes** containing small, barbed harpoonlike structures called **nematocysts** (fig. 36.10). Notice the transition from intracellular digestion within choanocytes of poriferans to a combination of intra- and extracellular processing in cnidarians. The phyla discussed in upcoming exercises rely almost exclusively on extracellular digestion.

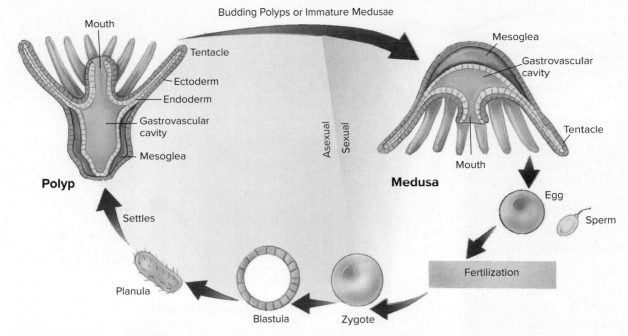

**Figure 36.8** A generalized life cycle for Cnidaria. Cnidarians alternate between medusa and polyp forms. Male and female medusae use meiosis to produce gametes that, in water, are fertilized. These gametes are the only haploid (*n*) stage; all other stages are diploid (2*n*). After a short period of free swimming, the planula larva settles to the substrate and forms a polyp. When the polyp buds (an asexual process), additional polyps and medusa buds form. Medusae separate from the polyp and swim away. The polyp or medusa stage has been lost or reduced in many cnidarians, such as anemones and corals.

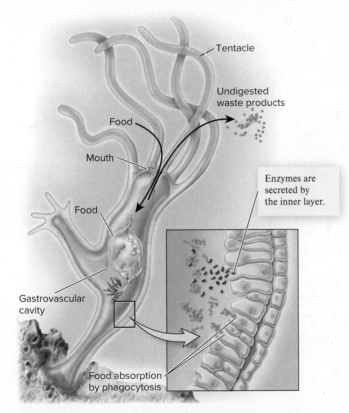

**Figure 36.9** Extracellular digestion. In animals with gastrovascular cavities, such as the cnidarian *Hydra,* most digestion occurs extracellularly. Food sources are trapped by tentacles and delivered to the mouth. Digestion occurs within the gastrovascular cavity. Digested food products are phagocytosed directly into the cells that line the cavity, and wastes are excreted out the same opening in which food entered.

### Question 5

*a.* Consider Learning Objective 1 listed at the beginning of this exercise. Are cnidocytes significant to fundamental processes for cnidarians? In what ways?

*b.* Consider Learning Objective 1 listed at the beginning of this exercise. How could polymorphism contribute to the evolutionary success of cnidarians in their environment?

## Class Hydrozoa

The polyp stage dominates the hydrozoan life cycle, although both polyps and medusae occur in most species. The outer layer of cells, the ectoderm, and the inner layer, the endoderm, surround the gastrovascular cavity; these layers are separated by the gelatinous, acellular mesoglea. Amoeboid cells circulate in the mesoglea. Ectodermal cells include cnidocytes and muscular contractile cells. Endodermal and glandular cells secrete enzymes into the gastrovascular cavity for extracellular digestion. The gastrodermis lacks cnidocytes.

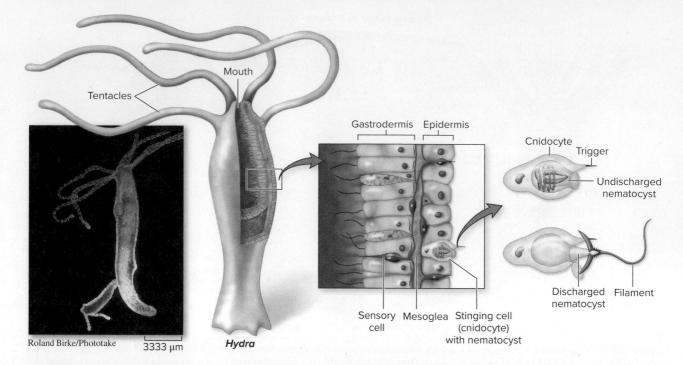

Figure 36.10   Phylum Cnidaria: cnidarians. The cells of a cnidarian such as this *Hydra* are organized into specialized tissues. The interior gut cavity is specialized for extracellular digestion—that is, digestion within a gut cavity rather than within individual cells. Cnidarians are radially symmetrical, with parts arranged around a central axis like the petals of a daisy. The epidermis includes stinging cells called cnidocytes, and each cnidocyte can discharge a harpoonlike nematocyst. Cnidocytes are scattered on the body wall and dense on the tentacles.

## *Hydra*

*Hydra* are small, common hydrozoans that live in shallow, freshwater ponds (figs. 36.9 and 36.10). They are usually less than 1 cm tall and prey on smaller invertebrates among the filaments and leaves of freshwater algae and plants. *Hydra* have no medusa stage.

### *Procedure 36.2*   Observe Hydra

1.  Obtain a living *Hydra;* observe it in a small petri dish with a dissecting microscope. Polyps of *Hydra* are solitary and occasionally hang from the water's surface with their **basal disks** adhering to the surface of the water. More often, they attach to a hard substrate with their basal disk. However, *Hydra* can detach themselves and move, not by swimming but by somersaulting along their substrate.

2.  Allow a few minutes for the animal to relax in the petri dish; then tap the edge and observe the animal's response.

3.  If small, living crustaceans such as *Daphnia* or *Artemia* (brine shrimp) are available, place some of these organisms near the tentacles of a *Hydra*. When a prey item touches a tentacle, it sticks tightly. The nematocysts help entrap the *Daphnia*.

4.  Observe the cellular structure of *Hydra* by studying a prepared slide of a cross section of the organism (figs. 36.10 and 36.11).

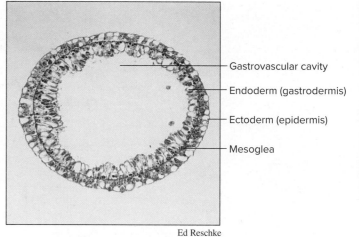

Figure 36.11   Micrograph of a cross section of a *Hydra,* a hydrozoan (100×).

### Question 6
*a.*  How do *Hydra* respond to a tap on their substrate?

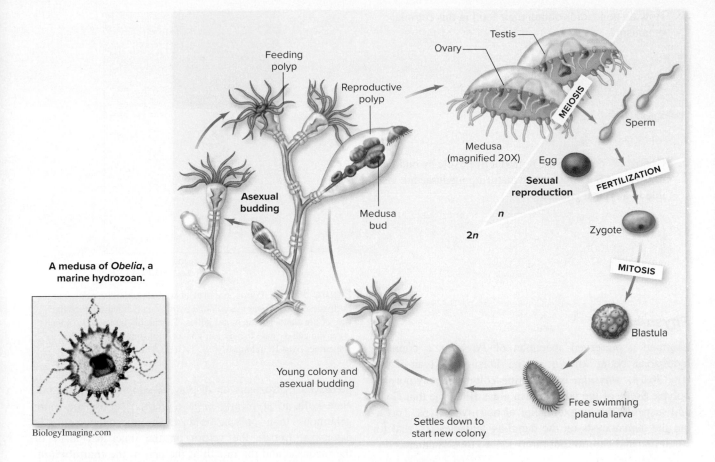

Feeding
polyp

Reproductive
polyp

Ovary

Testis

MEIOSIS

Sperm

Medusa
(magnified 20X)

Egg

Sexual
reproduction

FERTILIZATION

Asexual
budding

*n*

Medusa
bud

2*n*

Zygote

MITOSIS

Blastula

A medusa of *Obelia*, a
marine hydrozoan.

BiologyImaging.com

Young colony and
asexual budding

Settles down to
start new colony

Free swimming
planula larva

**Figure 36.12** The life cycle of *Obelia,* a marine colonial hydroid. Polyps reproduce by asexual budding, forming colonies. Reproductive polyps may also give rise to medusae, which reproduce sexually via gametes. These gametes fuse, producing zygotes that develop into planulae, which in turn settle down to produce polyps. The inset photo shows a young recently budded medusa.

*b.* What tissues must exist for this response?

*c.* *Hydra* are predators. How actively do *Hydra* stalk their food?

*d.* What specialized cells of tentacles aid in capturing prey?

## *Obelia*

Examine a prepared slide of another hydrozoan, *Obelia* (fig. 36.12). Examine a prepared slide of the small medusae of *Obelia* in addition to the polyp colony. *Obelia* typifies most hydrozoans because it has colonial polyps and free-swimming medusae. These colonial polyps appear plantlike and branch from a tube. Polyps of *Obelia* are polymorphic because some are specialized feeding polyps called **gastrozoids;** others are reproductive polyps called **gonozoids.**

### Question 7
*a.* What structures determine whether a polyp of *Obelia* is a gastrozoid (feeding polyp) rather than a gonozoid?

**b.** How do gonozoids obtain their food in this colonial organism?

**c.** Gonozoids continue the reproductive cycle by budding medusae. About how many maturing medusae are visible in a typical gonozoid?

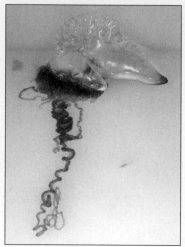

Image courtesy of Islands in the
Sea 2002, NOAA/OER

**Figure 36.13** Portuguese man-of-war, *Physalia utriculus*. The Portuguese man-of-war is a colonial hydrozoan that has adopted the way of life characteristic of sea jellies. This highly integrated colonial organism can ensnare fish by using its painful stings and tentacles, sometimes over 15 m long.

## Physalia

Examine a preserved specimen of *Physalia*, a common hydrozoan better known as the Portuguese man-of-war (fig. 36.13). *Physalia* is a floating colony of polymorphic polyps. Some of the polyps form a gas-filled sac that floats and suspends the long tentacles of nutritive polyps. Touching the nematocysts on the dangling tentacles is lethal for small fish and painful (but rarely lethal) for swimmers.

## Gonionemus

Examine a preserved specimen of *Gonionemus* with your dissecting microscope (fig. 36.14). Examine any other preserved hydrozoans on display. *Gonionemus* is a hydrozoan with an atypically large medusa. Medusa are more gelatinous than polyps because the mesoglea is more extensive. Locate the **velum** on the inner periphery of the medusae and the mouth at the end of the **manubrium.** The **gastrovascular cavity** radiates from the center as **ring canals** connected by a **circular canal** around the perimeter. The **gonads** (tissue that produces gametes) attach to the radial canals and appear similar in males and females. The **tentacles** have a rough surface.

## INQUIRY-BASED LEARNING

### How well do Hydra detect nearby prey?

Observations: *Hydra* are predators capable of relatively complex behavior. They readily detect vibration as well as changes in water chemistry produced by their potential prey.

Question: How do *Hydra* sense their prey?

**a.** Establish a working lab group and obtain Inquiry-Based Learning Worksheet 36 from your instructor.

**b.** Patiently observe *Hydra* movement and their responsiveness. Discuss with your group a well-defined question relevant to *Hydra* response to prey movement, to dissolved molecules from their prey, or both. Record the question on Worksheet 36.

**c.** Translate your question into a testable hypothesis and record it.

**d.** Devise a procedure to determine if *Hydra* are more responsive to prey movement, to dissolved molecules from their prey, or both.

**e.** Outline on Worksheet 36 your experimental design and supplies needed to test your hypothesis. Ask your instructor to review your proposed investigation.

**f.** Conduct your procedures.

**g.** Record your data, answer your question, and make relevant comments.

**h.** Discuss with your instructor any revisions to your questions, hypotheses, or procedures. Repeat your work as needed.

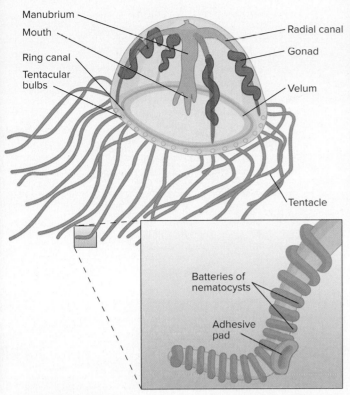

**Figure 36.14** Diagram showing the structure of the medusa of *Gonionemus*, a hydrozoan.

Manubrium
Mouth
Ring canal
Tentacular bulbs
Radial canal
Gonad
Velum
Tentacle
Batteries of nematocysts
Adhesive pad

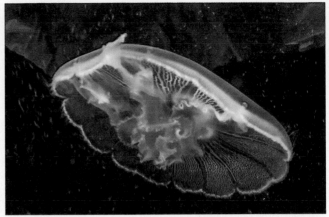

BiologyImaging.com

**Figure 36.15** *Aurelia*, the moon sea jelly (1×). The gelatinous medusa is the dominant body form in the life cycle of these cnidarians.

## Question 8

***a.*** Which cells give tentacles of *Gonionemus* their rough surface?

***b.*** What functions might radial and ring canals serve?

## Class Scyphozoa

Observe preserved medusae of *Aurelia* and *Cassiopeia*. Scyphozoans are commonly called sea jellies (jellyfish) because the gelatinous medusa dominates their life cycle. The polyp is reduced to a small larval stage. The mesoglea has amoeboid cells. The GVC is divided into four radiating pouches, and the gastrodermis has cnidocytes. This group also includes one of the largest invertebrates in the world, *Cyanea capillota*. *Cyanea* lives in the North Sea and can exceed 2 m in diameter. *Aurelia* (2–6 cm) is a more typical sea jelly for you to examine (fig. 36.15).

Examine prepared slides of (1) planula larvae produced by sexual reproduction of medusae, (2) the polyp stage called a **scyphistoma**, and (3) ephyra (immature medusae) budded from the polyp. Review the life cycle of *Aurelia* (fig. 36.16).

## Question 9

How do medusae of *Aurelia* and *Gonionemus* differ in size, arrangement of tentacles, and shape of manubrium?

## Class Anthozoa

Anthozoans (anemones and corals) form the largest class of cnidarians with more than 6000 species (fig. 36.17). Anthozoan polyps are solitary or colonial, and there is no medusa. The mouth leads to a tubular pharynx and to a gastrovascular cavity with septate compartments. Gonads are gastrodermal.

## *Metridium*

Obtain a specimen of the common anemone (pronounced ah-NEH-moh-nee) *Metridium*, and find its mouth and tentacles. Make a cross section through the body of *Metridium* and expose the gastrovascular cavity (this dissection may be on demonstration). Locate the structures shown in figure 36.18. Anemones are sessile and attach themselves to a substrate with their flat and sticky basal disk. However, this attachment is not permanent, and anemones can slowly slide on a film of mucus. When pieces of the basal disk tear

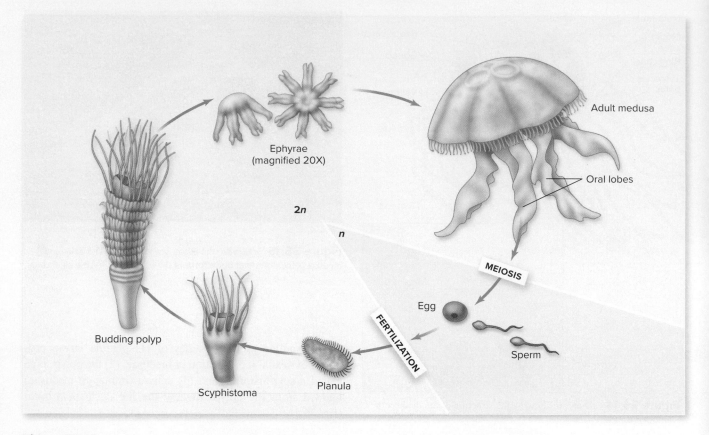

**Figure 36.16** *Aurelia* life history. For the dioecious *Aurelia,* like all scyphozoans, the medusa stage dominates the life cycle. Male and female medusae produce gametes for fertilization in the water. These gametes are the only haploid (*n*) stage of the life cycle; all other stages are diploid (2*n*). The zygote develops into a planula larva that settles to the substrate and forms a scyphistoma (polyp) that produces ephyrae (immature medusae) by budding. Medusae separate from the polyp and swim away.

BiologyImaging.com

**Figure 36.17** Sea anemone, a common anthozoan.

away from a moving anemone, the pieces form a new individual. This type of asexual reproduction is **fragmentation.**

The gastrovascular cavity of an anthozoan polyp is partitioned by thin septa. These septa distinguish anthozoan from scyphozoan and hydrozoan polyps. Along the edges of the septa are gonadal tissue and threads of tissue called

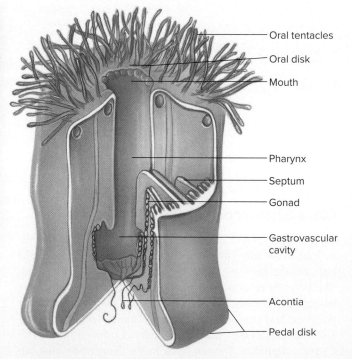

**Figure 36.18** Class Anthozoa. The structure of the anemone, *Metridium.* Anthozoans have no medusa stage.

BiologyImaging.com

**Figure 36.19** Calcium carbonate skeleton of a stony coral.

Depressions formed by polyps

BiologyImaging.com

**Figure 36.20** *Tubipora,* organ pipe coral.

acontia. Acontia threads bear dense batteries of cnidocytes that can protrude through the mouth or through pores in the body wall to overcome prey or provide defense.

## Corals

Examine a piece of dry, calcareous coral and look for small depressions (fig. 36.19). Coral polyps are structurally similar to anemones, but corals are usually colonial and much smaller. Most corals secrete a hard skeleton of calcium carbonate with many small cups surrounding the polyps. The small, fragile polyps are probably absent from the specimen you are examining, but the depressions in which they lived are numerous.

Examine a piece of *Tubipora* (fig. 36.20) and any other anthozoans on display. The tropical organ pipe coral, *Tubipora,* is organized differently. Long parallel polyps are encased in calcareous tubes connected at intervals by transverse plates. The calcareous tubes are impregnated with iron salts that give the colony an attractive color.

Unfortunately, the calcium skeletons forming the bulk of coral reefs are highly vulnerable to increased levels of atmospheric $CO_2$ produced by our growing use of fossil fuels. Increasing amounts of $CO_2$ dissolve in ocean water and form carbonic acid, which acidifies the water and weakens the calcareous coral structure. In addition, rising temperatures associated with global warming causes coral polyps to expel their symbiotic algae and die. This is known as "coral bleaching" because it leaves expanses of white, lifeless coral skeletons behind.

**Question 10**

***a.*** Locate the radial ridges within each depression on a piece of coral. What structures within the polyp did they support?

***b.*** What is the advantage of a partitioned gastrovascular cavity?

***c.*** Consider Learning Objective 1 listed at the beginning of this exercise. How does fragmentation contribute to the evolutionary success of anthozoans in their environment?

**Question 11**

*a.* In your notes draw and describe the life cycle of a cnidarian. What is the difference between a polymorphic life cycle and the typical life cycle of other animals?

*b.* How does the feeding method of a sponge compare with that of a coral?

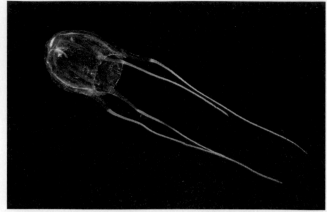

**Figure 36.21**  Class Cubozoa. *Chironex fleckeri,* a box jelly (1×).

## Class Cubozoa

As their name implies, cnidarians of class Cubozoa are box-shaped medusae; the polyp stage is inconspicuous and in many cases not known. Most medusae are only a few centimeters in height, although some are 25 cm tall. A tentacle or group of tentacles is found at each corner of the box (fig. 36.21).

Box jellies are strong swimmers and voracious predators of fish in tropical and subtropical waters. The stings of some species can be fatal to humans. Examine a preserved specimen of a box jelly. How does its shape compare to that of scyphozoans?

# Questions for Further Study and Inquiry

*1.* Which group within kingdom Protista probably gave rise to sponges? On what evidence do you base your answer?

*2.* Why are spicules used as a primary characteristic to classify sponges?

**3.** Sponges and cnidarians have no lungs or gills. How do they exchange gases with the environment? Are humans better off having lungs? If so, how?

**4.** Why are sponges considered to be an evolutionary dead end?

**5.** Explain how cnidocytes with their nematocysts function in food capture and defense.

**6.** Discuss how polymorphism in the cnidarians might have influenced adaptive radiation of the group.

7. How does digestion in cnidarians differ from digestion in sponges?

8. Prepare a simple table of all of the taxonomic groups, their common names, their distinguishing characteristics, and all representative genera covered in this exercise. Keep this table with your study notes.

**WRITING TO LEARN BIOLOGY**
What are the advantages and disadvantages to a solitary versus colonial existence?

# Survey of the Animal Kingdom

## Phyla Platyhelminthes and Mollusca

---

## Learning Objectives

By the end of this exercise you should be able to:

1. Describe how the prominent characteristics of platyhelminths and mollusks promote their survival and reproduction.
2. Describe the general morphology of flatworms in phylum Platyhelminthes and mollusks in phylum Mollusca.
3. List characteristics that phyla Platyhelminthes and Mollusca have in common with phyla Porifera and Cnidaria.
4. List characteristics of flatworms and mollusks more advanced than those of more primitive phyla.
5. List examples of each major class of flatworms and mollusks.
6. Understand the differences between acoelomates and coelomates and know which phyla are associated with each.
7. Discuss characteristics that have most likely contributed to flatworm and molluscan success over millions of years and relate the discussion to Learning Objective 1.

---

Please visit **connect.mheducation.com** to review online resources tailored to this lab.

---

**F**latworms of **phylum Platyhelminthes** and **mollusks** of **phylum Mollusca** are remarkably successful if we measure evolutionary success by high diversity and persistence in the environment for 400 million years. Both phyla occur in marine, freshwater, terrestrial, and parasitic environments. Their morphology is considerably more complex than that of sponges and sea jellies (table 37.1). For example, flatworms and mollusks are **triploblastic** meaning that they have three embryological germ layers: **ectoderm**, **mesoderm**, and **endoderm**. They also have **organs** made of interdependent tissues and flatworms are the most ancient animals having **bilateral symmetry** with distinct anterior and posterior ends (see fig. 36.7).

Flatworms are **acoelomate**, meaning without a coelomic cavity. Their mesoderm is a solid mass of tissue with no internal cavity (figs. 37.1 and 37.2). In contrast, mollusks and the remaining phyla we will discuss are **coelomate**. They have an internal cavity lined with mesoderm to accommodate complex internal organs. This spacious internal cavity likely enhances the ability of internal organs to move, contract, and otherwise function independently. Note that the coelom of mollusks is greatly reduced in modern species.

Coelomates are further divided into **protostomes** and **deuterostomes**. Protostomes have well-developed nervous, circulatory, excretory, reproductive, and digestive systems. Protostomes include phyla Platyhelminthes, Mollusca, and Annelida, all grouped as lophotrochozoa because they commonly have a trochophore larval stage (fig. 37.3).

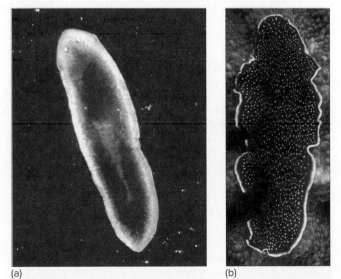

(a) blickwinkel/Hecker/Alamy Stock Photo (b) R. Aaron Raymond/Radius Images/
Getty Images

**Figure 37.1** Flatworms (phylum Platyhelminthes). (*a*) A common freshwater flatworm (*Dugesia dorotocephala*) (20×). (*b*) A free-living marine flatworm (*Thysanozoon nigropapillosum*) (5×).

Protostomes also include phyla Nematoda and Arthropoda grouped as ecdysozoa because they commonly molt (shed) their outer cuticle or exoskeleton (see Exercises 38 and 39). Deuterostome phyla include Echinodermata and Chordata (see Exercise 40). A detailed comparison of protostomes and deuterostomes is presented in Exercise 40.

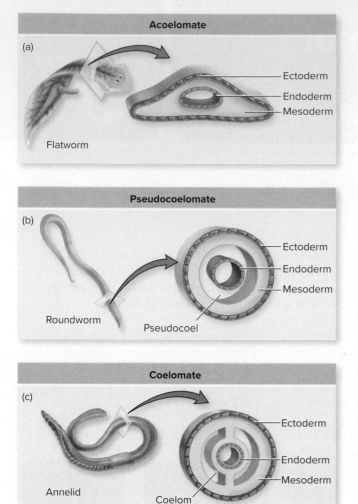

Acoelomate

(a)

Ectoderm
Endoderm
Mesoderm

Flatworm

Pseudocoelomate

(b)

Ectoderm
Endoderm
Mesoderm

Roundworm          Pseudocoel

Coelomate

(c)

Ectoderm
Endoderm
Mesoderm

Annelid          Coelom

**Figure 37.2**   Three body plans for bilaterally symmetrical animals. (*a*) Acoelomates (including flatworms) have no body cavity. (*b*) Pseudocoelomates (including nematodes) develop a body cavity between the mesoderm and endoderm. (*c*) Coelomates (including mollusks, annelids and more advanced phyla) have a body cavity bounded by mesoderm.

## PHYLUM PLATYHELMINTHES

Flatworms are dorsoventrally compressed, and free-living species have primitive sense organs (fig. 37.1). They also have a gastrovascular cavity with one opening that is both mouth and anus. Their nervous system is more advanced than that of cnidarians (see Exercise 36) and consists of a ladderlike arrangement of nerve cords extending the length of the body.

## Subphylum Turbellaria

Turbellarians (3000 species) are free-living flatworms inhabiting freshwater, saltwater, and moist terrestrial environments. Turbellarians scavenge and prey on small animals and are **hermaphroditic,** meaning that individuals have both male and female sex organs.

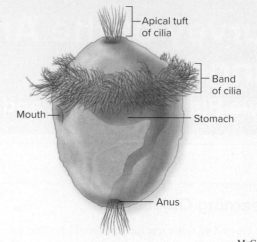

Apical tuft
of cilia

Band
of cilia

Mouth

Stomach

Anus

**Figure 37.3**   Trochophore larval form. Protostome phyla include Platyhelminthes, Mollusca, and Annelida. They share a close common ancestry as indicated by many of their species having a similar larval stage called a trochophore. These phyla are grouped as lophotrochozoans.

## Dugesia

*Dugesia,* often called planaria, is a common freshwater turbellarian with typical characteristics of flatworms (fig. 37.4). The head has lateral lobes and sensory organs called **eyespots.** *Dugesia* feed by sucking food through its mouth and into a tubular **pharynx** leading to the gastrovascular cavity. The muscular pharynx is usually retracted in the body but can be everted through an opening in the midventral epidermis (middle of the lower surface). Most digestion in the gastrovascular cavity is extracellular, but phagocytic cells line the cavity and complete digestion of small particles intracellularly.

The mesodermal tissue of planaria is the loose mass of cells between the ectoderm on the surface and the endoderm lining the gastrovascular cavity and pharynx (fig. 37.4). This body plan allows for remarkable powers of regeneration. A chopped planarian can regenerate a complete new head or regrow the entire right side of its body if cut in half lengthwise. Wouldn't it be remarkable if humans could rekindle the ability to regenerate lost appendages, an ability lost in more advanced phlya?

 **SAFETY FIRST**  Before coming to lab, you were asked to read this exercise so you would know what to do and be aware of safety issues. In the space below, briefly list the safety issues associated with today's procedures. If you have questions about these issues, contact your laboratory assistant before starting work.

### *Procedure 37.1*   Observe planaria

*1.*   Obtain a living *Dugesia* and examine its morphology with a dissecting microscope.

*2.*   Place the animal in the center of a petri dish and follow its movements for a few minutes.

Table 37.1

**Phyla Platyhelminthes and Mollusca**

| Phylum | Typical Examples | | Key Characteristics | Approximate Number of Named Species |
|---|---|---|---|---|
| Platyhelminthes (flatworms) | *Planaria*, tapeworms, liver flukes | | Solid, unsegmented, bilaterally symmetrical worms; no body cavity; digestive cavity, if present, has only one opening | 20,000 |
| Mollusca (mollusks) | Snails, oysters, octopuses, nudibranchs | | Soft-bodied coelomates whose bodies are divided into three parts: head-foot, visceral mass, and mantle; many have shells; almost all possess a unique rasping tongue, called a radula; 35,000 species are terrestrial. | 110,000 |

3. Gently touch the animal with a probe and watch how it responds.

4. If a spotlight is available, determine how *Dugesia* may respond to strong light.

5. Offer *Dugesia* a small piece of liver, boiled egg, or cat food; watch it eat.

6. After a few minutes, gently roll the animal from the food and find its protruded pharynx.

7. Examine a whole mount of a stained planaria and a prepared slide of a cross section through a planaria.

Specimens vary. Be sure to examine more than one slide, and do not rely on photographs.

8. Locate the structures shown in figure 37.4.

9. Examine cross sections taken through the region of the pharynx and away from the pharynx. Draw and label these cross sections, and have your instructor check them for accuracy.

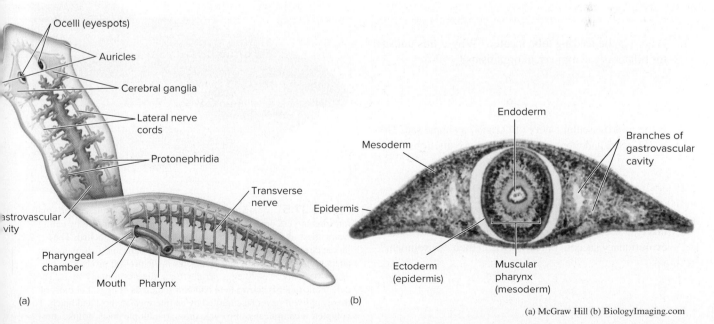

(a) McGraw Hill (b) BiologyImaging.com

**Figure 37.4** Anatomy of a free-living flatworm, *Dugesia*. (*a*) External structure. (*b*) Cross section of a planarian taken through the pharynx region (40×).

## Question 1

**a.** What features of *Dugesia* distinguish its head from its tail?

**b.** What is the difference between the eyes of most animals you are familiar with and the eyespots of *Dugesia?*

**c.** How does the head of *Dugesia* move differently from the tail?

**d.** Does *Dugesia* move randomly or in an apparent direction?

**e.** How is *Dugesia* adapted for directional movement?

**f.** How does a flatworm respond when touched with a probe?

**g.** Do the planaria move toward or away from light?

**h.** Where is the feeding tube located? Why is this unusual for bilaterally symmetrical organisms?

**i.** Is the gastrovascular cavity of *Dugesia* a simple sac? How is it divided and what advantage do these divisions offer?

**j.** Consider Learning Objective 1 listed at the beginning of this exercise. How could being monoecious contribute to evolutionary success of flatworms in their environment?

**k.** Planarians have a head. In biological terms, what constitutes a "head"? How does it relate to Learning Objectives 1, 4, and 7?

**l.** Planaria lack specialized gas-exchange organs. How do you think planaria accomplish this task?

## Subphylum Neodermata

All parasitic flatworms are in subphylum Neodermata. This name refers to "new skin" and refers to the animal's outer surface, termed the neodermis. Neodermata contains two subgroups: Trematoda, flukes, and Cercomeromorpha, tapeworms and their relatives.

## Subgroup Trematoda

Trematodes, commonly called flukes, are parasites, and their oval bodies are usually a few millimeters long. Flukes infect vertebrates and include both **endoparasites** (parasites inside their host) (fig. 37.5) and **ectoparasites** (parasites on the surface of their host). Trematodes lack an epidermis and are covered by an acellular but metabolically active **epicuticle.** This epicuticle is made of protein and lipids secreted by mesodermal cells and resists digestive enzymes. The epicuticle helps in respiration and absorbing nutrients. The ventral surface of a fluke usually has two adhesive organs (suckers). The oral sucker surrounds the mouth.

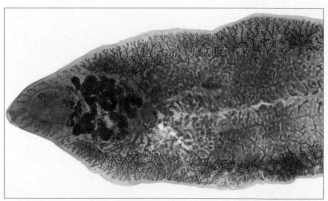

BiologyImaging.com

**Figure 37.5** *Fasciola* sp. is a parasitic fluke that infects the small intestine of vertebrates (10×). Humans acquire this parasite by eating water chestnuts and other aquatic plants with larvae attached. The larvae develop in intermediate hosts such as snails and fish. A major adaptation of flukes and other parasites is the prolific production of eggs—as many as 3000 eggs per day for a lifetime of many years. The dark, coiled uterus shown here contains hundreds of eggs, but most of these eggs will never be engulfed by an intermediate host and hatch. To complete a complicated life cycle through multiple hosts, a fluke must produce immense numbers of eggs to ensure that a few are successful. To appreciate the structure of the entire organism, see figure 37.8.

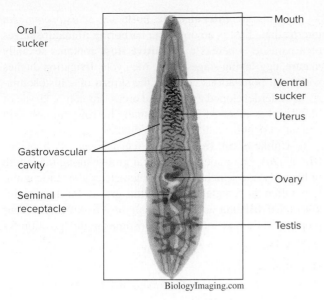

Oral sucker
Mouth
Ventral sucker
Uterus
Gastrovascular cavity
Ovary
Seminal receptacle
Testis

BiologyImaging.com

**Figure 37.6** Internal structure of *Opisthorchis*, the oriental liver fluke (30×).

that move through the digestive system of the host and exit with the feces. Larvae of flukes typically develop in snails and fish. Humans are infected when they eat raw or poorly cooked fish (fig. 37.7).

### Question 2

*a.* How does the shape of the digestive sac of *Opisthorchis* compare with that of *Dugesia?*

*b.* How does the position of the mouth of *Dugesia* and flukes compare?

## Opisthorchis

Use a dissecting microscope to examine a prepared slide of *Opisthorchis* and locate the structures shown in figure 37.6. *Opisthorchis* (*Clonorchis*) *sinensis,* the oriental liver fluke, often parasitizes humans in Japan and China. This hermaphroditic, adult fluke attaches to the bile duct and releases eggs

## Fasciola

Examine figure 37.5 and read its caption carefully. Use a dissecting microscope to examine a prepared slide or plastic mount of *Fasciola* and locate the features shown in figure 37.8. *Fasciola hepatica,* the sheep liver fluke, infects sheep, other vertebrates, and (rarely) humans. It is much larger than *Opisthorchis* (fig. 37.6) but similar in structure. *Fasciola* sucks food (blood, mucus, and cells) through a muscular pharynx located at the mouth.

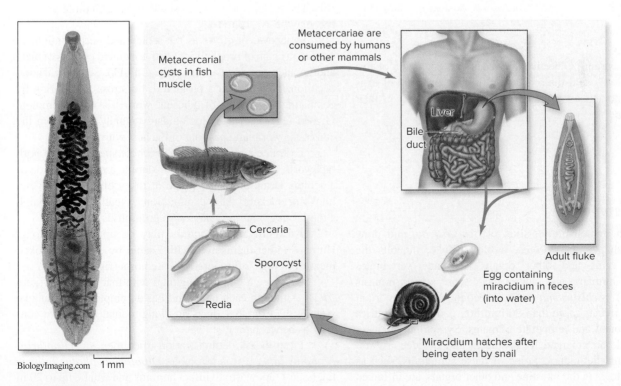

Metacercarial cysts in fish muscle

Metacercariae are consumed by humans or other mammals

Liver

Bile duct

Adult fluke

Cercaria

Sporocyst

Redia

Egg containing miracidium in feces (into water)

Miracidium hatches after being eaten by snail

BiologyImaging.com  1 mm

**Figure 37.7** Life cycle of the oriental liver fluke, *Opisthorchis sinensis.*

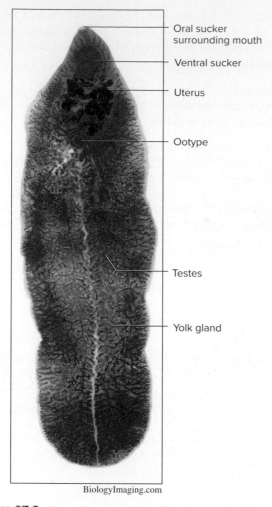

Oral sucker
surrounding mouth

Ventral sucker

Uterus

Ootype

Testes

Yolk gland

**Figure 37.8** Internal structure of *Fasciola,* the sheep liver fluke (2×).

## Question 3

Consider Learning Objective 1 listed at the beginning of this exercise. How could the production of large numbers of eggs contribute to the evolutionary success of flatworms in their environment?

## *Schistosoma*

Examine a demonstration slide of *Schistosoma,* and draw its body shape. *Schistosoma,* a blood fluke, inhabits the intestinal veins and other organs of many vertebrates (including humans) and causes the disease schistosomiasis (fig. 37.9). *Schistosoma* is socioeconomically important because it infects more than 200 million people in countries having tropical and temperate climates. Symptoms of infection include an enlarged liver, spleen, and bladder, as well as nutritional deficiency. Some of these symptoms, such as bleeding ulcers in the intestine and other organs, are inflamed by large deposits of eggs (up to 3500 per day from one female).

Snails are **intermediate hosts** for *Schistosoma.* An intermediate host is an organism harboring immature stages of a parasite, whereas a **definitive host** contains sexually mature, egg-laying stages of the life cycle. Irrigation ditches with snail populations enhance the spread of schistosomiasis in underdeveloped agricultural areas. Immature larvae of blood flukes released from the snails burrow through skin and infect people wading in these ditches.

Unlike most trematodes, *Schistosoma* is dioecious (fig. 37.9*a*). The male has a ventral groove along the length of his body into which the slender female cradles for copulation. Often the female remains in this groove for the remainder of her lifetime and extends slightly to lay eggs. Your prepared slide may include schistosomes in this position for copulation.

## Question 4

How does the shape of *Schistosoma* differ from that of other flukes you have studied?

## Subgroup Cercomeromorpha

Tapeworms are the most specialized platyhelminths. They are endoparasites of the gut of vertebrates and are covered by a cuticle similar to that of trematodes. However, tapeworms lack a mouth or digestive tract and have a unique body plan (fig. 37.10). Their cuticle efficiently absorbs nutrients from their host. The anterior end, or **scolex,** adheres to the host's intestinal wall with hooks or suckers. Behind the scolex is the **neck** followed by a series of segments called **proglottids.** The scolex and neck are small, but the chain of proglottids may be 10–15 m long.

A tapeworm grows as the scolex and neck continually produce a chain of proglottids as self-contained packets of male and female reproductive organs (fig. 37.11). Self-fertilization occasionally occurs in a proglottid, but cross-fertilization by copulating proglottids of adjacent worms is more common. **Gravid** (egg-carrying) proglottids eventually break from the end of the worm and pass out of the host with its feces.

Examine a prepared slide or whole specimen of a pork tapeworm, *Taenia solium.* Then examine a prepared slide of a scolex. Draw the basic shape of the scolex and note any hooks or suckers. Also examine and sketch from prepared slides young, mature, and gravid proglottids.

An adult *Taenia solium* may be up to 10 m long. Humans infect themselves with *Taenia* by eating uncooked meat from pigs, often intermediate hosts (fig. 37.12).

Compare a mature proglottid with that shown in figure 37.11. Although each proglottid has a complete reproductive system, the excretory ducts and longitudinal nerves are continuous between proglottids.

Examine a demonstration specimen of *Dibothriocephalus latus* if it is available. *Dibothriocephalus latus* is the largest tapeworm to infect humans and can be up to 20 m long. Its intermediate hosts are small crustaceans and fish.

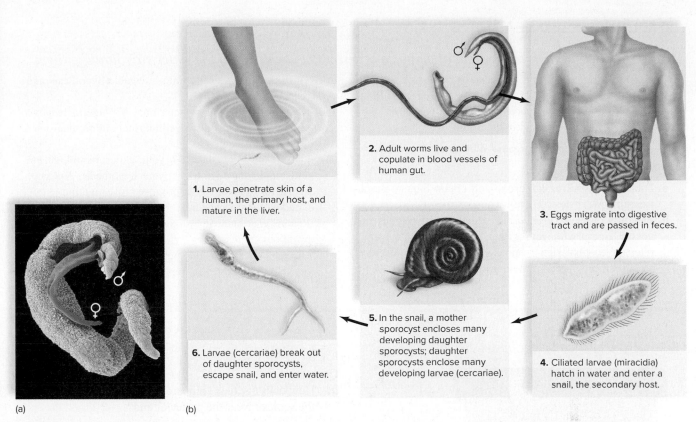

SPL/Science Source

**Figure 37.9** Life cycle of a blood fluke, *Schistosoma*. (*a*) Micrograph of *Schistosoma* (30×). (*b*) *Schistosomiasis*, an infection of humans caused by the blood fluke *Schistosoma*, is an extremely prevalent disease in Egypt—especially since the building of the Aswan High Dam. Standing water in irrigation ditches, combined with unsanitary practices, has created the conditions for widespread infection.

In panel (b):

1. Larvae penetrate skin of a human, the primary host, and mature in the liver.

2. Adult worms live and copulate in blood vessels of human gut.

3. Eggs migrate into digestive tract and are passed in feces.

4. Ciliated larvae (miracidia) hatch in water and enter a snail, the secondary host.

5. In the snail, a mother sporocyst encloses many developing daughter sporocysts; daughter sporocysts enclose many developing larvae (cercariae).

6. Larvae (cercariae) break out of daughter sporocysts, escape snail, and enter water.

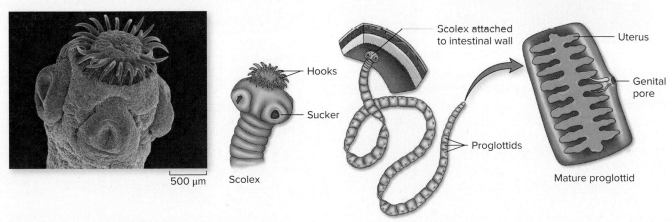

Dennis Kunkel Microscopy, Inc./Phototake

Labels: Hooks, Sucker, Scolex, Scolex attached to intestinal wall, Proglottids, Uterus, Genital pore, Mature proglottid

500 μm

**Figure 37.10** Body plan of a tapeworm (subgroup Cercomeromorpha). A mature proglottid has mature reproductive organs; a gravid proglottid contains many eggs.

*Can the primitive nervous system of planarians detect the chemistry of the water surrounding them?*

Observations: Detecting food is crucial to success of planaria. They must "taste" the water for compounds released from nearby food. Some classes of macromolecules (fats, carbohydrates, proteins, etc.) dissolve and disperse more readily than others.

Question: What types of organic molecules does *Dugesia* respond to most readily?

**a.** Establish a working lab group and obtain Inquiry-Based Learning Worksheet 37 from your instructor.

**b.** Patiently observe *Dugesia* movement and its responsiveness. Discuss with your group a well-defined question relevant to planarian response to dissolved molecules from their prey. Record the question on Worksheet 37.

**c.** Translate your question into a testable hypothesis and record it.

**d.** Devise a procedure to determine if *Dugesia* respond to some classes of dissolved nutrients more than other classes.

**e.** Outline on Worksheet 37 your experimental design and supplies needed to test your hypothesis. Ask your instructor to review your proposed investigation.

**f.** Conduct your procedures.

**g.** Record your data, answer your question, and make relevant comments.

**h.** Discuss with your instructor any revisions to your questions, hypotheses, or procedures. Repeat your work as needed.

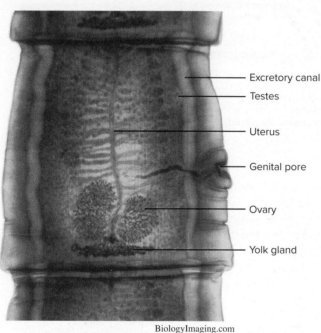

Excretory canal

Testes

Uterus

Genital pore

Ovary

Yolk gland

BiologyImaging.com

**Figure 37.11**   Mature proglottid of *Taenia pisiformis,* the dog tapeworm. Portions of two other proglottids are also visible at the top and bottom of the image (30×).

**Question 5**

**a.** How does a scolex compare in size to a proglottid near the scolex? Near the posterior end?

**b.** Tapeworms have no digestive system or mouth. How, then, do they obtain food?

**c.** Examine figure 37.10 and its caption carefully. Which proglottids, mature or gravid, occur closest to the scolex?

**d.** What is the difference between a mature and a gravid proglottid?

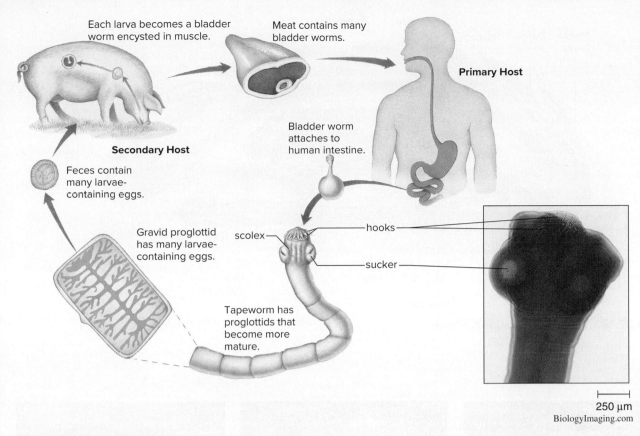

Each larva becomes a bladder worm encysted in muscle.

Meat contains many bladder worms.

**Primary Host**

Bladder worm attaches to human intestine.

**Secondary Host**

Feces contain many larvae-containing eggs.

Gravid proglottid has many larvae-containing eggs.

scolex

Tapeworm has proglottids that become more mature.

hooks

sucker

250 μm
BiologyImaging.com

**Figure 37.12** Life cycle of a tapeworm, *Taenia*. The life cycle includes a human (primary host) and a pig (secondary host). The adult worm is modified for its parasitic way of life. It consists of a scolex and many proglottids, which become bags of eggs.

*e.* Consider Learning Objective 1 listed at the beginning of this exercise. Is having a scolex significant to a fundamental process for tapeworms? For which process, and how is it significant?

*f.* Tapeworms are specialized. Would you expect specialists to have a unique morphology? Why?

## PHYLUM MOLLUSCA

Mollusks such as snails, clams, octopuses, and squids are soft-bodied animals with a specialized layer of epidermal cells called a **mantle** that secretes a **shell** (fig. 37.13). They have evolved a remarkable diversity of forms built on a consistent ancestral body plan (fig. 37.13). Molluskan forms are so diverse that a clam and a squid belong to the same phylum!

Mollusks produce many kinds of external shells (fig. 37.14*a,b*)—some mollusks have only a remnant internal shell (fig. 37.14*c*), whereas others have no shell at all. This phylum's diversity of 110,000 species is surpassed only by arthropods and probably nematodes. Although mollusks are coelomate, their coelom is often reduced to a small chamber surrounding the heart. The circulatory system is open (except in cephalopods), meaning that blood pools in sinuses and bathes the organs directly. **Open circulatory systems** have a few large vessels and a heart but no smaller vessels and capillaries. Mollusks have a **complete digestive tract** with a mouth and anus. This allows one one-way movement of food and specialized digestive processes along the tract.

The basic body-plan of a mollusk shows little segmentation but consistently includes (1) a **visceral mass** of organ systems (digestion, excretion, and reproduction) and

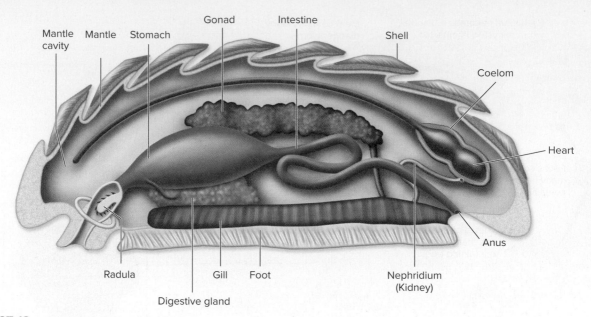

**Figure 37.13**  Generalized molluskan body plan.

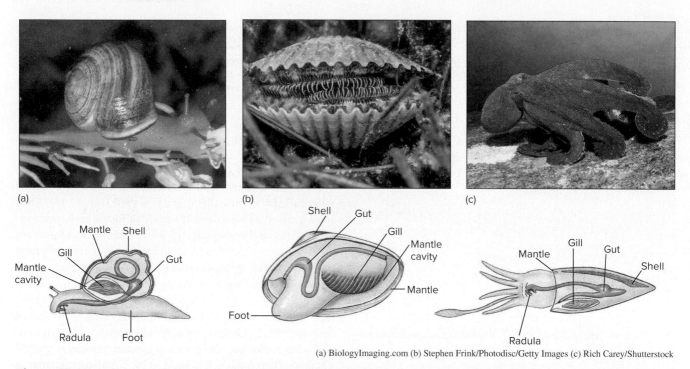

(a) BiologyImaging.com (b) Stephen Frink/Photodisc/Getty Images (c) Rich Carey/Shutterstock

**Figure 37.14**  The body plans of the three major classes of mollusks: (*a*) Gastropoda, (*b*) Bivalvia, and (*c*) Cephalopoda. A mollusk's shell has an outer layer overlying layers of densely packed crystals of calcium carbonate. A visceral mass includes organs of digestion, excretion, and reproduction and extends as a muscular foot adapted for locomotion, attachment, or food capture (in squids and octopuses). The radula is a scraping and feeding organ characteristic of mollusks (except bivalves, which obtain their food by filter feeding). Folds of tissue called the mantle arise from the dorsal body wall, line the shell, and enclose a cavity adjacent to the visceral mass. Within the mantle cavity are gills or lungs.

sensory structures; (2) a ventral, muscular, and often highly modified **foot** used for locomotion; (3) a calcium-based shell, though occasionally absent; and (4) a mantle that secretes the shell and may aid in respiration and locomotion in some species (fig. 37.14). Some mollusks also have a differentiated **head.**

## Class Polyplacophora

Obtain a preserved chiton and examine its external features. Polyplacophorans (*poly* = many, *placo* = plate, *phora* = move), commonly called chitons, are exclusively marine and have a primitive molluskan structure (fig. 37.15). The dorsal shell is divided into eight plates embedded in the mantle. The ventral foot is a broad oval muscle used to propel chitons slowly over the surface of rocks. The **radula,** a horny-toothed organ in the mouth, scrapes food (algae) from rocks.

## Class Gastropoda

Most gastropods (*gastro* = stomach, *poda* = foot) (snails) have a single shell that is often coiled and elaborate. Some gastropods, such as marine nudibranchs and the common garden slug, do not produce a shell (fig. 37.16). Most gastropod species are marine, but freshwater snails and land snails are common. Gastropods commonly feed with a rasping band of teeth called a radula (fig. 37.17)

Examine preserved gastropods and aquarium snails and compare the relative spiraling of the shells of different species. This spiral growth is common in mollusks and results from unequal growth of the two halves of the larva and mantle.

## Class Bivalvia

Clams, oysters, scallops, and mussels are bivalves (*bi* = two, *valve* = door or shell) having a dorsally hinged shell in two parts (figs. 37.14*b*, 37.18). The mantles of the left and right

valves join posteriorly to form a ventral **incurrent siphon** and a dorsal **excurrent siphon** that direct water through the clam (fig. 37.19*a*). When a clam burrows in sediment, the siphons extend to the water. The space between the

CK Ma/Shutterstock.com

**Figure 37.16**  Colorful nudibranchs, such as *Flaballina iodinae*, are gastropod mollusks without a shell. They have a rather mysterious defensive strategy—nudibranchs use the weapons of their prey. They eat sea jellies, hydrozoans, and corals, all of which have stinging structures called nematocysts. When a nudibranch attacks and eats sea jellies, it can swallow and digest these nematocysts without discharging them. The stingers pass through the digestive tract and are stored in feathery projections on the dorsal surface of the nudibranch. A predator taking a mouthful of nudibranch gets a nasty taste.

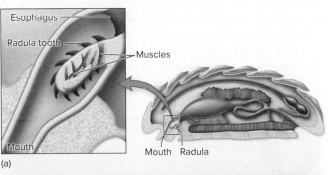

(a)

(b)

25 μm

(a) McGraw Hill (b) Eye of Science/Science Source

**Figure 37.17**  Radula of mollusks. (a) A radula is a unique rasping tongue of mollusks with rows of chitinous teeth covering its surface. (b) The membranous radula is stretched over a rigid cartilaginous rod pressed against a surface to be scraped. Snails use their radula to scrape algae from rocks or tear pieces from plant leaves.

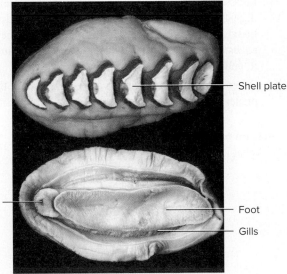

BiologyImaging.com

**Figure 37.15**  Chiton, *Chiton* (class Polyplacophora). Dorsal view (*above*) and ventral view (*below*) (1×). The shell of chitons has eight plates atop a mantle; visceral mass; and large, oval foot.

*Survey of the Animal Kingdom*   **429**

Figure 37.18 A giant clam (class Bivalvia, *Tridacna* sp.) of the Indopacific may weigh up to 900 lb (400 kg). Its mantle is often intensely colored by blue, green, and brown pigments, and it harbors rich and colorful colonies of symbiotic algae in the blood sinuses of the tissue. Photosynthesis by these algae provides extra food for the clam. Some clams have lenslike structures that focus light deep within their tissues and promote photosynthesis by the algae.

mantle and visceral mass is the **mantle cavity.** Water flows into the mantle cavity through the incurrent siphon, over the gills, and then dorsally into a space between the visceral mass where gills attach to the mantle. Water exits this chamber posteriorly through the excurrent siphon. As water flows over gills, suspended food particles are filtered by cilia and swept along grooves on the gills to the ventral edge. Cilia at the ventral edge move food to the labial palps that surround and direct food to the **mouth** (fig. 37.19*b*).

**SAFETY FIRST** Before coming to lab, you were asked to read this exercise so you would know what to do and be aware of safety issues. In the space below, briefly list the safety issues associated with today's procedures. If you have questions about these issues, contact your laboratory assistant before starting work.

## *Procedure 37.2* **Examine bivalve anatomy**

1. Obtain a preserved freshwater bivalve, such as *Anodonta.* The valves should be slightly separated by a wooden peg. Rinse the specimen thoroughly before dissection.

2. Locate the two valves and the anterior, posterior, dorsal, and ventral regions of the clam. The **hinge** and the **umbo** are on the dorsal surface (fig. 37.19*b*). The umbo is toward the anterior end.

3. Place the organism vertically and dorsal side (umbo) down on a dissecting pan. To further separate and loosen the valves, slip the flat end of a dull kitchen knife between the valves and rotate the knife. Be careful. Notice that a layer of mantle lies against the inside of each of the two valves.

4. Review figure 37.19 to find the general position of the **adductor muscles.**

5. When the valves are loose enough to be held slightly open (about 1.0 cm), slip a scalpel blade between the mantle tissue and shell of one side. When you have felt resistance of an adductor muscle against the tip of the blade, draw the blade snugly along the shell to cut the posterior and anterior adductor muscles. Try to cut them as close to the inner shell surface as possible. Always cut down toward the dissecting pan.

6. Pull the valves apart so that the organism lies on one side and the other valve lifts away. The mantle that once lined the removed shell now covers the visceral mass. Cut the mantle away to expose the visceral mass.

7. On the separated and empty valve notice the muscle's scars, hinge teeth, mantle line, and the inner, iridescent surface of the shell called **mother of pearl.**

8. Locate the **gills, labial palps,** tough muscular **foot,** and the softer, more dorsal visceral mass. The labial palps channel food to the open mouth.

9. To expose the internal organs of the visceral mass, slice the foot and visceral mass longitudinally (anterior-posterior plane) from the ventral surface toward the dorsal surface. This will produce left and right halves. Make the cut slightly off center to avoid cutting through all of the tough foot tissue.

10. Locate evidence of the coiled intestine, the green **digestive gland,** brown **gonad** tissue, and the dorsal **heart.** This sac is thought to be a remnant of the ancestral coelomic cavity. The intestine runs through it, and the sac contains the heart tissue.

11. When you finish your observations, dispose of the preserved tissue and shells in appropriate containers.

## Question 6
*a.* What is the texture of the mantle of *Anodonta?*

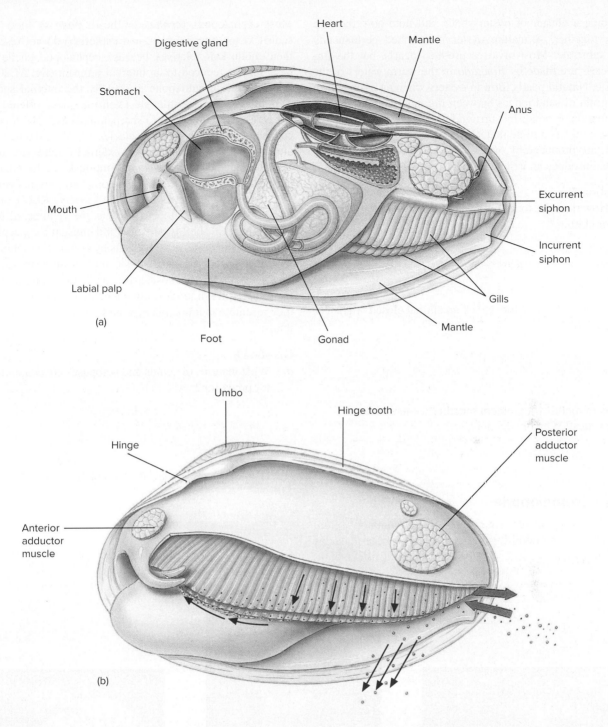

**Figure 37.19** (*a*) Clam (*Anodonta*) anatomy. The left valve and mantle are removed. (*b*) During filter feeding, water enters the mantle cavity posteriorly and is drawn forward by ciliary action to the gills and palps. As water enters the tiny openings of the gills, food particles are filtered out and caught in strings of mucus carried by cilia to the palps and directed to the mouth. Arrows show the path of food particles being moved toward the mouth by cilia on the gills. Sand and debris drop into the mantle cavity and are removed by cilia.

*b.* Are siphons in a preserved specimen of *Anodonta* as obvious as they are in figure 37.19?

*c.* Consider Learning Objective 1 listed at the beginning of this exercise. In what ways would having a shell contribute to the survival and reproductive success of mollusks in their environment?

Examine a cluster of oyster shells and note how they are stuck together. A mature oyster is attached permanently to a substrate. Most bivalves produce pearls, but the finest pearls are made by *Pinctada* in the warm waters of the Pacific. Natural pearls form in oysters when an irritant such as a grain of sand lodges between the mantle and the shell. The mantle responds by surrounding the irritant with layers of the same crystalline material used for the shell. Cultured pearls are produced by oysters about 3 years after aquaculturists introduce an irritant to their mantle.

## Question 7

**a.** How does the foot of a bivalve differ from that of snails or chitons?

**b.** How could predators attack an animal closed "tight as a clam"?

**c.** Is immobility a problem for filter feeders such as oysters? Why or why not?

## Class Cephalopoda

Examine a preserved specimen of the common squid, *Loligo*. Note the number and arrangement of the tentacles and the pointed, hardened beak surrounding the mouth.

Most cephalopods (*cephalo* = head, *poda* = foot) (e.g., squids, octopuses, nautilus, and cuttlefish) do not resemble their molluskan relatives because cephalopod shells may be absent or reduced to an internal remnant (fig. 37.20*a,b*). For example, a cuttlebone is actually the internal shell of a mollusk called a cuttlefish. (You may have offered your pet parakeet a cuttlebone to sharpen its beak.) The foot of a cephalopod is modified into tentacles. Squids and other cephalopods are predatory, and their external features are appropriately adapted. Unusual among mollusks is the relatively closed circulatory system of cephalopod arteries and veins.

Examine the eyes of *Loligo* closely. Also examine other displayed cephalopods such as the chambered nautilus; its external shell is beautiful and unusual for a cephalopod (fig. 37.20). Review mollusks by completing table 37.2 with brief descriptions of basic molluskan features. The eyes are probably the most surprising feature of a squid. The largest eyes in kingdom Animalia belong to the giant squid; they resemble mammalian eyes and use a lens to form clear images.

## Question 8

**a.** What features of squids and octopuses are adaptations for predation?

**b.** Do all of the tentacles of a squid have suckers?

(a) oceanbounddb/Getty Images
(b) ©atese/Getty Images
(c) bluehand/Shutterstock

**Figure 37.20** Cephalopod diversity. (*a*) An octopus. Octopuses generally move slowly along the bottom of the sea. (*b*) A squid. Squids are active predators, competing effectively with fish for prey. (*c*) Pearly nautilus, *Nautilus pompilius*.

**Table 37.2**

**A Comparison of Major Characteristics of Four Classes of Mollusks**

| | Polyplacophora | Gastropoda | Bivalvia | Cephalopoda |
|---|---|---|---|---|
| Shell | | | | |
| Mantle | | | | |
| Foot | | | | |
| Locomotion | | | | |
| Feeding | | | | |
| Sensory structures | | | | |

*c.* What are some functions of suckers?

*d.* Find the mouth at the base of the tentacles. What are the shape and consistency of the jaws? You may need to make an incision to expose the mouth and jaws.

*e.* Why are sensory organs more prominent in cephalopods than in other classes of mollusks?

*f.* Consider Learning Objective 1 listed at the beginning of this exercise. In what ways are image-forming eyes significant to fundamental processes for cephalopods?

# Questions for Further Study and Inquiry

*1.* Flatworms are the first organisms we have discussed with an anterior-posterior orientation. How does this affect their movement compared to the movement of more primitive organisms?

*2.* What are the disadvantages of a flatworm's digestive system having only one opening?

3. The complete digestive tract of mollusks and other phyla allows functional specialization. What specializations are common in the digestive tract of higher organisms such as humans?

4. Prepare a simple table of all of the taxonomic groups covered so far, their common names, their distinguishing characteristics, and all representative genera covered in this exercise. Keep this table with your study notes.

5. What is the advantage of radial symmetry for sessile animals such as hydras and bilateral symmetry for mobile animals such as planaria? What major evolutionary trends accompany bilateral symmetry?

6. Mollusks exhibit a variety of feeding methods. List at least four and discuss adaptations and examples for each type.

7. A snail shell is quite different from the familiar bony skeleton of a mammal. In what ways does a shell function as a skeleton?

8. Some land snails have formed a lunglike structure from a major layer of tissue. What is that layer?

9. Cephalopods are considered by many to be the most distinctive class of mollusks. What makes them seem almost out of place?

# Survey of the Animal Kingdom

## Phyla Annelida and Nematoda

Please visit **connect.mheducation.com** to review online resources tailored to this lab.

Annelids of phylum Annelida, commonly called segmented worms, share a close common ancestry with mollusks and platyhelminthes (see Exercise 37). They are coelomate protostomes with a free-living larval stage called a trochophore (see fig. 37.3). This shared type of larval stage indicates a close evolutionary relationship between annelids, mollusks, and platyhelminthes.

In contrast, nematodes of phylum Nematoda have ancestry more closely related to that of arthropods (see Exercise 40). Both of these phyla molt their outer covering. Nematodes molt their outer cuticle, just as arthropods molt their exoskeleton. As a nematode grows larger it forms a new, expandable cuticle underneath its existing one. Periodically the body swells, cracks, and sheds the old cuticle and replaces it with a new, expanded covering allowing for further growth. This periodic process of molting and shedding an outer covering is ecdysis.

Nematodes have a coelomic cavity, but contrary to that of mollusks and annelids it is not completely lined with mesoderm. This is why nematodes are called pseudocoelomates (fig 37.2).

## PHYLUM ANNELIDA

Annelids include earthworms, leeches, and many less familiar marine and freshwater species (Table 38.1). The most distinctive characteristic of this phylum is **segmentation.** The body of an annelid is divided into repetitive **segments** arranged on a longitudinal axis and divided by **septa.** Each segment contains parts of the circulatory, digestive, nervous, and excretory systems. The circulatory system of annelids is closed, with blood always retained in vessels. Annelids also have **setae,** small, bristlelike appendages often occurring in pairs on lateral and ventral surfaces. The degree of setal development is distinctive for each of the three classes of annelids.

## Class Polychaeta

Most polychaetes, such as the clam worm *Nereis,* are marine worms living in sediment (fig. 38.1). *Nereis* is distinctly segmented and each segment bears a pair of fleshy appendages called **parapodia.** These appendages have a large surface area, are highly vascularized with blood vessels, and help the polychaete move and respire. Protruding from the fleshy parapodia are many setae from which the class derives its name (*poly* = many, *chaeta* = setae). In some species the brittle, tubular setae are filled with poison and used for defense; in others the setae help filter food from the water.

Table 38.1

**Phyla Nematoda and Annelida**

| Phylum | Typical Examples | Key Characteristics | Approximate Number of Named Species |
|---|---|---|---|
| Nematoda (roundworms) | *Ascaris*, pinworms, hookworms, *Filaria* | Pseudocoelomate, unsegmented, bilaterally symmetrical worms; tubular digestive tract passing from mouth to anus; tiny; without cilia; live in great numbers in soil and aquatic sediments; some are important animal parasites | 20,000+ |
| Annelida (segmented worms) | Earthworms, polychaetes, beach tube worms, leeches | Coelomate, serially segmented, bilaterally symmetrical worms; complete digestive tract; most have bristles called setae on each segment that anchor them during crawling. | 20,000+ |

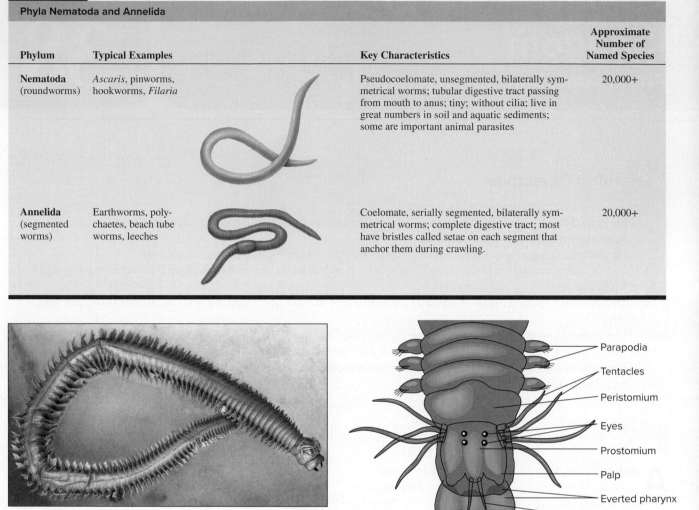

(a) BiologyImaging.com

(b)

Parapodia
Tentacles
Peristomium
Eyes
Prostomium
Palp
Everted pharynx
Prostomial tentacles
Jaw

**Figure 38.1** *Nereis virens,* a common polychaete (phylum Annelida). (*a*) External structure. (*b*) Anatomy of the anterior end.

## *Procedure 38.1*  Examine polychaetes

1. Examine a preserved *Nereis* and a prepared slide of a parapodium. Sketch the parapodium. Note the tufts of bristles and lobes of tissue.

2. Examine the well-developed head of *Nereis* and locate the features shown in figure 38.1b. The mouth and jaws of polychaetes are retractile, so you may have to pull out the jaws with a pair of forceps.

3. Examine other displayed polychaetes such as *Aphrodita*, the sea mouse, and *Chaetopterus,* the parchment worm. These worms have many setae and highly modified parapodia. The parchment worm gets its common name from the paperlike tube it builds.

## Question 1

*a.* The common name of the sea mouse refers to what external feature characteristic of polychaetes?

*b.* List several functions of parapodia and setae.

*c.* What is the probable function of the tentacles shown in figure 38.1*b*?

*d.* Consider Learning Objective 1 listed at the beginning of this exercise. Are parapodia significant to fundamental processes for polychaetes? In what ways?

*e.* What features of *Nereis* indicate that it is an annelid?

## Class Oligochaeta

A common oligochaete is *Lumbricus terrestris,* the familiar earthworm.

### *Earthworm Locomotion*

Movement of oligochaetes in their natural soil environment is not undulatory—instead, movement involves extension, anchoring, and contraction. These motions occur by alternating contractions of circular and longitudinal muscles. These contractions pull against a **hydrostatic** (water pressure) support system rather than against a rigid skeleton (fig. 38.2).

---

**Procedure 38.2** Examine locomotion in earthworms

*1.* Watch a living earthworm move on the hard surface of a pan, then compare and contrast this motion with that of a snake and a vinegar eel.

*2.* Place the worm on some loose soil, and describe its burrowing motion.

---

**Question 2**

*a.* Does the earthworm move randomly?

*b.* What do you suppose the worm is seeking or avoiding?

*c.* What muscles allow the worm to change its length and thickness?

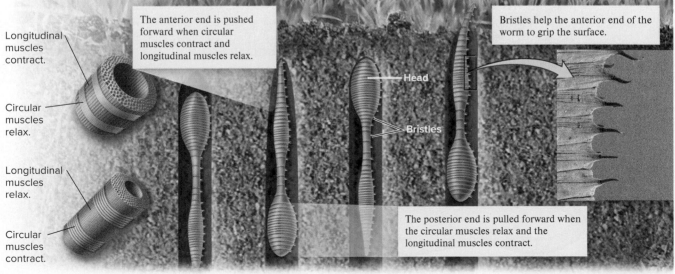

Longitudinal muscles contract.

Circular muscles relax.

Longitudinal muscles relax.

Circular muscles contract.

The anterior end is pushed forward when circular muscles contract and longitudinal muscles relax.

Head

Bristles

Bristles help the anterior end of the worm to grip the surface.

The posterior end is pulled forward when the circular muscles relax and the longitudinal muscles contract.

Cultura Creative RF/Alamy Stock Photo

**Figure 38.2** Hydrostatic skeleton in a worm. By alternately contracting and relaxing circular and longitudinal muscles, earthworms use hydrostatic pressure to achieve locomotion. Clinging bristles along the body surface help prevent backsliding.

**d.** How does an earthworm's motion differ from that of a snake and nematode?

**e.** What features of *Lumbricus* indicate that it is an annelid?

## External Anatomy of Earthworms

Oligochaetes (*oligo* = few or small, *chaeta* = setae), include many freshwater species besides the common earthworm. They lack parapodia and have few setae. The anus is on the terminal segment, but the mouth is preceded by a fleshy lobe called the **prostomium** (*pro* = before, *stoma* = mouth). Posterior to the mouth is the first body segment, the **peristomium** (*peri* = around, *stoma* = mouth) (fig. 38.3).

The most obvious external feature of an earthworm is the **clitellum,** a series of swollen segments at the anterior third of the body. Copulating worms attach at their clitella and exchange sperm (fig. 38.4). Earthworms are hermaphroditic, and each of the copulating worms produces egg and

sperm cells. Sperm mature in **seminal vesicles** and exit the worm through **male gonopores** on segment 15 (fig. 38.3). Sperm then pass to the adjacent worm and move along its body surface to openings of the **seminal receptacles** (each side of segment 10), where they are stored temporarily. After copulation, the worms separate. A few days later the clitellum secretes a mucous band that slides anteriorly and picks up eggs from the **female gonopores** (segment 14) and stored sperm from seminal receptacles. After the eggs are fertilized in the mucous band, the worm releases it as a **cocoon.**

### *Procedure 38.3* Examine the external features of an earthworm

1. Examine the external features of a preserved *Lumbricus,* and notice its segmentation.
2. Although the body lacks parapodia, if you touch the smooth epidermis and cuticle you can feel short setae on the ventral surface.
3. Use a dissecting microscope to determine the number and arrangement of these setae.
4. Locate the mouth and anus of the earthworm.
5. Use a dissecting microscope to locate the male gonopores and seminal receptacle openings on your specimen (fig. 38.3).
6. Locate the female gonopores on your specimen.

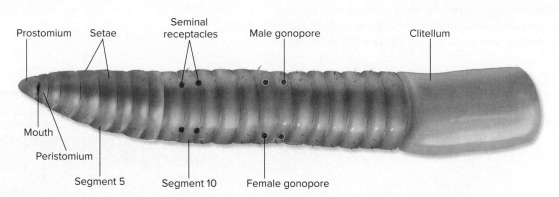

**Figure 38.3** External features of an earthworm's ventral surface.

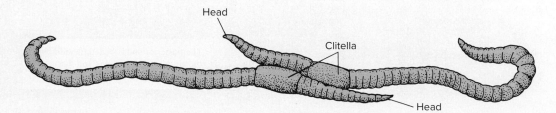

**Figure 38.4** Position for copulation and transfer of sperm in earthworms (class Oligochaeta).

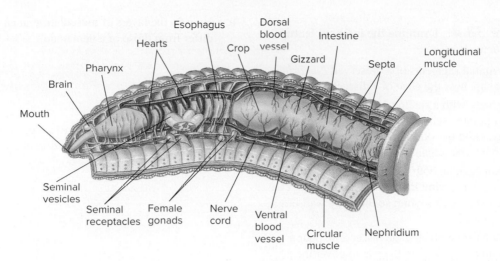

**Figure 38.5** Internal anatomy of an earthworm.

## Question 3

***a.*** How many setae are on each segment of the earthworm?

***b.*** Are they paired?

## Internal Anatomy of Earthworms

Segmentation is best appreciated by looking at internal anatomy (fig. 38.5). Some of the repeating segments are not identical. Segments of the digestive tract are fused and specialized to form a muscular **pharynx** (for suction and ingestion of food), **esophagus** (for transport of food), **crop** (for food storage and some digestion), **gizzard** (for maceration or crushing of food), and **intestine** (for absorption of nutrients). Recall that nematodes also have a linear digestive tract, but it lacks the specialization of that of annelids. Reproductive organs cluster around the anterior segments. Locating small structures such as the female gonads and seminal receptacles will require careful work. A rudimentary **brain** is just anterior and dorsal to the pharynx and is continuous with the **ventral nerve cord.**

Some structures such as lateral branches of the ventral nerve cord and paired **nephridia** occur in each segment (fig. 38.6). Nephridia, which are small, white, convoluted tubes, are found on the inner surface of each segment. Nephridia function like kidneys—they collect and release excretory wastes. Ciliated, funnel-shaped **nephrostomes** on the ends of the nephridia gather waste products from the coelom. The waste is released through external pores called **nephridiopores.**

The **dorsal blood vessel** and **ventral blood vessel** (above and below the intestine) are the main vessels of the closed circulatory system. They are connected by five lateral **"hearts."** However, these hearts promote blood circulation no more than the dorsal and ventral vessels.

The digestive tract has an internal fold of tissue called the **typhlosole** arising from the dorsal wall. This creates a **U**-shaped intestinal lumen and doubles the surface area for absorption.

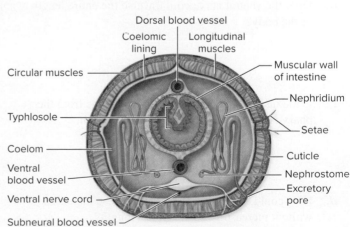

**Figure 38.6** Cross section and internal anatomy of an earthworm.

## Procedure 38.4  Examine the internal features of an earthworm

1. Pin the terminal segments of a preserved *Lumbricus*, dorsal side up, near the edge of a dissecting pan.
2. Open the body with a shallow, longitudinal incision. Use sharp-pointed scissors, and be careful not to cut too deeply. Hold the skin up with forceps to prevent damage to internal organs.
3. Pin the skin back on both sides of the incision, and expose the organs lying in the coelomic cavity. You may have to cut each septum so the body wall lies flat when pinned.
4. When you have completed the dissection, pour a small amount of water into the pan to cover the exposed worm.
5. Locate the structures shown in figure 38.5. You may need to use a dissecting microscope for close examination.
6. Locate the hearts, dorsal blood vessel, and ventral blood vessel.
7. Locate the organs of the digestive system. Slit open the digestive tract and observe its contents.
8. Locate a nephridium in each segment of your dissected specimen.
9. When you have completed your observations, dispose of your specimen in an appropriate container.
10. Examine a prepared slide of a cross section of an earthworm; locate the structures in figure 38.6.

### Question 4
a. How many segments of an earthworm have a heart?



b. Does the ventral nerve cord traverse the entire length of the body?



c. Is the inside of the digestive tract the same from the pharynx to the end of the intestine? Explain.



d. How could absorption in the intestine be increased without increasing the intestine's length?

e. How do the layers of musculature in an earthworm differ from those of a nematode?



f. List two or three features of an earthworm cross section that distinguish the dorsal and ventral surfaces.



## Class Hirudinea

Hirudineans include leeches, which are primarily freshwater ectoparasites (fig. 38.7). Leeches are not segmented as distinctly as are other annelids. Leeches lack setae, are dorsoventrally flattened, and have anterior and posterior suckers that hold prey. Blood sucking leeches eat infrequently, but species such as *Hirudo medicinal* can quickly consume five to ten times their body weight in blood. Many years ago physicians used these leeches for bloodletting from sick patients thought to have "too much bad blood." Today, laboratory-grown leeches are occasionally used to extract fluid that has accumulated around injuries and surgical incisions to enhance healing. Leeches extract fluid more efficiently and with less damage than does hypodermic suction (fig. 38.8).

Leeches reproduce sexually and individuals are hermaphroditic. Two leeches intertwine, and species with copulatory organs inject a packet of sperm called a **spermatophore** into the female gonopore. Other species have no copulatory organ; these leeches copulate by injecting a spermatophore directly through the epidermis of their partner (the spermatophore may have tissue-dissolving enzymes to aid penetration). The introduced spermatophore releases sperm into the coelomic cavity, and the cells move to the ovaries to fertilize the eggs. In two days to many months after copulation, the leech secretes a nutrient-rich cocoon to protect the eggs, and extrudes eggs into the cocoon. The cocoon is brooded by the leech or attached to submerged objects or vegetation.

## Procedure 38.5  Observe leeches

1. Observe living leeches; compare their movement with that of earthworms.
2. Examine the external anatomy of a preserved leech and locate the two suckers.
3. Open the body with a pair of scissors and look for signs of segmentation. A dissected leech may be on demonstration.

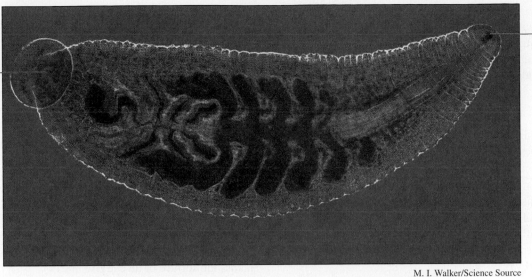

Posterior
sucker

Anterior
sucker

M. I. Walker/Science Source

**Figure 38.7**  External view of a leech (20×). Leeches are primarily freshwater ectoparasites. The mouth is within the anterior sucker.

BiologyImaging.com

**Figure 38.8**  Leech. These advanced annelids have external signs of segmentation but have little repetition of internal parts or internal compartments.

## Question 5

***a.*** What is the difference in general body shape of leeches compared to oligochaetes or polychaetes?

***b.*** What function other than feeding do suckers serve?

***c.*** Are setae visible on a leech?

***d.*** Is internal segmentation of a leech as distinct as that of an oligochaete?

***e.*** Consider Learning Objective 1 listed at the beginning of this exercise. How could production of a packetlike spermatophore contribute to the evolutionary success of leeches in their environment?

## Question 6

*a.* Now that you have examined the unifying characteristics of mollusks and annelids, can you list three or four characteristics that they share with flatworms and nematodes as their close ancestors?

*b.* Draw and label three coelomic body plans and list which of the phyla in Exercises 37 and 38 are associated with each plan.

## PHYLUM NEMATODA

Nematodes (fig. 38.9), commonly called roundworms, are everywhere and often occur in great numbers—a single decomposing apple may contain 100,000 nematodes of different species. If we removed everything but nematodes from the environment, we would still see a ghostly outline of our entire biosphere. Estimates of the global number of species living in all aquatic, terrestrial, and parasitic environments often exceed one million, most of which have not yet been formally described and named. Almost all feeding types (parasitic, predatory, etc.) are represented. Reproductive morphologies include dimorphic species, hermaphroditic species, and even species with males as well as hermaphrodites. Nematode diversity is extraordinary.

Many nematodes cause diseases in humans, other animals, and plants. One of the most serious of these diseases is elephantiasis, the grotesque swelling of an arm or leg resulting from nematodes (commonly *Filaria*) clogging the lymphatic system that drains the host's appendage (fig. 38.10). Fluid accumulates, and the appendage swells. Another parasitic nematode is the eye worm, *Loa loa,* which lives under the skin of humans and occasionally crawls across the surface of an eye (fig. 38.11).

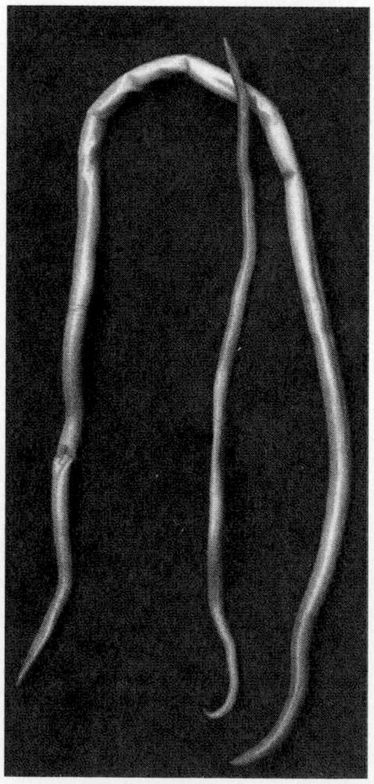

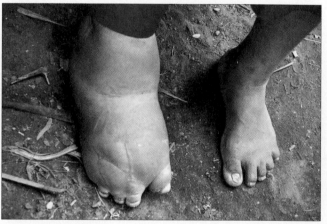

Rungtip Chatadee/age fotostock

**Figure 38.10**  Elephantiasis of a leg caused by adult filarial worms that live in lymph passages and block the flow of lymph. Tiny juveniles, called microfilariae, are picked up in a blood meal of a mosquito, where they develop to the infective stage and are transmitted to a new host.

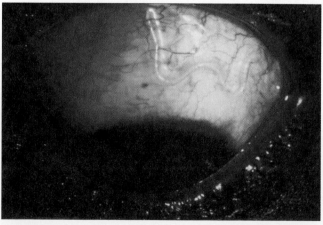

Sue Ford/Science Source

**Figure 38.11**  The nematode *Loa loa* appears in the eye of its host. This one is being surgically removed from someone's eye.

BiologyImaging.com

**Figure 38.9**  *Ascaris lumbricoides,* a common roundworm, inhabits the intestines of pigs and humans (0.5×). Male ascarid worms are smaller than females and have a curved posterior end.

Nematodes are slender and long with a rather featureless exterior. They lack flagella and cilia and are covered with a somewhat flexible and chemically complex **cuticle.** The cuticle resists digestive enzymes and is permeable only to water, dissolved gases, and some ions. Nematodes have a **complete digestive tract** with a mouth and anus. In addition to their digestive cavity, roundworms have a body cavity called a **pseudocoelom** consisting of a fluid-filled space between the body wall and digestive tract (see fig. 37.2). Internal organs are suspended in this cavity.

**Question 7**

***a.*** How does the number of body cavities of nematodes compare with that of flatworms?

***b.*** What are the advantages of a digestive tract having a separate entrance and exit?

***c.*** Female *Ascaris* are more numerous than males. Why might this be adaptive?

### Rhabditus and Turbatrix

**Procedure 38.6**   **Examine living nematodes**

*1.* Examine a culture of living *Rhabditus, Turbatrix* (the vinegar eel), or *Cephalobus.*

*2.* Using a toothpick, put a small piece of the culture medium on a slide with a drop of water and coverslip, and focus with low magnification of your microscope.

*3.* Watch the nematodes move.

The flexible cuticle and hydrostatic pressure of fluid in a nematode's pseudocoel aid locomotion by resisting antagonistic muscle contraction. Nematodes have only longitudinal muscles and lack circular or diagonal muscles. This combination of features produces a characteristic motion.

**Question 8**

***a.*** How would you describe the motion of a nematode?

***b.*** How is this movement related to the movement of its muscle layers?

### Ascaris

*Ascaris* is a large nematode that infects the intestinal tract of humans and other vertebrates (fig. 38.9). Males are smaller than females and have a hooked posterior end. The opening in the posterior end is the anus.

**Procedure 38.7**   **Dissect Ascaris**

*1.* Obtain a preserved *Ascaris lumbricoides* and examine its external features using a dissecting microscope. Compare the external features of males and females.

*2.* At the anterior end, locate the mouth surrounded by three lobes of tissue.

*3.* Prick the cuticle with a dissecting probe (teasing needle) and determine its consistency.

*4.* Pin the ends of a female specimen near the edge of a dissecting pan to permit viewing with a dissecting microscope. Slit the body wall longitudinally with a dissecting needle or sharp-pointed scissors.

*5.* Pin the body wall open and locate the internal organs shown in figure 38.12. The excretory pore may be small and difficult to find. A dissection of a male may be on demonstration.

*6.* Examine a prepared slide of a cross section of both a male and female *Ascaris.*

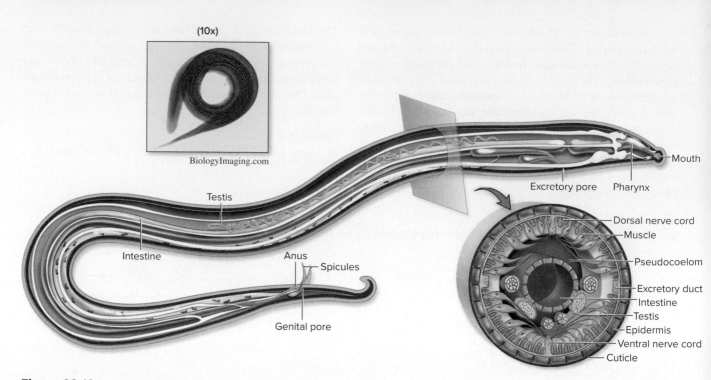

(10x)

BiologyImaging.com

Mouth

Excretory pore    Pharynx

Testis

Dorsal nerve cord
Muscle

Intestine

Anus
Spicules

Pseudocoelom

Excretory duct
Intestine
Testis
Epidermis
Ventral nerve cord
Cuticle

Genital pore

**Figure 38.12**    Internal anatomy of nematodes, which are commonly called roundworms.

7. Locate the features shown in figure 38.13 and determine from where along the length of your dissected specimen this section was taken.

8. Examine cross sections from different areas of the body if appropriate slides are available. Gametes mature as they move along the length of the tubular reproductive organs.

9. Dispose of waste in labeled containers.

**Question 9**

*a.* The cuticle of *Ascaris* is flaky and tough. Consider Learning Objective 1 listed at the beginning of this exercise. What might be an adaptive advantage of a thick and tough cuticle?

*b.* Where do the internal organs of *Ascaris* attach to the body wall?

*c.* How does the diameter of the female reproductive tract change?

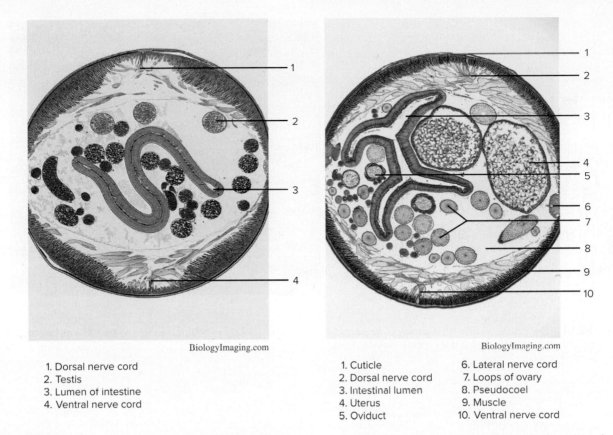

1. Dorsal nerve cord
2. Testis
3. Lumen of intestine
4. Ventral nerve cord

1. Cuticle
2. Dorsal nerve cord
3. Intestinal lumen
4. Uterus
5. Oviduct
6. Lateral nerve cord
7. Loops of ovary
8. Pseudocoel
9. Muscle
10. Ventral nerve cord

BiologyImaging.com

BiologyImaging.com

**Figure 38.13**  Cross sections of male (left) and female (right) *Ascaris,* a large nematode that infects the intestinal tract of a variety of vertebrates (30×).

*d.* Are any sensory organs evident in *Ascaris?* Why would this be adaptive?

## Trichinella

*Trichinella spiralis* causes the disease trichinosis. Adult females of *Trichinella* live in the intestine of their host and release larvae. These larvae migrate through the body to striated muscles, especially in the diaphragm and tongue, where they form painful calcified cysts (fig. 38.14). The larvae remain encysted until the muscle tissue is eaten by another host, where the larvae mature. Humans infect themselves by eating poorly cooked pork containing encysted larvae. However, the occurrence of trichinosis is decreasing. In the United States fewer than 50 cases of human trichinosis are reported annually to the Centers for Disease Control and Prevention, compared with 500 cases a year in the 1940s.

**Procedure 38.8**  Examine *Trichinella, Necator, Enterobius,* and *Dirofilaria*

*1.* Examine a slide of muscle tissue containing encysted larvae of *Trichinella.*

*2.* If slides are available, examine specimens of *Necator* (hookworm), *Enterobius* (pinworm), and *Dirofilaria* (heartworm).

*3.* Note variation in the morphology of the anterior and posterior ends of the organisms.

*4.* Sketch and note the body sizes of *Trichinella, Necator, Enterobius,* and larval *Dirofilaria.*

Hookworms cause anemia in infected livestock, which causes great economic losses. Pinworms, although less dangerous, irritate many children because the pinworms infect the intestine and inflame the anus (fig. 38.15). Children

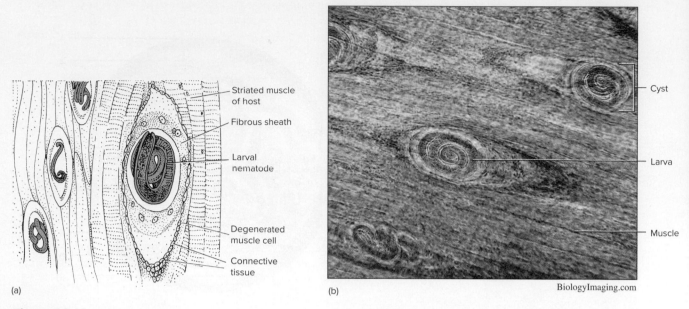

**Figure 38.14** *Trichinella spiralis.* (*a*) Anatomy of larvae encysted in muscle. (*b*) Larvae in muscle section (400×). The disease trichinosis is acquired by eating poorly cooked meat that contains encysted larvae.

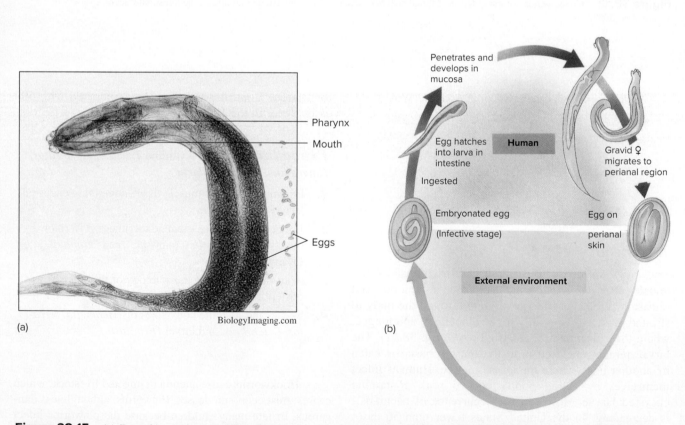

**Figure 38.15** (*a*) *Enterobius*, a pinworm (60×). Pinworms are the most common parasite in the United States—30% of children and 16% of adults are infected. (*b*) The life cycle of the pinworm, *Enterobius vermicularis*. The adult crawls from the intestine and lays eggs on the skin just outside the anus. The host, usually a human child, is irritated by the pinworms and scratches. Eggs are then unknowingly eaten and the larvae hatch in the intestine to complete the life cycle.

often scratch themselves, put their fingers in their mouths, and thereby reinfect themselves. Pinworms infect about 30% of children and 16% of adults in the United States. Heartworms (*Dirofilaria immitis*) are among the most significant parasites of dogs and cats. The larvae are transmitted by mosquitoes and large, mature stages (9 to 16 in. long) reside in the heart and lungs, where they disrupt blood flow.

## INQUIRY-BASED LEARNING

### *Of what use could leeches be to medicine?*

Observations: Many years ago, physicians used leeches for "bloodletting" from sick patients thought to have "too much bad blood." Today, laboratory-grown leeches are occasionally used to enhance healing by extracting fluid that accumulates around incisions following microsurgery and injuries. Leeches extract fluid more efficiently and with less damage than hypodermic suction (fig. 38.16). Medicine is a business as well as a science that must be effective and delivered safely to large populations.

Question: Are leeches appropriate as a viable medical treatment?

a. Establish a literature search group according to your instructor's guidelines, and obtain Inquiry-Based Learning Worksheet 38 from your instructor.

b. Discuss among your group and record on Worksheet 38 the requirements of effective use of leeches in medicine. Include necessary characteristics of the organism and the logistics of delivering treatment.

c. Determine with your group the specific questions that doctors must answer concerning leeches and their effective, safe use. Record these questions on Worksheet 38.

d. Research the current literature and outline information about the use of leeches in medicine.

e. Record your findings on Worksheet 38.

f. Discuss your findings with your instructor and with other students.

# Questions for Further Study and Inquiry

*1.* You have examined at least three phyla commonly referred to as "worms." How would you define this term?

*2.* Of what economic importance are earthworms?

3. Earthworms have no lungs or gills. Do they "breathe"? If not, how do they obtain oxygen to survive?

4. Do you suppose bloodletting by leeches was a good technique to cure psychosomatic illnesses? Why or why not?

5. Prepare a simple table of all of the taxonomic groups covered so far, their common names, their distinguishing characteristics, and all representative genera covered in this exercise. Keep this table with your study notes.

## WRITING TO LEARN BIOLOGY
Are leeches predators or parasites? What reasons would you give for both possibilities?

# Survey of the Animal Kingdom
## Phylum Arthropoda

## Learning Objectives

By the end of this exercise you should be able to:

1. Describe the structures that contribute significantly to the survival of arthropods in their environments.
2. Describe the general morphology of organisms of phylum Arthropoda.
3. List characteristics that arthropods share with the phyla discussed previously.
4. Discuss those characteristics of arthropods that were newly derived from those of their ancestral phyla.
5. List examples of the major classes of arthropods.
6. Describe modifications of the exoskeleton and paired appendages of arthropods.

Please visit **connect.mheducation.com** to review online resources tailored to this lab.

The success and predominance of the arthropod body plan in marine, freshwater, and terrestrial environments is extraordinary. **Phylum Arthropoda** is the most diverse as well as most abundant phylum of animals—estimates range from 2–10 million species. Their tremendous success in all major habitats is due mainly to the adaptive value of their rigid external skeleton with **jointed appendages** (*arthro* = jointed, *poda* = foot or appendage) (table 39.1, fig. 39.1). Their appendages are extensions of the main body and highly adapted for locomotion, feeding, reproduction, defense, and sensing the environment. Bodies of arthropods are segmented, as are annelids (fig. 39.2), although segments of arthropods often fuse during development. The classification of arthropods is typically based on the arrangement of these segments and the structure of their appendages. Arthropods are coelomate, their circulatory systems are open, and all organ systems are well developed.

The **exoskeleton** of arthropods is made of **chitin,** a long chain of nitrogen-containing sugar molecules arranged in strong fibers. Chitin may be as soft as the body of a butterfly or, if impregnated with calcium carbonate, as hard as the shell of a lobster. This exoskeleton provides protection, a moisture barrier, and a place for muscle attachment. Although this tough covering limits growth, arthropods periodically shed their exoskeleton (ecdysis) and quickly enlarge before the new exoskeleton hardens.

Each body segment of ancestral arthropods had a single pair of appendages. Recall that this condition of paired appendages also occurs in annelids. Appendages of modern arthropods are covered by the rigid exoskeleton and usually have flexible joints. This provides arthropods with tremendous strength and flexibility. The variation in structure and function of appendages is enormous.

BiologyImaging.com

**Figure 39.1** Scorpions are the oldest known terrestrial arthropods. Typical of all arthropods, scorpion bodies are segmented, have jointed appendages, and are covered with a hard, chitinous exoskeleton. Shown here is the world's largest scorpion (18 cm), the Emperor scorpion (*Pandinus emperator*) from west Africa. Some extinct species were five-times larger than the Emperor. The sting and neurotoxic venom of most species are equivalent to those of a hornet sting. The most notorious stingers are *Androctonus* of North Africa and species of *Centruroides* in Mexico, Arizona, and New Mexico. Children and weak adults may die from the neurotoxin of a scorpion in 6–7 hours by paralysis of respiratory muscles or cardiac failure. Scorpions locate their prey by detecting vibrations, often through sand rather than air. Desert species can locate and dig out a burrowing cockroach in just a few seconds. Desert species can also withstand temperatures of 46°C (115°F) and tolerate water loss equivalent to 40% of their body weight.

Table 39.1

**Major Groups within the Traditional Classification of the Phylum Arthropoda**

| Group | Characteristics | Members | Approximate number of named species |
|---|---|---|---|
| Crustaceans 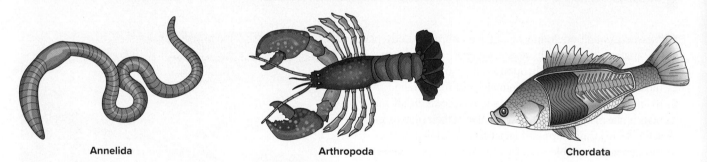 U.S. Fish and Wildlife Service | Mouthparts are mandibles (biting jaws); appendages are biramous ("two-branched"). | Lobsters, crabs, shrimp, isopods, barnacles | 67,000 |
| Chelicerates | Mouthparts are chelicerae (pincers or fangs). | Spiders, mites, ticks scorpions, daddy long-legs, horseshoe crabs | 78,300 |
| Centipedes and millipedes | Mouthparts are mandibles: bodies consist of a head and numerous body segments bearing paired uniramous ("single-branched") appendages. | Centipedes, millipedes | 20,000 |
| Insects | Mouthparts are mandibles; appendages are uniramous. | Beetles, bees, flies, fleas, true bugs, grasshoppers, butterflies, termites | 900,000 |

**Annelida**          **Arthropoda**          **Chordata**

**Figure 39.2**  Annelids, arthropods, and chordates all exhibit segmentation.

During the evolution of arthropods, their prominent characteristics (e.g., an exoskeleton and jointed appendages) were highly adaptive and probably allowed them to replace other species that weren't equipped to deal as well with their environment. These adaptations allowed primitive arthropods to exploit new environments, microhabitats, and niches. Taxonomists currently recognize four extant classes of arthropods: Chelicerata, Crustacea, Myriapoda, and Hexapoda. Review the introductory information in this exercise (as well as in your textbook) and complete table 39.2 for six or more arthropod characteristics.

## Who Cares about Horseshoe Crabs?

Blood from the horseshoe crab, *Limulus polyphemus,* is more valuable than you might think! It annually is used for tests worth $50–60 million to the medical industry. Amoebocytes circulating in the blood of horseshoe crabs have an ancient immune response that releases coagulogen from granules in the cells in response to bacterial toxins. The reaction forms a "clot" around a site of invading bacteria. The medical industry catches the crabs, bleeds about one-third of their blood, then releases them. Mortality is about 7.5% to 10%. Limulus Amoebocyte Lysate (LAL) is extracted from the blood, marketed, and used to detect any trace of Gram-negative bacterial endotoxin present on newly manufactured equipment, needles, vaccines, prosthetic limbs, or even blood samples from patients. As required by the Food and Drug Administration (FDA), injectable drugs and vaccines must pass a Limulus Test with LAL before they are approved. There is no synthetic substitute for LAL, so conservation and management of horseshoe crab populations is critical—and may save your life.

Courtesy U.S. Fish and Wildlife Service

Table 39.2

## The Major Characteristics of Arthropods and Their Adaptive Advantage

| Characteristic | Adaptive Advantage |
| --- | --- |
| | |
| | |
| | |
| | |
| | |
| | |

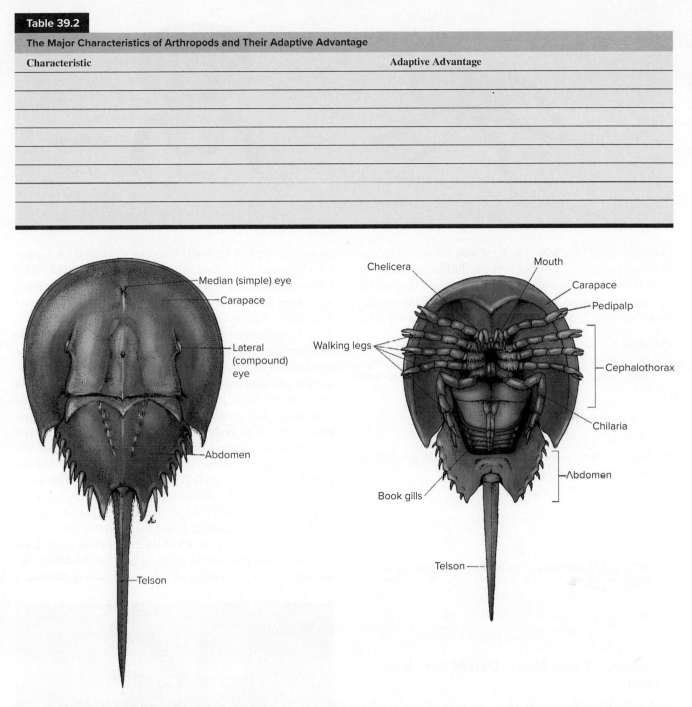

**Figure 39.3** Dorsal and ventral views of a horseshoe crab, *Limulus* sp. Horseshoe crabs are ancient chelicerates abundant on the Atlantic Coast and Gulf of Mexico.

## CLASS CHELICERATA

Chelicerates such as spiders, mites, scorpions, and horseshoe crabs are arthropods with the appendages of their most anterior segment modified into feeding structures called **chelicerae**. The second pair of appendages are **pedipalps** and are modified for capturing prey, sensing the environment, or copulating. Body segments of chelicerates are fused into two body regions: a **cephalothorax** consisting of a fused head and thoracic segments and an **abdomen** as the most posterior body region. Chelicerates lack antennae.

## Horseshoe Crabs

Horseshoe crabs, commonly *Limulus,* are ancient marine chelicerates existing since the Cambrian period (500 million years ago). They are abundant on the Atlantic Coast and Gulf of Mexico.

Examine a preserved *Limulus* and find the structures illustrated in figure 39.3. A horseshoe-shaped **carapace** covers the cephalothorax and a flexible joint seen easily from a dorsal view separates the cephalothorax from the abdomen. The ventral surface includes five pairs of appendages

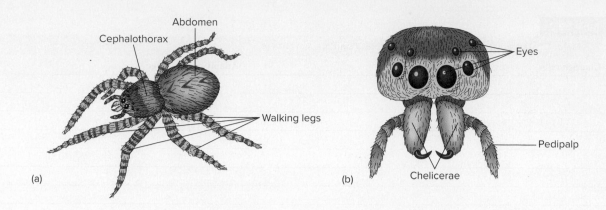

**Figure 39.4**   Anatomy of a jumping spider. (*a*) External view. (*b*) Anterior view of head.

modified as walking legs. The most anterior pair of legs are modified pedipalps. Anterior to the pedipalps are the chelicerae. Posterior to the walking legs are a pair of degenerated legs called **chilaria.** The appendages of the abdominal segments are modified as **book gills** and promote gas exchange.

### Question 1

*a.* Why do we call the external chitinous covering of an arthropod a skeleton even though it is not made of bone?

*b.* What external features of a horseshoe crab are used for burrowing and scavenging in sand?

*c.* Why are appendages of abdominal segments called "book" gills?

## Scorpions, Ticks, Mites, Daddy Longlegs, Spiders

These diverse groups, commonly referred to as arachnids, are diverse (85,000 species) and mostly terrestrial. The cephalothorax has chelicerae modified as fangs to pierce prey, has pedipalps to manipulate food and sense the environment, and has four pairs of walking legs.

Examine a preserved scorpion, noting the many distinct segments of its thorax and abdomen. Compare the segmentation of scorpions with that of other available preserved chelicerates such as mites, ticks, and daddy longlegs. Scorpions were the first terrestrial arthropods and have been around since the Silurian period (425 million years ago). Their lack of fused segments is unusual for chelicerates and indicates the scorpion's ancient origins (fig. 39.1). These secretive

carnivores are 1–8 cm long. Stings of *Centruroides,* a common scorpion in Mexico, Arizona, and New Mexico, have killed many humans, mostly children. Its venom is neurotoxic and causes convulsions, paralysis of respiratory muscles, and heart failure. However, most scorpion stings are not fatal.

Examine the external features of a preserved spider and compare them with the features shown in figure 39.4. Spiders are the most familiar arachnids and comprise 30,000 species. They are terrestrial and prey mostly on insects and other small invertebrates (figs. 39.5, 39.6). Spiders are often confused with insects, but spiders have two body regions and eight legs. Insects have three body regions and six legs. The spinnerets of spiders are independently moving nozzles that release silk from internal silk glands. Web silk is made primarily of polypeptides of the amino acids glycine, alanine, and serine. When this fluid is emitted it hardens, not from exposure to air but from polypeptide cross-linkages that form during release. The fluid is not forced out under pressure; rather, it is drawn out by the hind legs or by the weight of the body falling through the air. Spiders can produce silk as strong as nylon, highly elastic, and dry or sticky depending on the construction of the web.

BiologyImaging.com

**Figure 39.5**   Tarantulas, sometimes called bird spiders, may have a 25-cm leg span. They live on the ground, dwell in trees, and sometimes burrow. The hairy body of this brown tarantula is extremely sensitive to vibration. Many hairs contain chemoreceptors as well as mechanoreceptors, and hairs on the tips of the feet are often iridescent. Captive tarantulas can live 20 years or more.

BiologyImaging.com

**Figure 39.6** Black widow spiders (*Latrodectus mactans*) are shiny black with a red hourglass on the abdomen's ventral surface. As with other spiders, their chelicerae have poison glands opening at the tip of the fangs. Toxins and enzymes are injected into prey, liquefying the prey's tissue. The nutritious broth is then eaten. Black widows live in most parts of the world, and their venom causes nausea, muscular spasms, respiratory paralysis, and pain in the abdomen and legs. Human death from their bite is rare. Black widows, like other spiders, can withstand months of starvation by reducing their metabolic rate up to 40%. A male is only half the size of the female and is killed soon after mating.

## Question 2

*a.* What external features make scorpions appear so menacing?

*b.* Which are larger, a scorpion's chelicerae or pedipalps?

*c.* What is the shape of a spider's chelicerae?

*d.* Do you see any evidence that a spider's body is segmented?

*e.* How many eyes do most spiders have? Are they paired and similar in size?

*f.* Many spiders are hairy. How might this feature be adaptive?

## CLASS CRUSTACEA (CRAYFISH, CRABS, SHRIMPS)

Crustaceans (65,000 species) live in marine and freshwater. Only a few species are terrestrial. Crustaceans differ from the other classes because they have fundamentally **biramous,** or double-branched, appendages (fig. 39.7). Crustaceans have two pairs of antennae and usually have **compound eyes** with multiple lenses. Crustaceans along with insects are commonly called mandibulates because they have opposing mandibles derived from an anterior pair of appendages.

### Crustacean Anatomy

The body of a crustacean such as a crayfish usually has two regions: a cephalothorax covered by a carapace and an abdomen (fig. 39.8). The five anterior pairs of crustacean appendages are modified into **first antennae, second antennae, mandibles, maxillae,** and **maxillipeds.**

Obtain and examine a preserved crayfish. Appendages 3 (**mandible**) through 9 (**cheliped**) are used for feeding. These appendages have different shapes. The shape of some appendages such as the cheliped indicate an obvious function.

### Question 3

How might mouthparts of various shapes be adaptive for crayfish?

The four pairs of walking legs attach to the thorax at the ventral edge of the carapace. Abdominal appendages are much smaller than walking legs and are called **swimmerets** or **pleopods.** If you study a male crayfish, you'll notice that each of the first pair of swimmerets has an odd spatulate shape, modified to transfer packets of sperm to a female during reproduction. The most posterior pair of appendages are broad, flat **uropods.** They surround the terminal abdominal segment called the **telson.** All crustacean appendages are fundamentally biramous (i.e., double-branched).

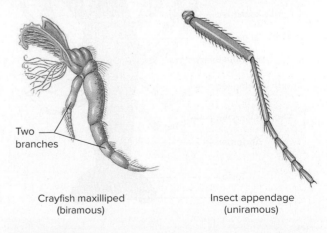

Two branches

Crayfish maxilliped (biramous)

Insect appendage (uniramous)

**Figure 39.7** Branched and single appendages. A biramous leg in a crustacean (crayfish) and a uniramous leg in an insect.

By now you've probably guessed correctly that studying arthropods consists mostly of examining body segments and their paired appendages. Appendages of two different species may have similar functions but different embryological origin. Such structures are **analogous.** For example, chelicerae of a spider are analogous to mandibles of a crayfish because they have similar functions even though they are not derived from the same body segment. In contrast, **homologous** structures of two different species have similar developmental origin but may or may not serve the same function. Chelicerae of a spider are homologous to the most anterior pair of antennae of a crayfish.

## *Procedure 39.1* **Study the external anatomy of a crayfish**

1. Obtain a preserved *Cambarus* (the common crayfish) and locate the external features shown in figure 39.8.
2. To appreciate the variation in appendages in *Cambarus,* use forceps to remove one of the first antennae, a cheliped, a walking leg, an anterior swimmeret, a posterior swimmeret, and a uropod.
3. Arrange them in order in a dissecting pan or on a piece of paper and relate each one's structure to its function.
4. Examine the appendages you've removed and note which ones are biramous.
5. Observe a live crayfish (if one is available) and determine its two major methods of locomotion.

6. Examine other available crustaceans such as a crab. The cephalothorax of a crab is obvious, but the abdomen is highly modified.

**Question 4**
***a.*** Which body region of a crayfish is most obviously segmented?

***b.*** What structures are located just under the carapace and attached to each leg of a crayfish? What is the adaptive advantage of these structures being attached to legs?

***c.*** How many legs are **chelate** (pincerlike with opposing claws)?

***d.*** What is the function of the uropods and telson, and what feature indicates this function?

***e.*** Are the anterior swimmerets different from the posterior pair? Is your crayfish a male or female?

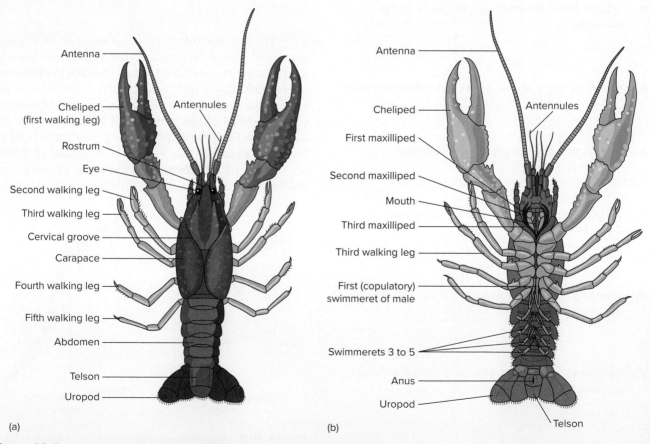

(a)

Antenna
Cheliped (first walking leg)
Antennules
Rostrum
Eye
Second walking leg
Third walking leg
Cervical groove
Carapace
Fourth walking leg
Fifth walking leg
Abdomen
Telson
Uropod

(b)

Antenna
Cheliped
Antennules
First maxilliped
Second maxilliped
Mouth
Third maxilliped
Third walking leg
First (copulatory) swimmeret of male
Swimmerets 3 to 5
Anus
Uropod
Telson

**Figure 39.8** External structure of the crayfish. (*a*) Dorsal view. (*b*) Ventral view.

**f.** Which set of legs (swimmerets or walking legs) appears best adapted to carry an incubating egg mass delicately and in a protected place on the body?

**g.** Crayfish can walk. Can they also "swim"? How would the uropods and telson help them do this?

**h.** How does the shape of a crab's abdomen differ from that of a crayfish?

The internal anatomy of a crayfish features a diamond-shaped **heart** surrounded by a thin **pericardial sac.** The heart lies on the dorsal midline just anterior to the abdominal segments (fig. 39.9). In the open circulatory system of a crayfish, blood flows from the heart through large arteries to the gills and to sinuses surrounding and bathing the internal organs. Blood returns to the pericardial sac and enters the heart through small openings called **ostia.**

The **gonads** (testes or ovaries) are lateral and just anterior to the heart. Testes are usually white, and ovaries are orange. Sperm from testes exit the body through a pore at the base of the fifth pair of walking legs, and eggs are released at the base of the third pair of walking legs. During reproduction, males and females copulate and sperm cells are passed to the genital openings of the female. Fertilization is internal, and fertilized eggs are extruded and retained for maturation on swimmerets of the female.

The stomach is a continuous, membranous structure along the dorsal midline of the cephalothorax. It is surrounded by muscle, reinforced with ridges of tissue, and receives food from the **esophagus** and **mouth.** Food moves from the stomach, through the intestine, and out through the anus at the base of the telson. A large **digestive gland** that secretes enzymes and stores food lies just posterior and lateral to the stomach.

Beneath the internal organs lies the anterior part of the **ventral nerve cord.** A pair of nerves from the ventral nerve cord pass around the esophagus and come together anteriorly as a **brain** between and beneath the eyestalks. Posterior to the esophagus the nerve cord extends the length of the body as a series of swellings or **ganglia,** each of which controls organs in the immediate segment.

### Procedure 39.2 Study the internal anatomy of a crayfish

1. Obtain a preserved crayfish.
2. Make an incision on each side of the cephalothorax. To make this incision, cut with scissors beginning at the posterior edge of the carapace halfway up the side (i.e., midlateral). Cut along the side toward the eye. Finish each lateral incision on each side of the rostrum.
3. Cut across the base of the rostrum and carefully remove the dorsal portion of the carapace.
4. Locate three pairs of ostia on the heart. Remove the heart.
5. Locate the stomach and digestive gland.
6. Carefully remove the internal organs (stomach, digestive gland, gonads). This exposes the end of the torn esophagus and the anterior part of the ventral nerve cord.
7. Remove a considerable amount of musculature on the floor of the abdomen (tail) to see the nerve cord more clearly.
8. Locate the **antennal glands.** They are excretory organs opening at the base of each antenna.
9. When you have finished examining your specimen, dispose of the material as directed by your instructor.

## CLASS MYRIAPODA

Myriapods have one pair of antennae on the head and three pairs of appendages that are modified as mouth parts, including mandibles that act like jaws.

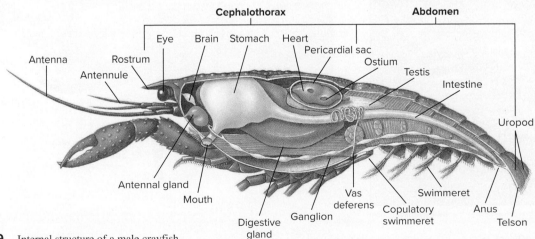

**Figure 39.9** Internal structure of a male crayfish.

*How can the same embryonic appendage pattern give rise to so many different appendage structures and functions?*

Observations: Crustaceans are highly diverse but share fundamental similarities retained throughout their evolution. Much diversity among crustaceans stems from variation in their appendages, yet consistent patterns emerge in appendage morphology and body plan.

Question: What variations and similarities are apparent among appendages of crustaceans?

a. Establish a working lab group and obtain Inquiry-Based Learning Worksheet 39 from your instructor.
b. Discuss with your group and instructor specific observations to be made during dissection of two closely related arthropods relevant to the preceding observations and question. Record them on Worksheet 39.
c. Translate your proposed observations into a testable hypothesis and record them on Worksheet 39.
d. Obtain a large, whole shrimp and crayfish for dissection.
e. Carefully remove appendages from one side of the specimens and tape the appendages side by side for comparison.
f. Complete Worksheet 39.

## Centipedes

Centipedes (3000 species) live in soil under logs and stones where they prey on small arthropods. Centipedes move rapidly and are dorsoventrally flattened. Some of the larger centipedes such as *Scolopendra* can inflict a painful bite but are not lethal to humans (fig. 39.10).

Examine a preserved centipede. Each body segment bears a pair of legs. The large fangs (sometimes called poison claws) on the head are not mandibles but are maxillipeds, which are appendages modified for feeding. The mandibles are smaller and lie between the maxillipeds.

### Question 5
*a.* Do centipedes have 100 legs as their name suggests?

*b.* What structures of chelicerates are analogous to the antennae of chilopods (centipedes) and other mandibulates?

## Millipedes

Examine a preserved millipede. Millipedes (8000 species) live in the same environment as centipedes but feed mainly on decaying plant material (fig. 39.11). Millipedes move slowly and are round in cross section. A disturbed millipede will frequently roll up to protect its soft underside. The number of legs on each segment distinguishes millipedes from centipedes.

### Question 6
*a.* How many pairs of legs does a millipede have?

*b.* Each apparent segment is actually two fused segments. How many legs are on each apparent segment?

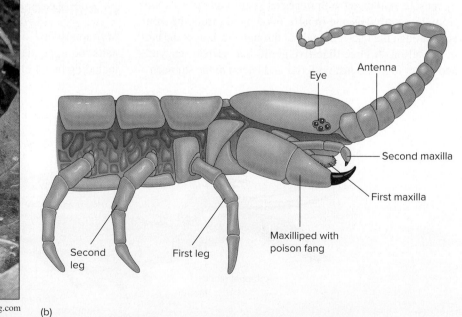

(a) BiologyImaging.com  (b)

**Figure 39.10**  Centipede, *Scolopendra* sp. (*a*) Most segments have one pair of appendages each. The first segment bears a pair of poison claws (maxillipeds), which in some species can inflict serious wounds. Centipedes are carnivorous. (*b*) The head of a centipede.

Michael P. Gadomski/Science Source

**Figure 39.11** Millipede, *Sigmoria.* Apparent body segments are actually fused pairs of body segments. Each apparent segment has two pairs of legs.

## CLASS HEXAPODA (Insects)

Hexapods are six-legged arthropods. Most are insects, but there are a few earlier-diverging noninsect hexapods, including soil-dwelling groups representing a separate but related lineage.

Insects are by far the largest group of organisms on Earth and probably include 1 to 10 million species. At least 70% of all known species of animals are insects, and they dominate virtually all terrestrial habitats. Three separate body regions and six thoracic legs are the major diagnostic features of insects. Insects possess all the major characteristics of arthropods, but their ability to fly is probably the key to their great success. Although other groups of organisms can fly, insects were the first fliers. There are many advantages to flying, such as avoiding predators. Can you name some others?

Most insects have two pairs of wings. However, insect wings are not modified appendages. Rather, they are evaginations (outgrowths) of the thoracic exoskeleton.

The success of insects on land is further enhanced by an efficient system of respiratory tubes called **tracheae** that conduct air throughout the body.

### External Anatomy of a Grasshopper

The grasshopper, *Romalea*, has characteristics typical of insects (fig. 39.12). Insects have three body regions: head, thorax, and abdomen; one pair of antennae; and six legs. The thorax is divided into the **pro-, meso-,** and **metathorax,** each having a pair of legs. Mouthparts are covered by the **labrum,** an extension of the head. Beneath the labrum are the **mandibles,** followed by a pair of **maxillae** with segmented extensions called **palps,** and then the **labium** with palps (fig. 39.13).

*Romalea* has 10 abdominal segments, each with a **spiracle** or breathing pore opening to the respiratory system of tracheal tubes (fig. 39.14). The terminal abdominal segment bears the reproductive genitalia. The terminal segment of males is blunt, whereas that of females is modified to lay eggs and is called an **ovipositor.**

> ### Procedure 39.3  Examine the external features of a grasshopper
>
> 1. Obtain a preserved specimen and locate the external features shown in figure 39.12.
> 2. Examine the head and find simple eyes called ocelli as well as compound eyes (fig. 39.13).
> 3. Remove the labrum and locate the remaining mouthparts.
> 4. Use your probe to locate the stout mandibles. Press the mandibles with your probe to test their rigidity.

### Question 7
*a.* How many ocelli does a grasshopper have?

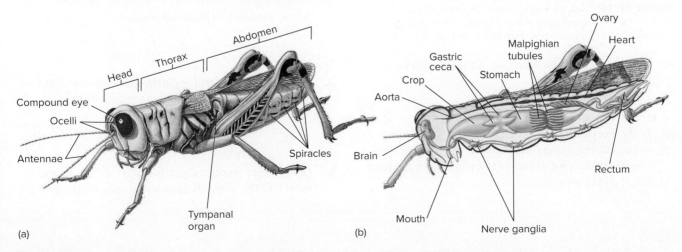

**Figure 39.12**  Grasshopper, *Romalea,* a member of class Insecta. (*a*) External anatomy. (*b*) Internal anatomy.

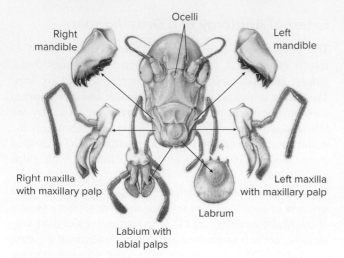

**Figure 39.13**  Head and mouthparts of a grasshopper.

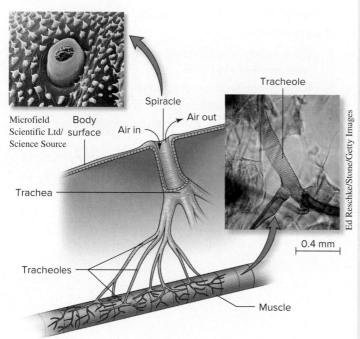

**Figure 39.14**  The tracheal system of an insect. Tracheae and tracheoles are connected to the exterior by specialized openings called spiracles and carry oxygen to all parts of a terrestrial insect's body.

**b.**  What is the probable function of the maxillary and labial palps?

**c.**  Wing morphology varies among species and orders of insects. Spread the wings of the grasshopper that you are examining. Do all of the wings have the same shape and consistency?

**d.**  Do you suspect that the fore- and hindwings have different functions? Why or why not?

**e.**  Consider Learning Objective 1 stated at the beginning of this exercise. Are wings significant to fundamental processes for arthropods? In what ways?

**Procedure 39.4**  **Examine the internal anatomy of a grasshopper**

1.  Use scissors to clip the wings away from the grasshopper's body.

2.  Make two lateral incisions (one on each side) from the posterior end. Cut the length of the abdomen and thorax along an incision line just dorsal to the spiracles.

3.  Connect the two lateral incisions by cutting across the dorsal surface just behind the head. Then remove the dorsal strip of exoskeleton.

4.  Pin the insect to the dissecting pan. Remove the legs if necessary.

5.  Locate the heart and aorta just below the strip of exoskeleton that you removed (fig. 39.12b).

6.  Notice the body cavity between the body wall and digestive tract. It is the hemocoel and is filled with hemolymph, a colorless blood.

7.  The internal organs may be covered with a yellow fat body. Remove it and the heart to expose the digestive tract.

8.  Search for tubules of the respiratory tracheal system. These tubules conduct air from the spiracles to the tissues.

9.  Locate reproductive organs (ovaries and testes) that lie on either side of the digestive tract.

10.  Examine the digestive tract and locate the major structures labeled in figure 39.12b.

11.  Remove the digestive tract to examine the ventral nerve cord and its series of swollen nerve ganglia.

**Question 8**

**a.**  How many abdominal segments does a grasshopper have? Is there a nerve ganglion for each segment?

**b.**  Gastric ceca produce digestive enzymes and secrete them into the stomach. How many ceca are there?

**c.** The spiracles can be opened or closed. Why might this be adaptive?

**d.** The brain is in the dorsal part of the head. Where does the ventral nerve cord encircle the digestive tract and join with the brain?

(a) Chewing (grasshopper)

The success of insects is also enhanced by a diversity of highly modified mouthparts (fig. 39.15). For example, mouthparts of grasshoppers grind coarse plant tissue, whereas mosquito mouthparts puncture and suck fluid from animals. Vision by insects is also remarkable. The structure of the arthropod compound eye allows extraordinary sensitivity and rapid detection of motion (fig. 39.16).

Examine all the insects available in the laboratory and note any differences from the basic anatomy of a grasshopper. If you find differences, ask your instructor to explain the variation among these insects. If you look carefully, you should keep your instructor busy for a long time discussing the most diverse animals in the world.

To review phylum Arthropoda, outline in table 39.3 the differences among the classes you have examined.

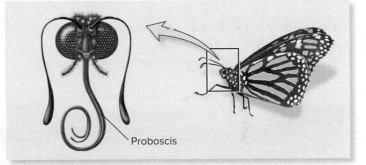

(b) Piercing and blood sucking (mosquito)

### Question 9

**a.** Are the chelicerae of horseshoe crabs homologous or analogous to (1) scorpion pedipalps? (2) grasshopper mandibles? Explain.

**b.** Describe four specialized appendages of insects and their function.

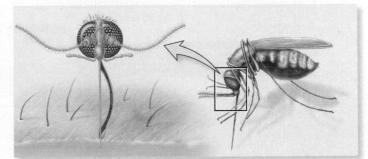

Proboscis

(c) Nectar sucking (butterfly)

**c.** Now that you have examined the unifying characteristics of arthropods, list three or four characteristics that they share with annelids, their most likely ancestors.

## DICHOTOMOUS KEY

A common tool for identifying organisms is a **dichotomous key,** such as the one presented in Exercise 31. Dichotomous keys list and describe pairs of opposing traits, each of which leads to another pair of traits until a level of classification of the specimen being identified is reached. By using a key you'll learn the characteristics that distinguish each of the groups identified by the key.

The diversity of insects is immense. They are classified into 26 orders distinguished mainly by the structure of

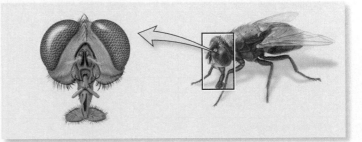

(d) Sponging liquid (housefly)

**Figure 39.15** Insect mouthparts have become modified in ways that allow insects to feed by a variety of methods, including (*a*) chewing (Orthoptera, Coleoptera, and others), (*b*) piercing and blood sucking (Diptera), (*c*) nectar sucking (Lepidoptera), and (*d*) sponging liquid (Diptera).

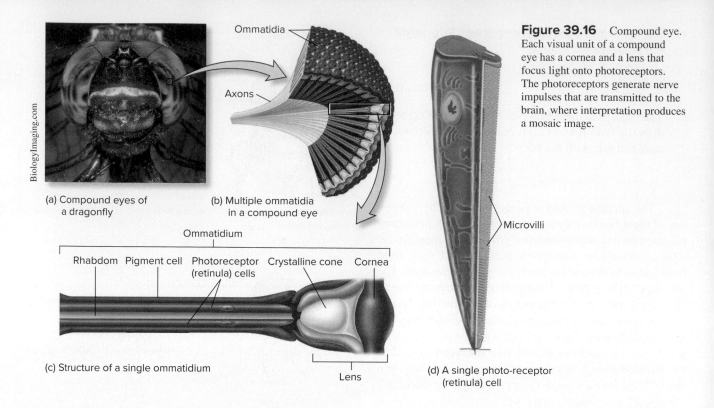

(a) Compound eyes of a dragonfly

(b) Multiple ommatidia in a compound eye

**Figure 39.16** Compound eye. Each visual unit of a compound eye has a cornea and a lens that focus light onto photoreceptors. The photoreceptors generate nerve impulses that are transmitted to the brain, where interpretation produces a mosaic image.

Ommatidium

Rhabdom  Pigment cell  Photoreceptor (retinula) cells  Crystalline cone  Cornea

(c) Structure of a single ommatidium

Lens

Microvilli

(d) A single photo-receptor (retinula) cell

wings, mouthparts, and antennae. You will be provided with five to eight preserved insects. Use this key to identify the order of each specimen.

## Procedure 39.5   Use a dichotomous key to identify the order of an insect

The key will eventually lead you to the order of the specimen.

1.  Select a specimen and read the first pair of characteristics.

2.  Choose the one that best describes your specimen.

3.  Proceed according to the number at the end of your choice to the next pair of characteristics.

### Table 39.3

**A Comparison of Major Characteristics of Organisms of Phylum Arthropoda**

|  | Horseshoe crabs | Arachnids | Crustaceans | Centipedes | Millipedes | Insects |
|---|---|---|---|---|---|---|
| Names of body regions |  |  |  |  |  |  |
| Number of legs |  |  |  |  |  |  |
| Arrangement of legs |  |  |  |  |  |  |
| Segmentation |  |  |  |  |  |  |
| Number of antennae |  |  |  |  |  |  |
| Names of major sensory organs |  |  |  |  |  |  |
| Names of major mouthparts |  |  |  |  |  |  |

## DICHOTOMOUS KEY TO SOME MAJOR ORDERS OF INSECTS

1. Insects with two wings .........................(flies) Diptera

   Insects with four wings, a pair of forewings, and a pair of hindwings.................................. 2

2. Fore- and hindwings are not alike in texture and color. One pair may be hard and dense while the other may be light and transparent ....................... 3

   Fore- and hindwings similar, usually clear, thin, and transparent ....................................... 5

3. Forewings thick and leatherlike at base, tips much thinner and may be transparent; mouthparts pointed and beaklike to puncture prey and suck body fluids ................................(bugs) Hemiptera

   Forewings same texture throughout, biting mouthparts with opposing mandibles .................... 4

4. Forewings leathery and with veins ............ (grasshoppers, crickets) Orthoptera

Forewings hard, without veins .......................................(beetles) Coleoptera

5. Wings of same length, antennae usually shorter than head ...............................................6

   Wings not of same length, antennae long or enlarged toward end ...............................................7

6. Large insects (usually > 3 cm), wings long, transparent, and with many strong veins; abdomen long and slender ................................(dragonflies) Odonata

   Smaller insects, wing venation faint, wings extending posterior to the abdomen ...................................(termites) Isoptera

7. Wings covered with fine, opaque scales; tubular, coiled, sucking mouthparts ...........................(butterflies, moths) Lepidoptera

   Wings thin, transparent, and not covered with scales; mandibles well developed ...........................(ants, bees, wasps) Hymenoptera

# Questions for Further Study and Inquiry

*1.* Arthropods usually have a distinct head. How would you define a "head"? What are the advantages and disadvantages of having such a body region?

*2.* Does an insect's exoskeleton limit growth? Why or why not?

*3.* Diagram the arrangement of muscles necessary to bend a joint with an exoskeleton versus a joint supported by an endoskeleton.

*4.* Arthropod body segments are sometimes distinct, sometimes indistinct, and sometimes fused as groups to form body regions. Which groups of arthropods appear the most distinctly segmented? Which appear the least segmented?

5. What effect would 2.5 million spiders per acre have on the insect community?

6. Do you suspect that each eye of a spider provides the same sensory input to the brain? Why or why not?

7. What activities and body functions of arthropods require the most specialized appendages?

8. Do beetles have wings? If so, where are they?

9. What other group of organisms you have studied thus far has chitin as part of its outer covering?

10. What group of arthropods dominates the sea?

11. Prepare a simple table of all of the taxonomic groups, their common names, their distinguishing characteristics, and all representative genera covered in this exercise. Keep this table with your study notes.

12. Does a crayfish have an open or closed circulatory system? Review detailed references and summarize how the circulatory systems vary among animal phyla.

13. A major feature of annelids and arthropods is segmentation. Speculate on the adaptive advantages of segmentation.

**WRITING TO LEARN BIOLOGY**

Do you think that arthropods constitute a single phylum, or should they be divided into multiple phyla? Describe what divisions you would make and your reasons for them.

# Survey of the Animal Kingdom
## Phyla Echinodermata and Chordata

Please visit **connect.mheducation.com** to review online resources tailored to this lab.

Members of the two phyla remaining in our survey of animals—that is, Echinodermata and Chordata—are **deuterostomes.** Deuterostomes are a major departure from the phylogenetic line of **protostomes** such as annelids, mollusks, and arthropods. The fundamental basis for the separation of deuterostomes from protostomes involves different morphological patterns of embryonic development (fig. 40.1); most notably, the blastopore of deuterostomes gives rise to an anus rather than the mouth. The blastopore is the opening to the first cavity formed in a developing embryo and is discussed more in Exercise 50.

Examine figure 40.1 carefully. Refer to your textbook for a more detailed comparison of protostomes and deuterostomes.

## Question 1

Describe three major differences between deuterostome and protostome development.

## PHYLUM ECHINODERMATA

Echinoderms (6000 species) are marine bottom-dwellers and include sea stars, brittle stars, sea urchins, sand dollars, sea cucumbers, and sea lilies (table 40.1, fig. 40.2). These organisms are called echinoderms (*echino* = spiny, *derm* = skin) because their internal skeleton of calcareous plates,

called **ossicles,** usually has spines protruding through a thin layer of skin. The five classes of echinoderms are distinguished primarily by the arrangement of their ossicles.

Adult echinoderms are radially symmetrical, and their bodies typically consist of a ring of five repetitive parts (i.e., they are pentaradial). In contrast, larvae of echinoderms are bilaterally symmetrical (fig. 40.3). This indicates that radial symmetry is secondarily derived and not directly related to the symmetry of more ancient phyla such as Cnidaria.

Echinoderms have a unique **water vascular system** consisting of a series of coelomic water-filled canals ending in hollow projections called **tube feet.** Muscle contractions and hydrostatic pressure in the water vascular system extend and move the tube feet and other parts of the system and thereby move the animal (fig. 40.4).

## Class Asteroidea (Sea Stars)

Ossicles of class Asteroidea, including the common sea star *Asterias,* are arranged loosely under the skin, and spines are small and blunt. Arms of sea stars are continuous with the central disk. The mouth is at the center of the lower, **oral** surface, and the anus is on the upper, **aboral** surface. Surrounding the blunt spines of sea stars are **dermal gills** for respiration by diffusion and pincerlike **pedicellariae** used to remove debris from the surface (fig. 40.5). Also on the aboral surface is the **madreporite,** a sieve connecting the water vascular system with the environment.

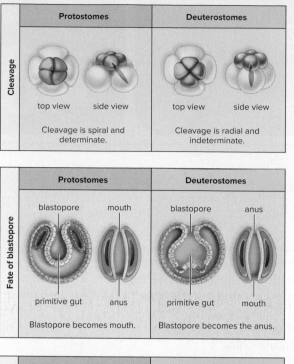

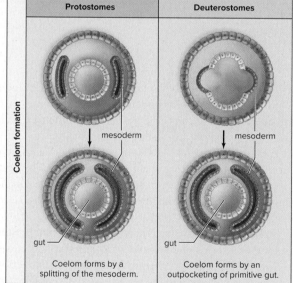

**Figure 40.1** Patterns of embryonic development of protostomes compared to deuterostomes. *Left:* In the embryo of protostomes, cleavage is spiral and determinate—new cells form at an angle to old cells—and each cell has limited potential and cannot develop into a complete embryo. The blastospore is associated with the mouth, and the coelom, if present, develops by a splitting of the mesoderm. *Right:* In deuterostomes, cleavage is radial and indeterminate—new cells sit on top of old cells—and each one can develop into a complete embryo. The blastopore is associated with the anus, and the coelom, if present, develops by an outpocketing of the primitive gut.

Sea stars often prey on oysters and clams by using their arms and tube feet to grip the shell and persistently apply pressure to pry it open. Sea stars then evert their stomach inside the clam to digest and engulf the tissue.

**SAFETY FIRST** Before coming to lab, you were asked to read this exercise so you would know what to do and be aware of safety issues. In the space below, briefly list the safety issues associated with today's procedures. If you have questions about these issues, contact your laboratory assistant before starting work.

## Procedure 40.1 Examine the external anatomy of a sea star

1. Examine a prepared slide of the bipinnaria larval stage of starfish development. Note the bilateral symmetry.
2. Examine a preserved sea star and locate the external and internal features shown in figure 40.6.
3. Examine the oral surface of the sea star, and locate the central mouth. Locate the tube feet protruding from the **ambulacral grooves.**
4. Use a dissecting microscope to examine the spines on the aboral surface (fig. 40.5).
5. Touch the madreporite with a probe, and note its consistency.
6. Examine figure 40.4 and trace the path of water from the madreporite to the tube feet.
7. If living sea stars are available, observe their locomotion.

### Question 2
*a.* How many tube feet would you estimate are on the oral surface of a sea star?

*b.* Are sea star spines movable?

*c.* What is the consistency of the madreporite?

*d.* How fast does a living sea star move?

*e.* Do tube feet of a living sea star move in unison?

Table 40.1

**Phyla Echinodermata and Chordata**

| Phylum | Typical Examples | | Key Characteristics | Approximate Number of Named Species |
|--------|------------------|---|---------------------|-------------------------------------|
| Echinodermata (echinoderms) | Sea stars, sea urchins, sand dollars, sea cucumbers | | Deuterostomes with radially symmetrical adult bodies; endoskeleton of calcium plates; five-part body plan and unique water vascular system with tube feet; able to regenerate lost body parts; marine | 6000 |
| Chordata (chordates) | Mammals, fish, reptiles, birds, amphibians | | Segmented coelomates with a notochord; possess a dorsal nerve cord, pharyngeal slits, and a tail at some stage of life; in vertebrates, the notochord is replaced during development by the spinal column; 20,000 species are terrestrial; deuterostomes | 42,500 |

## *Procedure 40.2*  **Examine the internal anatomy of a sea star**

1. Obtain a preserved sea star, and cut off at least 2 cm of one arm.
2. Examine the severed end and compare the structures with those shown in figure 40.7.
3. Cut along both sides of the arm up to the central disk. Then cut across the aboral (upper) surface to join the two lateral incisions.
4. Remove the upper body wall and expose the coelomic cavity of the arm. Locate the internal organs shown in figure 40.8.
5. Remove one of the digestive glands so you can examine the gonad lying underneath.
6. Cut around the perimeter of the central disk. Then cut from the perimeter up to and around the madreporite. This incision will allow you to remove the upper body wall without tearing away the madreporite.
7. Remove the upper body wall. As you lift it, try to see the delicate connection between the anus and the surface of the thick-walled pyloric stomach below.
8. Locate the structures shown in figure 40.8.

### Question 3

*a.* How many tube feet would you estimate are in one arm?

*b.* What part of the water vascular system extends into each arm?

*c.* What other phyla that you have examined rely on "hydraulics" as part of their locomotion system?

*d.* Does the stomach wall appear highly folded and extensible? How does that relate to the feeding method of most sea stars?

## Class Ophiuroidea (Brittle Stars)

Examine a preserved brittle star such as *Ophioderma*. Brittle stars have slender, sometimes branched arms clearly demarcated from the central disk (fig. 40.2*d*). Ossicles of brittle stars are typically thick and have attached musculature. They may form "shields" on the surface. As the name "brittle" star implies, their arms detach easily, allowing escape from predators.

The ambulacral grooves are closed in brittle stars, and the reduced tube feet are not used for locomotion. The thin flexible arms of brittle stars allow them to crawl rapidly like an octopus rather than creep slowly like sea stars. Brittle stars eat suspended food particles captured with their tube feet and passed to their mouth.

(a)

(b)

(c)

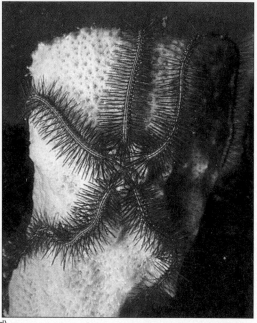

(d)

Bill Ober

(e)

**Figure 40.2** Diversity in echinoderms (phylum Echinodermata). (*a*) Sea star, *Dermasterius imbricata* (class Asteroidea), in the Gulf of California. (*b*) California sea cucumber, *Parastichopus californicus* (class Holothuroidea). (*c*) Feather star, *Florometra serrastissima* (class Crinoidea), uses its highly branched arms in filter feeding. Although filter feeding probably reflects the original use of echinoderm appendages, most modern echinoderms use their arms for locomotion, capturing prey, and scavenging the substrate for food. (*d*) Brittle star, *Ophiothrix* (class Ophiuroidea). (*e*) Sea urchin, class Echinoidea. Urchins, such as this red sea urchin, *Strongylocentodus franciscanus,* defend themselves with long and sometimes barbed spines, often filled with toxins.

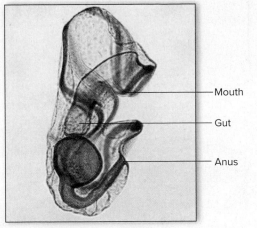

BiologyImaging.com

**Figure 40.3** The free-swimming larva of the common sea star, *Asterias rubens*. Such bilaterally symmetrical larvae suggest that the ancestors of the echinoderms may not have been radially symmerical. Bilateral larva is shown in side view (100×).

**Question 4**

*a.* Between brittle stars and sea stars, which have the most apparent ossicles? Do they overlap?

*b.* Are tube feet visible in *Ophioderma*?

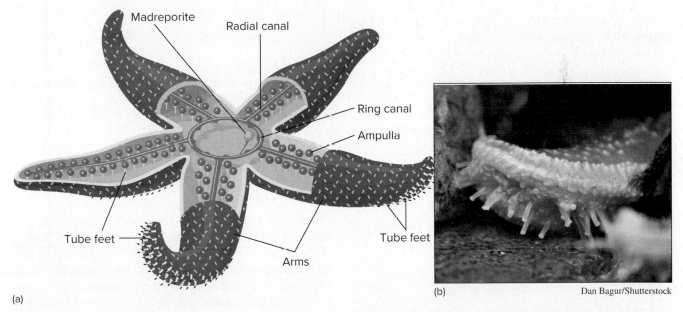

(a)

(b)                                                                 Dan Bagur/Shutterstock

**Figure 40.4** Echinoderms. (*a*) The echinoderm body plan of a sea star, emphasizing the water vascular system. (*b*) The extended tube feet of a sea star, *Ludia magnifica* (10×). Tube feet are used for locomotion and for gripping and pulling apart prey such as clams.

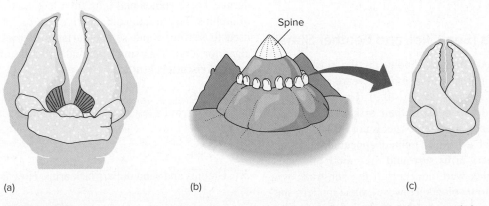

(a)                                  (b)                                  (c)

**Figure 40.5** Pedicellariae of sea stars. (*a*) Forceps-type pedicellaria of *Asterias*. (*b*) Spine surrounded by (*c*) scissors-type pedicellaria of *Asterias*.

| Oral View | Aboral View |
| --- | --- |

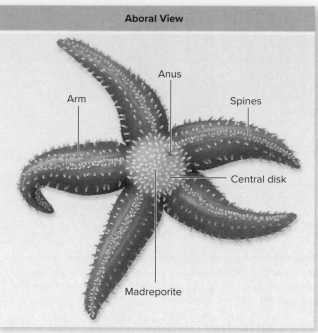

**Figure 40.6** External features of sea stars.

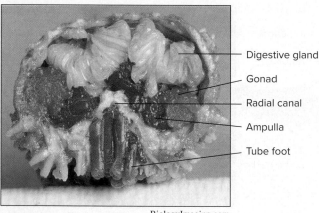

BiologyImaging.com

**Figure 40.7** Cross section of an arm of a sea-star showing internal anatomy (5×).

## Class Crinoidea (Sea Lilies and Feather Stars)

Examine a preserved crinoid. Crinoids are the most ancient echinoderms and flourished 450 million years ago; only a few genera live today (figs. 40.2*c*, 40.9). They differ from other living echinoderms because their oral surface (mouth and anus) usually faces up. Their ossicles are well developed and give the animal a coarse, jointed appearance. Highly branched and feathery arms surround the mouth and anus. Most ancient crinoids were attached to the substrate by a stalk and appeared to be plants. However, most modern species are not stalked or permanently attached. Crinoids filter feed by capturing food particles on the mucus of their tube feet.

### Question 5
How does the position of the mouth and anus of a crinoid relate to a primitive sessile existence?

## Class Echinoidea (Sea Urchins and Sand Dollars)

Examine an urchin such as *Arbacia* and locate the features shown in figure 40.10. Also examine a dissected Aristotle's lantern if one is available. In addition, examine a sand dollar and compare its test of fused ossicles to an urchin's test. Urchins lack distinct arms, and their ossicles are fused into a solid shell called a **test** (fig. 40.2*e*). Holes in the test allow long tube feet to protrude. Spines of sea urchins are jointed, movable, and longer than those of other classes of echinoderms. These spines and long tube feet control locomotion of urchins. The mouth contains five ossified plates, or teeth, used to scavenge and scrape surfaces of rocks and gather algae for food. This small internal structure of five teeth is called **Aristotle's lantern.**

### Question 6
*a.* Is an urchin's test pentaradially symmetrical?

*b.* Urchins and sand dollars lack arms. How do they move?

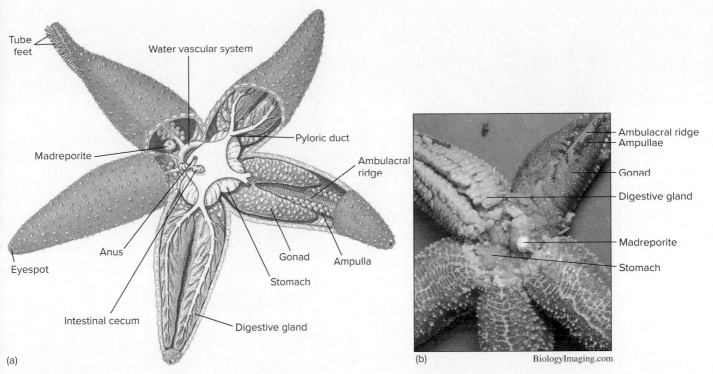

**Figure 40.8** Sea star internal anatomy. (*a*) Schematic with aboral surface removed. (*b*) Partially dissected sea star.

## Sea Urchins and RAGs

The vertebrate immune system is astoundingly complex, even though fewer than 1% of human genes encode antibodies that recognize invaders. Underlying this versatile immune system is the ability to shuffle and reshuffle DNA fragments that encode antibodies that recognize a virtually limitless variety of invading antigens. Surprisingly, scientists have learned much about the vertebrate immune system and its encoding system for antibodies by studying sea urchins.

By studying the genome of the purple sea urchin, *Strongylocentrotus purpuratus*, scientists have revised our understanding of the vertebrate immune system. In 2006, the purple sea urchin became the first echinoderm to have its DNA sequenced. When immunologist Jonathan Rast and his colleagues searched the echinoderm's DNA for evidence of genes required for vertebrate-style adaptive immunity, they found genes encoding a pair of enzymes that shuffle the echinoderm's DNA! The enzymes are called RAG proteins, for **r**ecombination **a**ctivating **g**ene. In vertebrates, two RAG proteins (called Rag1 and Rag2) must be present for lymphocytes to mature properly; these two proteins interact

to recognize sites where a DNA fragment is to be cut and spliced. The origin of our vertebrate immune system may have begun taking shape in ancestors of our distant deuterostome cousins, the echinoderms!

gcammino/iStock/Getty Images

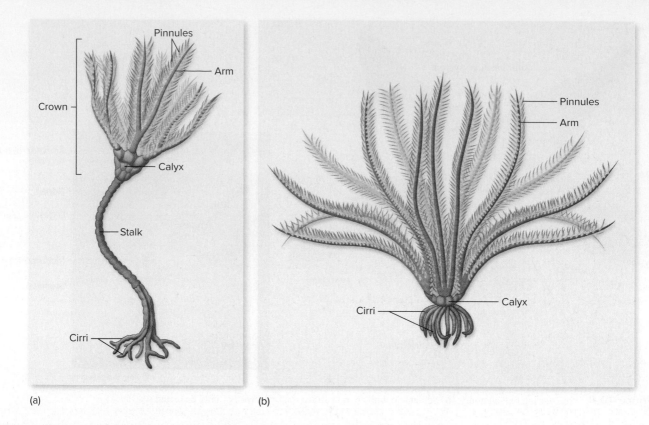

(a)                                    (b)

**Figure 40.9** Class Crinoidea. (*a*) A sea lily (*Ptilocrinus*). (*b*) A feather star (*Neometra*).

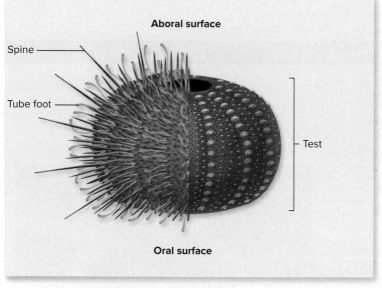

**Figure 40.10** External anatomy of a sea urchin. Spines and tube feet are removed on the right half of the diagram to show the test.

## Class Holothuroidea (Sea Cucumbers)

Examine a sea cucumber *Cucumaria* and determine the orientation of its radial symmetry. Find the mouth at one end; it is surrounded by modified tube feet called **tentacles.** Sea cucumbers look different from other echinoderms because they have soft bodies with reduced ossicles and few if any spines (figs. 40.2*b*, 40.11). Radial symmetry is less evident

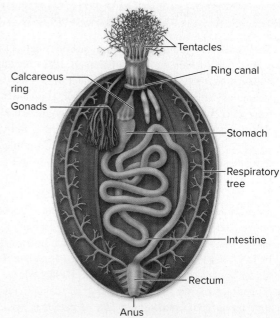

**Figure 40.11** Internal structure of a sea cucumber, *Thyone*. The mouth leads to a stomach supported by a calcareous ring. The calcareous ring is also the attachment site for longitudinal retractor muscles of the body. Contractions of these muscles pull the tentacles into the anterior end of the body. The stomach leads to a looped intestine.

Table 40.2

**A Comparison of the Major Characteristics of the Classes of Echinoderms**

| | Class | | | | |
|---|---|---|---|---|---|
| | **Asteroidea** | **Ophiuroidea** | **Crinoidea** | **Echinoidea** | **Holothuroidea** |
| Shape of arms | | | | | |
| Development of tube feet | | | | | |
| Development of ossicles | | | | | |
| Feeding method | | | | | |
| Spine structure | | | | | |

in sea cucumbers and their body axis is oriented horizontally. This orientation gives sea cucumbers a semblance of cephalization. The tentacles secrete a mucus that captures small floating organisms, which they eat. Interestingly, some sea cucumbers respond to stress by rupturing anteriorly and rapidly expelling their pharynx, digestive tract, and other organs. This process is called evisceration; because of it, the animal must regenerate the lost parts of the organs. Some gourmets consider sea cucumbers a delicacy.

Examine other preserved echinoderms and review in your textbook the major characteristics of each class. Then complete table 40.2.

**Question 7**
*a.* Are tube feet visible on the sea cucumber?

*b.* Hydras, octopuses, and sea cucumbers have tentacles. Do tentacles have a single universal function or varied functions? What functions are common?

*c.* Describe the functions of pedicellariae, madreporite, dermal gills, Aristotle's lantern, tube feet, water vascular system.

## PHYLUM CHORDATA

Chordates include 42,500 species of fish, amphibians, reptiles, birds, and mammals. They all are characterized by (1) a **dorsal hollow nerve cord;** (2) a **notochord,** a cartilaginous rod that forms on the dorsal side of the gut in the embryo; (3) **pharyngeal slits,** openings in the throat that filter water that has entered through the mouth, and (4) a **postanal tail** (fig. 40.12). An internal, bony skeleton is also common and provides sites for muscle attachment for efficient movement.

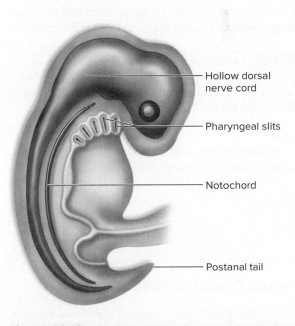

Hollow dorsal nerve cord

Pharyngeal slits

Notochord

Postanal tail

**Figure 40.12** The four principal features of the chordates, shown in a generalized embryo.

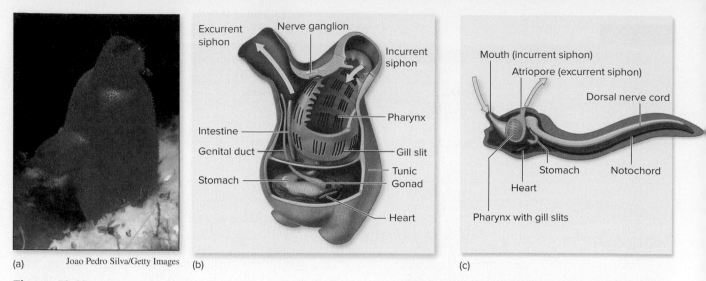

**Figure 40.13** Tunicates (phylum Chordata, subphylum Urochordata). (*a*) Living adult. (*b*) Structure of adult tunicate. (*c*) Larval structure.

## Subphylum Urochordata (Tunicates or Sea Squirts)

Examine a preserved adult tunicate. Urochordates, sometimes called tunicates, are sessile or planktonic marine organisms whose larvae possess the general chordate form—that is, they are elongated with a notochord and dorsal nerve cord (fig. 40.13). In contrast to larvae, the structure of an adult is highly modified to include a sievelike basket perforated with pharyngeal gill slits and surrounded by a cellulose sac called a **tunic.** Water enters through an incurrent siphon, is filtered by the pharyngeal basket, and exits through an excurrent siphon. Water is actively filtered; some tunicates only a few centimeters long can filter 170 liters of water per day. Food collected by mucus on the pharyngeal basket is moved by cilia to the stomach and intestine. The intestine empties into the body cavity near the excurrent siphon.

Examine a prepared slide of larval tunicates. A larval tunicate has bilateral symmetry, a dorsal nerve cord, a notochord, and a postanal tail but loses these features when it settles for adult life (fig. 40.13*c*).

### Question 8

What other group of organisms has cellulose in its supporting structures? Does this shared feature surprise you?

## Subphylum Cephalochordata (Lancelets)

Examine either a preserved lancelet or a slide of a whole mount and compare the specimen with that shown in figure 40.14. Also examine a cross section through the pharynx and try to visualize the paths of food and water (fig. 40.15).

(a) Paulo Oliveira/Alamy Stock Photo

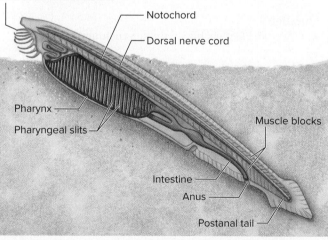

(b)

**Figure 40.14** (*a*) A lancelet, *Branchiostoma lanceolatum* (phylum Chordata, subphylum Cephalochordata), protruding from shell gravel. The muscle segments are visible; the square, pale yellow objects along the side of the body are gonads, indicating that this is a male lancelet. (*b*) Internal structure of a lancelet. This bottom-dwelling cephalochordate has the four distinctive features of chordates: notochord, dorsal nerve cord, pharyngeal gill slits, and a postanal tail. The vertebrate ancestor probably had a similar body plan.

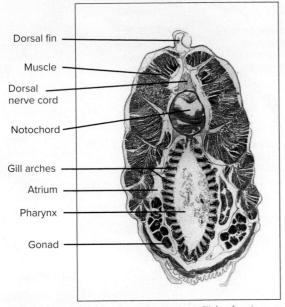

**Figure 40.15** Cross section through the pharynx of a lancelet (20×).

Lancelets are small, fishlike, marine chordates that burrow in sand or mud. They are commonly called amphioxus, but the most common genus is *Branchiostoma*.

The dorsal nerve cord and notochord extend the length of the animal. The buccal cavity surrounds the mouth followed by a long pharynx with many gill slits (openings) separated by gill arches of reinforced tissue. As seawater enters the mouth and exits through the slits, it must pass over the surfaces of the arches that form the sides of the slits. As this occurs, food particles are caught on the arches and are eventually swept to the intestine. The anus is not terminal. Lancelets and vertebrates have a post-anal tail, another diagnostic trait of chordates. After water passes by the gill arches, it moves into a surrounding chamber called an **atrium** and then leaves the body through the **atriopore.**

## Subphylum Vertebrata (Fish, Birds, Amphibians, Reptiles, and Mammals)

Vertebrates have a distinct head and a vertebral column that replaces the notochord in adults and surrounds the dorsal nerve cord (fig. 40.16). Three classes of vertebrates discussed here are fishes and four are terrestrial tetrapods.

## Class Cephalaspidomorphi (Lampreys)

Examine a preserved lamprey, *Petromyzon* (fig. 40.17). Also examine a prepared slide of an **ammocoete,** the larva of a lamprey. Living lampreys (38 species) descended from representatives of the earliest stages in the evolution of vertebrates. They lack jaws typical of other vertebrates but have a cartilaginous endoskeleton and a notochord. Seven pharyngeal gill slits are evident near the head. The gill arches separating the gill slits are reinforced with cartilage. The mouth is at the center of the round **buccal funnel** and is armed with horny teeth and a rasping tongue. They attach their buccal funnel to the side of a fish, rasp a hole in the body with their tongue, and feed on the body fluids of the fish.

### Question 9
Which closely related subphylum of chordates does an ammocoete resemble?

## Class Chondrichthyes (Sharks, Skates, and Rays)

Sharks and their relatives (1000 species) are abundant in oceans as predators and scavengers. Their **endoskeleton** is cartilaginous and the anterior gill arches are modified into jaws. Jaws are a significant adaptation and modification of strong, anterior gill arches to process food (fig. 40.18). Like agnathans, their cartilaginous skeleton is not necessarily primitive but is probably derived secondarily from an ancestral bony skeleton.

Examine a preserved specimen of *Squalus*, the dogfish shark (fig. 40.19). Its external anatomy illustrates some advanced features appropriate for a predator. Fin structure

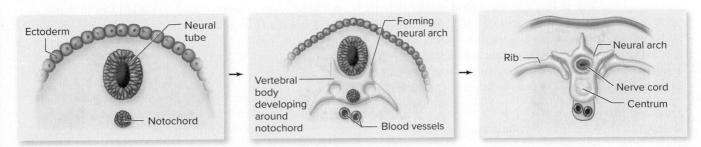

**Figure 40.16** Embryonic development of a vertebra. During the course of evolution, or of development, the flexible notochord is surrounded and eventually replaced by a cartilaginous or bony covering, the centrum. The neural tube is protected by an arch above the centrum. The vertebral column is a strong, flexible rod that the muscles pull against when the animal swims or moves.

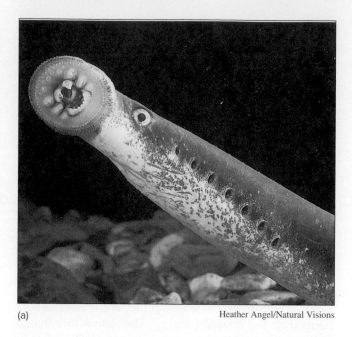

(a)

Heather Angel/Natural Visions

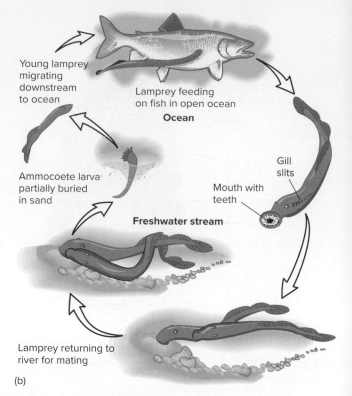

Young lamprey migrating downstream to ocean

Lamprey feeding on fish in open ocean
**Ocean**

Gill slits

Mouth with teeth

Ammocoete larva partially buried in sand

**Freshwater stream**

Lamprey returning to river for mating

(b)

**Figure 40.17** Lamprey (class Cephalaspidomorphi *Petromyzon marinus*). (*a*) Note the sucking mouth attached to aquarium glass and teeth used to feed on other fish. (*b*) External structure and life history of a sea lamprey. Sea lampreys feed in the open sea; toward the end of their lives lampreys migrate into freshwater streams, where they mate. Females deposit eggs in nests on the stream bottom, and the young larvae hatch 3 weeks later.

Skeletal elements   Cranium   Gill slits

Early jawless fish       Early jawed fish       Modern jawed fish

(a)

(b)

Jeff Rotman Photography

**Figure 40.18** (*a*) Jaws and support structures likely evolved from skeletal elements forming gill arches that separate gill slits of agnathans. (*b*) Head of sand tiger shark, *Carcharias* sp., showing a series of successional teeth on strong jaws.

includes paired **pelvic fins** (on the ventral surface near the anus) and **pectoral fins** (behind the gill slits) for stabilization and maneuvering. Jaws are large and powerful, and receptors in the nostrils and epidermis are sensitive to smells and electrical currents. A **lateral line** runs along each side of the body and contains sensory cells to detect slight vibrations.

**Question 10**
*a.* Which fins of sharks provide power and speed?

*b.* Why is the number of pharyngeal gill slits in sharks fewer than that in lampreys?

*c.* Consider Learning Objective 1 listed at the beginning of this exercise. Is a lateral line system significant to fundamental processes for sharks and bony fish? How so?

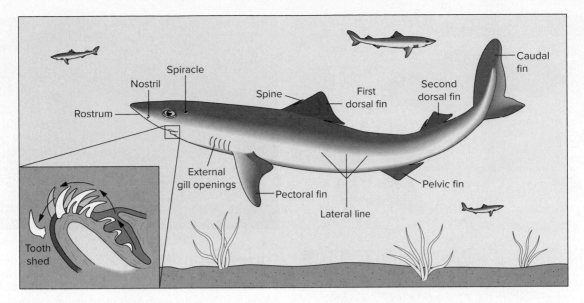

**Figure 40.19**   Sharks (class Chondrichthyes). Dogfish shark, *Squalus acanthias*. Section of lower jaw (inset) shows new teeth developing inside the jaw. These teeth move forward to replace lost teeth. The rate of replacement varies in different species.

## Class Actinopterygii (Ray-finned Fish)

> ### *Procedure 40.3*   Examine the anatomy of a bony fish
>
> *1.* Examine the external anatomy of a preserved fish (fig. 40.20).
> *2.* Although sharks are built for speed, the maneuverability of bony fishes is much greater. If living fish are available, observe their swimming and "breathing."
> *3.* Examine a dissected perch and locate the structures shown in figure 40.20.

Ray-finned or "bony" fish are the most diverse class of vertebrates (30,000 species). Advanced features of ray-finned fish include a bony endoskeleton, modified gill arches, and internal air bladders for balance and buoyancy. Gills are protected by a movable gill cover called an **operculum.** Along each side and branching over the head of most fishes is a **lateral-line system** consisting of sensory pits in the skin. These pits detect water currents and predators or prey that may be moving near the fish.

**Question 11**

*a.* How do the number and shape of fins of a ray-finned fish differ from those of a shark?

*b.* Fins of a ray-finned fish are flexible and diverse in shape. Describe the location of a fin present in ray-finned fish but not in sharks.

*c.* How does the symmetry of the tail of a fish compare with that of a shark?

*d.* Does most of the power for movement by a fish come from the tail or from other fins?

*e.* Can fish move water over their gills without moving through the water? What role does the operculum play in this movement?

*f.* How does the buoyancy of an air bladder affect the motion of a fish compared to that of a shark?

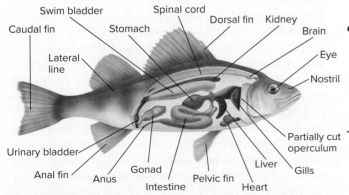

**Figure 40.20**   Anatomy of a bony fish, class Actinopterygii.

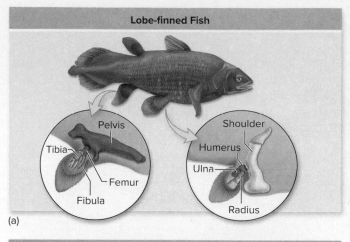

Lobe-finned Fish

Pelvis
Tibia
Femur
Fibula

Shoulder
Humerus
Ulna
Radius

(a)

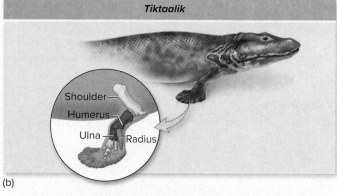

*Tiktaalik*

Shoulder
Humerus
Ulna
Radius

(b)

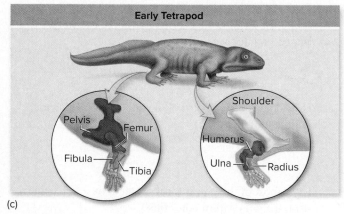

Early Tetrapod

Pelvis
Femur
Fibula
Tibia

Shoulder
Humerus
Ulna
Radius

(c)

**Figure 40.21** Transitional forms in the tetrapod lineage. This figure shows two early tetrapod ancestors, a lobe-finned fish and the transitional form *Tiktaalik roseae,* as well as a descendant, an early amphibian. Skeletal modifications show ancestral continuity as well as adaptation to brief periods out of the water by early amphibians.

Recent fossil evidence strongly supports our understanding of the evolutionary transition of ancestral bony fish to a tetrapod body plan of amphibians (fig. 40.21).

## Class Amphibia (Frogs, Toads, and Salamanders)

Examine preserved amphibians on display. Amphibians were the first land vertebrates, arising from fish with stout, fleshy fins. Most amphibian adults are terrestrial, but they

BiologyImaging.com

**Figure 40.22** A frog (class Amphibia). This poison arrow tree frog (*Dendrobates* sp.) exhibits strong coloration. These colors advertise its powerfully toxic secretions to predators, which quickly learn the frog's noxious taste. Natives of South America use the frog's toxins as a weapon; they kill the frog by piercing it with a sharp stick and holding it over a fire. The heat causes the cutaneous glands to secrete drops of venom, which are scraped into a container and allowed to ferment. Arrows dipped into the poison can paralyze birds and small monkeys. One poison arrow tree frog contains enough toxin to kill about 20,000 mice.

lay eggs in water (fig. 40.22). The eggs are fertilized externally and each hatches into an aquatic larval stage called a **tadpole.** Tadpoles undergo a dramatic metamorphosis of body shape as they become adults.

Development of legs and the development of lungs in amphibians were major evolutionary events. However, primitive lungs had already developed in some fish. In addition to lungs, the soft moist skin of some amphibians is highly vascularized and accounts for as much oxygen diffusion as the lungs.

### Question 12
How are the legs of a frog different from the fins of a fish to enable movement on land?

## Class Reptilia (Turtles, Snakes, and Lizards)

Examine preserved reptiles and note their morphological diversity (fig. 40.23). Reptiles, unlike their ancestors, are independent of aquatic environments and have developed structures for internal fertilization (fig. 40.24). Most reptiles also lay watertight eggs that contain a food source (the yolk) and a series of four membranes—the chorion, the amnion, the yolk sac, and the allantois (fig. 40.25). Each membrane plays a role in making the egg an independent life-support system. The outermost membrane of the egg, the **chorion,** allows oxygen to enter the porous shell but retains water within the egg. The **amnion** encases the developing embryo within a fluid-filled cavity. The **yolk sac** provides food from the yolk for the embryo via blood vessels connecting to the embryo's gut. The **allantois** surrounds a cavity into which waste products from the embryo are excreted.

BiologyImaging.com

**Figure 40.23** Pit vipers are venomous reptiles; they have a pair of heat-detecting pit organs, one on each side of the head. Pit organs are visible between the eye and the nostril of this copperhead snake. These vipers can locate and strike a motionless warm animal in total darkness by sensing heat from its body. Pit organs are highly sensitive to infrared wavelengths and are especially sensitive to sudden changes of temperature. Pit organs can detect temperature differences of 0.2°C or less, allowing effective hunting of small animals at night.

Reptiles have a dry skin covered with scales that retard water loss. This dry skin does not aid respiration, but the lungs are well developed. Reptiles, fish, and amphibians are *poikilothermic,* meaning that their body temperature depends on the environment.

## Question 13

*a.* What is the adaptive significance of internal fertilization and a watertight egg?

BiologyImaging.com

**Figure 40.24** Internal fertilization. The male injects sperm-containing semen into the female's body during copulation. Ancestors of reptiles such as these lizards were the first terrestrial vertebrates to develop this form of reproduction, which is particularly suited to terrestrial existence.

*b.* How do the legs of different reptiles vary in number, size, and function?

*c.* Would you expect the legs of a terrestrial tetrapod to be more robust than those of an aquatic organism? Why or why not? Is this true for the reptiles and amphibians that you examined?

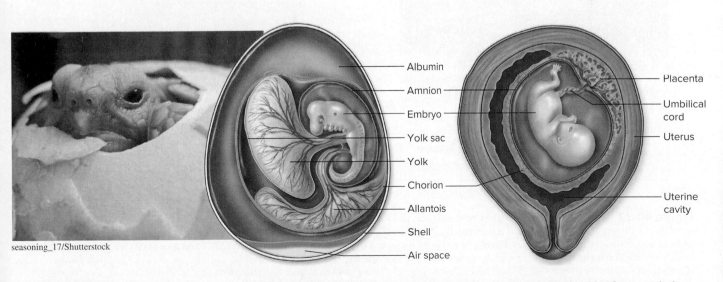

seasoning_17/Shutterstock

**Figure 40.25** The watertight amniotic egg enables reptiles to live in a wide variety of habitats. This sea turtle is hatching from a typical reptilian egg. The amnion of vertebrate eggs is a sac that encloses the developing embryo of a reptile, bird, or mammal. In an amniotic egg, the embryo is encased in a hard, protective shell, and is supported internally by three membranes—the amnion, allantois, and chorion. Placental mammals also enclose their embryos in an amnion.

**d.** Consider Learning Objective 1 listed at the beginning of this exercise. How could poikilothermy contribute to the evolutionary success of reptiles in their environment?

## Class Aves (Birds)

Examine a prepared slide and whole mount of a feather (fig. 40.26). Notice the interlocking structures. Also examine specimens of birds (fig. 40.27). Birds are the only animals with feathers, and they share the ability to fly with

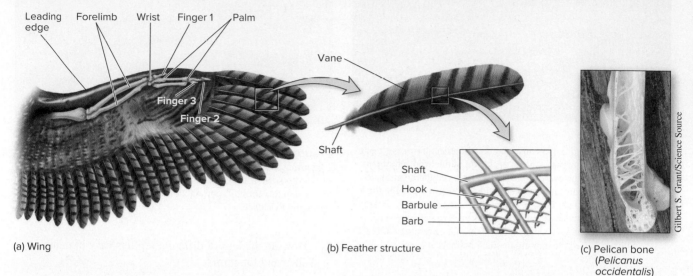

(a) Wing

(b) Feather structure

Shaft
Hook
Barbule
Barb

(c) Pelican bone (*Pelicanus occidentalis*)

Gilbert S. Grant/Science Source

**Figure 40.26** Features of the bird wing and feather. (*a*) The wing is supported by an elongated and modified forelimb with extended fingers. (*b*) Each feather has a hollow shaft that supports many barbs, which in turn support barbules that interlock with hooks to give the feather its form. (*c*) The bones of a pelican (*Pelicanus occidentalis*) are hollow but crisscrossed with a honeycomb structure that provides added strength (8×).

(a)                                              BiologyImaging.com

(b)                                              kojihirano/iStock 360/Getty Images

**Figure 40.27** Birds (class Aves). (*a*) The flightless cormorant, *Phalacrocorax harrisi,* lives only on the Galápagos islands and is the only cormorant in the world (out of 30 species) that cannot fly. Ancestors arriving at the islands had no predators and little competition for their feeding niche of bottom-fish, eels, and octopuses. Flight was no longer adaptive. Over time, natural selection favored a streamlined body and strong legs for swimming. Flight muscles atrophied through the generations, and their sparsely feathered wings have become vestigial. (*b*) The California condor (*Gymnogyps californianus*) is the largest land bird in North America. Young condors acquire full adult plumage after 6 years and may live 50 years. They are efficiently adapted to soaring effortlessly in search of carrion. Their bald heads are adapted for reaching deep within the carcass and tearing pieces of meat. Unfortunately, they are in danger of becoming extinct. The remaining three or four wild individuals were captured in 1987. Offspring have been raised in captivity and have been periodically reintroduced into their dwindling habitat. Even efficient survival adaptations of the condor have not prevented a dramatic population decline. Condors are extremely sensitive to human disturbances, and we have steadily encroached on their habitat.

only a few groups. Eyes of birds are always prominent, and vision is one of their most highly developed senses. Birds are *homeothermic,* meaning that they maintain a constant body temperature. Other adaptations to flight include a high body temperature for high metabolism, a lightweight skeleton, an efficient respiratory system, and heavy musculature at the breast to move the wings. The evolutionary origin of feathers is hotly debated, but biologists agree that their lineage likely traces back to dinosaur ancestry (fig. 40.28).

**Question 14**

*a.* What are wings of flying animals other than birds made of?

*b.* Why might birds use keen vision more than reptiles or amphibians do?

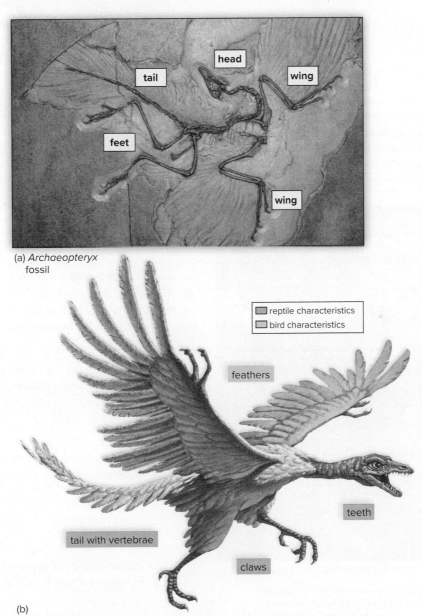

(a) *Archaeopteryx* fossil

(b)

(a) Jason Edwards/National Geographic/Getty Images; (b) Joe Tucciarone, Interstellar Illustrations

**Figure 40.28** Transitional fossils. (*a*) *Archaeopteryx* was a transitional link between dinosaurs and birds. Fossils indicate that it had feathers and wings with claws, and teeth. Most likely, it was a poor flier. (*b*) *Archaeopteryx* also had a feather-covered, bony reptilian-type tail that shows up well in this artist's representation.

*c.* Consider what you have learned about enzymes in
Exercise 11. What might be the adaptive advantage of
homeothermy?

*d.* Describe six adaptations of birds to flight.

## Class Mammalia

Examine some preserved mammals. Mammals are covered
with insulating body fat and **hair** and maintain a constant
body temperature as birds do (fig. 40.29). Mammals are active
and have a well-developed circulatory system with a four-
chambered heart. The circulatory system distributes oxygen,
nutrients, and heat. Mammals nourish their young with milk
produced by the mother's **mammary glands** (fig. 40.30).

Although you are already familiar with the external
anatomy of *Homo sapiens*, you will study the anatomy of a

Magdalena Biskup Travel Photography/Getty Images

**Figure 40.29** This tarsier (*Tarsius syrichia*) is a primitive
primate (class Mammalia) found in the Philippines. It is the size of
a rat, lives in a tree, and eats insects. The position of its eyes in the
front of its head allows full stereoscopic vision. It has nails instead of
claws, which indicates a common ancestry with higher primates. Its
large eyes are efficient adaptations for nocturnal activity. The retinas
lack cones for detecting color but are extra rich in rods for black/white
sensitivity.

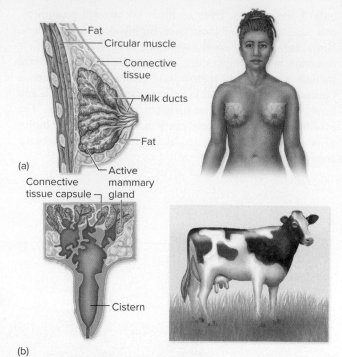

**Figure 40.30** Mammary glands are specialized to secrete milk
following the birth of young. (*a*) Many ducts lead from the glands to
a nipple. Parts of the duct system are enlarged to store milk. Suckling
by an infant initiates a hormonal response that causes the mammary
glands to release milk. (*b*) Some mammals (e.g., cattle) have teats
formed by the extension of a collar of skin around the opening of
mammary ducts. Milk collects in a large cistern prior to its release.

rat, another representative mammal, in Exercises 47, 48, and
49. As you examine preserved and living mammals, search
for common features such as hair distribution, body orientation, and structures for locomotion.

### Question 15

*a.* What factors govern the distribution of hair on
species such as the human or rat?

*b.* How do the mammals that you are examining vary in
body orientation (resting stance and position during
movement)?

*c.* What characteristics of mammals help explain how
they can occupy a variety of habitats?

Examine other preserved chordates and review in your
textbook the major characteristics of each chordate group.
Then complete table 40.3.

**Table 40.3**

**A Comparison of the Characteristics of the Major Chordate Groups**

| | Descriptive Characteristics | | | | | |
|---|---|---|---|---|---|---|
| | Appendages description | Feeding methods | Skeletal elements | Diagnostic features | Gill slit development | Notochord/ spine condition |
| Tunicates | | | | | | |
| Lancelets | | | | | | |
| Lampreys | | | | | | |
| Sharks | | | | | | |
| Bony fish | | | | | | |
| Amphibians | | | | | | |
| Reptiles | | | | | | |
| Avians | | | | | | |
| Mammals | | | | | | |

## INQUIRY-BASED LEARNING

*Would you argue that skeletons of all major groups of vertebrates are similar? Different?*

Observations: Natural selection shapes available genetic variation into adaptations that boost fitness. Over many generations, characteristics with no adaptive advantage for survival and reproduction may decrease in frequency, while those that confer reproductive advantages become increasingly frequent. External features interface an organism with its environment and are subject to strong selective pressures. External features serving multiple functions also vary a great deal among the various classes of vertebrates.

Question: To what extent are vertebrate external morphologies adapted for various functions?

a. Establish a working lab group and obtain Inquiry-Based Learning Worksheet 40 from your instructor.
b. Obtain and examine the external features of a preserved fish, amphibian, bird, reptile, and mammal.
c. Discuss with your group and instructor specific comparisons to make among these vertebrates. Record them on Worksheet 40.
d. The table on Worksheet 40 lists four broad functions of an adaptation. Can you think of others? List at least two other functions in the worksheet table.
e. Complete the investigation directed by Worksheet 40.

## Questions for Further Study and Inquiry

*1.* Does it surprise you that echinoderms are more closely related to our own phylum (Chordata) than are other phyla? Why would you have thought otherwise?

*2.* Why are embryological features important for distinguishing the major groups of phyla?

*3.* Echinoderms lack cephalization. What characteristics of this group deemphasize the need for a head?

**4.** What problems were associated with colonizing land during the evolution of vertebrates?

**5.** Why do you suppose four rather than five or six appendages is the rule for vertebrates?

**6.** A cuticle occurs on the surface of organisms of many phyla and appears to be an advantageous feature. Why have higher organisms not retained this structure?

**7.** What is the difference in the developmental derivation of mandibles among insects, jaws of vertebrates, and the beak of an octopus?

**8.** Although other groups of vertebrates are more numerous and have existed longer than mammals, mammals are often called the most advanced form of life. Why?

**9.** Locomotion in mammals is varied. Do you believe that their powers of locomotion are superior to those of birds? Why or why not?

**10.** Compare the origin and function of reptile scales, bird feathers, and mammal hair. How are they similar? How do they differ?

**11.** Prepare a simple table of all of the taxonomic groups, their common names, their distinguishing characteristics, and all representative genera covered in this exercise. Keep this table with your study notes.

**12.** Vertebrates have a closed circulatory system, meaning that the blood is always enclosed within vessels and does not fill body cavities. Mollusks (Exercise 38) and arthropods (Exercise 39) have open circulatory systems, meaning that blood is pumped by a heart into body cavities, where tissues are surrounded by the blood. What are the advantages and disadvantages of each type of circulatory system?

**13.** Two classes of vertebrates (Aves and Mammalia) are *endothermic*. What is meant by this term? Hypothesize some evolutionary advantages of being endothermic. What are some of the costs?

## WRITING TO LEARN BIOLOGY
What external anatomical features of amphibians are associated with their dual life on land and in water?

# Vertebrate Animal Tissues

## Epithelial, Connective, Muscular, and Nervous Tissues

## Learning Objectives

By the end of this exercise you should be able to:

1. Understand the general classification scheme for vertebrate tissues.
2. List examples, functions, and distinguishing features of each type of tissue.
3. Associate structure with function for each type of tissue that you examine.

Please visit **connect.mheducation.com** to review online resources tailored to this lab.

**C**ells with similar structure and function constitute a **tissue,** such as muscle tissue or nervous tissue. Tissues and their functions integrate to form **organs.** Organs are structures such as the stomach, lungs, and liver composed of several different tissues grouped together and having an integrated function. Organs work together as **systems,** such as the respiratory system or digestive system (fig. 41.1). Thus, we can define an organism at various levels of biological organization. At the cellular level, vertebrates contain between 50 and several hundred different kinds of cells, depending on how finely you differentiate cell types. These diverse cells are traditionally grouped into four tissue types, based on structure and function: **epithelial, connective, muscular,** and **nervous tissues** (fig. 41.2). The study of the microscopic anatomy of tissues is called **histology.**

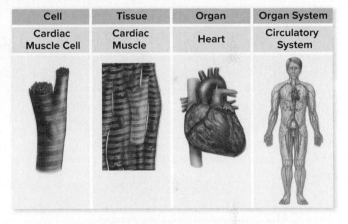

| Cell | Tissue | Organ | Organ System |
|------|--------|-------|--------------|
| Cardiac Muscle Cell | Cardiac Muscle | Heart | Circulatory System |

**Figure 41.1** Levels of organization within the body. Similar cell types operate together and form tissues. Tissues functioning together form organs. Several organs working together to carry out a function for the body are called an organ system. The circulatory system is an example of an organ system.

## EPITHELIAL TISSUE

Epithelial cells protect the body. They cover the exterior of an organism, line the gut and other cavities, and line the coelomic cavity. Specifically, epithelial cells (1) protect underlying tissues from dehydration and mechanical damage, (2) provide a selectively permeable barrier that facilitates or impedes passage of materials, (3) provide sensory surfaces, and (4) secrete fluids.

Epithelial cells are often classified by their layers and shape. **Simple** refers to a tissue that occurs in one layer. **Stratified** refers to multiple layers. **Pseudostratified** refers to a single layer of cells that appear stratified because the nuclei appear in different positions within the columnar cells (see fig. 41.6). **Squamous** cells are flat like fried eggs, **cuboidal** cells are shaped roughly like cubes, and **columnar** cells are tall and narrow. You will examine three classes of epithelial tissue in lab: **simple epithelium,**

**stratified epithelium,** and **glandular epithelium.** As you view each of the slides, summarize what you've learned in table 41.1 at the end of this exercise.

### Simple Epithelium

Simple epithelial tissues are a single cell layer thick and are classified according to the shapes of their cells (fig. 41.3).

- **Squamous** epithelial cells are irregular and flattened. Thus, one cell layer of simple squamous epithelium is a minimal barrier to diffusion. Squamous cells line the alveoli of the lungs, the filtration system of the kidneys, and the major cavities of the body. These cells are relatively inactive and are associated with sites of passive movement of water, electrolytes, and other substances.

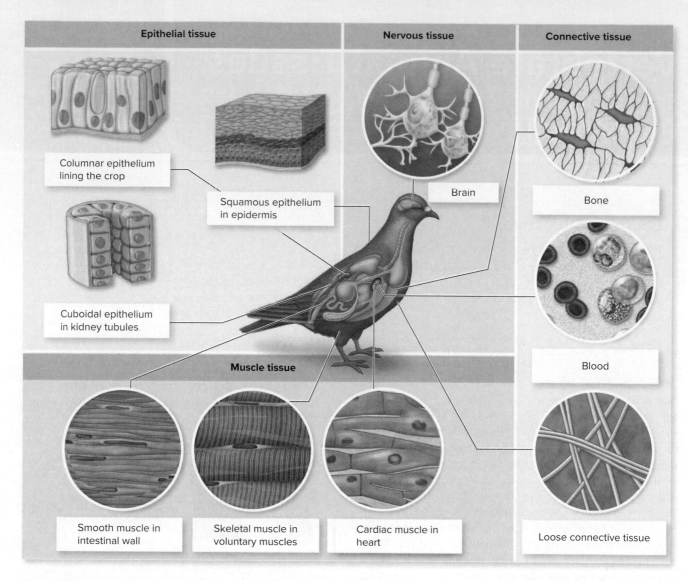

**Figure 41.2** The types of tissues in vertebrates.

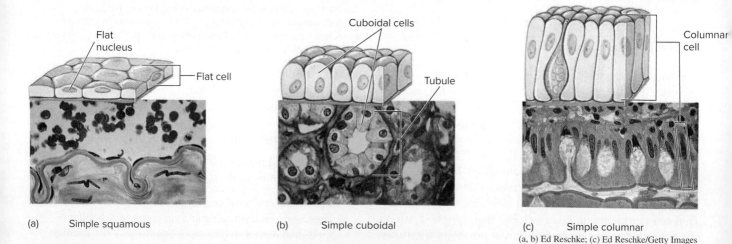

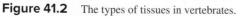

(a)  Simple squamous      (b)  Simple cuboidal      (c)  Simple columnar

(a, b) Ed Reschke; (c) Ed Reschke/Getty Images

**Figure 41.3** Types of epithelial tissue, based on shape. (*a*) Squamous epithelium lines the artery shown here (200×). The nuclei are flat. The round cells above the epithelium are blood cells in the interior of the artery (also see fig. 41.4). (*b*) Cuboidal epithelium forms the walls of these kidney tubules, seen in cross section (200×) (also see fig. 41.4). (*c*) Columnar epithelium forms the outer cell layer of this human intestine. Interspersed among the epithelial cells are goblet cells, which secrete mucus. They are clear in most slide preparations (200×) (also see fig. 41.5).

- **Cuboidal** and **columnar** epithelial cells appear fuller than do squamous cells and are shaped as their names imply. They line the respiratory and intestinal tracts and ducts such as kidney tubules. Cuboidal and columnar epithelial cells often have cilia and secrete fluids.

Let's examine some of your living epithelial cells.

## Procedure 41.1  Examine squamous and cuboidal epithelial cells

1. Discuss with your instructor the safety issues associated with this procedure.

2. Gently scrape the inside of your cheek with the tip of a clean toothpick and stir the tip in a small drop of water on a microscope slide.

3. Lay a coverslip over the preparation and place a small drop of methylene blue at the edge of the coverslip. This stain will add color and contrast to the cells.

 Be careful not to spill methylene blue—it will stain your skin and clothes! Do not scrape your cheek forcefully. Be gentle.

4. If the stain does not readily diffuse under the coverslip, pull the fluid under by touching a dry paper towel to the opposite edge of the coverslip.

5. Examine your cells under low, then high, magnification.

6. The cells may be clumped, so scan the slide to find cells that have floated free.

7. Sketch a few cells.

## Question 1
Which cell structures can you identify?

8. Examine a prepared slide of a cross section of a kidney. Simple squamous and cuboidal cells are both common in vertebrate kidneys.

9. Refer to figure 41.4. Simple squamous epithelium surrounds Bowman's capsules, and simple cuboidal cells compose kidney tubules. The wall of a tubule cut in cross section will appear as a ring or rough circle of cells.

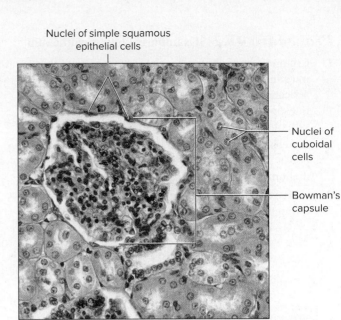

Nuclei of simple squamous epithelial cells

Nuclei of cuboidal cells

Bowman's capsule

BiologyImaging.com

**Figure 41.4**  Kidney tissue showing simple squamous and cuboidal epithelia (200×). These cuboidal cells form the walls of kidney tubules (see fig. 41.3b). The squamous cells are seen on edge because they line a narrow cavity.

## Question 2
*a.* How thin are the cells, and how does this relate to function?

*b.* What term describes the shape of cheek cells?

*c.* What is the approximate diameter (in micrometers) of these epithelial cells? If necessary, refer to the exercise on microscopy (Exercise 3) for instructions on measuring cells.

*d.* How does a tissue differ from a cell? From an organ?

*e.* The cuboidal cells in figure 41.4 are arranged in circles. How can these circles be interpreted as tubes?

## Procedure 41.2   Examine columnar epithelium

1. Columnar epithelium lines the inner surface of the intestinal tract and trachea. Examine under low magnification a prepared slide of a cross section of small intestine and locate the relatively large, fingerlike villi of the inner intestinal wall.

2. Increase the magnification to medium, then high power, focusing on the single layer of columnar epithelium covering the villi (fig. 41.5).

3. Also examine a prepared slide of a cross section through a trachea (fig. 41.6).

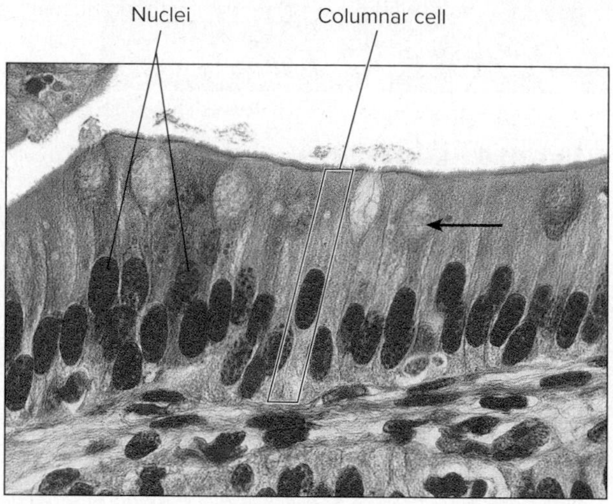

### Question 3

*a.* Are epithelial cells of the trachea similar in size and structure to those lining the intestine?

*b.* What is the approximate ratio of length to width for epithelial cells of the trachea?

*c.* Tracheal cells have many small, hairlike projections that move and sweep debris from the surface of the trachea. What are these projections called?

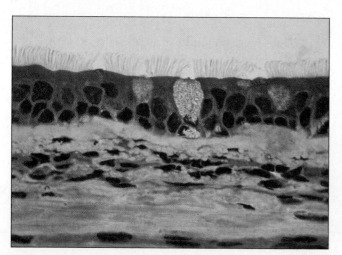

Nuclei    Columnar cell

BiologyImaging.com

**Figure 41.5**   Simple columnar epithelium of frog intestine consists of a single layer of elongated cells. The arrow points to a specialized goblet cell that secretes mucus (400×).

## Stratified Epithelium

Examine a slide of a cross section of skin and locate the stratified epithelium (fig. 41.7). **Stratified** tissues are several layers thick. Typically, the upper layer is squamous, the middle one cuboidal, and the basal (bottom) layer

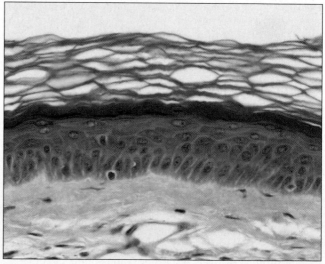

Lutz Slomianka

**Figure 41.7**   Stratified squamous epithelium, such as this cross section of skin, consists of many layers of cells (120×). The surface of the skin is toward the top.

Ed Reschke

**Figure 41.6**   Pseudostratified ciliated columnar epithelium taken from the trachea. Notice the tuft of cilia at the top of each cell (500×).

columnar. The skin is the most obvious example of stratified epithelium, although "skin" includes tissues other than epithelium.

### Question 4

*a.* Are layers of skin cells distinct, or is there a gradual change in cellular shape from the basal to the surface layers?

*b.* Skin cells produce **keratin,** a strong fibrous protein found in hair and fingernails. What is the function of keratin?

*c.* List several functions of skin that relate to the shape and toughness of these epithelial cells.

### Glandular Epithelium

Some glands of the body consist of highly modified epithelial cells that do not function as a protective covering. These cells are more active metabolically than is simple epithelium. For example, **exocrine glands** are derived from tubular invaginations of epithelial layers and include the liver, mammary glands, and pancreas. Cellular secretions of exocrine glands move to the surface and away from the organ via ducts.

Examine a prepared slide of liver. Liver contains many sinuses that carry blood. Note the large nucleus in each **hepatocyte** (liver cell).

### Question 5

*a.* What fluid do hepatocytes secrete and store in the gallbladder?

*b.* What is the dark spot in the nucleus of a hepatocyte?

## CONNECTIVE TISSUE

Connective tissues support the body, fight pathogens, and store energy. These cells are not tightly packed (as are epithelial cells) and are typically suspended in an extracellular matrix of fibers. Some connective tissues are dispersed and flow in the circulatory system. Classification of connective tissue cells is based as much on function and the nature of the extracellular matrix as on cellular morphology. There are two major classes of connective tissue: (1) **connective tissue proper,** further divided into loose and dense connective tissues with an abundance of fibers, and (2) **special connective tissues,** which include blood, cartilage, and bone, each having a characteristic extracellular matrix.

### Connective Tissue Proper

**Loose connective tissue** consists of cells scattered within an amorphous mass of proteins that form a **ground substance.** Examine a slide of subcutaneous tissue, sometimes called areolar tissue, and note the irregular arrangement of **fibroblasts** and fibers (fig. 41.8). Fibroblasts, which are widely dispersed in vertebrate bodies, are irregular branching

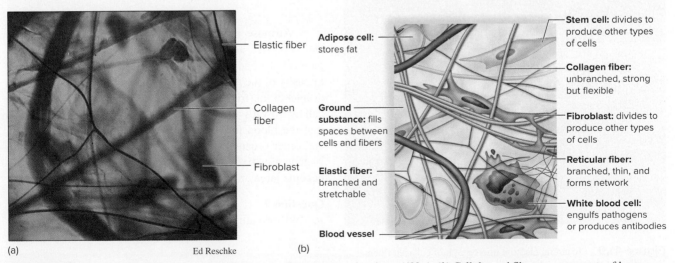

(a)                                    Ed Reschke                    (b)

**Figure 41.8** Loose connective tissue. (*a*) Subcutaneous, fibrous connective tissue (400×). (*b*) Cellular and fibrous components of loose connective tissue.

cells that secrete an extracellular matrix of strong fibrous proteins. The most commonly secreted protein is **collagen,** which represents 25% of all vertebrate protein (fig. 41.9). If all components of the body except collagen were removed, a ghostly mesh of fibers would remain as the framework of the body and its organ systems. Collagen is not the only fiber produced by fibroblasts. **Elastin** fibers have protein molecules arranged so that the fibers can stretch.

Examine a slide of tissue having reticulin fibers taken from a lymph gland (fig. 41.10). **Reticulin** is a protein that forms thin, branching, reticular fibers that support glands such as the spleen and lymph nodes. Reticular fibers also compose junctions between several other kinds of tissues.

Loose connective tissue also includes **macrophages,** the immune system's first defense against invading organisms. Macrophage connective tissue consists of many relatively small, round cells. Macrophages are defensive cells that engulf and digest cellular debris, invading bacteria, and foreign particles (fig. 41.11). Macrophages may move individually in the circulating fluids of the body or remain fixed in an organ such as the liver or spleen. Examine a prepared slide of liver tissue. Macrophages of this tissue may contain small black particles of India ink that they engulfed before the tissue was fixed and preserved.

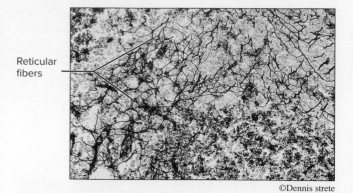

Reticular fibers

©Dennis strete

**Figure 41.10**   Reticular connective tissue in a lymph node (300×).

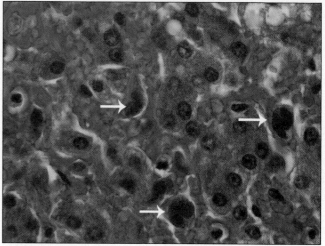

Dennis K. Burns, MD, Travis G. Brown, MD, Walter L. Kemp, MD/McGraw Hill

**Figure 41.11**   Macrophage ingesting cells (400×). Arrows indicate individual macrophages.

### Question 6
Review lysosomes in Exercise 4, The Cell. Why are macrophages rich in lysosomes?

**Adipose cells** are found in loose connective tissue and comprise **adipose tissue** (fig. 41.12). Examine a prepared slide of adipose tissue. For best contrast, keep the light intensity of the microscope low. Each adipose cell contains a droplet of fat (triglycerides). To generate energy, the adipose cell hydrolyzes its stored triglyceride and secretes fatty acids into the blood for oxidation by cells of the muscles, liver, and other organs. The number of adipose cells in an adult is generally fixed. When a person gains weight, the cells become larger, and when weight is lost, the cells shrink.

### Question 7
*a.*   Why do adipose cells appear empty?

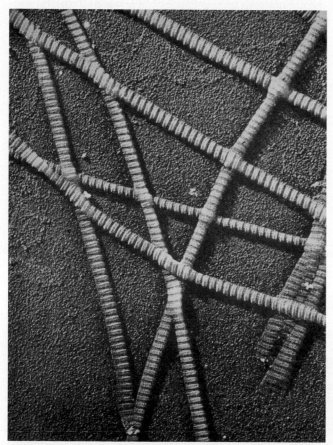

J. Gross/Biozentrum, University of Basel/Science Source

**Figure 41.9**   Scanning electron micrograph of collagen fibers. Each fiber of this structural connective tissue consists of many individual collagen strands and is very strong (5000×).

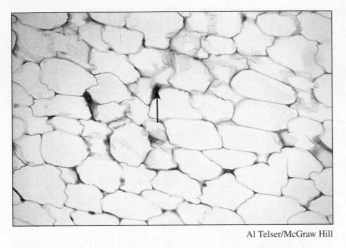

Al Telser/McGraw Hill

**Figure 41.12** Adipose tissue cells contain large droplets of fat that push the nuclei close to the plasma membranes. The arrow points to a nucleus (200×).

*b.* Of what use is the reserve of oil in adipose cells?

**Dense connective tissue** contains tightly packed collagen fibers; these fibers make dense connective tissue stronger than loose connective tissue. Examine a prepared slide of longitudinal section through a tendon (fig. 41.13). Examine a prepared slide of a ligament if one is available. Tendons connect muscle to bone and derive their strength from this regular, longitudinal arrangement of bundles of collagen fibers. Ligaments bind bone to bone and are similar

in structure to tendons. Notice the lack of apparent blood vessels in your prepared slide of tendon. Low vascularization makes tendons and ligaments slow to heal.

**Question 8**
*a.* How can a fibroblast produce a fiber many times its own length?

*b.* In what areas of the body would the elasticity of elastin fibers be advantageous?

*c.* Which tissues in the body require the greatest strength? Explain your answer.

*d.* Are all the fibers in a tendon oriented in the same direction? Of what importance is this?

## Special Connective Tissues

**Blood cells** and their intercellular fluid matrix called **plasma** perform a variety of tasks in the vertebrate body,

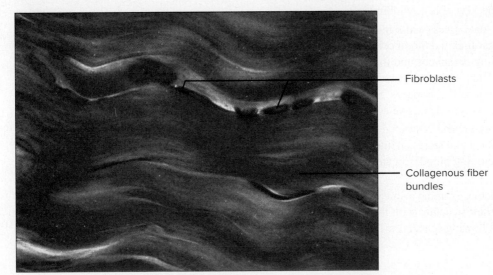

Fibroblasts

Collagenous fiber bundles

Ed Reschke

**Figure 41.13** Tendon. Dense fibrous connective tissue is a strong tissue that forms tendons, which attach muscle to bone. Bundles of collagen fibers are oriented in the same direction to increase strength (400×).

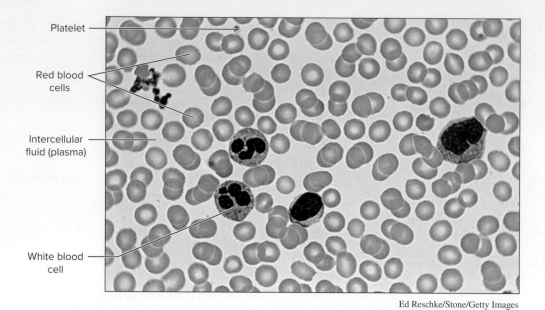

Ed Reschke/Stone/Getty Images

**Figure 41.14** Blood tissue consists of red blood cells, white blood cells, and platelets suspended in an intercellular fluid called plasma (800×).

including maintaining proper pH and transporting oxygen and carbon dioxide. Practically every type of substance used by cells is dissolved in plasma. Blood cells are classified as **erythrocytes** (red blood cells), **leukocytes** (white blood cells), or **platelets** (enucleated fragments of large bone-marrow cells).

## Procedure 41.3   Examine blood components

1. Examine a prepared slide of a human blood smear (fig. 41.14).

2. The most numerous blood cells are erythrocytes, or red blood cells. They produce **hemoglobin** (fig. 41.15), which binds and transports oxygen. Notice the uniform shape of erythrocytes; these cells have lost their nuclei and become packets of hemoglobin.

3. Locate some leukocytes (white blood cells), which are much larger than red blood cells and defend against invading organisms and foreign substances. Inerestingly, leukocytes can leave blood vessels and migrate into intercellular spaces. Some biologists classify leukocytes as defensive connective tissue.

4. Platelets, which appear as small dark fragments, sequester (collect and store) chemicals and enzymes essential for clotting blood. Locate platelets in your blood smear.

5. If one is available, examine a smear of frog blood cells and contrast their structure with that of human blood cells. In the following space, draw some of these frog cells.

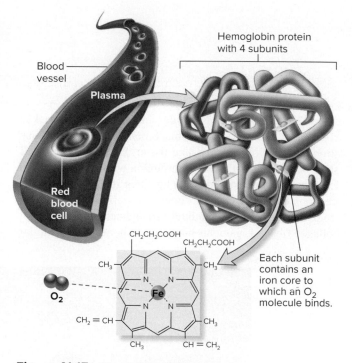

**Figure 41.15** Hemoglobin. Erythrocytes contain large amounts of the protein hemoglobin. Oxygen binds reversibly to iron atoms at the core of each subunit of hemoglobin.

## Question 9

*a.* Is a nucleus visible in each blood cell? Are nuclei apparent in white blood cells?

*b.* The shape of the nucleus is often used to subclassify a leukocyte. What variation do you see in the shape of nuclei in leukocytes?

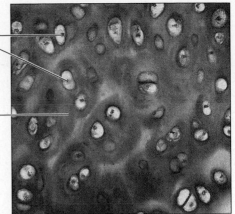

Chondrocytes in lacunae

Chondrin matrix

BiologyImaging.com

*c.* What is the ratio of leukocytes to erythrocytes?

*d.* Do frog blood cells have nuclei?

**Figure 41.16** Hyaline cartilage cells called chondrocytes are located in lacunae surrounded by a chondrin martrix of intercellular material and fine collagenous fibers (250×).

**Cartilage** is found in skeletal joints and derives its resilience and support from an extracellular gelatinous matrix of **chondrin.** Chondrin of cartilage may be impregnated with fibers of collagen. This matrix is secreted by cells called **chondrocytes.** As in most connective tissue, cells of cartilage are rather isolated within the extracellular matrix. Chondrocytes in cartilage reside in cavities called **lacunae.**

Examine a slide of hyaline cartilage, which cushions bone surfaces between joints (fig. 41.16). Also examine a slide of elastic cartilage, which commonly occurs in the external ear and in the epiglottis of the voice box (larynx) (fig. 41.17). Elastic cartilage is more flexible than is hyaline cartilage and is rich in fibers.

**Question 10**
What are some general characteristics of chondrin that make it adaptive to the function of cartilage?

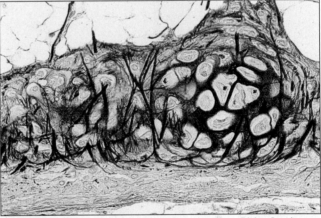

BiologyImaging.com

**Figure 41.17** Elastic cartilage contains fine collagenous fibers and many dark elastic fibers in its intercellular material (100×).

**Question 11**
Why is an elaborate system of canals needed in bone more than in cartilage?

**Bone** and its properties are also derived from a strong extracellular matrix with fibers (fig. 41.18). Collagen fibers of bone are also surrounded by hard crystals of calcium salts rather than the flexible matrix of chondrin in cartilage. This fibrous and crystalline matrix is maintained by bone cells called **osteocytes.**

Examine a prepared slide of bone. Bone forms in thin concentric layers called **lamellae** forming **osteons,** sometimes called **Haversian systems.** Lamellae form a series of tubes around narrow channels called **central canals,** which align parallel to the long axis of the bone. Central canals surround blood vessels and nerve cells throughout bone and communicate with bone cells in lacunae through **canaliculi.**

## MUSCLE TISSUE

The distinctive feature of muscle is its ability to contract, which results from the interaction of **actin** and **myosin** filaments. These proteins occur in other eukaryotic cells but not in such abundance and uniform orientation. Bundles of these contractile filaments, called **myofibrils,** occur within

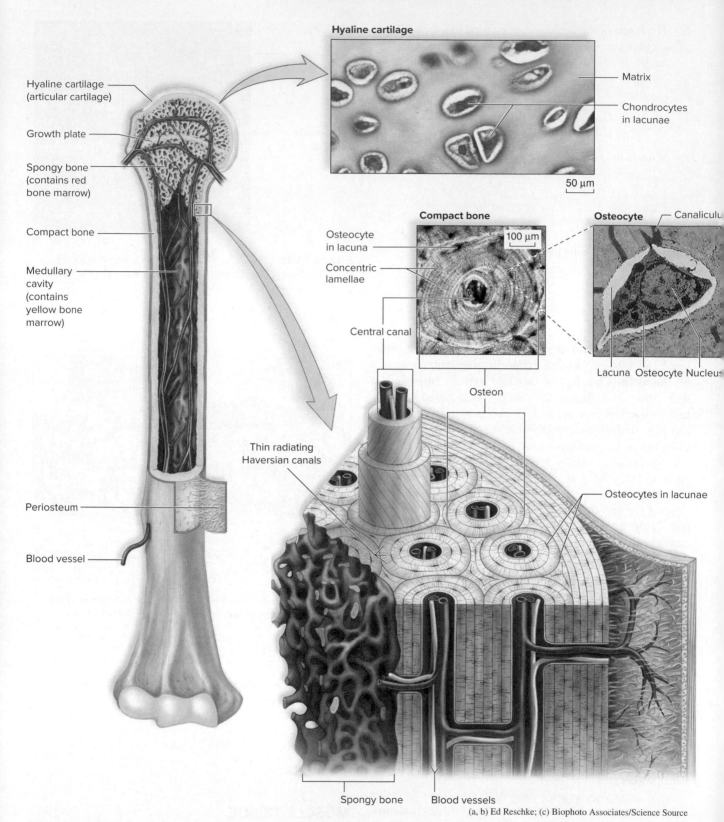

**Hyaline cartilage**

Matrix

Chondrocytes
in lacunae

50 µm

Hyaline cartilage
(articular cartilage)

Growth plate

Spongy bone
(contains red
bone marrow)

Compact bone

Medullary
cavity
(contains
yellow bone
marrow)

Periosteum

Blood vessel

**Compact bone**

Osteocyte
in lacuna

Concentric
lamellae

Central canal

100 µm

**Osteocyte**

Canaliculu

Lacuna   Osteocyte Nucleus

Osteon

Thin radiating
Haversian canals

Osteocytes in lacunae

Spongy bone

Blood vessels

(a, b) Ed Reschke; (c) Biophoto Associates/Science Source

**Figure 41.18**   Anatomy of a long bone. Bone shown at three levels of detail. Some parts of bones are dense and compact, providing great strength. Other parts, such as marrow, are spongy and have a more open lattice. Most red blood cells are formed in marrow. New bone is formed by cells called osteocytes, which secrete collagen fibers as sites for deposition of hard calcium-phosphate crystals. Bone is deposited in thin, concentric layers called lamellae. Lamellae form a series of tubes around narrow channels called central canals, which run parallel to the length of the bone. Central canals are interconnected and contain nerves and blood vessels.

a single muscle cell, and their uniform contraction produces considerable force and movement (fig. 41.19). Vertebrates have three kinds of muscle: **skeletal, smooth,** and **cardiac** (fig. 41.20).

## Skeletal Muscle

Review in your textbook how skeletal muscle contracts. Then examine a prepared slide of skeletal muscle (fig. 41.20a). Skeletal (striated) muscles are attached to the skeleton and are controlled voluntarily (i.e., they

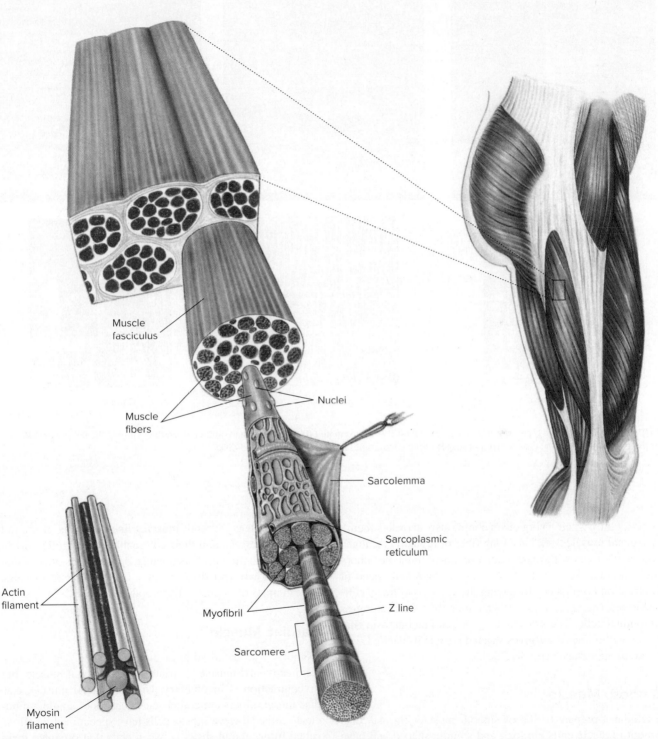

**Figure 41.19**   Striated muscle. Regular bands of actin and myosin are microfilaments arranged as sarcomeres that slide along their lengths and contract the muscle. Units of these microfilaments form myofibrils, which are bundled into muscle fibers. These fibers are surrounded by a membranous sarcolemma having many nuclei scattered along the length of the fiber. The sarcoplasmic reticulum is a membranous network that distributes the stimulus for contraction along the surface of the muscle fiber.

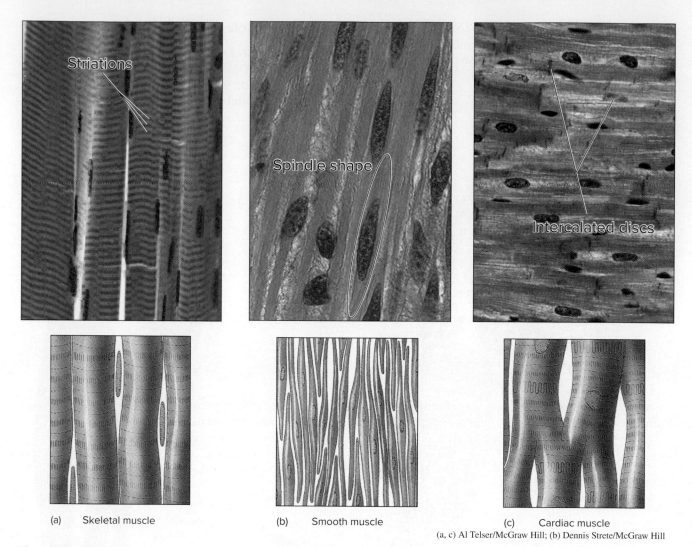

| (a) | Skeletal muscle | (b) | Smooth muscle | (c) | Cardiac muscle |

(a, c) Al Telser/McGraw Hill; (b) Dennis Strete/McGraw Hill

**Figure 41.20** Three types of muscle. (*a*) Skeletal (400×). (*b*) Smooth (400×). (*c*) Cardiac (400×). In cardiac muscle, the larger wave lines of the illustration (and the dark horizontal stripes of the photograph) indicate the intercalated disks.

contract only when an organism consciously activates them). A skeletal muscle "cell" is a long fiber of regularly arranged contractile units with many nuclei scattered at the periphery (outer boundaries) of the fiber. The strength and speed of contraction is enhanced by having the contents of many cells coalesced (merged) into a fiber rather than functioning as individual cells. The stacked array of actin and myosin filaments within the fibers gives striated muscle its banded (i.e., striated) appearance (fig. 41.20*a*).

## Smooth Muscle

Examine a prepared slide of smooth muscle (fig. 41.20*b*). Smooth-muscle cells are long and spindle-shaped and have a single nucleus. Smooth muscles line the walls of the gut and blood vessels, and their contraction is controlled involuntarily. Smooth muscle is organized into sheets of cells that contract slowly and rhythmically. The uterus and intestine are examples of organs with smooth muscle.

## Cardiac Muscle

Examine a prepared slide of cardiac muscle (fig. 41.20*c*). Cardiac (heart) muscle is striated, as is skeletal muscle, but its contraction is involuntary. Unlike skeletal muscle, cardiac muscle fibers are composed of chains of single, uninucleate cells. However, these cells have specialized junctions called **intercalated disks** between cells that organize them

Axon

Cell body

Dendrites

Steve Gschmeissner/Science Photo Library/Alamy Stock Photo

**Figure 41.21**  Anatomy of a neuron. Neurons are specialized to transmit nerve impulses. A neuron axon receives nerve impulses from other neurons, whereas a dendrite transmits the nerve impulse to subsequent neurons (600×).

*c.* Of the three types of muscle, which contracts without voluntary thought?

into rather continuous functional fibers similar to those of skeletal muscle. Thus, cardiac cells depolarize and contract more as a unit than do sheets of loosely associated cells of smooth muscle.

### Question 12

*a.* Can you distinguish the small striations perpendicular to the axis of a muscle cell?

*b.* Of the three types of muscle, which contracts most swiftly?

## NERVOUS TISSUE

The fourth major class of vertebrate tissue is nervous tissue. Nervous tissue consists of (1) **neurons,** cells specialized for transmitting nerve impulses, and (2) supporting cells called **glia,** including **Schwann cells,** which help propagate the nerve impulse and provide nutrients to neurons.

| Table 41.1 | | | | | |
|---|---|---|---|---|---|
| **A Comparison and Organization of Vertebrate Animal Tissues** | | | | | |
| Tissue Examined | Location in the Vertebrate Body | Function | Tissue Examined | Location in the Vertebrate Body | Function |
|  |  |  |  |  |  |
|  |  |  |  |  |  |
|  |  |  |  |  |  |
|  |  |  |  |  |  |
|  |  |  |  |  |  |
|  |  |  |  |  |  |
|  |  |  |  |  |  |
|  |  |  |  |  |  |

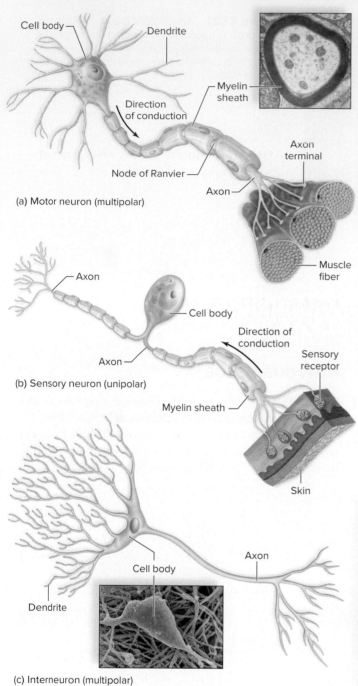

(a) Motor neuron (multipolar)

(b) Sensory neuron (unipolar)

(c) Interneuron (multipolar)

(a) Scott Camazine/Alamy Stock Photo; (c) Don W. Fawcett/Science Source

**Figure 41.22** Neuron anatomy. (*a*) Motor neuron. Note the branched dendrites and the single, long axon, which branches only near its tip. (*b*) Sensory neuron with dendrite-like structures projecting from the peripheral end of the axon. (*c*) Interneuron (from the cortex of the cerebellum) with very highly branched dendrites.

Examine a prepared slide of a smear of nervous tissue (figs. 41.21, 41.22). Neurons consist of (1) a **cell body** containing a nucleus and (2) cytoplasmic extensions that conduct nerve impulses. **Dendrites** are short extensions of a neuron that usually carry impulses toward the cell body from other cells or sensory systems. **Axons** are long extensions that usually carry impulses away from the cell body. An axon may carry an impulse to a muscle to make it contract or to the dendrites of another neuron. Because cell bodies occur only in the brain and spinal cord, some axons and dendrites must be a meter long to reach distant parts of the body. Axons and dendrites extending from neurons are often bundled as nerves that look like thin spaghetti among the muscles and organs of the vertebrate body (fig. 41.23).

To review all of the available examples of animal tissues, briefly reexamine each prepared slide. As you view each type of tissue, record in table 41.1 the tissue's location and function.

### Question 13
What is the difference between a nerve, such as that found in an arm or leg, and a neuron?

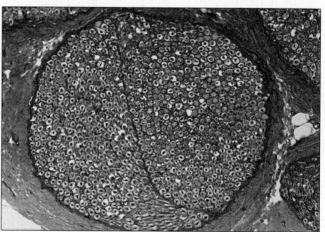

Ed Reschke/Getty Images

**Figure 41.23** Cross section of a nerve. A nerve is a bundle of axons bound together by connective tissue. In this micrograph of a nerve cross section, many myelinated neuron axons are visible, each looking somewhat like a severed hose (400×).

With the right stimulation, **stem cells** can produce a variety of tissues with great therapeutic value. Even a whole organism can be produced from properly treated stem cells! The fate of stem cells is undetermined until their genetic program and an appropriate microenvironment stimulate their differentiation into specialized tissue. But not all stem cells are the same. Human embryonic stem cells are derived from the inner cells of the blastocyst, a developmental stage occurring several days after fertilization. Since these blastomeres will soon begin to specialize, stem cells of therapeutic value must be harvested from early embryos. These **embryonic stem cells** are totipotent (i.e., can form all embryonic and adult tissues), but the bioethics of harvesting them from human embryos is controversial. Alternatively, **adult stem cells** found in areas of rapid cell division (such as bone marrow and reproductive organs) in mature organisms also have great utility, but they do not have the potential to develop into as many different cell types as do embryonic stem cells.

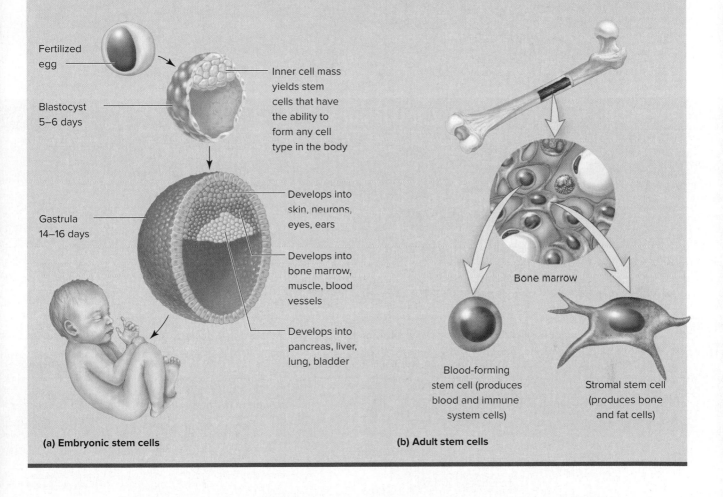

Fertilized egg

Blastocyst 5–6 days

Inner cell mass yields stem cells that have the ability to form any cell type in the body

Gastrula 14–16 days

Develops into skin, neurons, eyes, ears

Develops into bone marrow, muscle, blood vessels

Develops into pancreas, liver, lung, bladder

Bone marrow

Blood-forming stem cell (produces blood and immune system cells)

Stromal stem cell (produces bone and fat cells)

**(a) Embryonic stem cells**

**(b) Adult stem cells**

## INQUIRY-BASED LEARNING

### *Can the visibility of cellular structures be enhanced?*

Observations: Cells are so small and transparent that their structures are difficult to see, even at high magnifications. Stains with dissolved pigments will adhere to some organelles more than others and thereby increase contrast and visibility.

Question: What stains best enhance the visibility of cellular structures?

a. Establish a working lab group and obtain Inquiry-Based Learning Worksheet 41 from your instructor.

b. Discuss with your group a well-defined question relevant to the preceding observations and question. Record your question on Worksheet 41.

c. Ask your instructor which common cellular stains are available in the lab.

d. Outline on Worksheet 41 your experimental design and supplies needed to answer your question. Ask your instructor to review your proposed investigation.

e. Conduct your procedures, record your data, answer your question, and make relevant comments.

f. Discuss with your instructor any revisions to your questions, hypotheses, or procedures. Repeat your work as needed.

## Questions for Further Study and Inquiry

*1.* Why are vertebrate animal tissues difficult to classify into a single consistent system?

*2.* If a compound microscope produces an image in only two dimensions (length and width), how could you determine the three-dimensional shape of a cell?

*3.* How many animal tissues and cell types might be in a typical hamburger?

*4.* All living cells maintain a polarized membrane, meaning that positive and negative ions are separated on either side of the membrane. What role does this polarization play in the function of neurons?

*5.* How does the structure of bone tissue resemble woody eudicots in form and function?

*6.* What aspects of a neuron's structure enhance its function?

*7.* What structural components of cartilage promote its function as a pliable, yet tough, material?

### WRITING TO LEARN BIOLOGY

Does the presence of nuclei in frog blood cells indicate that the lack of nuclei in mammalian blood cells is an "advanced" characteristic? Why or why not?

# Human Biology
## The Human Skeletal System

## Learning Objectives

By the end of this exercise you should be able to:
1. Identify the major bones of the human skeleton and the different types of joints.
2. Understand how bones are held together.
3. Understand what bones form structures such as the elbow and knuckles.
4. Understand human morphology and how age influences bone structure.

Please visit **connect.mheducation.com** to review online resources tailored to this lab.

The human body consists of 206 bones that make up about 15% of our body-weight. Most bones are parts of larger structures and systems; for example, the skull has 22 bones, the vertebral column 26 bones, and the ribcage 24 bones.

The skeletal system, along with the muscular system (see Exercise 43), determines the shape of an organism, supports other organs, and allows for movement. Bones are the main structural element of the skeletal system. Bones are held together by **ligaments,** which are made of dense connective tissue that is slightly elastic. Bones are attached to muscles by **tendons.** Tendons have little elasticity.

In this exercise, you will study bones of the human body. Before you begin this exercise, review the structure of bone tissue in Exercise 41.

### Procedure 42.1 What bones compose the human body?

1. Examine the articulated skeleton in the lab. The human skeleton consists of the axial skeleton (skull, vertebrae, sternum, and ribs) and appendicular skeleton (shoulder, arm, hip, and legs) (fig. 42.1).

2. Use your textbook or other books in lab to identify the bones of these parts of the skeleton. Note the geometry of the skeleton, paying particular attention to the shapes of the bones, the textures of the bones, and the planes in which the joints can move (figs. 42.2, 42.3).

3. Identify and label the bones in figure 42.4. For each bone, list whether it is part of the axial or appendicular skeleton. Also identify these bones in your body.

### The Skeletal System by the Numbers

At birth, our bodies have more than 270 bones. During development, however, several of these bones fuse; that's why adults have only 206 bones. Of course, there is always phenotypic variation from one human to the next.

In adults, the skeleton accounts for 15% of our body weight.

The smallest bone in humans is the stapes, which is a stirrup-shaped bone in our ears that is critical for hearing. The stapes is approximately 3 mm × 2.5 mm. Unlike other bones, the stapes is full size at birth.

We have 12 pairs of ribs. Approximately 4% of people have an extra pair. Ribs are attached to the sternum (or breastbone), which is a long, flat bone shaped like a necktie in the center of the chest.

Bones in our bodies are approximately 48% water. For comparison, our heart and liver are both 72% water, our lungs are 84% water, and our adipose (i.e., fat) tissue is 14% water.

## THE APPENDICULAR SKELETON (126 BONES)

The appendicular skeleton consists of the shoulders, the upper limbs (in humans, commonly called arms, forearms, and hands), the pelvis (coxal bones including illium, ischium, and pubis), and lower limbs (in humans, commonly called legs, ankles, and feet). The appendicular skeleton enables movement and protects the major organs of digestion, reproduction, and excretion (fig. 42.1).

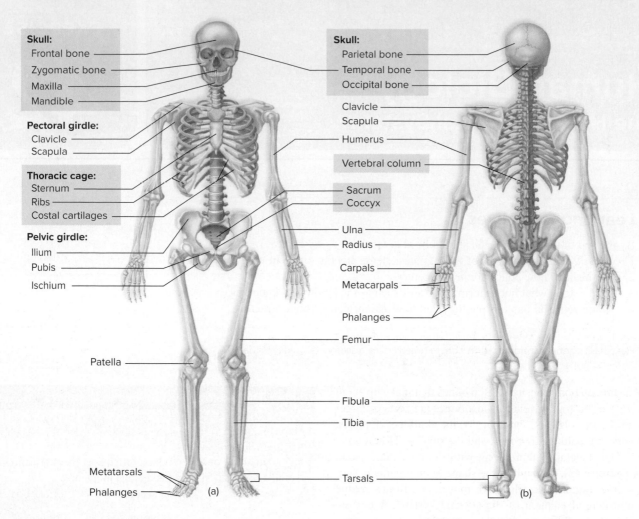

**Figure 42.1** The human skeleton. (*a*) Anterior view. (*b*) Posterior view. The bones of the axial skeleton are in blue, and the rest is the appendicular skeleton. Not all bones are shown.

## Shoulder

    **Clavicle**—collarbone

    **Scapula**—shoulder blade

## Arm

    **Humerus**—upper arm; is the longest and largest bone of the upper limb

    **Ulna**—longer of the two bones in the forearm; is on the side of the little finger

    **Radius**—shorter of the two bones of the forearm; is on the side of the thumb

    **Carpals**—eight bones in the wrist bound by strong connective tissue

    **Metacarpals**—five bones in the hand

    **Phalanges**—bones of the fingers

## *Procedure 42.2*    Identifying bones in your body

*1.*   Clench your fist.

### Question 1

What bones form the raised knobs of your knuckles?

*2.*   Flex your forearm.

### Question 2

*a.*   What bones form the elbow?

*b.*   What is your "funny bone"? How did it get this name? Is it really bone that gets hit and causes sensation?

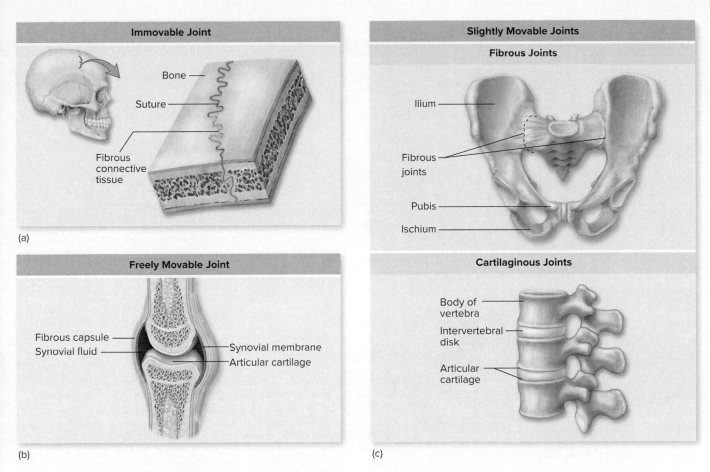

**Figure 42.2** Joints are classified functionally on the extent of movement they permit. (*a*) The sutures of the skull are immovable joints. (*b*) Freely movable joints are synovial joints, such as a finger joint. (*c*) Slightly movable joints include the sacroiliac joint of the mammalian pelvis, which is fibrous (*top*), and the cartilaginous joints between the vertebrae (*bottom*).

*3.* Use your left hand to hold your right forearm near the elbow. Now rotate your right wrist from palm up to palm down. What is this motion called? Refer to movements listed in procedure 42.3.

**Question 3**
What bone is stationary, and which bone rotates?

**Leg**

**Femur**—thigh bone; above the knee

**Fibula**—smaller, more slender of the two leg bones below the knee

**Tibia**—shin bone; larger of the two leg bones below the knee

**Patella**—kneecap

**Tarsals**—seven bones of the ankle

**Metatarsals**—five long bones of the feet

**Phalanges**—bones of the toes; two in the big toe and three in each of the other toes

*4.* Bend your leg at the knee and feel your patella. Then feel the lump just below your patella.

**Question 4**
What bone forms that lump? What might be an adaptive function for that lump?

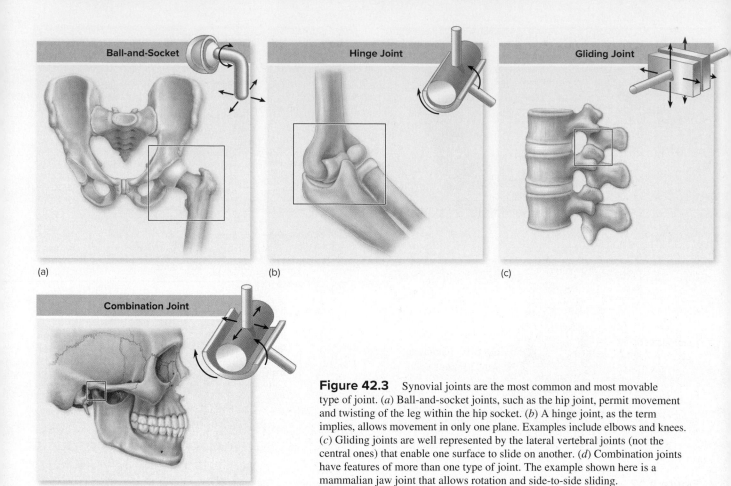

**Figure 42.3** Synovial joints are the most common and most movable type of joint. (*a*) Ball-and-socket joints, such as the hip joint, permit movement and twisting of the leg within the hip socket. (*b*) A hinge joint, as the term implies, allows movement in only one plane. Examples include elbows and knees. (*c*) Gliding joints are well represented by the lateral vertebral joints (not the central ones) that enable one surface to slide on another. (*d*) Combination joints have features of more than one type of joint. The example shown here is a mammalian jaw joint that allows rotation and side-to-side sliding.

**Question 5**

What are the functions of your legs? Ankles? Feet?

**Vertebrae**—26 bones, including the sacrum (forms part of the hip) and the coccyx (tailbone) (sacrum and coccyx are made of fused vertebrae)

**Skull**—22 bones, including six bones of the middle ear. Most bones of the skull are fused with immovable joints called sutures that appear as wavy lines (the head also includes six auditory ossicles)

**Hyoid**—a single, small, U-shaped bone suspended at the front of the throat above the pharynx. The hyoid bone does not articulate with (i.e., touch) any other bone.

**Question 6**

If the human skull is made of bone, how does it enlarge as we age?

## AXIAL SKELETON (80 BONES)

The axial skeleton is formed by the vertebral column, part of the rib cage, and the skull. In humans, the axial skeleton maintains our upright posture.

**Sternum**—breastbone

**Ribs**—normally 24 bones (i.e., 12 pairs of ribs); increase in length from the first through seventh ribs, then decrease in length to the twelfth rib.

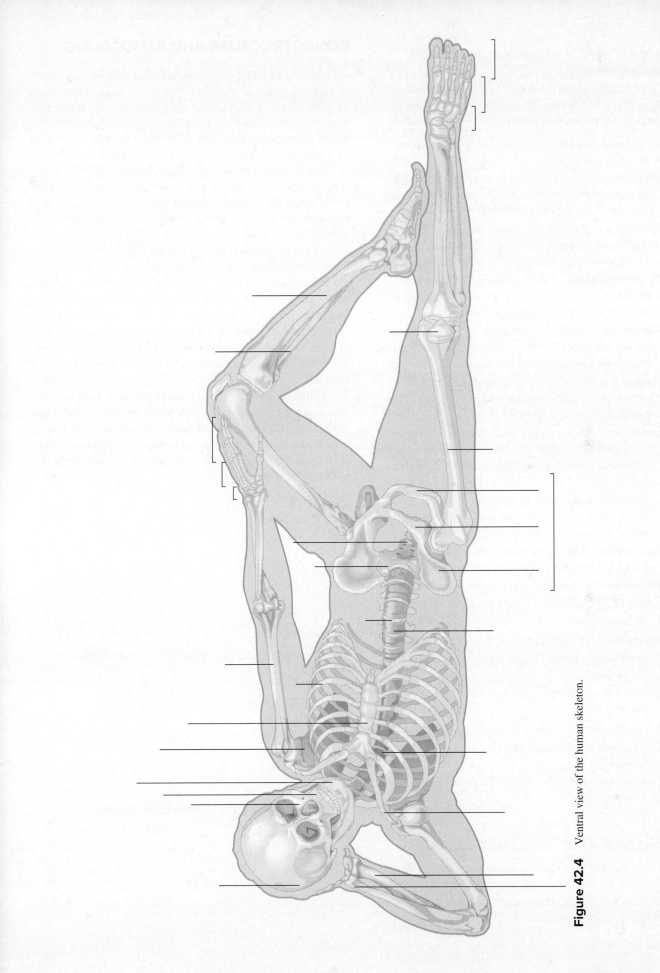

**Figure 42.4** Ventral view of the human skeleton.

## Procedure 42.3  Vocabulary of joints and skeletal movement

Bodies of human adults have about 300 joints, which are points at which bones meet (fig. 42.3). Most of these joints are synovial joints, such as elbows, knees, and knuckles.

1. Examine the following terms describing the movement of synovial joints.

   **Flexion**—bending parts at a joint so that the angle between them decreases and the parts come closer together (bending the lower limb at the knee)

   **Extension**—straightening parts at a joint so that the angle between them increases and the parts move farther apart (straightening the lower limb at the knee)

   **Hyperextension**—excess extension of parts at a joint, beyond the anatomical position (bending the head back beyond the upright position)

   **Abduction**—moving a part away from the midline (lifting the upper limb horizontally to form a right angle with the side of the body)

   **Adduction**—moving a part toward the midline (returning the upper limb from the horizontal position to the side of the body)

   **Rotation**—moving a part around an axis (twisting the head from side to side). Medial rotation involves movement toward the midline, whereas lateral rotation involves movement in the opposite direction.

   **Circumduction**—moving a part so that its end follows a circular path (moving the finger in a circular motion without moving the hand)

   **Supination**—turning the hand so the palm is upward or facing anteriorly (in anatomical position)

   **Pronation**—turning the hand so the palm is downward or facing posteriorly (in anatomical position)

   **Eversion**—turning the foot so the sole faces laterally

   **Inversion**—turning the foot so the sole faces medially

   **Protraction**—moving a part forward (thrusting the chin forward)

   **Retraction**—moving a part backward (pulling the chin backward)

   **Elevation**—raising a part (shrugging the shoulders)

   **Depression**—lowering a part (drooping the shoulders)

2. Demonstrate and draw one or two examples of each movement.

## Procedure 42.4  Assemble a human skeleton

1. Examine the disarticulated human skeleton in the lab. Note the relative sizes and shapes of the major bones.

2. Work with your lab partner(s) to assemble the skeleton.

# BONE STRUCTURE AND REMODELING

Re-examine the basic structure of bone (fig. 41.18). Parts of bones are dense and strong, whereas other parts, such as marrow, are spongy. Most red blood cells are formed by bone marrow. Bones are built by cells called **osteoblasts,** which secrete collagen fibers as sites for the deposition of hard calcium-salt crystals. These salts are deposited in thin, concentric layers called lamellae (fig. 41.18). When muscles are developed by exercise, the bones they pull against also become thicker and stronger. This is why exercises such as weight lifting increase the mass of bone as well as muscle.

Strong, healthy bone is continually maintained by **bone remodeling,** the ongoing replacement of old bone tissue by new bone tissue. As osteoblasts produce collagen and other organic components they become trapped in these secretions. Soon they mature and are called **osteocytes.** Bone is broken down by large cells called **osteoclasts** in a process called **bone resorption.** In humans, bone remodeling replaces bones as many as 10 times during an average lifetime.

Bone remodeling is a balance between bone deposition and bone resorption. If too much mineral is deposited in the bone, the surplus bone tissue often forms thick bumps called bone spurs that can interfere with movement of joints. If too much bone is resorbed, the bones become weak and overly susceptible to fracture.

**Question 7**

*a.* What causes osteoporosis?

*b.* What factors increase a person's chances of getting osteoporosis?

*c.* What can help prevent the development of osteoporosis?

Osteoporosis is a progressive bone-disease that reduces both the mineral and organic portions of bone, thereby increasing the risk of fracture (fig. 42.5). This disease affects four times as many women as men and occurs when the normal balance between bone formation and bone breakdown is disrupted. A common cause of osteoporosis is prolonged disuse of muscles. The force produced by active skeletal muscle contractions helps maintain bone mass. When muscles are not used due to paralysis or illness, bone mass declines.

Osteoporosis can also result from hormonal imbalances. Hormones such as estrogen stimulate bone formation. When a woman's reproductive cycles cease (menopause), estrogen levels decline and bone density may decrease and heighten the risk of bone fractures. In contrast, some hormones demineralize bone to maintain normal mineral homeostasis in the blood. An excess of these hormones can demineralize enough bone to cause osteoporosis.

Osteoporosis itself has no symptoms; its main consequence is the increased risk of bone fractures. Osteoporosis is slowed by adequate calcium and vitamin D intake along with weight-bearing exercise and, in some cases, hormone replacement therapy for postmenopausal women. Osteoporosis is the most prevalent bone disease in the United States, affecting 55% of Americans older than 50 years of age, and costing our nation approximately $15–$20 billion in hospital and other expenses.

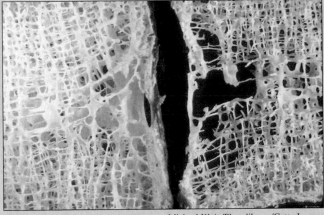

Michael Klein/Photolibrary/Getty Images

**Figure 42.5** In osteoporosis, the loss of calcium weakens bones. The vertebra on the left is normal, and the vertebra on the right has been weakened by osteoporosis.

## Spinal Curvatures

Chronic pain can result from misalignment and curvature of the spine. **Lordosis** (hollow back) is an exaggeration of the convex curve of the lumbar region, resulting in a swayback condition. **Kyphosis** (hump back) is an exaggeration of the concave curve of the thoracic region, resulting in a hunchback condition. **Scoliosis** (see figures below) is an abnormal lateral and rotational curvature of the vertebral column, which is often accompanied by secondary abnormal curvatures, such as kyphosis.

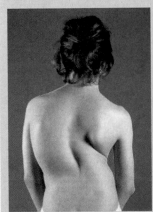

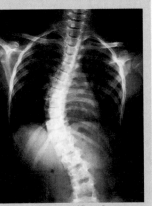

Princess Margaret Rose Orthopaedic
Hospital/Science Source

Lester V. Bergman/Getty Images

## HOW DO THE SKELETONS OF HUMANS COMPARE WITH THOSE OF OTHER MAMMALS?

Examine the skeletons of other mammals available in lab. Give special attention to the relative sizes, locations, and functions of each bone. Many bones—for example, those in our arms and in birds' wings—are homologous, meaning that they have the same basic structure and origin (because they were inherited from a common ancestor).

### Question 8
Do homologous bones always have the same function? Explain your answer.

### How are vertebrates' skeletons adapted for strength versus protection?

Observations: No organ system is more versatile in function than the skeleton and its adaptive features. Strength and protection are critical for highly mobile vertebrates.

Question: To what extent are vertebrate skeletons adapted for strength versus protection?

**a.** Establish a working lab group and obtain Inquiry-Based Learning Worksheet 42 from your instructor.

**b.** Discuss with your group and instructor specific skeletal/bone characteristics that would indicate adaptation for strength, protection, or both. Record them on Worksheet 42.

**c.** Obtain and examine the features of a human skeleton. Record your observations and assessments of skeletal features.

**d.** Complete the investigation directed by Worksheet 42.

## Questions for Further Study and Inquiry

*1.* How many bones are in the body of an adult human? How many are in a human infant? Why is there a difference?

*2.* What is the difference in locations of tendons and ligaments?

*3.* How does a tibia differ from a femur, especially in regards to location and range of motion?

*4.* What bones form your ankle? Your neck? Your chest?

*5.* What are the functions of bones?

*6.* Osteoporosis affects 35 million elderly and middle-aged people in the United States. About 80% of these people are women. Why do older women suffer more from osteoporosis than men?

*7.* What are some examples of bone disease?

## WRITING TO LEARN BIOLOGY

What health problems affect joints? Describe how these ailments are treated.

# Human Biology
## Muscles and Muscle Contraction

## Learning Objectives

By the end of this exercise you should be able to:

1.  Identify the major muscles of your body.
2.  Describe how muscles can flex or extend a joint.
3.  Describe how the use of a muscle causes fatigue.
4.  Distinguish between isotonic and isometric contractions.

Please visit **connect.mheducation.com** to review online resources tailored to this lab.

**M**uscles are structures specialized for contraction. Contrary to what many people think, muscles cannot actively lengthen; they can only contract (shorten). Some other force (e.g., gravity or the contraction of another muscle) is necessary to return the muscle to its original (uncontracted) length. Muscles are ineffective without something rigid or antagonistic to pull against. Sometimes the antagonistic structure is another muscle, but usually bone is the rigid surface against which a muscle can pull. Imagine how you might move if your muscles had no bones to pull against.

The force exerted on a muscle by an object is **muscle load.** For example, a 4-kg object would exert a muscle load of 4 kg when being lifted. **Muscle tension** is the force a contracting muscle exerts on an object. Muscle load and muscle tension are opposing forces. If the tension produced by a muscle does not exceed the load on the muscle by the object, then the object will not move. To lift a load, muscle tension must exceed muscle load.

There are two primary types of muscle contractions: isotonic and isometric. **Isotonic contractions** are contractions in which the muscle shortens while the load is constant. For example, if you use your arm's biceps muscle to bend your arm and lift a 10-lb weight, then (fig. 43.1) your biceps muscle shortens as you lift, but the weight (load) is 10 lb at all times. This is an isotonic contraction. **Isometric contractions** occur when muscles develop tension but do not shorten. For example, if you push against an immovable wall, your muscles develop tension but do not shorten significantly. This is an isometric contraction.

## Question 1

*a.* What common activities involve isotonic contractions? Isometric contractions?

*b.* Which type of contraction develops more muscle tension: isotonic or isometric? Explain why.

In both types of contractions, the amount of tension generated by the muscle is proportional to the number of muscle fibers that contract; the more fibers are stimulated by a nerve impulse to contract, the greater the tension.

**Muscle tone** is a state of moderate to slight tension that is maintained continuously by a muscle or group of muscles. For example, our posture relies on the muscle tone of a variety of skeletal muscles.

## Question 2
Why is muscle tone important?

Most joints are movable; they will move in one, two, or three planes, depending on the joint. Movement results from contraction of a skeletal muscle that connects a nonmoving

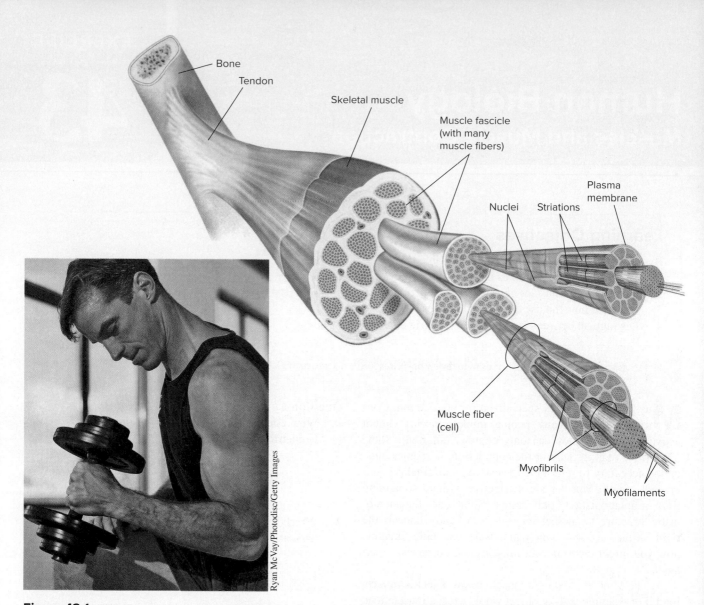

Figure 43.1 The diagram shows muscle attachment to the bone, fascicle arrangement of muscle fibers, and a large-to-small view of myofilaments in muscle fibers.

bone (i.e., the origin) to a moving bone (i.e., the insertion) across a joint.

In this exercise, you will study how muscles flex or extend joints. The extension of a joint, such as when you straighten your arm, increases the angle between two bones. Flexing a joint, such as when you bend your arm at the elbow, decreases the angle between two bones. More specifically, the contraction of your biceps muscle flexes your forearm, whereas the contraction of your triceps muscle extends your forearm. In this example, as is true throughout most of your body, skeletal muscles are arranged in **antagonistic pairs:** when one contracts, the other relaxes.

In this exercise, you will also study the major groups of skeletal muscles. You'll also study the mechanics of skeletal muscle contraction. Before doing this lab, review the introduction to muscle cell biology in Exercise 41.

## Muscles by the Numbers

Our body has 639 skeletal muscles accounting for 42% of a male's body mass and 36% of a female's mass. Muscle tissue is 15% denser than fat.

Thirty of our muscles are facial muscles that provide us with remarkable subtlety of expression.

Eye muscles are the busiest of our muscles; we blink as much as 100,000 times a day! They are also notably large and strong considering how small our eyeballs are. Research shows them to be 100-times stronger than needed.

Our strongest muscle is the masseter or jaw muscle. It can bite with a force of 4300 Newtons. In comparison, a uterus during childbirth rhythmically contracts with a force of about 300 Newtons. A Newton is the force that gravity exerts on approximately 102 g (equal to a small apple).

# MAJOR GROUPS OF MUSCLES

The following terms help biologists describe the structure and function of muscles (fig. 43.2):

> **Extensor**—muscle that straightens a joint
>
> **Flexor**—muscle that bends a joint
>
> **Insertion**—where a muscle attaches at its more movable end
>
> **Origin**—where a muscle attaches to a relatively fixed position

Use your textbook and other materials available in the lab to identify the following groups of muscles. After locating these muscles on yourself or your partner, label figures 43.3 and 43.4.

## Shoulder and Trunk

> **Deltoid**—inserts on humerus; originates on clavicle. When arm is at rest, the deltoid is the outer muscle along the upper third of the humerus. When the arm is raised, the deltoid is the hard mass of muscle above the shoulder joint. The short, thick deltoid raises the upper arm to horizontal or slightly higher. This motion away from the body midline is **abduction.** Muscles causing this motion are abductors.
>
> **Pectorals**—large, triangular muscle covering the upper part of the chest. Inserts on the humerus; originates from the clavicle, upper ribs, and sternum. Contractions move the arms toward the body midline (i.e., **adduction**). If your arm is fixed, such as during climbing, the pectoral helps pull the chest upward. The pectorals are the "breast" of poultry and are the main flight muscles of birds.
>
> **Trapezius**—inserts on the clavicle and along the scapula; originates along the upper dorsal midline. Aids in lifting with the arms or carrying loads on shoulders; braces and shrugs the shoulders.
>
> **Latissimus**—large sheet of muscle in back. Inserts on the upper part of the humerus, and originates along the middorsal line. Moves the arm downward. The latissimus is a primary muscle used in a swimming stroke or in bringing the arm forcibly downward.

## Arm

> **Triceps brachii**—inserts on ulna and originates from scapula and humerus. The triceps extends the arm at the elbow and is the primary muscle for doing a pushup.
>
> **Biceps brachii**—inserts on the radius; originates on the scapula. The biceps flexes the forearm at the elbow and is the primary muscle for doing a biceps curl or raising a glass to your mouth.

---

### Procedure 43.1 What muscles flex and extend the forearm?

1. Find a partner.
2. Position your or your partner's elbow comfortably on the table top with the palm up.
3. Feel the muscles of the upper arm as the forearm is extended and flexed.
4. Repeat this exercise with a heavy weight in the hand.
5. Repeat this exercise with the elbow pointed at the ceiling.

---

### Question 3

*a.* What muscle flexes the forearm?

*b.* What is its origin? Its insertion?

*c.* Which muscle extends the forearm?

*d.* What is its origin? Its insertion?

> **Wrist extensors**—muscles on the dorsal side of the lower arm that raise the wrist upward.
>
> **Wrist flexors**—set of muscles on the ventral side of the lower arm that bends the hand at the wrist.

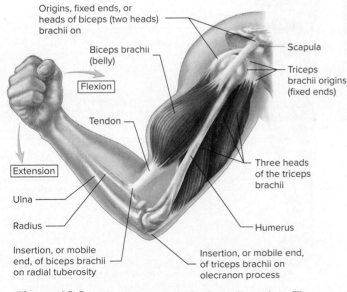

Origins, fixed ends, or heads of biceps (two heads) brachii on

Biceps brachii (belly)

Scapula

Triceps brachii origins (fixed ends)

Flexion

Tendon

Three heads of the triceps brachii

Extension

Ulna

Radius

Humerus

Insertion, or mobile end, of biceps brachii on radial tuberosity

Insertion, or mobile end, of triceps brachii on olecranon process

**Figure 43.2** Muscles are attached to bones by tendons. The elbow is flexed. The biceps brachii has two heads, which originate on the scapula. The triceps brachii has three heads, which originate on the scapula and humerus. The biceps brachii inserts onto the radial tuberosity and onto nearby connective tissue. The triceps brachii inserts onto the olecranon process of the ulna.

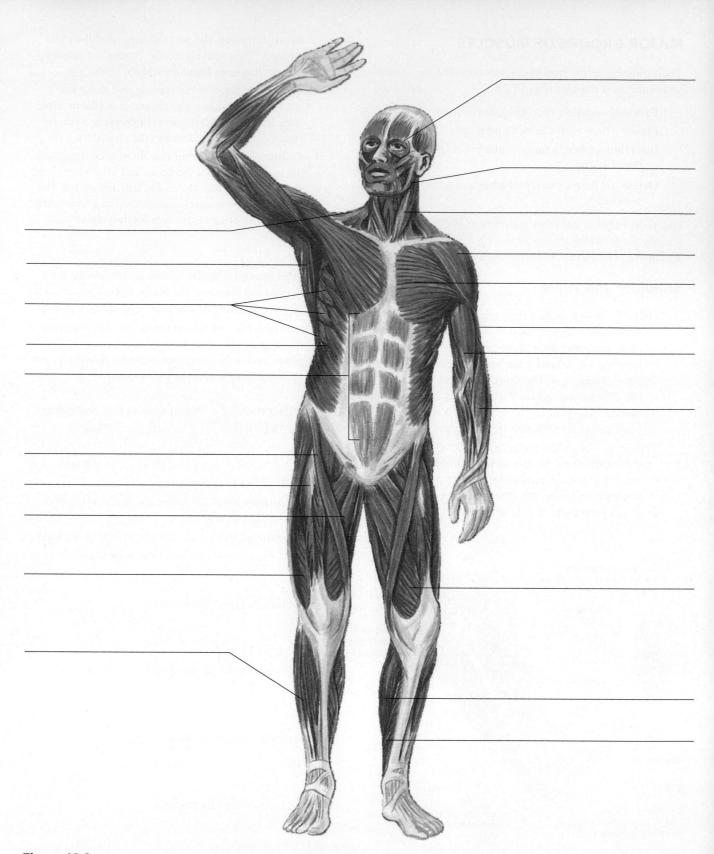

**Figure 43.3** Ventral view of superficial muscles of the human body.

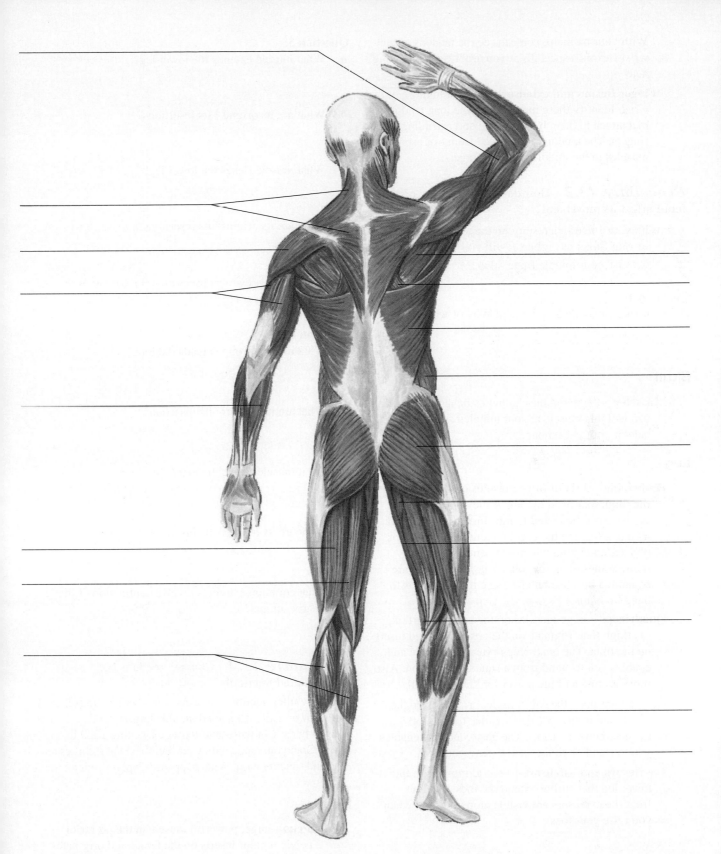

**Figure 43.4**  Dorsal view of superficial muscles of the human body.

With your palm up, you can see the tendons of the wrist flexors, especially if you are lifting something heavy.

**Finger flexors and extensors**—similar to those of the wrist. Identify these muscles and the long tendons that attach to them by clenching and extending your fingers. The tendons are crossed from extensor muscles to the middle and ring fingers.

## *Procedure 43.2* How does the structure of your hand affect its movement?

*1.* With your fingertips resting on the table, raise only your ring finger as high as it will go.

*2.* Now raise your middle finger also.

### Question 4
Did your ring finger also go higher? Why or why not?

## Skull

**Masseter**—the main muscle that clenches the jaw. You can feel this muscle at your temples and on either side of your cheekbone.

## Leg

**Hamstring**—a set of three muscles on the back of the thigh that bend the leg at the knee. Originate on the coxal bone and femur; insert on the fibula. You can feel the hamstring's tendons at the back of your knee joint when you bend your knee while standing on the other leg. Hamstrings are so named because butchers use these tendons to hang up hams.

**Quadriceps**—large muscles on the anterior part of the thigh that originate on the coxal bone and insert on the tibia. The quadriceps extends the knee and enables you to stand from a squatting position. Also provides much of the power for kicking a ball.

**Gastrocnemius**—the calf muscle; originates on the femur and inserts (by the Achilles tendon) on the heel bone (a tarsal). The gastrocnemius enables you to stand on tiptoes and extend your foot.

**Toe flexors and extensors**—several muscles in the lower leg that curl or extend the toes. Tendons from the extensors are visible atop your foot when you raise your toes.

## *Procedure 43.3* What muscles flex and extend the lower leg and foot?

*1.* Feel the muscles of your thigh as you flex and extend the lower leg against an externally applied force.

### Question 5
*a.* What muscle extends the lower leg?

*b.* What are its origins? Its insertions?

*c.* What muscle flexes the lower leg?

*d.* What are its origins? Its insertions?

*2.* Repeat the previous observation for your or your partner's lower leg.

### Question 6
*a.* What muscle group extends the foot?

*b.* What are its origins? Its insertions?

*c.* What muscle flexes the foot?

*d.* What are its origins? Its insertions?

*e.* What common activities involve contraction of the gastrocnemius?

## *Procedure 43.4* Can any tendons be manipulated manually?

Your Achilles tendon connects your heel to muscles of your lower calf. This tendon, the largest tendon in the human body, can withstand forces exceeding 1000 lb.

Grab your ankle with your thumb on the Achilles tendon and squeeze hard. What happens? Explain your answer.

## Hip

**Gluteus**—large, powerful muscle in the posterior pelvic region. Inserts on the femur and originates from the coxal bone. The gluteus supports the pelvis and trunk on the femur (you can show this by standing on one leg and feeling the muscle). Used in climbing, cycling, jumping, and regaining an erect position after bending forward.

**Abdominals**—set of muscles below the chest that flatten and compress the abdomen. Abdominals bend the body forward and from side to side; also used in urination and defecation.

 Do not do the following exercises if you have heart or lung problems. Stop immediately if you feel faint.

## Procedure 43.5 How fast do muscles fatigue?

Work with two lab partners as you do this procedure.

1. Squeeze a tennis ball as rapidly as possible with one hand. While you squeeze, a partner will call out "time" every 15 sec for 3 min. Your other lab partner will count and record your number of contractions during each 15-sec period.

| | |
|---|---|
| 0–15 sec | _____ contractions |
| 15–30 sec | _____ contractions |
| 30–45 sec | _____ contractions |
| 45–60 sec | _____ contractions |
| 60–75 sec | _____ contractions |
| 75–90 sec | _____ contractions |
| 90–105 sec | _____ contractions |
| 105–120 sec | _____ contractions |
| 120–135 sec | _____ contractions |
| 135–150 sec | _____ contractions |
| 150–165 sec | _____ contractions |
| 165–180 sec | _____ contractions |

2. Wait 1 min and repeat the experiment with the same hand. Record your results:

| | |
|---|---|
| 0–15 sec | _____ contractions |
| 15–30 sec | _____ contractions |
| 30–45 sec | _____ contractions |
| 45–60 sec | _____ contractions |
| 60–75 sec | _____ contractions |
| 75–90 sec | _____ contractions |
| 90–105 sec | _____ contractions |
| 105–120 sec | _____ contractions |
| 120–135 sec | _____ contractions |
| 135–150 sec | _____ contractions |
| 150–165 sec | _____ contractions |
| 165–180 sec | _____ contractions |

3. Wait another minute and repeat the experiment. Record your results:

| | |
|---|---|
| 0–15 sec | _____ contractions |
| 15–30 sec | _____ contractions |
| 30–45 sec | _____ contractions |
| 45–60 sec | _____ contractions |

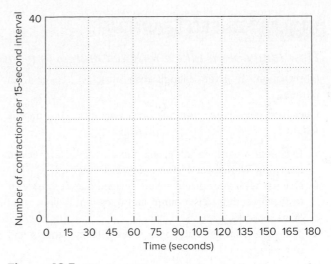

**Figure 43.5** Decline in the number of contractions as a muscle fatigues.

| | |
|---|---|
| 60–75 sec | _____ contractions |
| 75–90 sec | _____ contractions |
| 90–105 sec | _____ contractions |
| 105–120 sec | _____ contractions |
| 120–135 sec | _____ contractions |
| 135–150 sec | _____ contractions |
| 150–165 sec | _____ contractions |
| 165–180 sec | _____ contractions |

4. On figure 43.5, graph the number of contractions versus time for each trial. Connect the data points with a straight line for each trial.

### Question 7
*a.* Did you produce the same number of contractions in each of the three trials?

*b.* Why or why not?

### Question 8
*a.* Does the number of contractions increase or decrease during the experiment? Explain your answer.

*b.* Is the slope of the line constant in the experiments? What do you conclude from this?

*c.* What causes muscle fatigue?

## INQUIRY-BASED LEARNING

### *How long does it take a fatigued muscle to recover?*

Observation: If given enough rest, muscles recover from fatigue.

Question: What is the minimum recovery time from muscle fatigue?

**a.** Establish a working lab group and obtain Inquiry-Based Learning Worksheet 43 from your instructor.

**b.** Discuss with your group a well-defined question relevant to the preceding observation and question. Record it on Worksheet 43.

**c.** Translate your question into a testable hypothesis and record it.

**d.** Suggest one way to measure recovery time. Outline on Worksheet 43 your experimental design and supplies needed to test your hypothesis. Ask your instructor to review your proposed investigation.

**e.** Conduct your procedures, record your data, answer your question, and make relevant comments.

**f.** Discuss with your instructor any revisions to your questions, hypotheses, or procedures. Repeat your work as needed.

## Questions for Further Study and Inquiry

*1.* What would happen if both muscles of an antagonistic pair contracted simultaneously?

*2.* What is a "pulled" muscle?

*3.* What other types of muscles besides skeletal muscles are there?

*4.* What is an example of a muscle that is not part of an antagonistic pair?

*5.* Many people take dietary supplements to improve their strength and endurance. Do these supplements "work"? What is the evidence?

*6.* Review the information presented in Exercise 42. How do muscles and bones work together?

*7.* Do invertebrates without bones have something rigid for their muscles to pull against? How so?

## WRITING TO LEARN BIOLOGY

What is muscle fatigue? How could you delay the fatigue of a muscle?

# Human Biology

## Breathing

In Exercise 12 you studied how cells oxidize sugars to release energy for their activities. This process is called cellular respiration, and in humans and most other organisms it requires oxygen (fig. 44.1). To get this oxygen for respiration (and to get rid of respiratory waste products such as carbon dioxide), humans have a pulmonary respiratory system in which ventilation and gas exchange occur in specialized organs called lungs. We promote gas exchange by forcing air into and out of our lungs as we breathe. The importance of breathing can't be overestimated—every day we breathe about 25,000 times. If our rate of breathing slows too much, we either faint or suffocate.

Air moves in response to pressure gradients produced by a complex anatomy (fig. 44.2). This is true for large masses of air, such as cold and warm fronts that sweep into town, and for relatively small masses of air that move into and out of our lungs. Air always moves from areas of high pressure to areas of lower pressure. For example, when we prick a balloon, air quickly moves from inside the balloon, where pressure is high, to the outside of the balloon, where pressure is lower. Air moving into and out of our bodies as we breathe also moves in response to pressure gradients. As you'll learn in this exercise, we expend much energy to create pressure gradients that help us breathe.

Our lungs are in our thoracic (chest) cavity. We inhale by expanding our lungs, which creates a negative pressure (i.e., partial vacuum) in our lungs. This suction pulls air into our lungs. Expanding our lungs is more complicated than it seems because lung tissues lack skeletal muscle. Several muscles must be coordinated for inhalation and exhalation. To expand our lungs and create the negative pressure needed to inhale, we use our diaphragm and intercostal muscles.

- The **diaphragm** is a sheetlike muscle separating the abdomen from the chest cavity. It is the primary muscle used

in breathing; when the diaphragm contracts, it flattens and expands the chest cavity. This expansion of our chest creates a negative pressure (i.e., a partial vacuum), thereby pulling air into our nostrils, mouth, and lungs.

- **Intercostal muscles** are located between the ribs (fig. 44.3). When these muscles contract, they expand the chest cavity and negative pressure pulls air into our lungs.

Contracting your diaphragm and intercostal muscles expands your chest cavity, and decreases the pressure in the cavity. Your lungs adhere to the lining of the cavity, so as it expands, your lungs also expand. This expansion decreases the pressure in the lungs and causes air to move into the lungs. Relaxing the diaphragm and intercostal muscles shrinks the chest cavity, thereby increasing the pressure in the lungs.

Rex Brown/Getty Images Sport/Getty Images

**Figure 44.1** We humans promote gas exchange by forcing air into and out of our lungs as we breathe. In cold weather, we see condensed water vapor in air exhaled from our lungs.

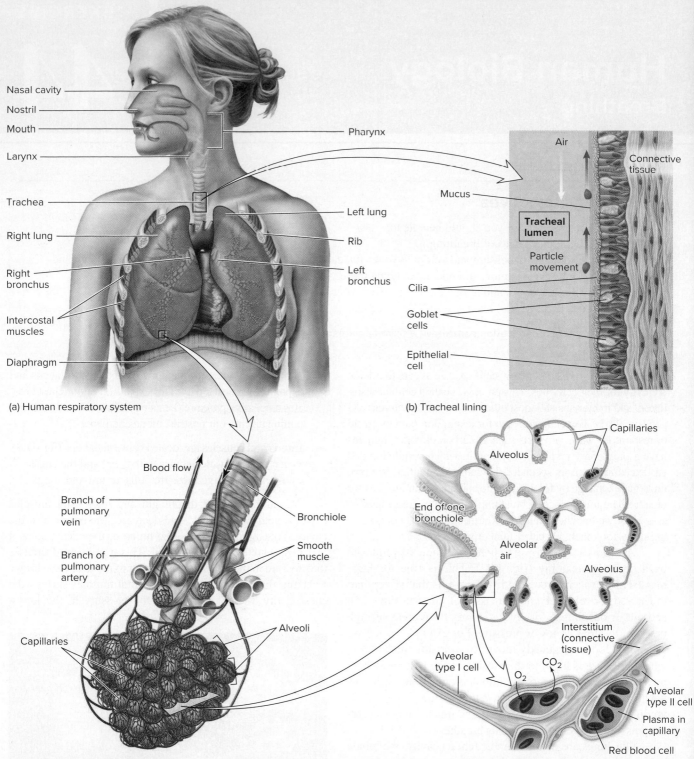

(a) Human respiratory system

Nasal cavity
Nostril
Mouth
Larynx
Trachea
Right lung
Right bronchus
Intercostal muscles
Diaphragm
Pharynx
Left lung
Rib
Left bronchus

(b) Tracheal lining

Air
Connective tissue
Mucus
**Tracheal lumen**
Particle movement
Cilia
Goblet cells
Epithelial cell

(c) Structure of a bronchiole and alveoli

Blood flow
Branch of pulmonary vein
Branch of pulmonary artery
Capillaries
Bronchiole
Smooth muscle
Alveoli

(d) Cross section of an alveolar cluster, with enlarged region

Capillaries
Alveolus
End of one bronchiole
Alveolar air
Alveolus
Interstitium (connective tissue)
Alveolar type I cell
$O_2$
$CO_2$
Alveolar type II cell
Plasma in capillary
Red blood cell

**Figure 44.2**   The mammalian respiratory system. (*a*) In this overview, the ribs have been removed in front, and the major airways have been drawn in the lungs. The thoracic cavity is bounded by the ribs and intercostal muscles and the muscular diaphragm. For simplicity, external and internal intercostal muscles are not indicated separately. (*b*) Ciliated epithelial cells and mucus-producing goblet cells line the trachea. The mucus traps inhaled particles and the cilia help move the mucus toward the mouth, where it can be swallowed. (*c*) The bronchioles deliver air to the clusters of alveoli. Note the smooth muscle cells around the bronchioles, which can cause the bronchioles to constrict or dilate. Capillaries surround the alveoli. Red represents oxygenated blood; blue represents partly deoxygenated blood. (*d*) Cross section through a cluster of alveoli. Note the single cell layer of alveoli cells and their close proximity to adjacent capillaries.

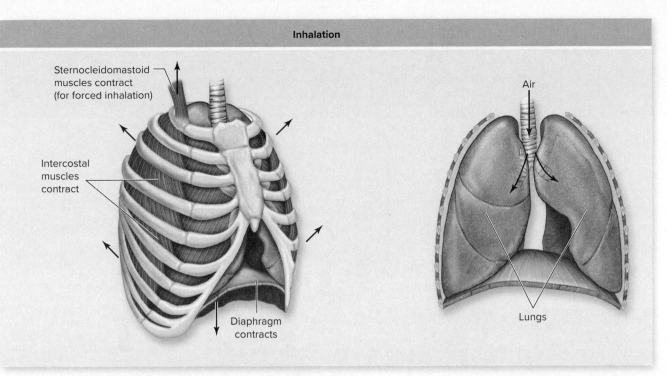

Inhalation

(a)

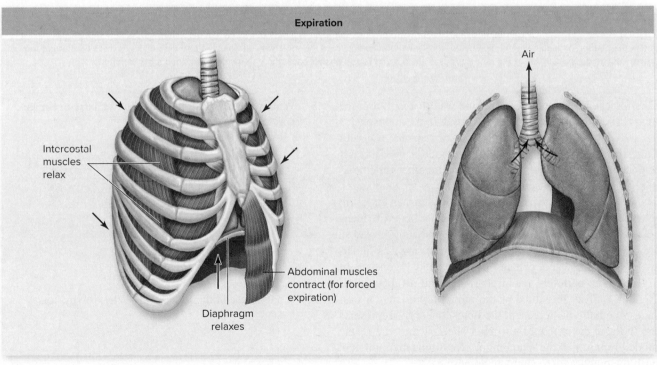

Expiration

(b)

**Figure 44.3** How a human breathes. (*a*) Inhalation. The diaphragm contracts and the walls of the chest cavity expand, increasing the volume of the chest cavity and the lungs. As a result of the larger volume, air is drawn into the lungs. (*b*) Exhalation. The diaphragm and chest walls return to their normal positions as a result of elastic recoil, reducing the volume of the chest cavity and forcing air out of the lungs through the trachea. Note that inhalation can be forced by contracting accessory respiratory muscles (such as the sternocleidomastoid), and exhalation can be forced by contracting the abdominal muscles.

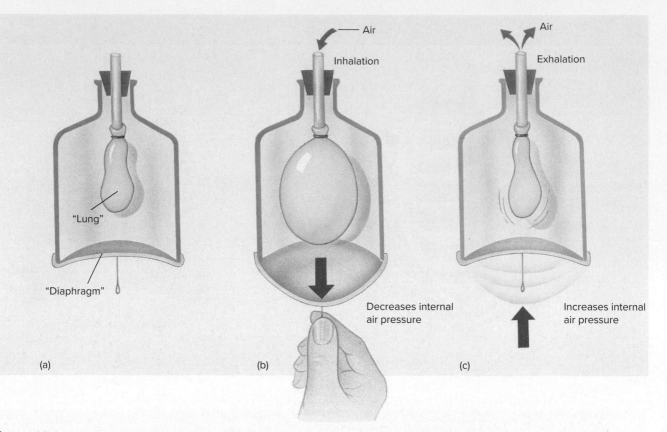

**Figure 44.4**   A simple experiment that shows how we breathe. In the jar is a balloon (*a*). When the diaphragm is pulled down, as shown in (*b*), the balloon expands; when it is relaxed (*c*), the balloon contracts. In the same way, air is taken into the lungs when the diaphragm moves down, expanding the volume of the lung cavity. When the diaphragm pushes back, the volume decreases and air is expelled.

This forces air with elevated $CO_2$ and water vapor out of the lungs because pressure there exceeds that in the atmosphere.

To better understand breathing movements, use your fingertips on your chest and then abdomen to feel the difference in your body movements as you breathe while holding your rib cage steady and using only your diaphragm. Then feel your body movements when using only rib intercostal muscles, and then when using both rib intercostals and diaphragm. Next, examine the lung model, a simple device that helps demonstrate the principles underlying inhalation and exhalation. The model consists of a glass tube, balloons, enclosed space, and a rubber sheet (fig. 44.4). The glass tube of the model represents air passages, the balloons represent the lungs, the space represents the thoracic cavity, and the rubber sheet at the base of the model represents the diaphragm. Rhythmically pull and push the diaphragm to simulate breathing.

**Question 1**

*a.*  What happens to the lungs when the diaphragm is pulled down? When it is pushed back? Why?

*b.*  What would happen if the seal at the base of the jar was broken?

*c.*  What causes a collapsed lung?

*d.*  Is a collapsed lung functional? Why or why not?

  Don't do the following exercises if you have heart or lung problems. Work with a partner in each of the following procedures, and stop immediately if you feel faint.

## Procedure 44.1 Measure differences in chest diameter during breathing

1. Wrap a tape measure snugly under your armpits and around the fullest part of your chest. Women should wrap the tape measure just below their breasts. Record your chest diameter: _____ cm

2. Now take a deep breath, hold it, and measure your chest diameter: _____ cm

### Question 2

*a.* How much did your chest enlarge?

*b.* What caused your chest to enlarge?

*c.* What is the significance of this change?

3. Place your hand at the bottom of your sternum (chest bone) and take a deep breath.

### Question 3

*a.* What direction did your hand move when you inhaled?

*b.* When you exhaled?

4. Place your hands on your abdomen and take five deep breaths. Describe what you feel.

5. Repeat this with your hands on your chest. Describe what you feel.

### Question 4

Can you take a deep breath without expanding both your chest and abdomen? Why or why not?

6. Exhalation occurs by relaxing your diaphragm and intercostal muscles. To force more air from your lungs you must use your abdominal muscles. Place your hand near your belly button and force as much air out of your lungs as possible.

### Question 5

*a.* What happened to your abdominal muscles when you exhaled?

*b.* What happened to your intercostal muscles when you exhaled?

## LUNG CAPACITY

Your lungs hold several liters of air. This volume of air has several subvolumes (fig. 44.5).

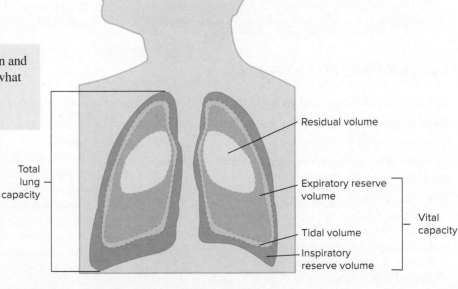

**Figure 44.5** Pulmonary volumes and capacities.

- **Tidal volume** (TV) is the volume of air inhaled or exhaled during a single breath and is the amount of air necessary to maintain the oxygen supply to your tissues. Tidal volume is about 500 mL for an average adult.

- **Expiratory reserve volume** (ERV) is the amount of air that can be exhaled after a normal, quiet exhalation. Expiratory reserve volumes range from 800 to 1300 mL.

- **Inspiratory reserve volume** (IRV) is the amount of air that can be inhaled after a person takes a normal breath. Inspiratory residual volumes range from 2500 to 3500 mL.

- **Residual volume** (RV) is the air that cannot be exhaled from the lungs. Residual volume is about 1200 mL.

- **Vital capacity** is the total of tidal volume plus inspiratory and expiratory reserve volumes, the maximum amount of air that can be inhaled after maximum exhalation. Vital capacity is sometimes used to indicate pulmonary function. A significant decrease in vital capacity is often associated with emphysema, pneumonia, and other lung diseases.

## Measuring Lung Capacity

The volume of air that enters and leaves the lungs can be measured with an instrument called a spirometer (fig. 44.6). Adequate respiratory volume is critical for good health, and is dramatically impacted by asthma and many pulmonary diseases. Respiratory volumes change as you age, and they vary from individual to individual. The normal adult breathes about 12 breaths per minute. In the following procedures, record all of your results in table 44.1.

Phipps & Bird, Inc., Richmond, VA. Used with Permission.

**Figure 44.6**    A spirometer is used to measure lung capacity.

---

### Procedure 44.2    Measure your tidal volume (TV)

*1.* Set the dial of the spirometer at zero. Place a sterile mouthpiece over the stem of the spirometer. Insert the mouthpiece into your mouth with the spirometer's dial facing upward.

*2.* Inhale through your nose. Exhale through your mouth and into the spirometer for five normal breathing cycles without removing the mouth piece from your mouth.

*3.* After five breathing cycles, observe the dial reading. Divide the reading by five to determine your tidal volume for one breath.

*4.* Record this volume, according to the units on the spirometer, as the first of three tidal volume measurements in table 44.1.

*5.* Repeat steps 2–3 twice and record these values in table 44.1.

---

## INQUIRY-BASED LEARNING

### Do the diaphragm and intercostal muscles contribute equally to breathing?

Observation: Breathing involves the combined action of the diaphragm and intercostal muscles.

Question: What are the relative contributions of the diaphragm and intercostal muscles to breathing?

**a.** Establish a working lab group and obtain Inquiry-Based Learning Worksheet 44 from your instructor.

**b.** Discuss with your group a well-defined question relevant to the preceding observation and question, and relate it to your inspiratory reserve volume and/or tidal volume. Record it on Worksheet 44.

**c.** Translate your question into a testable hypothesis and record it.

**d.** Outline on Worksheet 44 your experimental design and supplies needed to test your hypothesis. Ask your instructor to review your proposed investigation.

**e.** Conduct your procedures, record your data, answer your question, and make relevant comments.

**f.** Discuss with your instructor any revisions to your questions, hypotheses, or procedures. Repeat your work as needed.

6. Calculate the average of the resulting three values and record this average value in table 44.1.

## Procedure 44.3   Measure your expiratory reserve volume (ERV)

1. Reset the dial of the spirometer to zero.
2. Place your mouth on the tube. Inhale normally through your nose, then exhale maximally into the mouthpiece. Observe the reading for this maximal exhalation volume.
3. From this maximal exhalation volume subtract your average tidal volume (determined in procedure 44.2).
4. Record this value in table 44.1 as the first of three ERV values.
5. Repeat steps 2–4 twice and record these ERV values in table 44.1.
6. Calculate your average ERV and record it in table 44.1.

## Procedure 44.4   Measure your vital capacity (VC)

1. Reset the dial of the spirometer to zero.
2. Take several deep breaths and exhale completely after each. Then take as deep a breath as possible and exhale slowly and evenly through the spirometer.
3. Record this value in table 44.1 as the first of three VC values.
4. Repeat steps 2–3 twice, being sure to reset the dial after each exhalation. Record these volumes in table 44.1.
5. Calculate your average VC and record it in table 44.1.

## Procedure 44.5   Compute your inspiratory reserve volume (IRV)

Your instrument is not set up to measure your inspiratory reserve volume directly. However, you can compute it by using the following formula:

$$\text{Inspiratory reserve volume (mL)} = \text{vital capacity} - \text{tidal volume} - \text{expiratory reserve volume} = \underline{\quad} \text{ mL}$$

The normal adult's vital capacity varies with his or her age, sex, and height. However, a typical vital capacity for men is

about 5200 mL, whereas that for women is about 4000 mL. Tidal volume is affected strongly by exercise.

## Procedure 44.6   Measure the effect of exercise on your tidal volume

1. Exercise vigorously for 3 min by stepping onto and off of a small stool, or doing some other approved exercise. Then measure your tidal volume according to procedure 44.2.

**Tidal volume = _____ mL**

### Question 6

*a.* How do these values compare with your "at rest" values (table 44.1)?

*b.* How do you explain this difference?

2. Stop exercising and continue to measure your tidal volume every 30 sec for 10 min. Graph your results (fig. 44.7).

### Table 44.1

**Measurements of Lung Volumes**

| Tidal Volume (TV) | Expiratory Reserve Volume (ERV) | Vital Capacity (VC) | Inspiratory Reserve Volume (IRV) |
|---|---|---|---|
| 1st _____ | 1st _____ | 1st _____ | |
| 2nd _____ | 2nd _____ | 2nd _____ | |
| 3rd _____ | 3rd _____ | 3rd _____ | |
| Average _____ | Average _____ | Average _____ | Calculated value = _____ |

**Question 7**

What is your recovery time?

3. When you've finished the experiment, place the disposable mouthpiece in the collecting bag.

## BODY SIZE AND VITAL CAPACITY

A normal adult's vital capacity varies with body size. To show this, use a meterstick to measure your height: _____ cm

Write your height, gender, and vital capacity on the chalkboard in the lab. Graph all of the data for your class on figure 44.8. Your instructor may ask you to graph the data for males separately from those of females.

**Question 8**

*a.* What do you conclude about the relationship of height and vital capacity?

*b.* Is this what you would have predicted? Why or why not?

## BREATHING RATE

Your rate of breathing is controlled by many factors, the most important of which is the concentration of carbon dioxide in your blood. Slowing your breathing rate slows the release of carbon dioxide into the lungs, thereby increasing the amount of carbon dioxide in the blood.

The average adult breathes about 12 times per minute, compared to an elephant's 10 breaths per minute. A giraffe maintains alveolar ventilation by breathing about 8 to 10 times per minute. This slow ventilation rate helps reduce the number of times per minute the dead space, mostly in the long trachea, must be filled with air. The resting tidal volume of the giraffe is about 4 L.

### Procedure 44.7   Measure your breathing rate

1. Determine your breathing rate at rest.
2. Breathe deeply and rapidly for 12 breaths (i.e., hyperventilate). Try not to move your shoulders as you hyperventilate. Simply increase the rate and depth of breathing.

**Question 9**

Does deep breathing get easier or harder as time goes by? Why?

3. After hyperventilating, hold your breath for as long as you can. Record how long you held your breath: _____ sec
4. Exhale, take a breath, and hold your breath again as long as you can. Again record how long you held your breath: _____ sec
5. Repeat this exercise three more times.

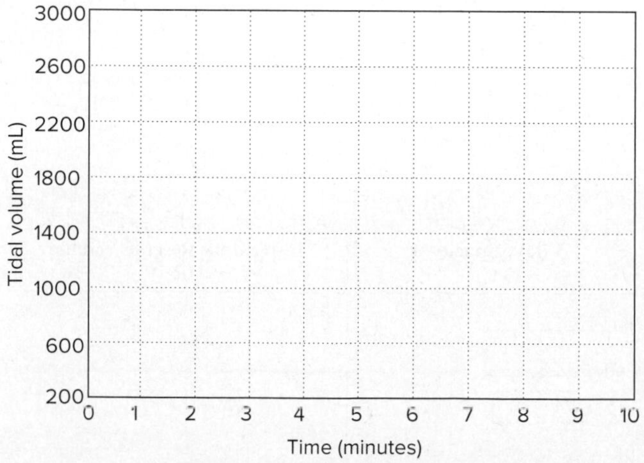

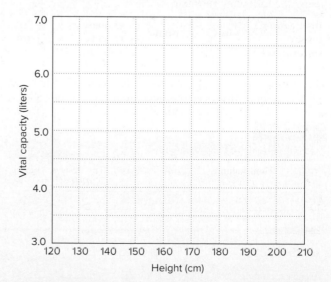

**Figure 44.7**   Change in tidal volume during recovery from exercise.

**Figure 44.8**   The relationship between lung vital capacity and a person's height.

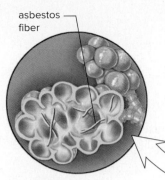
asbestos
fiber

**Pneumonia**
Alveoli fill with thick
fluid, making gas
exchange difficult.

**Pulmonary Fibrosis**
Fibrous connective tissue
builds up in lungs, reducing
their elasticity.

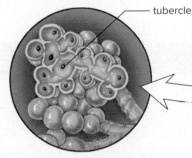
tubercle

**Pulmonary Tuberculosis**
Tubercles encapsulate bacteria,
and elasticity of lungs is reduced.

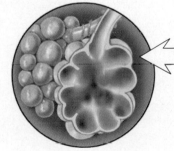

**Emphysema**
Alveoli burst and fuse into
enlarged air spaces. Surface area
for gas exchange is reduced.

**Asthma**
Airways are inflamed due
to irritation, and bronchioles
constrict due to muscle spasms.

mucus

**Bronchitis**
Airways are inflamed due
to infection (acute) or due to
an irritant (chronic). Coughing
brings up mucus and pus.

**Figure 44.9**    Common bronchial and pulmonary diseases. Exposure to infectious pathogens and/or polluted air, including tobacco smoke, causes the diseases and disorders shown here.

## Question 10

*a.* What pattern do you see in the results?

*b.* How do you explain these data? That is, how is this response adaptive under these conditions?

**6.** Rest until your respiratory rate returns to normal. Then run vigorously in place for 3 min. Stop and hold your breath for as long as possible. Record your time: _____ sec

**b.** What do you think causes the breathing rate to increase: the increase of $CO_2$ in the blood or the depletion of $O_2$? How could you test your answer?

**Question 11**

**a.** How does this time compare with others you recorded? Explain your results.

**c.** A large surface area is critical for efficient breathing. Review figure 44.9. Which diseases would most affect alveolar surface area?

# Questions for Further Study and Inquiry

**1.** What is the clinical significance of vital capacity?

**2.** How does smoking affect the various aspects of lung capacity?

**3.** How would you measure the effects of exercise on vital capacity?

**WRITING TO LEARN BIOLOGY**

Who do you think would have a shorter recovery time after exercising, a well-conditioned athlete or an out-of-shape professor? Explain your answer.

# Human Biology
## Circulation and Blood Pressure

### Learning Objectives

By the end of this exercise you should be able to:
1. Describe the path of blood flow in a four-chambered heart.
2. Describe the sounds made by a beating heart.
3. Describe the structure and function of red blood cells, white blood cells, capillaries, veins, and arteries.
4. Describe how exercise affects blood pressure and pulse rate.

Please visit **connect.mheducation.com** to review online resources tailored to this lab.

The cells of multicellular organisms are linked by an elaborate circulatory system. In humans, this circulatory system is based on a fast-flowing river of blood that delivers materials, food, and oxygen to cells (see fig. 49.7). Our circulatory system also removes waste products such as carbon dioxide from cells. The circulatory system in humans and other vertebrates is a closed system, meaning that blood is enclosed at all times within vessels and does not fill body cavities. The circulatory system in humans consists of a pumping heart, blood, and blood vessels.

In this exercise, you'll examine the structure and function of your circulatory system. You'll also make some diagnostic measurements of your circulatory system, including your pulse rate and blood pressure.

## HEART

Your heart is a muscular organ that weighs 200–400 g (7–15 ounces) and is slightly larger than your fist. Every day, a person's heart beats about 100,000 times and pumps about 7570 L (2000 gallons) of blood.

A human heart has four chambers: a left and right atrium, and a left and right ventricle. These chambers are separated by one-way valves that help control blood flow. The left and right sides of the heart are separated by an inner wall called the septum.

Study the path of blood flow through the mammalian heart (see fig. 45.1; also fig. 49.9).

- Oxygen-poor and $CO_2$-rich blood from the superior vena cava and inferior vena cava enter the right atrium.

- The right atrium pumps blood through the right atrio-ventricular (tricuspid) valve to the right ventricle.

- The right ventricle then pumps blood through the pulmonary semilunar valve into the pulmonary trunk and the two pulmonary arteries to the lungs.

- Oxygen-rich blood from the lungs travels in four pulmonary veins to the left atrium.

- The left atrium pumps blood through the left atrioventricular (bicuspid or mitral) valve to the left ventricle.

- The left ventricle pumps blood through the semilunar valve into the aorta to the body.

Examine a heart from a cow or available heart model. Use your fingers to trace the path of blood flow.

### Question 1
*a.* Which chamber has the thickest wall?

*b.* How does this relate to the function of that chamber?

### Circulation and Blood Pressure by the Numbers

Our heart beats approximately 100,000 times per day, during which it pumps more than 7500 L of blood throughout our bodies. This means that during an average lifetime, our heart pumps about 48 million gallons of blood, which would fill more than 70 Olympic-size swimming pools.

The human heart weighs about 300 g; that's comparable to the weight of an empty stomach, but is only one-fifth that of our liver. For comparison, the heart of a lion weighs about 750 g, while that of a sparrow weighs only 0.4 g.

If all of the veins, arteries, and capillaries in a human body were stretched end-to-end, they would span more than 60,000 miles and would circle Earth more than 2.5 times.

Our bodies produce more than 100 billion red blood cells per day; each of these cells lives about 120 days. For comparison, cells lining our stomach and small intestine live an average of only 2 days, and cells in our liver survive about 15 days.

Our pulmonary artery, which takes blood from our heart to our lungs, has a diameter of 2.4 cm. For comparison, our capillaries have a diameter of about 8 μm.

Typical blood pressure in humans is 120/80 mm Hg, which is similar to that in rabbits (110/80) and chimps (135/80). Frogs have a lower blood pressure (32/20), while giraffes have a higher blood pressure (340/230).

Each beat of the heart produces a characteristic sound. The first sound is a low-pitched "lub" made by the left atrioventricular (also called the bicuspid or mitral) and right atrioventricular (tricuspid) valves closing when the ventricles start to contract. Soon thereafter you hear a "dub" made by the pulmonary semilunar valve and aortic semilunar valve closing after the ventricles have contracted. In both cases, blood falling back on the cuplike valves snaps these valves closed, much like wind snaps open a parachute. If any of the valves do not close completely, there is a turbulence in the heart that can be heard as a **heart murmur.** Heart murmurs often sound like sloshing liquid.

Obtain a stethoscope. To best hear the "dub" sound of your heart, press the stethoscope against the fifth or sixth rib, slightly left of center. To hear the "lub" sound, press the stethoscope against the second rib.

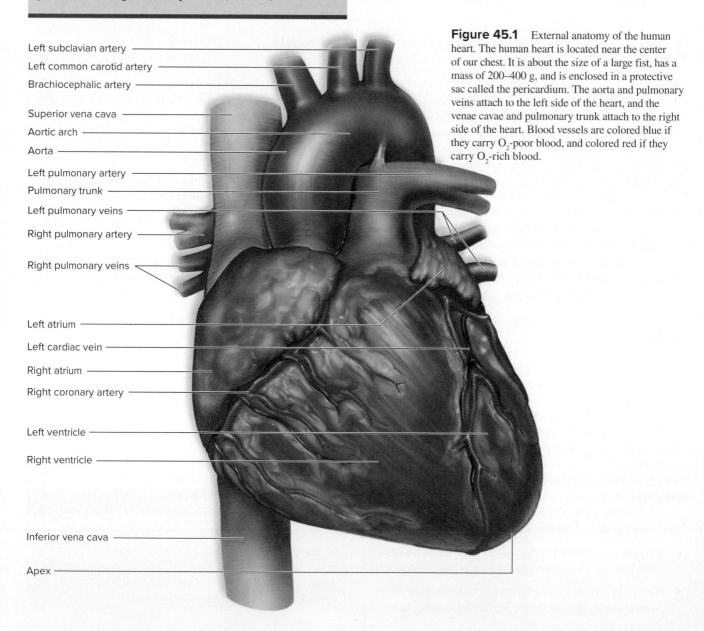

Left subclavian artery
Left common carotid artery
Brachiocephalic artery
Superior vena cava
Aortic arch
Aorta
Left pulmonary artery
Pulmonary trunk
Left pulmonary veins
Right pulmonary artery
Right pulmonary veins
Left atrium
Left cardiac vein
Right atrium
Right coronary artery
Left ventricle
Right ventricle
Inferior vena cava
Apex

**Figure 45.1** External anatomy of the human heart. The human heart is located near the center of our chest. It is about the size of a large fist, has a mass of 200–400 g, and is enclosed in a protective sac called the pericardium. The aorta and pulmonary veins attach to the left side of the heart, and the venae cavae and pulmonary trunk attach to the right side of the heart. Blood vessels are colored blue if they carry $O_2$-poor blood, and colored red if they carry $O_2$-rich blood.

## Question 2

*a.* About how fast is your heart beating?

*b.* Why can't you hear these sounds when you press the stethoscope against your neck or leg?

*c.* What sounds of a beating heart did you hear? What made those sounds?

## BLOOD

Recall from Exercise 41 that blood is a type of connective tissue (see fig. 41.14). About 55% of blood is a yellowish fluid called **plasma.** Suspended in plasma are cells, the most abundant of which are **red blood cells,** or **erythrocytes.** Red blood cells are biconcave disks 7–8 μm in diameter (fig. 17.4a). There are about 250,000 erythrocytes in a drop of blood. Red blood cells are made by red bone marrow.

Use the low, then high, magnification lenses of your microscope to examine a prepared slide of human blood. The pink cells lacking nuclei are red blood cells. The larger cells stained bluish-purple are **white blood cells,** or **leukocytes.** Leukocytes, like erythrocytes, are made by red bone marrow.

## Question 3

*a.* What's in plasma?

*b.* What is the function of plasma?

*c.* What is the shape of each kind of blood cell? How is this shape important to the function of the cells?

*d.* What is the function of each kind of blood cell?

*e.* What is the approximate ratio of red blood cells to white blood cells in human blood?

*f.* What are some diseases of blood? What are the symptoms of these diseases?

*g.* Red blood cells lack nuclei. How might this improve their ability to function?

## Blood Vessels

Blood vessels include arteries, capillaries, and veins (fig. 45.2). **Arteries** carry blood away from the heart. Arteries consist of four concentric layers: an outer layer of connective tissue, a middle layer of smooth muscle, an elastic layer, and an inner layer of epithelial cells. Examine a prepared slide of an artery.

## Question 4

*a.* Which of the layers is thickest?

*b.* What does this tell you about the function of the arteries?

*c.* Plaque is a buildup of cholesterol, white blood cells, calcium, and other substances in the walls of arteries. How would the deposition of plaque in an artery affect the function of the artery?

**Veins** carry blood to the heart. Veins have the same three layers that arteries have. Examine a prepared slide of a vein.

## Question 5

*a.* Which of the layers is thickest?

*b.* How is this different from an artery?

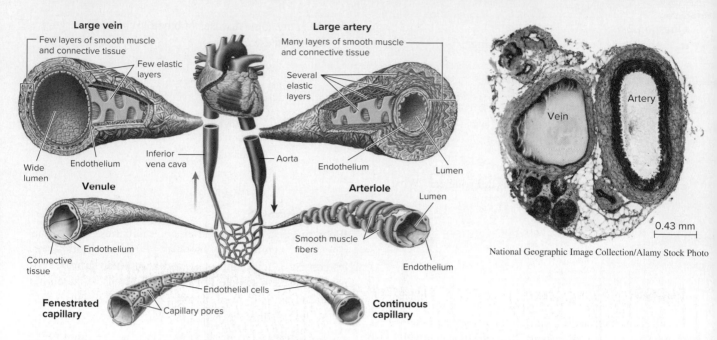

**Figure 45.2** Comparative features of blood vessels. Sizes are not drawn to scale. *Inset*: Light micrograph (12×) of a medium-size artery near a vein. Note the difference between the artery and vein in wall thickness and lumen diameter.

*c.* What does this tell you about the functioning of veins?

*c.* Which has the thinnest wall?

**Capillaries** connect arteries and veins and have a diameter slightly larger than that of a single red blood cell. Collectively, capillaries form an elaborate lattice of narrow, thin-walled tubes. Humans have about 97,000 km of capillaries with a total surface area of 800–1000 m² (an area greater than three tennis courts). All cells in our body are within 100 μm of a capillary. Examine a prepared slide of an artery, vein, and capillary.

*d.* What does this tell you about where gas exchange occurs?

**Question 6**

*a.* Which of the types of blood vessels has the largest diameter?

*e.* What does the absence of muscle tissue in capillaries suggest about the function of this part of the circulatory system?

### Blood Circulation in Goldfish

Use a net to catch a goldfish from the aquarium in the lab. Gently wrap the goldfish in cotton soaked in water from the aquarium. Let the tail protrude from the cotton. Place the fish in a petri dish and examine the tail with a dissecting microscope. Note the moving blood. Gently return the goldfish to the aquarium. Act quickly and decisively to minimize stress on the goldfish.

*b.* Which has the smallest diameter?

## Question 7

*a.* Is the blood moving at a steady rate and in the same direction?

*b.* What does this tell you?

*c.* Are you looking at an artery, vein, or capillary?

*d.* How do you know?

## HUMAN PHYSIOLOGY

 Do not do the following exercises if you have heart or lung problems. Stop immediately if you feel faint.

## Pulse

During an average lifetime, the human heart beats approximately 2.5 billion times. With each beat, the heart forces blood into arteries. This surge of blood stretches the artery coming from the heart. Surge after surge of blood from your beating heart produces waves of blood that pulse through your arteries. These pressure waves of blood are known as **pulses.** The pulse rate indicates the number of heart contractions per minute. Typical pulse rates usually range from 65 to 80 contractions per minute, but well-conditioned athletes may have rates as low as 40 contractions per minute.

Many arteries are well positioned for measuring your pulse (fig. 45.3).

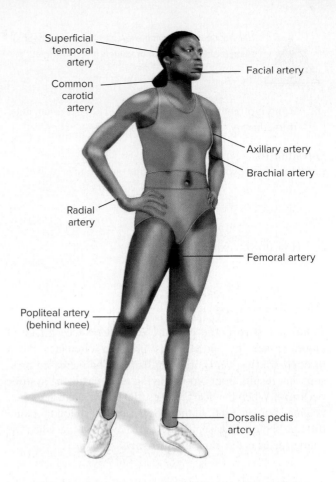

**Figure 45.3**   Many arteries in our bodies are well positioned for measuring a pulse.

---

### *Procedure 45.1*   **Measure the effect of exercise on pulse rate**

*1.* Find your pulse by placing your second and third fingers on the thumb side of your inner wrist (this is where the radial artery passes into the hand). Press down slightly. Count your pulse for 15 sec:
_____ beats in 15 sec

*2.* Multiply this number by 4 to convert this to beats per minute:

**Beats in 15 sec × 4 = _____ beats per min**

*3.* Repeat this measurement three times and average your results. Record your average resting pulse rate:
_____ beats per minute

---

**4.** Measure your pulse at your common carotid artery.

## Question 8

How does this pulse rate compare with that measured at your wrist?

---

**5.** Hold your breath for 15 sec. Then measure your pulse for another 15 sec while still holding your breath.

---

## Question 9

*a.* How does holding your breath affect your pulse rate?

*b.* How do you explain this?

Do not do the following exercises if you have heart or lung problems. Stop immediately if you feel faint.

**6.** Run vigorously in place for 5 min. Then sit down and immediately measure your pulse rate.

## Question 10
**a.** How does exercising affect your pulse rate?

**b.** How do you explain this?

## Blood Pressure

**Blood pressure** is the pressure exerted on the surface of blood vessels by blood. This pressure circulates blood through arteries, veins, and capillaries. The increased pressure that results from blood leaving the heart is the **systolic pressure.** When the heart relaxes, the arteries return to their original diameter and, in the process, squeeze blood forward through the cardiovascular system. This pressure when the heart relaxes is the **diastolic pressure.**

## INQUIRY-BASED LEARNING

### How do minor movements affect our pulse rate?

Observation: Pulse rate changes quickly and is highly sensitive to minor movements and exercise.

Question: Do minor movements such as sitting and standing affect pulse rate? How long does it take for pulse rate to recover from these changes in position?

**a.** Establish a working lab group and obtain Inquiry-Based Learning Worksheet 45 from your instructor.

**b.** Discuss with your group and instructor a specific question relevant to the preceding observation and question. When posing your question, consider magnitude of change in pulse rate as well as recovery time. Record your question on Worksheet 45.

**c.** Translate your question into a testable hypothesis and record it.

**d.** Outline on Worksheet 45 your experimental design and supplies needed to test your hypothesis. Ask your instructor to review your proposed investigation.

**e.** Conduct your procedures, record your data, answer your question, and make relevant comments.

**f.** Discuss with your instructor any revisions to your questions, hypotheses, or procedures. Repeat your work as needed.

## Question 11
**a.** Is blood pressure the same throughout the circulatory system?

**b.** How do you know?

Blood pressure, like barometric pressure, is measured in units called millimeters of mercury (mm Hg). This unit is based on a measuring device called a *manometer,* an inverted tube of liquid mercury. Pressure against the mercury reservoir at the base of the tube raises the column of mercury—the more pressure, the higher the column. Thus, the greater the number of millimeters of mercury, the greater the pressure. For example, a pressure of 100 mm Hg would raise a column of mercury 100 mm, whereas a pressure of 160 mm Hg would raise a column of mercury 160 mm.

Blood pressure is usually measured in the brachial artery just above the elbow (fig. 45.3, also see fig. 45.5). Systolic pressure there typically ranges from 100 to 140 mm Hg, with the average being near 120. Diastolic pressure typically ranges from 70 to 85 mm Hg, with an average of about 80. Blood pressure is reported as systolic pressure/diastolic pressure. Thus, a typical blood pressure is "120 over 80" (expressed as 120/80).

The difference between systolic and diastolic pressure is **pulse pressure**—this is what you feel in arteries when you touch your skin (fig. 45.4). Blood pressure is affected by many factors, including a person's intake of salt, the volume of blood, age, and the elasticity of blood vessels.

Blood pressure is measured with a stethoscope and a sphygmomanometer, which is a hollow, inflatable cuff attached to a pressure gauge. The principle is simple: You first tighten the cuff until blood flow through the artery stops. When this occurs, the pressure exerted by the inflated cuff exceeds the pressure in the artery. Then you slowly deflate (i.e., loosen) the cuff and use a stethoscope to determine when blood flow resumes through the artery (fig. 45.5).

### Procedure 45.2 Measure the effect of exercise on blood pressure

**1.** Do all of the following experiments in pairs, alternately serving as subject and experimenter. Have your lab partner lie down and relax for 2 min. Attach the cuff around his/her arm above the elbow. Tuck the flap of the bag under the fold.

**2.** Inflate the cuff to about 200 mm Hg. Because this pressure exceeds the force of the beating heart (i.e., its systolic pressure), the brachial artery in the arm

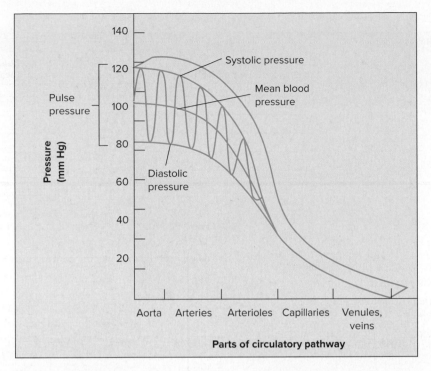

**Figure 45.4** The changing pressures associated with the pulse wave in the human circulatory system.

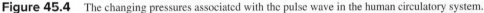

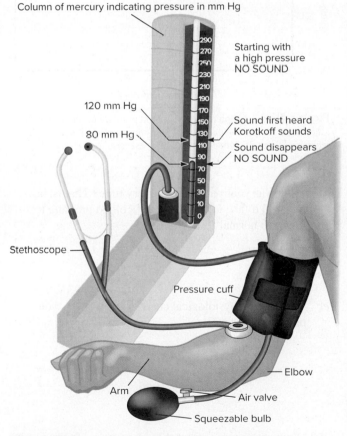

**Figure 45.5** Measurement of blood pressure with a sphygmomanometer. Blood pressure is recorded as two numbers separated by a slash; the first number is the systolic pressure, and the second number is the diastolic pressure. For a healthy 20-year-old college student at rest, a typical blood pressure is 120/80.

collapses and stops blood flow through the artery. You'll feel no pulse in your partner's wrist when pressure in the cuff is 200 mm Hg.

3. Place the bell of the stethoscope under the cuff and over the brachial artery just above the elbow. Again inflate the cuff to a pressure of about 200 mm Hg. Slowly release pressure in the cuff. When the cuff pressure falls below the systolic pressure, blood spurts through the artery. This rush of blood produces turbulence that you hear with a stethoscope as loud, tapping sounds. The pressure at which you hear these so-called Korotkoff sounds is the systolic pressure.

4. Continue to slowly release pressure from the cuff. As the pressure drops, the sounds become louder and more distinct as more blood flows through the artery. When the cuff pressure reaches the diastolic pressure, blood flow is normal (i.e., nonturbulent) and the sounds disappear. The pressure at which the sound disappears is the diastolic pressure.

5. Repeat this pressure until you obtain consistent measurements. However, do not keep the cuff inflated around your partner's arm for more than a minute or so at a time.

6. Record the average blood pressure: _____ mm Hg

7. When you stand up, gravity causes arterial pressure to decrease in the upper parts of your body and increase in the lower parts of your body. Indeed, standing up from a prone position has an effect on brachial artery blood pressure equivalent to losing about 500 mL of

### Table 45.1

**The Effect of Posture on Blood Pressure**

| Posture | Blood Pressure (mm Hg) (systolic pressure/diastolic pressure) | Pulse Rate (beats/sec) |
|---|---|---|
| Prone | _____ | _____ |
| Standing (10 sec after rising) | _____ | _____ |
| Standing (5 min after rising) | _____ | _____ |
| Standing (7 min after rising) | _____ | _____ |
| Standing (9 min after rising) | _____ | _____ |

blood. Measure your partner's pulse and blood pressure while he or she is reclined.

8. Have your partner stand up. Measure his or her pulse rate and blood pressure immediately and 5, 7, and 9 min after rising. Record your results in table 45.1.

**Question 12**

How did your partner's blood pressure change? Why?

9. Have your partner do bench-step exercises for 3 min at a rate of 60 steps per min. Have your partner lie down again and continue to measure your partner's blood pressure at 2-min intervals until it returns to normal. Plot these blood pressures on figure 45.6.

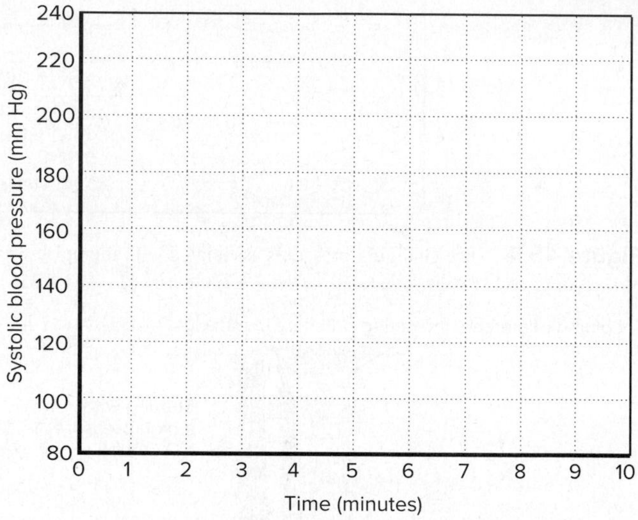

**Question 13**

*a.* How does exercise affect blood pressure?

*b.* Which is affected more, systolic pressure or diastolic pressure?

*c.* How do you explain this?

**Figure 45.6** Graph of systolic blood pressure on the *y*-axis and time on the *x*-axis. The data you plot on this graph will describe the recovery of systolic blood pressure after exercise.

*d.* What was your partner's recovery time? That is, how long did it take for your partner's blood pressure to return to normal?

*e.* What other physiological changes occurred as your partner recovered?

10. Temperature also affects blood pressure. To show this, have your partner put his or her hand in cold (5°C) water for 1–2 minutes. Then remeasure the blood pressure.

## Question 14

**a.** How does cold affect blood pressure?

**b.** Are systolic and diastolic pressures affected similarly?

**c.** Compare your data with those of your classmates. What is the average blood pressure for your class?

## Venous Blood Pressure

Blood moves slowly and at low pressure through capillaries. After exiting the capillaries, there is no mechanism (e.g., heart) to pump and increase the blood pressure so it remains low, and blood continues to move slowly through the veins. One-way valves prevent the blood under such low pressure from flowing backward (fig. 45.7).

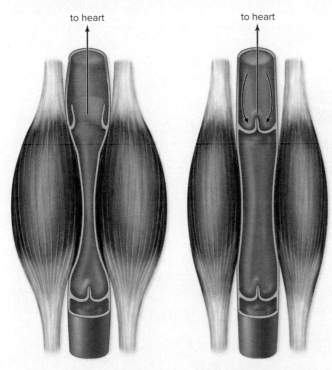

(a) Contracted skeletal muscle pushes blood past open valve.

(b) Closed valve prevents backward flow of blood as muscles relax.

**Figure 45.7** Valves regulate blood flow in veins. (*a*) Contracted muscles increase pressure and force valves open. (*b*) Relaxed muscles lower local pressure and back pressure forces the valves closed. This prevents blood from flowing backward.

---

### Procedure 45.3    Blood flow in veins

1. Hang your hands down at your side. Note the veins on the back sides of your hands.
2. Raise your hands above your head.

## Question 15
What happens to the veins? Why?

3. Have your partner hold his or her hand out to heart level and next to a meterstick taped to the wall. Slowly raise your partner's hand; the height where the veins disappear is the venous pressure measured in centimeters (cm) of water. Convert this measurement to mm Hg using the following formula:

$$mm\ Hg = cm\ water \times 0.73 =$$

## Question 16
**a.** How does blood pressure in veins compare with that in the brachial artery?

**b.** Why would you expect such a difference?

### Procedure 45.4    Locate valves in veins

1. Compress the vessels near your right elbow until the veins stand out.
2. Lay the index finger of your left hand on a vein near your wrist.
3. Move your thumb along and on top of the vein toward your elbow (i.e., toward your heart).
4. Lift your thumb and note what happens to blood in the vein. If blood refills all of the vein, repeat the experiment by placing your finger where the thumb reached. Continue until you reach a point at which the blood does not return toward the finger when the thumb is lifted.

## Question 17
What blocks the backflow of blood in veins?

## ANEMIA

Anemia is a decrease in the oxygen-carrying capacity of blood. Anemia can result from red blood cells that are too small, contain too little hemoglobin, are manufactured too slowly, or die too quickly. Iron deficiency is the most common cause of anemia; sickle cell disease (see fig. 17.4) is a type of inherited anemia.

## ATHEROSCLEROSIS

Fatty deposits inside coronary arteries reduce flow to the heart muscle. This "hardening of the arteries" is called atherosclerosis (*athero* is from the Greek word for "paste," and *sclerosis* meaning "hardness"). A cross section of a partially clogged artery is shown in the upper right corner.

## HEART ATTACK

Blocked blood flow in a coronary artery kills part of the myocardium, the heart muscle. This is a heart attack (myocardial infarction), and it may come on suddenly. A common treatment for a blocked coronary artery is a bypass operation. A surgeon creates a bridge around the blockage by sewing pieces of blood vessel taken from the patient's chest or leg onto the blocked artery.

## ARRHYTHMIA

An arrhythmia is an abnormal heartbeat. Some arrhythmias originate in the atria, causing transient flutters or racing that lasts only a few seconds. An electronic pacemaker implanted under the skin is a common treatment. In ventricular fibrillation, the ventricles contract wildly, causing sudden cardiac arrest. Death may occur within minutes.

## ANEURYSM

Atherosclerosis can so weaken the wall of an artery that a region of the vessel forms a pulsating, enlarging sac called an aneurysm. If it bursts, blood loss may be great.

## THE EFFECTS OF SMOKING ON CARDIOVASCULAR HEALTH

Smoking is the most common preventable cause of death. Cigarette smoke damages the lungs, impairing their ability to deliver $O_2$ to the heart (and increasing the chance of lung cancer). Nicotine stimulates the secretion of epinephrine and norepinephrine, increasing both heart rate and blood pressure. Nicotine also damages blood vessels and stimulates the formation of blood clots, increasing the risk of stroke.

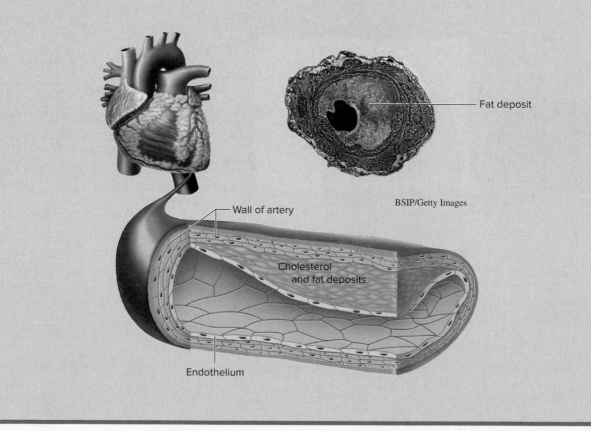

Fat deposit

BSIP/Getty Images

Wall of artery

Cholesterol and fat deposits

Endothelium

## Blood Pressure and Your Health

Your blood pressure is relatively constant when you are sitting down and resting. **Optimal blood pressure** occurs when systolic pressures are less than 120 mm Hg and diastolic pressures are less than 80 mm Hg. 118/75 is an example of an optimal blood pressure. **Prehypertension** occurs when systolic pressures are between 120 and 129 mm Hg, and/or when diastolic pressures are between 80 and 89 mm Hg. People with prehypertension will probably develop hypertension if they do nothing to lower their blood pressure. **Hypertension,** or **high blood pressure,** occurs when systolic pressures exceed 129 mm Hg and/or when diastolic pressures exceed 90 mm Hg. A blood pressure of 145/96 is an example of hypertension.

The American Heart Association estimates that approximately 50 million Americans have hypertension, and at least 30% of them do not know it. If left untreated, hypertension increases the risk of health problems such as stroke, heart attack, kidney damage, and heart failure. Smoking, obesity, stress, and a poor diet can contribute to hypertension, and regular exercise can help lower blood pressure (fig. 45.8). Although hypertension has no symptoms, it kills more than 50,000 people per year in the United States.

### Question 18

***a.*** Hypertension is the most important modifiable risk factor for stroke, the #3 killer, and a leading cause of severe, long-term disability in the United States. How could hypertension affect someone's health?

***b.*** Why is hypertension called a "silent killer"?

Organizations such as the American Heart Association provide much useful information about health and blood pressure. If your blood pressure is high, or if you'd like to learn more about how blood pressure affects your health and well-being, study their websites (e.g., http://www.americanheart.org) and visit your physician.

Gary He/McGraw Hill

**Figure 45.8** Regular exercise can help you lower your blood pressure.

## AN ANALYSIS OF YOUR RISK OF CARDIOVASCULAR DISEASE

A number of factors are suspected or proven to influence your risk of developing heart disease. Factors such as exercise, diet, persistent emotional stress, race, salt intake, obesity, age, family history, and smoking are clearly associated with the probability of suffering from cardiovascular disease later in life, even though no single factor is a guaranteed predictor. To assess your risk, complete the following questionnaire provided by the The University of Arizona's Sarver Heart Center. To complete the questionnaire, simply record the number of points assigned to each level of each risk factor. Compare your total points with the ranges associated with different levels of risk. Although this is not a definitive test, it will heighten your self-awareness of the consequences of your lifestyle.

**The University of Arizona's Sarver Heart Center's Heart Disease Risk Assessment**

DIRECTIONS: Note your score for the various risk factors. Total your points at the end and find your risk of having a heart attack or stroke during the next 10 years. BP = blood pressure

| RISK FACTOR | SCORE | YOUR SCORE |
|---|---|---|
| AGE | | |
| Male 0–44 | 0 | |
| Male 45–55 | 2 | |
| Male 56–65 | 3 | |
| Male 66–75 | 4 | |
| Male > 75 | 5 | |
| Female 0–54 | 0 | |
| Female 55–65 | 2 | |
| Female 66–75 | 3 | |
| Female > 75 | 5 | |
| **HYPERTENSION** (High BP) | | |
| Never hypertensive; BP below 130/80 | 0 | |
| Treated hypertension; BP average below 130/80 | 1 | |
| Treated hypertension; BP average above 130/80 | 2 | |
| Untreated hypertension; BP usually above 130/80 | 3 | |
| **SMOKING** | | |
| Never smoked or nonsmoker for 5 or more years | 0 | |
| Nonsmoker for less than 5 years | 1 | |
| Current smoker for <10 years | 2 | |
| Current smoker for more than 10 years | 3 | |
| **DIABETES** | | |
| Absent | 0 | |
| Type II, treated with diet alone | 1 | |
| Type II, treated with pills and/or insulin | 2 | |
| Type I (juvenile onset), insulin dependent | 3 | |
| **ACTIVITY LEVEL** (aerobic exercise, at least 30 minutes per session) | | |
| 4 or more times/week | 0 | |
| 2–3 times/week | 1 | |
| 0–1 time/week | 2 | |

©The University of Arizona Sarver Heart Center. Reprinted by permission.

| RISK FACTOR | SCORE | YOUR SCORE |
|---|---|---|
| **TOTAL CHOLESTEROL (mg/dL)** | | |
| Below 200 | 0 | |
| Above 200 | 1 | |
| **LDL (bad) CHOLESTEROL (mg/dL)** | | |
| below 70 | 0 | |
| 70–100 | 1 | |
| above 100 | 2 | |
| **HDL (good) CHOLESTEROL (mg/dL)** | | |
| Above 45 (female) or above 40 (male) | 0 | |
| Below 45 (female) or below 40 (male) | 1 | |
| **TRIGLYCERIDES (mg/dL)** | | |
| Below 150 | 0 | |
| Above 150 | 1 | |
| **WAIST CIRCUMFERENCE** | | |
| Male more than 40 inches | 2 | |
| Female more than 35 inches | 2 | |
| Male less than 40 inches | 0 | |
| Female less than 35 inches | 0 | |
| **STRESS LEVEL** (personal perception) | | |
| I don't feel stress | 0 | |
| I feel stressed 1–2 times per week | 1 | |
| I feel stressed most days | 2 | |
| I feel stress everyday | 3 | |
| **BODY MASS INDEX (BMI)** | | |

Calculate your body mass index (BMI) with the following formula: BMI = (mass in kg)/(height in m)$^2$ For example, a person with a mass of 75 kg and a height of 1.80 m has a BMI = $75/(1.8)^2 = 75/3.24 = 23.15$ (You can also find websites that will calculate your BMI from these measurements.)

| | | |
|---|---|---|
| Below 25 | 0 | |
| 25–29.9—overweight | 1 | |
| Above 30—obese | 2 | |
| Above 40—severe obesity | 3 | |

| RISK FACTOR | SCORE | YOUR SCORE |
|---|---|---|
| **FAMILY HISTORY** (heart attack or angina in **sibling or parent** under the age of 55 if male, 65 if female) | | |
| No family history | 0 | |
| One family member | 1 | |
| More than one family member | 2 | |
| **PERIPHERAL VASCULAR DISEASE** | | |
| None | 0 | |
| History of stroke, claudication, aortic aneurysm, carotid artery disease | 2 | |
| **CORONARY ARTERY DISEASE** | | |
| No known history | 0 | |
| Known plaque through calcium score or angiogram | 1 | |
| Prior heart attack or intervention (bypass, stent) | 3 | |
| **DIET/NUTRITION** | | |
| Vegan (no meat or dairy) | 0 | |
| Vegetarian (no meat) | 1 | |
| Eat everything, but no trans and minimum saturated fat | 2 | |
| No restrictions to what I eat | 3 | |

**MAXIMUM TOTAL POINTS: 38   YOUR TOTAL ___**

Your risk of having a heart attack or stroke during the next 10 years, based on your total score:

- **0–4:** Lowest risk (less than 3 percent)—Heart attack extremely unlikely.

- **5–9:** Low risk (3–8 percent)—You are doing most things right, but you can probably do some fine-tuning. If you pay attention, you will live a long, healthy, active, and productive life.

- **10–13:** Moderate risk (9–14 percent)—You do some things right, but you have work to do. Identify your risk factors and take corrective action.

- **14–16:** High risk (15–19 percent)—You stand a good chance of having a first heart attack before you are 55, if male, and 65, if female. If you have already had a heart attack, you are at high risk for a repeat performance.

- **Over 16:** Extreme risk (greater than 20 percent) — Unless you have a death wish, you have work to do. It will take time and effort to move to the lower risk categories. Don't bite off too much at one time. The process of change should be slow, committed, persistent, and permanent.

This analysis is modeled after risk analyses such as the Framingham Coronary Heart Disease Risk Score and the ACC Heart Disease Risk Assessment. For more information, see www.heartseries.org and consult your physician.

## Questions for Further Study and Inquiry

*1.* What is the value of increased blood pressure and pulse rate during exercise?

*2.* A well-conditioned athlete shows fewer changes in his or her circulation and breathing in response to exercise than does someone in poor condition. Why?

3. How would "hardening of the arteries" (arteriosclerosis) affect blood pressure? Why?

4. What is congestive heart failure?

5. Why is high blood pressure dangerous?

6. Vertebrates have a closed circulatory system, meaning that the blood is always enclosed within vessels and does not fill body cavities. Mollusks (Exercise 37) and arthropods (Exercise 39) have open circulatory systems, meaning that blood is pumped by a heart into body cavities, where tissues are surrounded by the blood. What are the advantages and disadvantages of each type of circulatory system?

7. Giraffes have blood pressures that often exceed 340/230, whereas salamanders have blood pressures as low as 22/12. How would you explain these differences?

8. Do all arteries carry $O_2$-rich blood? Explain.

9. Although red blood cells in humans have a lifespan of only about 4 months, we don't run out of them. Why?

10. Do some library research and list up to 10 diseases that directly affect the heart and circulatory systems.

**WRITING TO LEARN BIOLOGY**

How do you think that smoking affects blood pressure? Why?

# Human Biology
## Sensory Perception

## Learning Objectives

By the end of this exercise you should be able to:

1. Identify the location and basis for your blind spot.
2. Describe afterimages, the distribution of your touch and taste receptors, your type of eye dominance, adaptation to stimuli, and the effect of stimulus intensity on perception.
3. Measure your visual acuity, near point, astigmatism, and peripheral vision.
4. Distinguish between nerve deafness and conduction deafness.

Please visit **connect.mheducation.com** to review online resources tailored to this lab.

Like most animals, you have a complex nervous system that informs your brain of your body's condition and provides your brain with detailed information about the environment. The nervous system also coordinates movements and perceives, translates, and responds to environmental stimuli such as light, touch, and temperature. A deficiency in any part of your sensory systems would be dangerous because you could no longer respond to your ever-changing external and internal environment.

The purpose of this exercise is to increase your awareness of the senses that you use daily. Do all of these experiments with a partner; alternate with him or her as subject and experimenter.

## BLIND SPOT

The **retina** is a layer of photoreceptors on the back inner surface (i.e., the posterior, deepest, or innermost surface) of the eye (fig. 46.1). Most of this surface is covered by photoreceptor cells called rods and cones, which are modified epithelial cells (fig. 46.2). A human retina has roughly 160,000 photoreceptors per mm², whereas an owl retina may have 700,000 per mm². The central fovea of the retina is the region used for color vision and the region giving greatest visual acuity (sharpness of image). Other parts of the retina are important for peripheral vision, but images there are not in sharp focus.

The **optic disc** is the region of the retina where blood vessels and the optic nerves enter or leave the retina. This region lacks photoreceptors and is therefore a "blind spot" of the retina. We usually don't notice the blind spot because our brain fills in the blank area for us. However, it's still there.

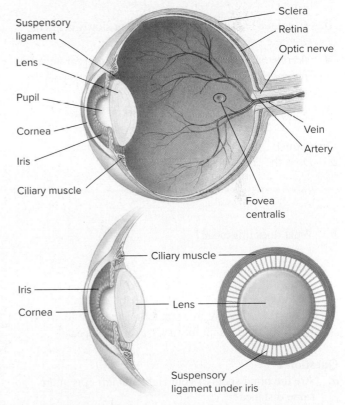

**Figure 46.1** Structure of the human eye. The transparent cornea and lens focus light onto the retina at the back of the eye, which contains the rods and cones. The center of each eye's visual field is focused on the fovea centralis, a 1.5-mm-wide depression on the surface of the retina that is responsible for the eye's sharpest vision. Focusing is accomplished by contraction and relaxation of the ciliary muscle, which adjusts the curvature of the lens.

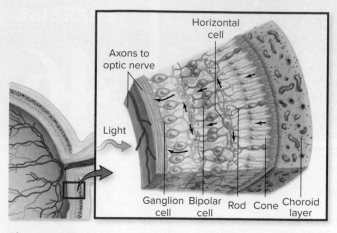

Figure 46.2 Structure of the retina. Note that the rods and cones are at the rear of the retina, not the front. Light passes through several layers of ganglion and bipolar cells before it reaches the rods and cones. Arrows show the direction of nerve impulses.

### Procedure 46.1 Discover your blind spot

1. Cover your left eye. Hold figure 46.3 about 50 cm (20 in) from your face and directly in front of your right eye.

2. Stare at the cross in figure 46.3. You can also see the circle.

3. Continue to stare at the cross as you slowly bring the figure closer to your eye.

4. At one point the circle will seem to disappear because its image has fallen on your blind spot. Have your partner record that distance.

5. Continue to move the figure closer to your face.

**Question 1**

*a.* Does the circle reappear?

*b.* What does this mean?

6. Test your other eye in a similar manner, but focus on the circle and watch for the cross to disappear.

**Question 2**

*a.* Are the blind spots of your right and left eyes at the same distance?

Figure 46.3 Images for detecting a blind spot.

*b.* What is the anatomical cause of the blind spot in your eye?

The fovea is sometimes invaded by blood vessels. When this occurs, the fovea also becomes a blind spot. This condition characterizes a disease called macular degeneration, the most common cause of legal blindness in people older than 65.

## EYE DOMINANCE AND BINOCULAR VISION

We're all familiar with preferences for using a particular hand for jobs such as writing and throwing. Our eyes also exhibit right-left dominance.

### Procedure 46.2 Determine eye dominance

1. With both eyes open, carefully focus on an object a few feet away.

2. Close one eye, then reopen it.

3. Close the other eye, then reopen it.

Which eye seems more directly in line with the object? If it's the right eye, you are right-eye dominant; if it's the left eye, you are left-eye dominant. If the object is in the middle of both eyes, you are central-eye dominant.

**Question 3**

*a.* Do you have right, left, or central dominance?

*b.* When you look at a distant object, why would you move your head and eyes to an off-center position when one eye is dominant?

Eye dominance is important for how we see and react to our world. For example, right-handed hitters in baseball have their right hand in the upper control position when they hit. Similarly, 65% of baseball players are right-eye and right-hand dominant. Only about 17% are crossed dominant (right hand–left eye or left hand–right eye), whereas another 18% have no eye dominance. These players see the world from a point halfway between both eyes. Interestingly, the best hitters (as judged by batting average) either are crossed dominant or lack dominance. The best pitchers (judged by their earned-run averages) have central-eye dominance.

## Question 4
How could eye dominance affect one's ability to hit a baseball?

Binocular vision refers to vision through two eyes. Each eye views an object at a different angle. The brain blends this information from each eye and produces a stereoscopic, three-dimensional image.

## Procedure 46.3   Binocular vision

1. Obtain a test tube and pencil.
2. Have your lab partner hold a test tube vertically approximately two feet from your face.
3. Close one eye. Try not to touch the rim of the test tube with the pencil. Lower the pencil into the test tube.
4. Based on your success, devise a simple experiment to test the value of binocular vision.

5. Ask your instructor to review your experimental design and data.

## NEAR POINT

The shortest distance at which an object is in sharp focus is called the **near point:** the closer the distance, the greater your eye's ability to accommodate for changes in distance. This distance gradually increases as we get older. By age 60, this distance is very large, a condition called presbyopia (table 46.1).

## Procedure 46.4   Determine your near point

Here's how to determine your near point. If you wear glasses or contacts, keep them on.

1. Hold this page in front of you at arm's length. Close one eye and focus on a word on this page.
2. Slowly move the page toward your face until the image of the word is blurred. Then move the page away until the image is sharp. Have your partner record the distance between your eye and the page. That distance is your near point.

### Table 46.1

**Typical Values for Distance to Near Point by Age Groups**

| Age | Distance to Near Point (cm) |
|---|---|
| 10 | 9 |
| 20 | 10 |
| 30 | 13 |
| 40 | 18 |
| 50 | 53 |
| 60 | 83 |
| 70 | 100 |

## Question 5
How does our near point affect what we do and see?

## AFTERIMAGES

Images sensed by your eyes do not disappear immediately after you close your eyes or shift your glance. Rather, an image lingers as an afterimage.

## Procedure 46.5   Demonstrate afterimage

1. Place a large sheet of white paper and another sheet of black paper on your laboratory bench a short distance apart.
2. Place a bright blue index card on the black paper. Stare at the card intently for 30 sec.
3. Quickly shift your gaze to the white paper. Record your observations.
4. Repeat this procedure with a yellow card and record your observations.
5. Repeat the experiment, this time placing the card initially on the white paper and shifting your stare to the black paper. In the following space, record and provide an explanation for your results.

## VISUAL ACUITY

Visual acuity is the sharpness of a visual image and is usually measured with a Snellen eye chart. The sizes of letters on the chart are such that you should be able to read the first line (i.e., the letter E) of the chart from 200 ft away. Conversely, line 8 of the chart is tall enough so that it should be read from 20 ft away. Thus, if a person can read line 8 from 20 ft away, his or her visual acuity is designated as 20/20,

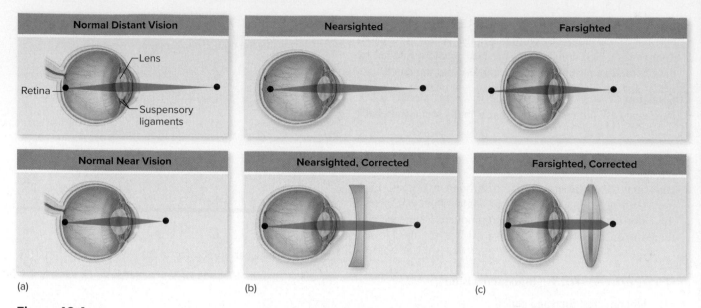

| Normal Distant Vision | Nearsighted | Farsighted |
| Normal Near Vision | Nearsighted, Corrected | Farsighted, Corrected |

(a)  (b)  (c)

**Figure 46.4** Focusing the human eye. (*a*) In people having normal vision, the image remains focused on the retina in both near and far vision because of changes produced in the curvature of the lens. When a person with normal vision stands 20 ft or more from an object, the lens is in its least convex form and the image is focused on the retina. (*b*) In nearsighted people, the image comes to a focus in front of the retina and the image thus appears blurred. (*c*) In farsighted people, the focus of the image would be behind the retina because the distance from the lens to the retina is too short.

or 1. If you can see line 8 from 10 feet away but not farther, your acuity is 20/40, or 1/2 of normal. Snellen charts typically have letter sizes for assessing acuity of 20/15, which is better than normal acuity, to 20/200, which is poor acuity.

The eyes of farsighted people focus the image behind their retina. These people are referred to as farsighted because they see distant objects clearer than close objects. Conversely, the eyes of nearsighted people focus the image in front of the retina; these people see close objects better than they do distant objects. Both conditions can be corrected with glasses or contact lenses provided that the fovea remains functional (fig. 46.4).

## Procedure 46.6 Determine visual acuity

1. Locate the Snelling eye chart in lab. Stand 20 ft (6.1 m) from the chart, facing it.

2. Cover one eye and read the letters that your partner points to on the chart. Begin at the top of the chart and work your way down.

3. Note the lowest row of letters that you can read accurately. Record the number printed next to that row: _____. This number is the farthest distance (measured in ft) that a person with normal vision can read the letters in that row. For example, if the number is 40, then the person has 20/40 vision, meaning that the person can see at 20 ft what a person with normal vision can see at 40 ft.

4. Test both eyes. If you wear glasses or contacts, test your eyes with and without your lenses.

**Question 6**
What is your visual acuity?

Right eye: _____      Left eye: _____

## ASTIGMATISM

Astigmatism usually results from abnormal curvature of the cornea. Test for astigmatism by facing an astigmatism test chart 10 ft (3 m) away. Cover one eye and stare at the

### Sensory Perception by the Numbers

Humans can detect more than 45,000 scents and see more than 7 million colors. Displaying this many colors would require a paint chart more than 21 miles long.

The human brain weighs about 1.4 kg, of which more than 1 kg is water. For comparison, an elephant's brain weighs 4.7 kg, a dolphin's brain weighs 700 g, a cat's brain weighs 31 g, and a mouse's brain weighs 0.4 g.

When we touch something, the touch-signal travels to our brain at 200 km h$^{-1}$ (124 mph). For comparison, reflex signals travel at 400 km h$^{-1}$ (250 mph), thought signals at 110 km h$^{-1}$ (70 mph), and pain signals at 5 km h$^{-1}$ (3 mph). On average, blood moves at a speed of approximately 5.6 km h$^{-1}$ (3.5 mph), and food moves through our esophagus at about 0.07 km h$^{-1}$ (0.75 inch per second).

circle in the center of the chart. Test both eyes; if you wear glasses or contact lenses, do the test with and without your glasses or contact lenses. If all lines radiating from the circle are straight, no astigmatism is present. If the lines appear crooked or wavy, you probably have astigmatism.

## PERIPHERAL VISION

Dim light is best perceived by rods, whereas bright light and colors are best perceived by cones. Most cones are in an area called the central fovea, a small part of the retina we use when we focus directly on an object and look at its color and detail. In this area of the retina, almost all of the receptors are cones. Rods dominate the periphery of the visual field. This demonstration will show the monochrome nature of peripheral vision.

### Procedure 46.7    Test peripheral vision

1. Have your lab partner stare forward.
2. Slowly bring a piece of colored paper into your partner's visual field from behind his or her head.
3. Stop when your partner tells you that the paper has just entered his or her visual field.
4. Ask what color the paper is. If you did the test properly and if your partner didn't cheat, he or she will probably be unable to determine the color of the paper. Why?

## COLOR BLINDNESS

Color blindness is a color vision deficiency that usually is inherited (see Exercise 17). The most common type of color blindness is red-green color blindness that results from a deficiency of red- and green-sensitive cones. People with this deficiency have difficulty distinguishing shades of red and green. A totally color-blind person sees everything as a shade of gray.

Test for color blindness with the *Ishihara Color Test Book* available in the lab. Hold the plates about 60 cm from your partner's face in bright, natural light. Give your partner about 5 sec to view each plate. Without touching the figure, ask your partner to name the number in each mosaic.

If the *Ishihara Color Test Book* is not available, examine the image in figure 46.5 in bright light.

## HEARING LOSS

There are two kinds of hearing loss: sensorineural (nerve) deafness and conduction deafness. Nerve deafness results from damage to the sound receptors or neurons that transmit impulses to the brain (fig. 46.6). Such damage is usually caused by exposure to loud sound; modern surgical implants may improve, but not fully restore, hearing. Conduction deafness results from damage that prevents sound vibrations from reaching the inner ear, especially damage to the ear canal, ear drum, or bones of the middle ear. Causes include middle ear infections

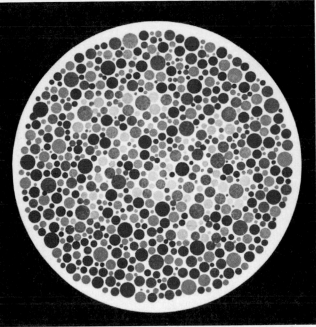

Steve Allen/Brand X Pictures/Getty Images

**Figure 46.5**    Color-blindness test. A person with normal color vision should see the number 74 here. (This is only an example and is not meant for diagnosis.)

and abnormal bone growth near the middle ear. Conduction deafness is usually correctable by surgery or hearing aids.

You can use a tuning fork to distinguish between nerve and conduction deafness.

### Procedure 46.8    Testing for hearing loss

1. Have your lab partner sit down and plug one ear with cotton.
2. Strike the tuning fork against the heel of your hand (do not strike the fork against a hard object).
3. Hold the fork about 25 cm (10 in) away from your partner's ear, with the edge of the fork pointing toward the ear (fig. 46.7). If your partner can hear the fork, he or she has normal hearing or only minimal hearing loss.
4. As the sound fades, have your partner tell you when he or she can no longer hear the fork. Then quickly and without touching the prongs of the vibrating fork place the base of the fork against the temporal bone behind the ear. If a slight hearing loss exists and the sound reappears, some conduction deafness is present. You may want to simulate a conduction impairment by having your partner wear an earplug.

**Question 7**
How does an earplug interrupt sound?

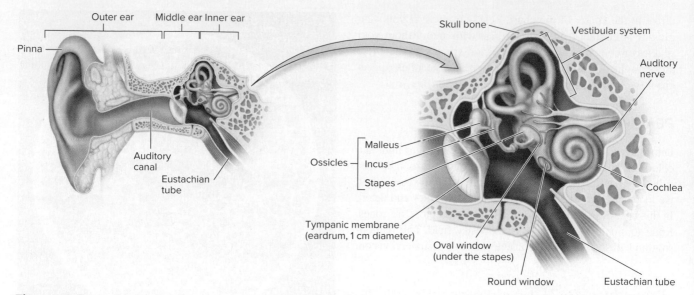

**Figure 46.6** Structure of the human ear. Sound waves passing through the ear canal produce vibrations of the tympanic membrane, which cause movement of the middle ear ossicles (the malleus, incus, and stapes) against an inner membrane called the oval window. Vibration of the oval window sets up pressure waves that travel through the canals of the cochlea.

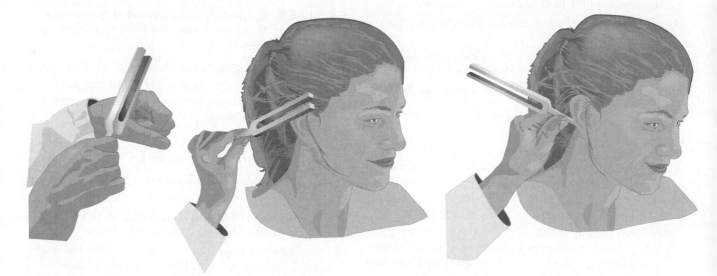

**Figure 46.7** Using a tuning fork to test for hearing loss.

Someone with severe hearing loss will not hear the vibrating fork (or will hear it only briefly). That person suffers conduction deafness if the sound reappears when the end of the fork is placed against his or her temporal bone. If the sound does not reappear, the person is nerve deaf.

## DISTRIBUTION OF TOUCH RECEPTORS

Touch receptors are varied and distributed unevenly in the skin (fig. 46.8). To perceive two stimuli as separate sensations, the stimuli must be far enough apart to stimulate two touch receptors. Thus, the distribution of touch receptors affects our ability to distinguish different stimuli.

---

*Procedure 46.9*   **Test for distribution of touch receptors**

1.  Have your lab partner close his or her eyes.

2.  Touch your partner's skin on the back of the calf with one or two points of a scissors (or caliper) spread far apart. Take care while touching your lab partner with scissor points. Ask your partner to report the sensation as "one" or "two." Randomly alternate the two-point touches with one-point touches to keep your partner honest.

3.  Decrease the distance between the two points until your partner reports a one-point stimulus about 75% of the time. Measure this distance as the minimum distance evoking a two-point sensation.

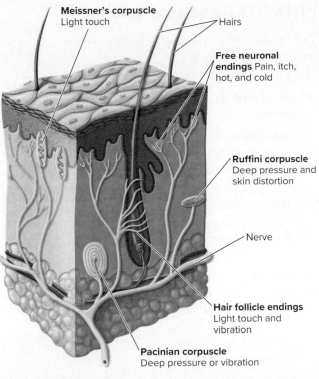

**Meissner's corpuscle**
Light touch

Hairs

**Free neuronal
endings** Pain, itch,
hot, and cold

**Ruffini corpuscle**
Deep pressure and
skin distortion

Nerve

**Hair follicle endings**
Light touch and
vibration

**Pacinian corpuscle**
Deep pressure or vibration

**Figure 46.8**   Examples of sensory receptors in the skin of mammals. Organisms use sensory receptors to obtain information about the environment. External cues are converted into information that the organisms use to survive.

---

*4.* Repeat this procedure on the following parts of your partner's body: (a) inside the forearm, (b) back of the neck, (c) palm of the hand, and (d) tip of the index finger.

**Question 8**

*a.* Which parts of your body are most sensitive to touch?

*b.* Which parts are least sensitive?

*c.* What is the significance of these differences in sensitivity?

*d.* How would someone's sensitivity to touch by their fingertips affect their ability to read Braille?

*e.* What common touches would stimulate each of the receptors shown in figure 46.8?

---

## SENSING TASTE

Taste is a poorly understood sense because our final perception of flavor arises from a combination of other sensory inputs. The texture, smell, and appearance of foods strongly influence our sense of taste. Strictly speaking, the sense of taste refers to sensations originating from taste cells in the mouth and has traditionally been described as a combination of sensations of salty, bitter, sour, sweet, and possibly other sensations. To cause a taste sensation, food dissolves in saliva, moves through taste pores in taste buds (fig. 46.9), and interacts with neural

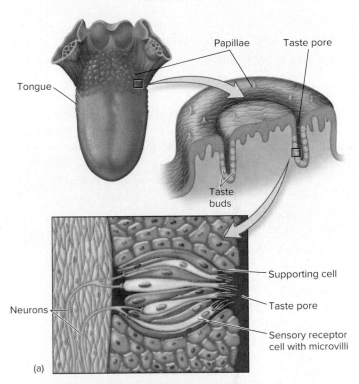

Papillae

Taste pore

Tongue

Taste
buds

Supporting cell

Taste pore

Neurons

Sensory receptor
cell with microvilli

(a)

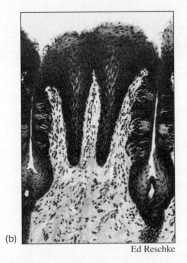

(b)

Ed Reschke

**Figure 46.9**   Taste. (*a*) Structures involved in the sense of taste. This sense occurs in taste buds, which contain the sensory receptor cells that respond to dissolved food molecules. (*b*) Individual taste buds are bulb-shaped collections of chemical receptor cells that open into the mouth through a pore (200×).

receptor cells. When food interacts with receptor proteins or with ion channel proteins on the surface of these cells, an electrical impulse is generated and sent to the brain.

Biologists once thought that each of the four areas of the tongue were exclusively responsible for a particular sensation (sweet, sour, salty, bitter). However, research shows that neurons in all areas of the tongue are widely responsive to a variety of tastes. The final perception of taste is an integration of signals from receptors and is more complex than was first thought. An integration of impulses from taste receptors combined with other senses (vision, touch, and smell) all combine to produce flavor.

## *Procedure 46.10*   Test for taste sensitivity

1. Before doing this procedure discuss with your lab instructor if you or your partner have any food allergies.
2. Cut two identical small pieces of a raw onion, raw apple, and raw potato. Have your partner close his or her eyes, clamp the nostrils shut, and open his or her mouth.
3. Put a piece of onion, apple, or potato on your partner's tongue. Ask your partner to identify it without swallowing.
4. Repeat the procedure with the other food.
5. Record results and then repeat the procedure with the nostrils open.

## Question 9
For enjoyment of which foods is aroma particularly important? For which foods is it less important?

## INTENSITY OF SENSATIONS

The intensity of a sensation is usually proportional to the intensity of the stimulus.

## *Procedure 46.11*   Investigate intensity of sensations

1. Fill large pans with ice water, warm water, and water at room temperature.
2. Sequentially place your index finger in the water in each pan. Can you sense the changes in temperature?
3. Repeat the experiment (1) with different fingers and (2) with pans containing water having only slight differences in temperature.

## Question 10
Would you expect your toes to detect narrower ranges of temperature than your fingers would detect? Why?

4. Place your left hand in the ice water and your right hand in the warm water (be sure that the warm water has not cooled). Leave your hands in these pans until you "get used to" the temperatures.

## Question 11
*a.*   Does the sensation change with time? Why or why not?

## INQUIRY-BASED LEARNING

### *How is the nervous system adapted to sense stimuli?*

Observation: Your nervous system can "tune out" stimuli so that you are not constantly bombarded with trivial sensations such as clothing, wristbands, and background noise. Imagine your world without that adaptation!

Question: Does the time needed for your skin to adapt to stimuli depend on the nature of the stimulus?

a. Establish a working lab group and obtain Inquiry-Based Learning Worksheet 46 from your instructor.
b. Discuss with your group a well-defined question relevant to the preceding observation and question. Review procedure 46.10 as a starting point for your investigation. Record your question on Worksheet 46.

c. Translate your question into a testable hypothesis and record it. Consider hypotheses about temperature, pressure, texture, or composition (metal, wood, paper, etc.) of the stimulus.
d. Outline on Worksheet 46 your experimental design and supplies needed to test your hypothesis. Ask your instructor to review your proposed investigation.
e. Conduct your procedures, record your data, answer your question, and make relevant comments.
f. Discuss with your instructor any revisions to your questions, hypotheses, or procedures. Repeat your work as needed.

**b.** Would air at the same temperature feel the same? Why or why not?

**c.** In terms of receptors, what does it mean to "get used to" a sensation?

**d.** Would your perceptions have changed if the cold water was 0°C and the warm water was 80°C?

**e.** Why would it be important to be able to perceive the difference between 45°C and 80°C?

**f.** After at least 1 minute, place both hands in the pan of water at room temperature. Note the sensation. Does it change with time? Why or why not?

**g.** What do temperature receptors really respond to?

**h.** Were your receptors giving you useful information in both situations even though the information they gave you changed?

## OPTICAL ILLUSION

An optical illusion is an experience of seeming to see something that does not exist or that is different than it appears. The optical illusion presented here "confuses" certain neurons in your brain that sense both the orientation and direction of movement of visual curves and lines.

> ### Procedure 46.12    Optical illusion
> 1. Focus on the dot in the center of figure 46.10.
> 2. Slowly move your head toward, and then away from, the image.

**Question 12**
What happened? What did you see? Explain your results.

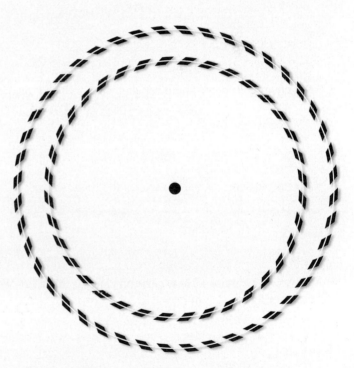

**Figure 46.10**    An optical illusion.

# Questions for Further Study and Inquiry

1. Are you aware of all activity within your nervous system? Why is this advantageous?

2. How could you show whether there are specific receptors in your skin that respond to heat and cold?

3. Are all regions of your body equally sensitive to touch and temperature?

4. Do animals differ in how they perceive their environment? Explain your answer and list some examples.

5. What extra receptors would help humans survive in today's world?

6. How is our sense of balance related to vision?

7. Review in Exercise 17 how color blindness is inherited. Would you expect color blindness to be more common in males or females? Why?

## WRITING TO LEARN BIOLOGY
What are the advantages and disadvantages of receptor adaptation to an organism?

# Vertebrate Anatomy
## External Features and Skeletal System of the Rat

## Learning Objectives

By the end of this exercise you should be able to:
1. Define the terms used to describe the general planes, surfaces, and anatomy of a vertebrate.
2. Identify the external mammalian features of the male and female rat.
3. Identify the bones of a rat skeleton.
4. Compare the skeletal system of a rat with the skeletal system of a human.
5. Discuss the relationship of structure and function for the external features and skeletal system of a rat.

Please visit **connect.mheducation.com** to review online resources tailored to this lab.

**S**tudying anatomy involves much more than just memorizing and recognizing parts of an organism. It requires understanding the relationship of the function of an anatomical feature to the structure and location of that feature. As you study rat anatomy, remember that your objective is not just to memorize parts of the rat but to understand the overall design of the rat's anatomy and how this structure relates to function. We have chosen a rat (*Rattus norvegicus*) for dissection because it is a small and familiar vertebrate.

### Question 1
What features of rats make them a good experimental model to learn about factors that affect humans?

Before you begin handling your specimen, review the safety information presented on pages xiii-xiv. Chemically preserved specimens must always be treated with care. Protect your eyes, and be aware of allergic reactions.

Before you begin dissection, familiarize yourself with the following terms. They describe the location of structures and their relationship to other structures (fig. 47.1).

**Dorsal**—toward the upper surface (back)

**Ventral**—toward the lower surface (belly)

**Anterior**—toward the head

**Posterior**—toward the tail

**Cranial**—toward the head

**Caudal**—toward the tail

**Medial**—toward the midline of the body

**Proximal**—toward the end of the appendage nearest the body

**Lateral**—away from the midline of the body; toward the side

**Distal**—toward the end of an appendage farthest away from the body

**Frontal plane**—divides the body into dorsal and ventral halves; two-dimensional plane parallel to the anterior-posterior axis and perpendicular to the dorsal-ventral axis

**Transverse plane**—a cross section; two-dimensional plane perpendicular to the anterior-posterior axis of the body

**Sagittal plane**—divides the body into left and right halves; two-dimensional plane parallel to the anterior-posterior axis and parallel to the dorsal-ventral axis

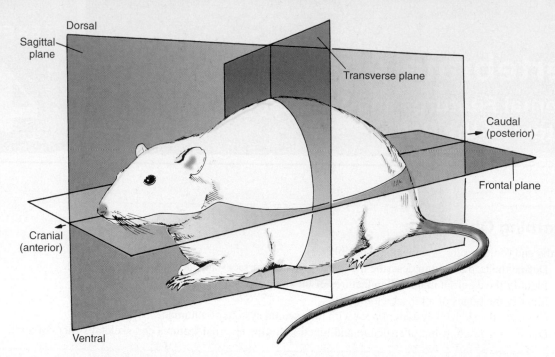

**Figure 47.1**  Surfaces and planes of section of a rat.

## EXTERNAL ANATOMY

Rinse your rat with cold water to remove excess preservative. Then lay it in a small dissecting pan and observe its general characteristics.

The rat's body is divided into six anatomical regions: the **cranial region** or head, the **cervical region** or neck, the **pectoral region** or the area where forelegs attach, the **thorax** or chest region, the **abdomen** or belly, and the **pelvic region** or area where hind legs attach. The terms *right* or *left* refer to the organism's right or left side. Notice in figure 47.1 that the *left* side of the rat is in full view.

Locate and examine the structures noted in bold print in the following descriptions. A hairy coat called the **pelage** covers most of the rat's body (fig. 47.2). The tactilely (relating to touch) sensitive "whiskers" on the face are called **vibrissae.** The **nares** (nostrils) are at the anterior end of the head. The **subterminal mouth** has a **cleft** in the upper lip that exposes large front teeth called **incisors.** The eyes bulge from the head and enable the rat to see in almost all directions at once. Eyes equipped with large **pupils** are one of many adaptations to nocturnal life. The **nictitating membrane** is located at the inside corner of the **eye.** This translucent membrane may be drawn across the eyeball for protection. Locate the **eyelids**—they are similar to those found in humans. **Ears** are located posteriorly on the head and composed of a **pinna** (external ear) and an **auditory meatus** (ear canal).

On the ventral surface of the rat and just lateral to the midline are six pairs of nipples or **teats.** The **tail** is located posteriorly (caudally) and is covered with hair and scales. Ventral to the base of the tail is the **anus.** Just caudal to the last pair of teats of a female is a small protuberance, the

**clitoris,** from which the **urinary aperture** opens. Just caudal to the clitoris is the **vaginal orifice** in a small depression called the **vulva.**

In males, a pair of large **scrotal sacs** lie on each side of the anus and contain the **testes,** which produce sperm cells. The testes may be withdrawn into the abdomen during nonreproductive periods. Just anterior to the scrotal sacs is the **prepuce,** a bulge of skin surrounding the **penis.** The penis can be extended from the prepuce through the **preputial orifice.** The end of the penis has a **urogenital orifice** where both urine and sperm cells exit the body. Be sure to examine rats of both sexes.

The rat is **quadruped** (four legs) with four digits and a vestigial thumb on each foot. Both the sole and heel of a rat's foot contact the ground as indicated by the lack of hair from the toes to the heel. This stance is **plantigrade.** Some other animals are **digitigrade** and walk on their toes.

**Question 2**

*a.* Which is more cranial, the shoulder of the forelegs or the hips of the hind legs?

*b.* A slice through which plane would divide the brain into left and right halves?

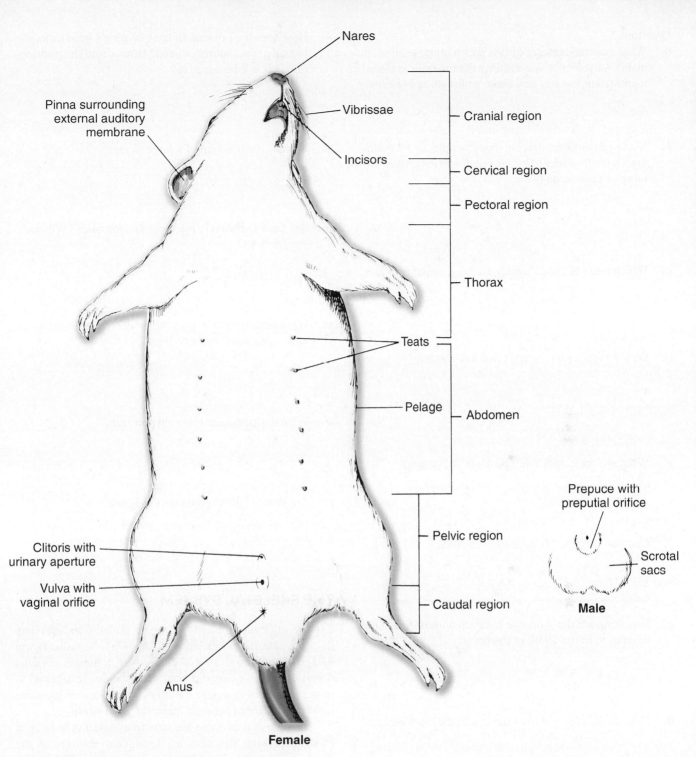

**Figure 47.2** External anatomy of male and female rats.

*c.* A slice through which plane would separate the upper from the lower jaw?

*d.* Which is more proximal: the tip of the tail or the base of the tail?

## Question 3

**a.** What external features of rats are common to all mammals? As you continue to study the rat, refer to these mammalian features and make additions and deletions.

**b.** Mammals are very diverse. For example, do all mammals have four legs? Do you recall any mammals with more or fewer than two eyes?

**c.** Which areas of the rat's body are not covered with hair?

**d.** Why do these parts of the body have no hair?

**e.** What are some additional functions of pelage?

**f.** Can you suggest a function for vibrissae?

**g.** How long are the vibrissae in relation to the head and in relation to the width of the body?

**h.** How do the upper incisors mesh with the lower incisors?

**i.** Are any teeth other than incisors present?

**j.** Why is the mouth referred to as *subterminal?*

**k.** How does the position of eyes on a rat's head differ from that on a human's head? How would the position of a rat's eye be adaptive?

**l.** What is the function of a large pinna?

**m.** Are teats exclusively mammalian structures? Why do you think so?

**n.** What term similar to *quadruped* describes animals such as humans with two legs?

**o.** What digitigrade animals do you recall?

**p.** Are humans plantigrade or digitigrade?

## THE SKELETAL SYSTEM

Examine a cleaned and articulated rat skeleton and compare its parts with those labeled in figure 47.3. See also figure 42.1. Then compare a rat skeleton with a human skeleton if one is available. Record your observations in table 47.1. In the margin of this page list a few words describing some general differences between these skeletal systems.

Locate and examine the structures noted in bold print in the following descriptions. Prominent features of the **skull** include the **mandible** (lower jaw), **maxilla** (upper jaw), the **zygomatic arch,** and the **cranium** (braincase). The mandible has a broad, flat surface for muscle attachment, and some of the jaw muscles pass under the zygomatic arch and attach to the lateral surface of the cranium. The skull and vertebral column form the **axial skeleton.**

Examine the **vertebral column.** It is divided into five groups of vertebrae. From anterior to posterior these groups are the **cervical, thoracic, lumbar, sacral,** and **caudal vertebrae.** Count the number of vertebrae in each group. The shapes of the anterior two cervical vertebrae, the **atlas** and **axis,** differ from the other cervical vertebrae. The atlas and axis form

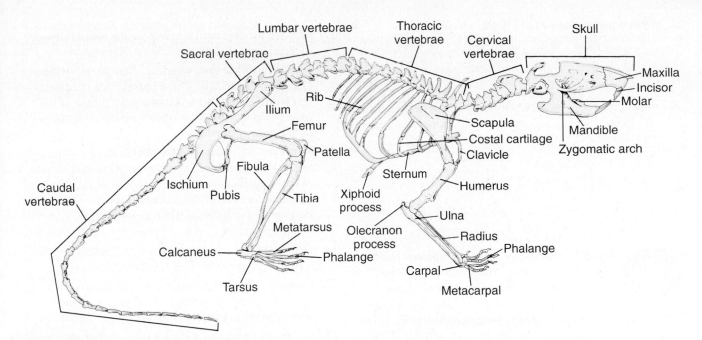

**Figure 47.3** The skeleton of a rat.

**Table 47.1**

| Skeletal Features of Rats and Humans | | |
|---|---|---|
| | Rat | Human |
| Number of bones | | |
| Arrangement of bones | | |
| Presence of unique bones | | |
| Orientation of bones | | |

**Table 47.2**

| Features of the Vertebral Columns and Rib Cages of Rats and Humans | | |
|---|---|---|
| | Rat | Human |
| Number of vertebrae in each group | | |
| Number of ribs | | |
| General shape of vertebrae | | |
| Prominence of spines | | |
| Degree of fusion of caudal and sacral vertebrae | | |
| Attachment of vertebrae to other bones | | |

a versatile joint and allow for a variety of head movements. Extending from the thoracic vertebrae are **ribs** that form an enclosure, the **rib cage.** The caudal vertebrae form a long tail.

Compare the vertebral columns and rib cages of rats and humans. List your observations in table 47.2.

Bones of the anterior appendages (forelegs) attach to the **pectoral girdle,** and bones of the posterior appendages (hind legs) attach to the **pelvic girdle.** These bones compose the **appendicular skeleton.** Compare the variation between rat and human bones of the pectoral girdle with the anterior appendages and the pelvic girdle with the posterior appendages. You will find many similarities between rat and human skeletons.

## Question 4

*a.* How does contraction of muscles passing under the zygomatic arch move the mandible?

*b.* What are the relative sizes of braincases in humans and rats?

*c.* Approximately how many vertebrae are in a rat's tail?

*d.* Does a human have a tail? What is it called?

*e.* What does the similarity between the skeletal systems of rats and humans suggest about the ancestry of these two mammals?

**SAFETY FIRST** Before coming to lab, you were asked to read this exercise so you would know what to do and be aware of safety issues. In the space below, briefly list the safety issues associated with today's procedures. If you have questions about these issues, contact your laboratory assistant before starting work.

## Procedure 47.1 **Skin your rat**

*1.* Lay your rat on its dorsal surface with the tail toward you. You will need scissors, a scalpel, and patience to remove the skin without damaging the underlying muscles and organs. An incision may have already been made in the throat or abdomen to inject colored latex into the circulatory system.

*2.* If no throat incision is apparent, then loosen the skin in the throat and pull it slightly away from the body. Snip carefully through the skin to make a small hole as an incision. Do not penetrate any internal organs.

*3.* Snip anteriorly from this incision toward the lower lip (fig. 47.4). Do not cut deeply. Cut only through the **integument** (skin).

*4.* Hold the skin up with forceps to prevent cutting muscles or internal organs. Next, snip posteriorly along the ventral midline to the **genitals** (external sex organs).

*5.* Cut through the skin in a circle around the genitals and continue the midventral incision to the tail.

*6.* Cut a circle around the base of the tail.

*7.* Make four more incisions on the ventral surface from the midline to the two ankles of the hind legs and the two wrists of the forelegs.

*8.* Then cut through the skin in a circle around the wrists and ankles.

*9.* To separate the skin from the body, grasp the flap of skin extending from the chest under a forelimb and force the tip of a blunt probe between the skin and the connective tissue holding the skin to the body. This connective tissue is called **fascia** and is tough yet flexible.

*10.* Work slowly and gently peel the skin from the body.

*11.* Remove the skin until you expose the ventral surface of the chest and abdomen.

*12.* Work toward the dorsal surface. Thin sheets of muscles attach to the skin. These muscles are the **cutaneous trunci,** which arise near the ventral base of the foreleg and spread to the lateral and dorsal surfaces of the chest and abdomen.

*13.* When you have detached the skin to the dorsal midline, begin to peel the skin anteriorly toward the base of the skull. You may notice a deposit of fat between the shoulder blades. This is **brown fat** and is specialized for heat production if the animal is stressed by low temperatures.

*14.* Do not remove the skin from the face unless instructed to do so by your instructor.

*15.* Leave the skin attached at the base of the skull and use it as a protective cape to wrap your specimen when you are not studying it.

*16.* If available, spray the rat with preservative, then wrap the skin around its body. Place the rat in a plastic bag, seal the bag tightly, and label the bag.

## INQUIRY-BASED LEARNING

### *How does the structure of the human skeleton accommodate walking on two legs?*

Observation: Efficient bipedal locomotion likely requires an adapted skeleton.

Question: Which features in the skeletal system are associated with bipedal locomotion?

**a.** Establish a working lab group and obtain Inquiry-Based Learning Worksheet 47 from your instructor.

**b.** Discuss with your group and instructor characteristics likely associated with walking on two legs, and record them on Worksheet 47.

**c.** How will you provide evidence to support your predictions of likely characteristics? Design an approach to study of the skeletal system to document morphology associated with walking on two legs. Record the design of your investigation and needed supplies on Worksheet 47.

**d.** Ask your instructor to review your proposed investigation.

**e.** Conduct your procedures, record your findings, answer your questions, and make relevant comments on Worksheet 47.

**f.** Discuss with your instructor any revisions to your questions, hypotheses, or procedures. Repeat your work as needed.

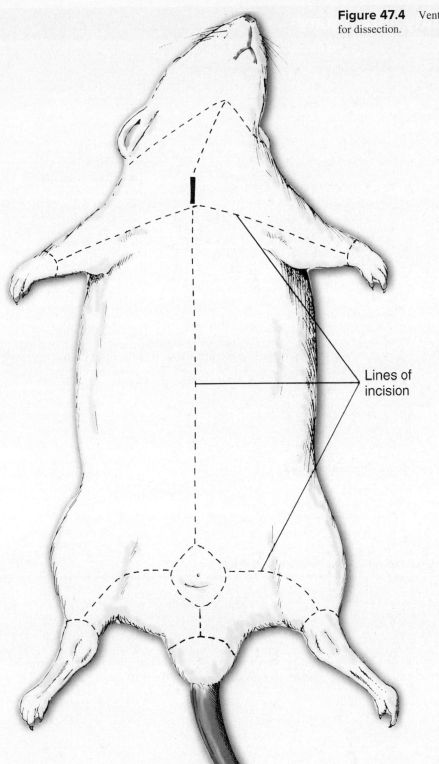

Lines of incision

*1.* What modifications of the skeletal system would be adaptive for a bipedal vertebrate?

*2.* What area of the axial skeletal system of a human is weakest? Why?

*3.* What is the value of a tail to a rat? To a monkey? To a deer?

*4.* Of what adaptive significance is a ball-and-socket joint at the hip of mammals rather than a hinge joint such as the knee?

*5.* Why is the rat a good model system for studying human anatomy and physiology?

*6.* What modifications to a rat's skeleton would be needed to achieve bipedalism?

*7.* Is the arrangement of bones and joints of rats and human skeletal systems perfect? Can you imagine a different arrangement that would be more efficient? What would be the advantages and disadvantages of the differences you propose?

**WRITING TO LEARN BIOLOGY**
What features of the skeletal system diversify movement in humans as compared to movement of a rat?

# Vertebrate Anatomy
## Muscles and Internal Organs of the Rat

---

### Learning Objectives

By the end of this exercise you should be able to:
1. Identify the major muscles of a rat.
2. Define terms describing the location, structure, and movement of muscles.
3. Describe the origin, insertion, and action of selected muscles of a rat.
4. Identify and state the function of major organs of the thorax and abdomen.

Please visit **connect.mheducation.com** to review online resources tailored to this lab.

**D**issection of muscles and internal organs requires patient concentration. Use a blunt probe rather than a scalpel or a needle to separate muscles and to move internal organs. Keep your specimen moist with dilute preservative so tissue won't be damaged as you move muscles and organs to reveal hidden structures.

### MUSCULAR SYSTEM

Muscles are usually arranged on the skeleton in **antagonistic** (opposing) pairs. Contracting one member of the antagonistic pair moves the body, whereas contracting the other muscle restores the body and first muscle to their original positions. For example, after you contract a muscle to bend your arm at the elbow you must contract an opposing muscle to straighten your arm and stretch the first muscle.

When muscles contract, they use bones as levers for resistance and motion. As you examine the muscles of a rat, you should refer often to a rat skeleton and locate the bones where each muscle attaches.

Muscles are attached to bones by connective tissue called **tendons** that adhere to spines, knobs, ridges, and depressions on bones. The end attached to the bone that does not move during contraction is called the **origin.** The end of the muscle attached to the bone that moves during a contraction is called the **insertion.** Movement caused by the contraction of a muscle is the muscle's **action** (see fig. 43.2). Examine preserved rat and human skeletons and find likely surfaces for attachment of tendons.

### Question 1
*a.* Where on a human skeleton is a likely origin and insertion for a muscle that lifts your knee?

*b.* Could a muscle have both its origin and insertion on the same bone? Why or why not?

The following terms describe muscles that move the body in a particular fashion. Familiarize yourself with these terms and demonstrate each of these terms with movements of your own body and the body of a rat. Also be sure to review the descriptive anatomical terms presented in the introduction of the previous exercise on vertebrate anatomy.

**Extensors**—increase the angle of a joint

**Flexors**—decrease the angle of a joint

**Abductors**—move an appendage away from the midline of the body

**Adductors**—move an appendage toward the midline of the body

**Retractors**—move an appendage backward

**Protractors**—move an appendage forward

**Supinators**—rotate the palm or bottom of foot upward

**Pronators**—rotate the palm or bottom of foot downward

---

### Procedure 48.1    Examine and identify muscles

1. Begin examining and identifying the muscles on your skinned rat by picking the stringy fascia away from the surface of the body. The more connective fascia that you remove, the more clearly you will see the muscles.

2. Use a blunt probe to gently separate the various muscles. Work cautiously until you develop a sense of where two muscles overlap. Notice the direction of the fibers within each muscle.

3. Refer to the figures frequently as you separate each muscle.

4. Answer the questions that will guide your observation as you separate muscles in each region of the body.

---

Figures 48.1, 48.2, and 48.3 illustrate the muscles of a rat in dorsal, lateral, and ventral views. Not all of the muscles of a rat are shown here, but you can easily understand the muscular system by studying those illustrated.

When separating and dissecting the muscular system (as well as organ systems) avoid needlessly cutting away muscle tissue. Remember, you are dissecting the rat, not disassembling it. The best dissection will bring all structures into view without removing any of them. Never cut or remove a structure just to "get it out of the way."

## Muscles of the Pectoral Girdle and Appendages

**Question 2**

*a.* Which of the muscles on the dorsal surface are shaped like a trapezoid?

*b.* What bones along the dorsal midline are points of attachment for the **acromiotrapezius, spinotrapezius,** and **latissimus dorsi?**

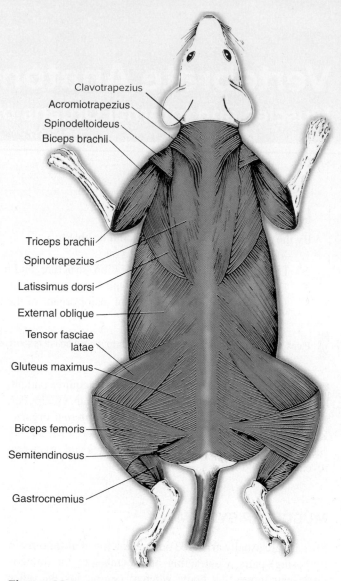

**Figure 48.1**    Dorsal view of musculature of a rat.

*c.* Are the bones named in your previous answer sites of the origin or insertion of these muscles?

*d.* Which muscle is antagonistic to the **biceps brachii?**

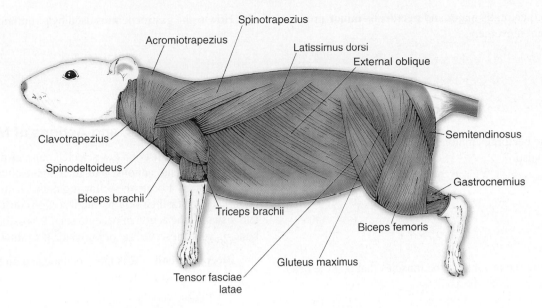

**Figure 48.2**  Lateral view of musculature of a rat.

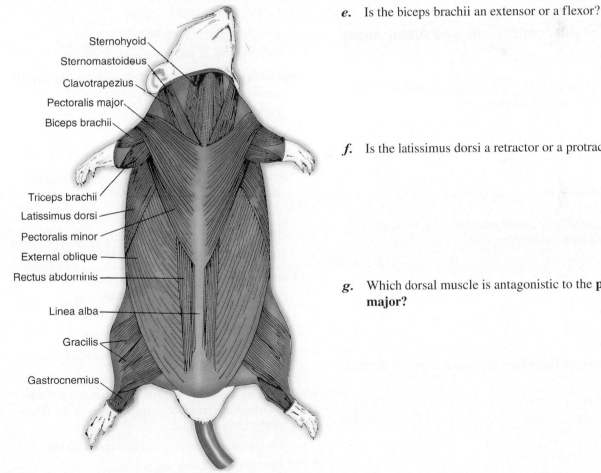

**Figure 48.3**  Ventral view of musculature of a rat.

*e.*  Is the biceps brachii an extensor or a flexor?

*f.*  Is the latissimus dorsi a retractor or a protractor?

*g.*  Which dorsal muscle is antagonistic to the **pectoralis major?**

*h.* Do the pectoralis major and **pectoralis minor** produce different movements?

*i.* Are the pectoralis major and pectoralis minor antagonistic?

*j.* What are the origins for the muscles that pull the head down toward the chest?

## Muscles of the Pelvic Girdle and Appendages

### Question 3

*a.* Which of the large lateral muscles are used in power movements of the hind legs?

*b.* Which of the large lateral muscles extends the knee joint and pulls the leg forward?

*c.* Which muscle flexes the knee joint and retracts the leg?

*d.* How is the gastrocnemius used by a running rat?

## Origins, Insertions, and Actions of Muscles

The following is a list of a few selected muscles that illustrate origins, insertions, actions, and their associations with the skeletal system. Locate the following muscles on your specimen. Then locate their origins and insertions on a rat skeleton and visualize the action of each muscle. If possible, move the bone on your preserved rat or on yourself to show its action.

**Biceps brachii**—This muscle is located on the anterior surface of the humerus.

Origin: scapula

Insertion: radius

Action: flexes the lower arm

**Triceps brachii**—This muscle is located on the sides and back of the upper arm.

Origin: humerus and scapula

Insertion: elbow (olecranon)

Action: extends the lower arm

**Spinotrapezius**—This muscle lays across the dorsal thoracic region of the rat.

Origin: thoracic vertebrae

Insertion: spine of the scapula

Action: moves the scapula upward and backward

**Acromiotrapezius**—This is the central muscle of the trapezius group.

Origin: cervical vertebrae

Insertion: scapula

Action: moves the scapula dorsally

**Latissimus dorsi**—This muscle is posterior to and partially covered by the spinotrapezius.

Origin: thoracic and lumbar vertebrae

Insertion: medial side of the humerus

Action: moves the humerus dorsocaudad

**External oblique**—This muscle covers the sides of the abdominal cavity from the hip to the rib cage. Its fibers are arranged at right angles to the fibers of the latissimus dorsi.

Origin: lumbodorsal fascia and posterior ribs

Insertion: linea alba and pelvis

Action: compresses and retains the viscera and acts with the rectus abdominis to form a flexible ventral and lateral body wall

**Cutaneous trunci**—This muscle attaches to the skin on the lateral and dorsal sides of the body.

Origin: under the upper front leg

Insertion: skin

Action: moves the skin

**Biceps femoris**—This large muscle is in two bundles; it is easily located on the side of the thigh.

Origin: ischium

Insertion: distal portion of femur and proximal portion of tibia

Action: abducts the thigh and flexes the shank

**Gastrocnemius**—This muscle is easy to locate because it forms the bulk of the calf muscle in the lower leg.

Origin: distal end of femur

Insertion: heel (calcaneus) by means of the tendon of Achilles

Action: extends the foot

## INQUIRY-BASED LEARNING

### How are muscle systems modified to accommodate walking on two legs?

Observation: Efficient bipedal locomotion likely requires an adapted muscular system.

Question: Which modifications in the muscular system are associated with bipedal locomotion?

a. Establish a working lab group and obtain Inquiry-Based Learning Worksheet 48 from your instructor.

b. Discuss with your group and instructor characteristics likely associated with walking on two legs, and record them on Worksheet 48.

c. How will you provide evidence to support your predictions of likely characteristics? Design an approach to study of the muscular system to document morphology associated with walking on two legs. Record the design of your investigation and needed supplies on Worksheet 48.

d. Ask your instructor to review your proposed investigation.

e. Conduct your procedures, record your findings, answer your questions, and make relevant comments on Worksheet 48.

f. Discuss with your instructor any revisions to your questions, hypotheses, or procedures. Repeat your work as needed.

## INTERNAL ANATOMY

### Organs of the Head and Neck

Locate and examine the structures noted in bold print in the following descriptions. To locate the **salivary glands** (**sublingual, parotid,** and **submaxillary glands**), first locate the medial sternohyoid muscle. The salivary glands lie to the side of the anterior part of this muscle. Salivary glands are rather soft, spongy tissue. They secrete **saliva** that lubricates food and contains **amylase** that degrades starch to maltose. Do not mistake lymph glands for salivary glands; the lymph glands are darker, more compact, circular glands that lie anterior to the salivary glands and are pressed against the powerful jaw muscles. The submaxillary glands are large, roughly oval, and elongate. The sublingual glands touch the submaxillary glands on their anterolateral surface. Parotid glands are best seen in lateral view. Parotid glands extend toward and behind the ear; they are beneath the ear and between the jaw and points of the shoulders. Infected parotid glands in humans produce swellings commonly associated with mumps.

---

*Procedure 48.2* **Examine organs of the head and neck**

1. After you have located the submaxillary salivary glands, remove them to see structures deeper in the neck.

2. The **thyroid gland** is a gray or brown swelling on either side of the **trachea.** The swellings are quite small. You may need help finding them. A thin band (isthmus) transverses the trachea. To locate the thyroid gland, expose the trachea by cutting through the **sternohyoid** muscles that cover the ventral surface of the throat.

---

The two lobes of the thyroid gland lie on each side of the trachea just ventral to the **larynx.** The major secretion of the thyroid gland, **thyroxin,** regulates the metabolic rate of the body. The two lobes are connected by a narrow band of tissue known as the **isthmus of the thyroid.**

The larynx is a rigid, cartilaginous chamber containing the vocal cords. Just caudal to the larynx is the trachea, which conducts air to and from the lungs. The rigid trachea is composed of a series of incomplete rings of cartilage forming a tube.

**Question 4**
Why is the trachea reinforced with cartilage?

## Procedure 48.3   Open the body cavity

1. Make a small incision with your scalpel just caudal to the posterior end of the sternum. Cut no more than 4–5 mm into the body cavity.

2. With your scissors, snip posteriorly for about 1–2 cm through the abdominal wall (toward the pelvis) until you encounter the transversely attached **diaphragm** (fig. 48.4). Try to preserve the attachment of the diaphragm as you follow the next few steps. This is difficult. You may ask your instructor for a demonstration.

3. Do not cut through the diaphragm. Instead, cut laterally in both directions (fig. 48.4) slightly anterior to the diaphragm and along the line of attachment of the diaphragm to the body wall.

4. To open the thoracic cavity, use scissors to cut the ribs one at a time from the posterior to anterior. This will eventually open the chest cavity.

5. To begin opening the abdominal cavity, make a second abdominal, midline incision just posterior to the diaphragm. Cut laterally in both directions (fig. 48.4) slightly posterior to the diaphragm, and along its attachment in a similar manner to step 3.

6. From the second incision, cut posteriorly slightly to the right of the white **linea alba** on the midventral surface of the abdomen until you reach the genitals.

7. Cut from the genitals laterally to the junction of each hind limb with the body.

8. After you have opened the animal, take it to the sink and rinse the two body cavities. Be sure that a colander covers the sink drain. After you have rinsed your rat, remove and discard all waste material from the sink.

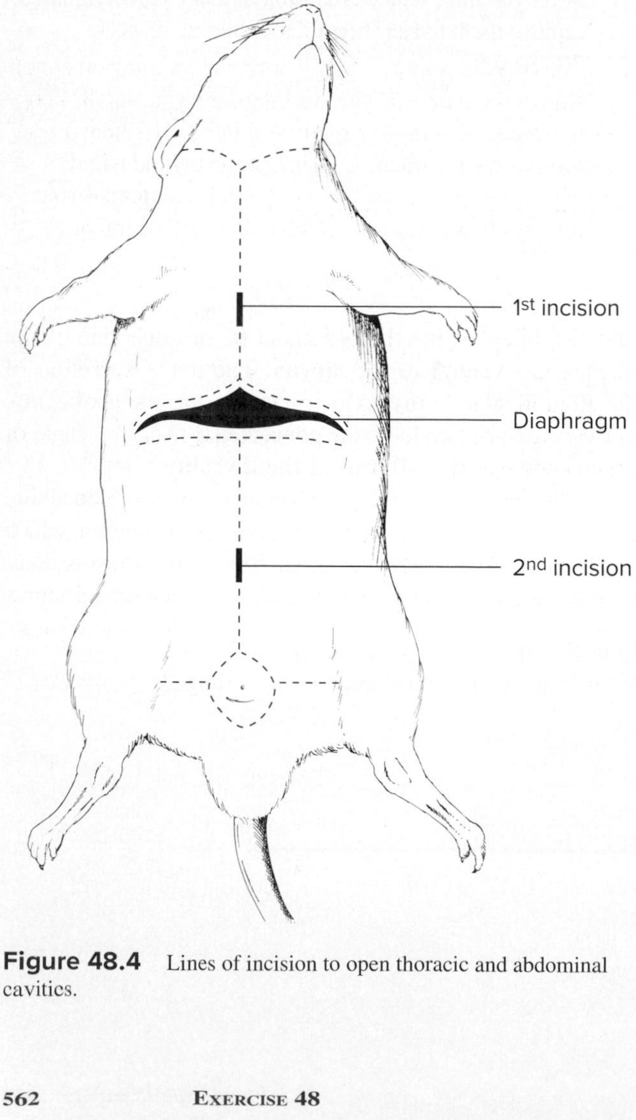

## The Thoracic Organs

Locate and examine the structures noted in bold print in the following descriptions. The diaphragm, a thin layer of muscle, separates the **thoracic cavity** from the **abdominal cavity.** The **heart** is centrally located in the thoracic cavity (fig. 48.5). The two dark-colored chambers at the top of the heart are the **atria** (sing., *atrium*) that receive blood from veins. The light brown muscular areas below the atria are the **ventricles,** which are the pumping chambers of the heart. The heart is covered by a thin membrane known as the **pericardium.** Section the heart to see its chambers.

**Figure 48.4** Lines of incision to open thoracic and abdominal cavities.

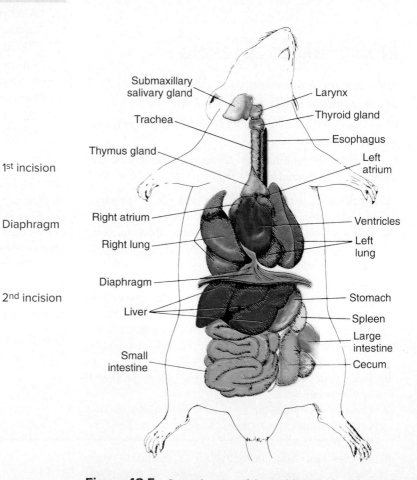

**Figure 48.5** Internal organs of the rat. Move the intestines to the right or left to see the dispersed pancreas tissue. Muscles cover the esophagus, thyroid, and larynx.

The **thymus gland** lies directly over the upper part of the heart. The thymus functions in the developing immune system and is larger in young rats than in old rats. Compare your rat with other specimens.

Air enters the lungs from the **bronchial tubes** branching from the trachea, which terminate into thin-walled air sacs, the **alveoli** (sing., *alveolus*), where gas exchange occurs. The bronchial tubes and alveoli are embedded in the lung tissue and are not visible (see fig. 44.2). Alveoli are very small; this is important because the smaller the individual alveoli, the larger the overall surface area for gas exchange. Posterior to the lungs and heart lies the diaphragm. Cramps of the muscular diaphragm sometimes occur when you run; these cramps can cause a sharp pain in your side when you try to breathe rapidly. The diaphragm can also develop a twitch, which may occur every 10 or 20 sec and be rather embarrassing. Normal rhythmic contractions of the diaphragm help you breathe.

## Question 5

*a.* What is the name of this common condition of a twitching diaphragm?

*b.* Is your rat young or old? How can you tell?

*c.* The lungs surround the heart on both sides. Are the lungs the same size? Are they both lobed?

*d.* Observe the texture of the lungs. Are the lungs muscular? If the lungs were muscular, would the contraction of the muscles aid the function of the lungs?

*e.* If the lungs are not muscular, how is air brought into the lungs?

*f.* Is the diaphragm a flat sheet of muscle or does it curve? If it curved the other way, how would its function be impacted?

*g.* How does the diaphragm function in breathing?

*h.* Which chambers of the heart have the heaviest musculature? How does that difference relate to function of the atria versus ventricles?

## The Abdominal Organs

Locate and examine the structures noted in bold print in the following descriptions. The **coelom** is the body cavity within which the **viscera** (internal organs) are suspended. The abdominal cavity and viscera are covered by a membranous tissue called the **peritoneum,** formed from mesoderm. Recall that the coelomic cavity of evolutionarily advanced invertebrates and vertebrates is always lined by mesodermal tissue. The peritoneum is extensive and occurs in four regions:

- **Parietal peritoneum** covers the walls of the abdominal cavity.

- **Visceral peritoneum** covers the internal organs.

- **Mesenteries** are thin membranes continuous between the parietal and viscera peritoneums. They attach the internal organs to the dorsal body wall.

- **Omenta** (sing., *omentum*) are thin membranes extending from the visceral peritoneums and connect organ to organ.

Note the continuous nature of the peritoneum as you examine the abdominal organs. Locate the **liver,** the dark-colored organ suspended just under the diaphragm. The many functions of the liver include producing **bile** to help digest fat, storing glycogen (a high-energy molecule), transforming nitrogenous wastes into less harmful substances, and transforming digested nutrients into various molecules needed by the body. Rats do not have a gallbladder. For these functions, the liver is strategically located near the digestive system to receive a rich supply of blood directly from the intestinal tract. The liver has four parts:

- **Median** or **cystic lobe** is located atop the organ. There is a cleft in its central part.

- **Left lateral lobe** is large and partially covered by the stomach.

- **Right lateral lobe** is partially divided into an anterior and posterior lobule, is smaller than the left lateral lobe, and is hidden from view by the median lobe.

- **Caudate lobe** is small, folds around the esophagus and stomach, and appears to be in two small sections seen most easily when the liver is raised.

The **esophagus** pierces the diaphragm and conducts food from the mouth to a muscular enlargement of the digestive tract called the **stomach.** Locate the stomach on the left side just under the diaphragm. The functions of the stomach include food storage, physical breakdown of food, and initial enzymatic digestion of protein. The entrance of the esophagus to the stomach is guarded by the **cardiac (or lower esophageal) sphincter.** Sphincters are circular muscles that control the flow of enclosed fluids by contracting and closing the opening between cavities. The outer margin of the curved stomach is known as the **greater curvature;** the inner margin is the **lesser curvature.** Slit open the stomach and notice ridges, called rugae, that line the stomach. The stomach is attached to the intestine via the **pyloric sphincter.**

The **spleen** is similar in color to the liver and is attached to the greater curvature of the stomach. It is associated with the circulatory system and functions in the formation, storage, and destruction of blood cells.

The **pancreas** is the brownish, flattened gland suspended in membranous tissue near the junction of the stomach and small intestine (fig. 48.6). This gland has exocrine and endocrine functions. Review these terms in your textbook. The pancreas produces digestive enzymes conducted to the intestinal tract through small ducts. The pancreas also secretes and releases **insulin** into the bloodstream to regulate glucose metabolism. The **greater omentum** is the membranous curtain of tissue that hangs from the stomach and contains lymph nodes, blood vessels, and fat.

The **small intestine** is the slender coiled tube that receives partially digested food from the stomach. The small intestine continues digestion and nutrient absorption. It consists of three sections: duodenum, jejunum, and ileum. These sections are distinguishable histologically but may be

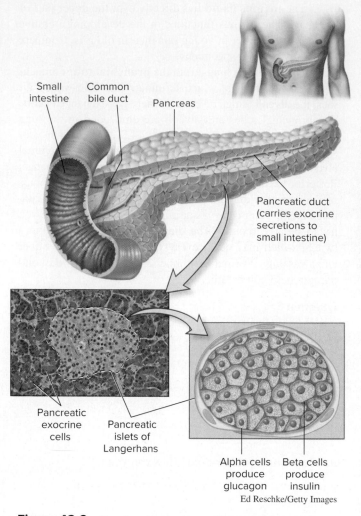

**Figure 48.6**   Location, appearance, and internal structure of the mammalian pancreas. Within the pancreas are scattered islets of Langerhans, which are endocrine tissue. Only the exocrine products (e.g., digestive enzymes) are secreted into the intestine; the hormones from the islets of Langerhans are secreted into the blood (not shown).

hard to differentiate anatomically. The **duodenum** receives enzymes from the digestive glands; it begins at the **pyloric sphincter** of the stomach and continues under the stomach. The duodenum has small, fingerlike folds of intestinal wall called **villi,** and the exposed surface of each cell has cytoplasm extensions called **microvilli** (fig. 48.7).

### *Procedure 48.4*   Location and function of internal organs

1. Design a three-column organizational table.
2. Complete the table with Organ Name in column 1, Organ Location in column 2, and Organ Function in column 3.

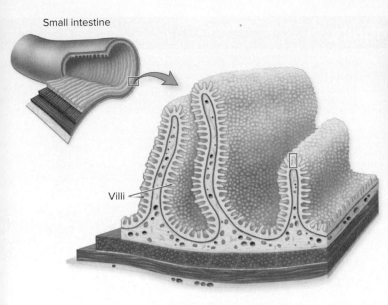

Small intestine

Villi

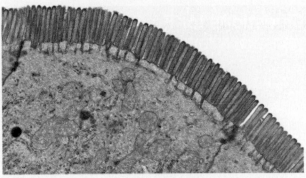

SCIENCE PHOTO LIBRARY/Science Source

**Figure 48.7** The small intestine. Successive enlargements show folded epithelium studded with villi that increase the surface area. The micrograph above shows an epithelial cell with numerous blue-colored microvilli.

## Question 6

*a.* Which structure would prevent the long tubular intestine from tangling and knotting?

*b.* Is the membrane attaching the spleen to the stomach a mesentery or omentum?

*c.* What is the function of villi? How does structure relate to function of the villi?

---

### Procedure 48.5    Examine abdominal organs

*1.* Locate the liver.

*2.* Locate the stomach, slit it open, and notice the rough folds called rugae.

*3.* Slit open the duodenum and examine the villi.

---

Locate and examine the structures noted in bold print in the following descriptions. The **jejunum** is the central portion of the small intestine. The **ileum** is the posterior portion of the small intestine that empties into the **colon** (large intestine). The **ileocecal valve** controls flow of food into the colon.

The colon is the large greenish tube extending from the small intestine to the **anus.** The colon is where the final stages of digestion and water absorption occur. It contains a rich flora of bacteria that help digestion. The colon also secretes a lubricant of mucus for the feces. The colon consists of five continuous sections identified by their appearance and location.

- **Cecum** is a large, flattened sac in the lower third of the abdominal cavity. It is a blind-ending pouch located caudal to the junction of the ileum and is often green. The cecum is homologous to the appendix in humans and is especially rich in microorganisms that aid in digestion.

- **Ascending colon** leads toward the right lateral lobe of the liver on the right side of the body cavity.

- **Transverse colon** crosses the body cavity under the liver and stomach.

- **Descending colon** descends the left side of the body cavity. You may need to move the small intestine aside to see it clearly.

- **Rectum** is the short, terminal section of the colon between the descending colon and the anus. The rectum temporarily stores feces before they are evacuated from the body.

Review the internal organs by listing all of the structures and areas of the digestive system through which food passes. Refer to your textbook and add a phrase describing the nature of digestion occurring in each area.

*1.*  Which muscles of a quadruped are most modified to achieve an upright, bipedal stance?

*2.*  Where on the human body are muscles attached to the skin most developed? What is the function of the movement of these muscles?

*3.*  Why might a human need a gallbladder, but a rat does not?

*4.*  Why is the cecum relatively larger in rats than it is in humans?

**DOING BIOLOGY YOURSELF**

On your body, test what factors limit motion of the vertebrate body. Explain your findings.

# Vertebrate Anatomy
## Urogenital and Circulatory Systems of the Rat

## Learning Objectives

By the end of this exercise you should be able to:
1. Describe the structural relationship of the excretory and reproductive systems of a rat.
2. Identify the major features of the male and female reproductive systems.
3. Identify the major arteries and veins of a rat.
4. Trace the flow of blood through the heart, arteries, and veins.
5. Describe and list the parts of the "pulmonary circuit" and "systemic circuit."
6. Describe the function and structure of the hepatic portal system.
7. Trace the pathway of oxygen-rich and oxygen-poor blood through the blood vessels of the respiratory system.

Please visit **connect.mheducation.com** to review online resources tailored to this lab.

All systems of vertebrates are dynamic. The excretory system filters nitrogenous waste from blood and removes it from the body. The reproductive system produces gametes, brings them into contact for fertilization, and nurtures the resulting embryo. The circulatory system distributes fluids, dissolved gases, hormones, and nutrients so that they come into close contact with each cell of the body. As you examine the various structures of these systems, note the sequence of the vessels that cells and fluids pass through as they exit or move through the body.

## UROGENITAL SYSTEM

The excretory and reproductive systems of vertebrates are closely integrated and are usually studied together as the urogenital system. However, they have different functions. The excretory system removes toxic wastes. The reproductive system produces gametes, an environment for the developing embryo, and hormones associated with sexual development. Before you begin dissecting the rat, review figures 49.1, 49.2, 49.3, and 49.4, which show various aspects of the human urogenital systems. These images of human anatomy are likely familiar to you. Their review will remind you of familiar vocabulary and help transition your learning to a less familiar vertebrate, the rat. As you study the rat's urogenital system, compare its anatomy with that of a human.

### Procedure 49.1 Observe the excretory and reproductive organs of a rat

1. Locate and observe the organs as described.
2. Do not remove organs; just move them to see other structures.

### Excretory Organs

Locate and examine the structures noted in bold print in the following descriptions. List in your notes the structures you are unable to locate. Your instructor will review your list and note which structures may or may not be fully visible in your specimen. Examine a model or demonstration dissection of a vertebrate kidney available in the lab (fig. 49.4). The primary organs of the excretory system are the **kidneys.** Kidneys of a human will process 423 gallons (1601 L) of blood per day and produce about 1.5 quarts (1.42 L) of urine. A sagittal section through the kidney reveals the **renal artery,** which transports blood to the kidney for filtering, and a **renal vein,** which transports filtered blood away from the kidney. The **ureter** is a tube that conducts urine from the kidney to the **urinary bladder.** The base of the ureter is expanded on the kidney to form a small chamber called the **renal pelvis,** which collects urine produced by the many tubules of the kidney. The outer area of the kidney is the **cortex** and the inner area is the **medulla.** Review in your textbook the fine structure and

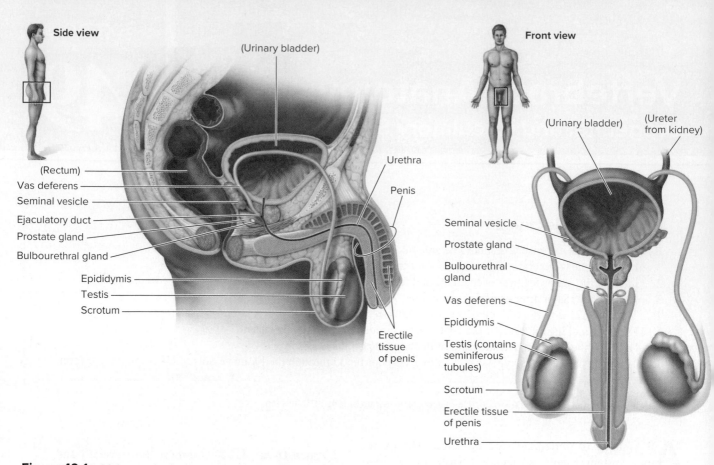

**Figure 49.1**  Male reproductive structures in humans (nonreproductive structures are identified in parentheses for orientation purposes).

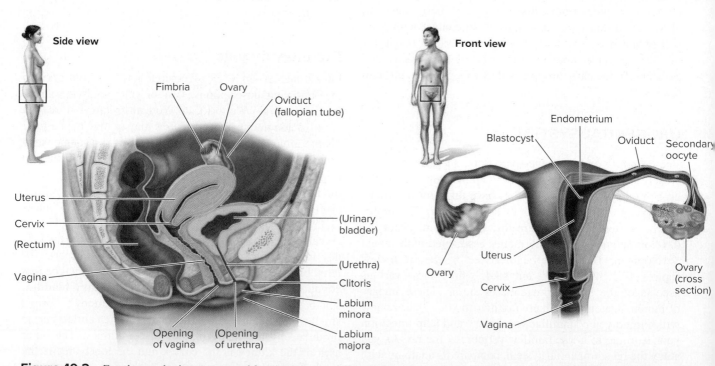

**Figure 49.2**  Female reproductive structure and function in humans. Side and front views of the female reproductive system (nonreproductive structures are identified in parentheses for orientation purposes). An oocyte moves from the ovary into the oviduct (also called the fallopian tube), where it may be fertilized and develop into a blastocyst. Subsequently, the blastocyst enters the uterus, where it may implant in the endometrium, the inner lining of the uterus.

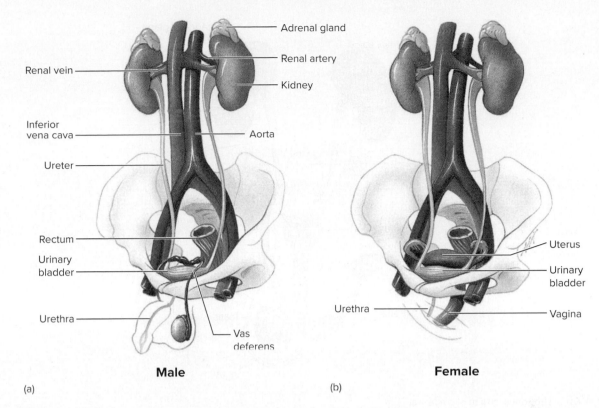

**Figure 49.3** (*a*) Human male and (*b*) female urinary systems shown in relation to the pelvis.

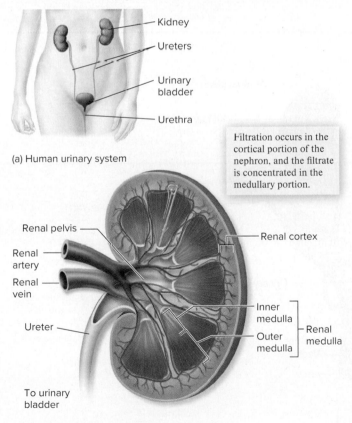

(a) Human urinary system

Filtration occurs in the cortical portion of the nephron, and the filtrate is concentrated in the medullary portion.

(b) Internal structure of a kidney

**Figure 49.4** (*a*) The human urinary system in a female. In the male, the urethra passes through the penis. (*b*) View of a section of a kidney showing major internal structures. By varying the concentration of solutes around capillaries in different regions, the kidney extracts fluid from the blood. Dilute filtrate derived from blood circulates through tubules looping through the medulla. Water is reabsorbed in response to the surrounding osmotic concentrations, and waste products are routed (as urine) to the renal pelvis and ureter.

function of the tubules that form the bulk of the kidney. If you pull the intestines to either side, you can see a pair of kidneys on the dorsal abdominal wall (figs. 49.5 and 49.6). Locate the delicate ureters by wiggling the kidneys and watching for movement on the body wall posterior to the kidneys. Remove one kidney and make a longitudinal section. Observe its internal structure; compare it to the model (fig. 49.4).

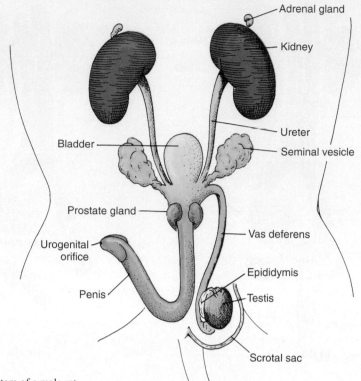

**Figure 49.5**   Urogenital system of a male rat.

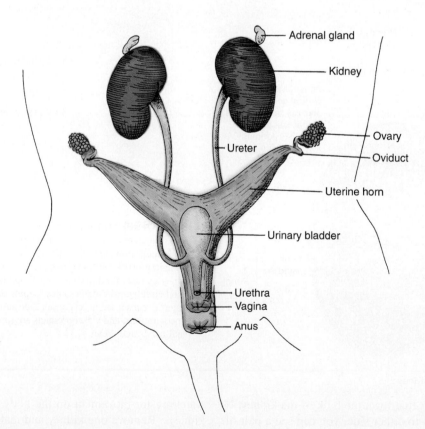

**Figure 49.6**   Urogenital system of a female rat.

The **urethra** carries urine from the bladder to the **urethral orifice,** where urine is expelled. The small, yellowish glands embedded in fat atop the kidneys are the **adrenal glands.**

### Question 1

What hormone do the adrenal glands produce? Why have two kidneys?

## Reproductive Organs of the Male Rat

Locate and examine the structures noted in bold print in the following descriptions. The major reproductive organs of the male are the **testes,** in the **scrotal sac** (fig. 49.5). Cut through the sac to expose a testis. On the surface of the testis is a coiled tube, the **epididymis,** which collects and stores maturing sperm cells produced within the testis. The tubular **vas deferens** conducts sperm from the epididymis to the **urethra,** which carries sperm through the penis and out of the body. Review gamete production in Exercise 15.

### Question 2

Where in the body do sperm and urine share the same passages? What male structures serve only for excretion, which only for reproduction, and which for both?

The lumpy, brown glands located to the right and left of the urinary bladder are the **seminal vesicles.** The white gland below the bladder is the **prostate gland.** The seminal vesicles and prostate gland secrete materials forming much of the **seminal fluid** (semen) that activates and transports sperm cells.

## Reproductive Organs of the Female Rat

Locate and examine the structures noted in bold print in the following descriptions. The short gray tube lying dorsal to the urinary bladder is the **vagina** (fig. 49.6). The vagina divides into two **uterine horns** extending toward the kidneys against the dorsal body wall. This **duplex uterus** (two uterine horns) is common in some groups of nonprimate mammals. It will accommodate more than one offspring (a litter) during a reproductive cycle. In contrast, a **simplex uterus** common in primates such as humans has a single, medial chamber for development of an embryo. At the tips of the uterine horns are small, lumpy glands called **ovaries.** The tiny tubes between the ovaries and the uterine horns are the **oviducts.** Oviducts capture eggs produced by the ovaries and conduct the eggs to the uterine horns. You may need a dissecting microscope to help locate the oviducts.

## CIRCULATORY SYSTEM

Locate and examine the structures noted in bold print in the following descriptions. The general structure of the circulatory system of the rat is almost identical to that of humans. Blood circulates in two general systems. **Pulmonary circulation** carries blood through the lungs for oxygenation and back to the heart. **Systemic circulation** moves blood through the body after it has left the heart. As you examine the blood vessels notice that the arterial system, especially near the heart, is distinctly asymmetrical. In future courses on comparative anatomy you will learn the derivation of this asymmetry. Also remember, the word "right" or "left" before a structure's name refers to the right or left side of the rat (or human) specimen's body—not yours.

Review the procedures in Exercise 45 and figures 45.1, 45.2, 49.7, 49.8, 49.9, and 49.10, which show aspects of the human circulatory system. As you study the rat's circulatory system, compare its anatomy with that of a human.

*Procedure 49.2* **Observe the circulatory system of a rat**

*1.* Locate organs of circulation.

*2.* Follow the path of circulation as described.

Begin your study of the circulatory system at the heart. The arteries and veins have been injected with colored latex so that you can see them more clearly. Do not cut through any arteries or veins as you dissect muscle tissue from around these vessels.

## Circulation of Blood through the Heart and Pulmonary System

You have already located the heart in an earlier lab. Examine it again; also examine the vertebrate heart shown in figures 49.9 and 49.10.

Blood from the posterior portion of the body enters the right atrium of the heart through the **inferior vena cava.** The terms *superior* and *inferior* are often used to describe the vena cavae in humans, who are bipeds (i.e., we have two legs) with a vertically oriented body axis. The terms *cranial* and *caudal* are sometimes used for other mammals, especially quadrupeds whose bodies are oriented horizontally. Blood from the anterior portion of the body enters through the **right** and **left superior venae cavae.** Blood flows from the right atrium, through the **tricuspid valve** to the **right ventricle.**

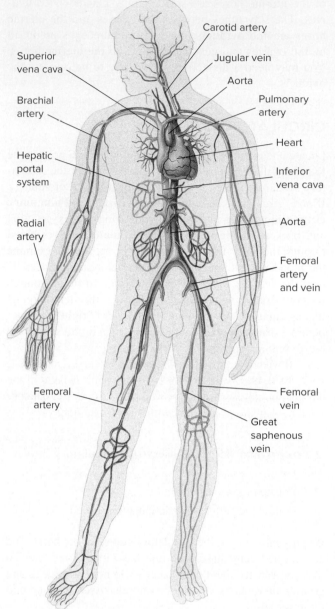

**Figure 49.7** Human circulatory system. A muscular heart pumps blood through arteries, veins, and capillaries. Arteries carry blood away from the heart to all parts of the body and veins carry blood to the heart. Blood passes from arteries into small, thin-walled capillaries where nutrients, oxygen, and wastes can diffuse into and out of tissues. Capillaries coalesce to form veins, which route blood to the liver and kidney for filtration and back to the heart. The circulatory systems of all vertebrates are similar.

It is then pumped through the pulmonary **semilunar valve** and into the **pulmonary trunk.** The pulmonary trunk divides dorsally into the right and left **pulmonary arteries;** these are the only arteries in the body that carry **deoxygenated** blood. Blood then flows through the pulmonary arteries to the lungs, where it is **oxygenated.** Blood returns from the lungs to the **left atrium** of the heart via four **pulmonary veins** and flows through the **bicuspid** (or **mitral**) **valve** to the **left ventricle.**

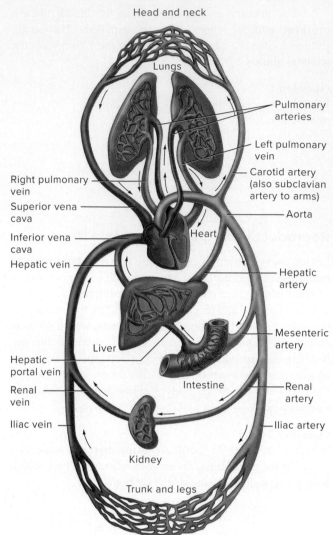

**Figure 49.8** Ventral view of the pulmonary and systemic circulation—the journey of blood.

The pulmonary veins are best traced by beginning at the left atrium and proceeding to the lungs.

## Circulation through the Heart and Systemic Arteries

Blood leaves the **left ventricle** of the heart through the **aortic semilunar valve** and flows into the **aorta** (fig. 49.9). The aorta has four general areas. Label the first three of these areas in figure 49.9. The (1) **ascending aorta** begins at the semilunar valve of the left ventricle and passes outside and over the left and right atria. The aorta then bends to the left and passes dorsally to the heart. This bend is the (2) **aortic arch.** The aorta continues as the (3) **descending aorta** and passes along the middorsal wall from the thorax through the diaphragm to the abdomen. Caudal to the diaphragm the aorta continues as the (4) **abdominal aorta.** You must move the stomach, liver, and intestines to the side to see the abdominal aorta as it passes along the vertebral column (fig. 49.11).

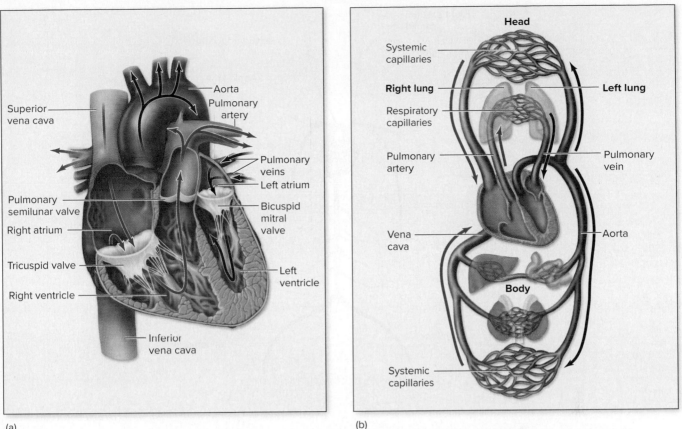

(a)

(b)

**Figure 49.9** Path of blood through the human heart. (*a*) The mammalian heart is divided into four chambers. Blood enters the right atrium from the superior vena cava and passes into the right ventricle through a valve that prevents backflow of blood. From the right ventricle, blood moves through the pulmonary artery and the lungs. Oxygenated blood from the lungs returns to the heart via the pulmonary veins. This blood enters the left atrium and then the left ventricle, from which it enters (*b*) the general circulatory system of the body through the aorta.

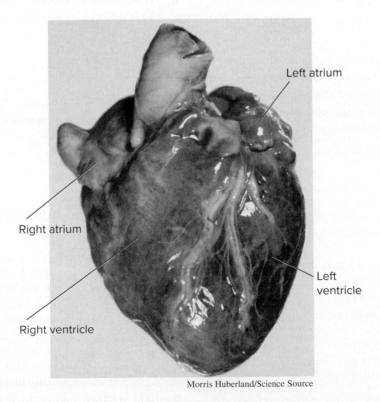

Morris Huberland/Science Source

**Figure 49.10** The human heart.

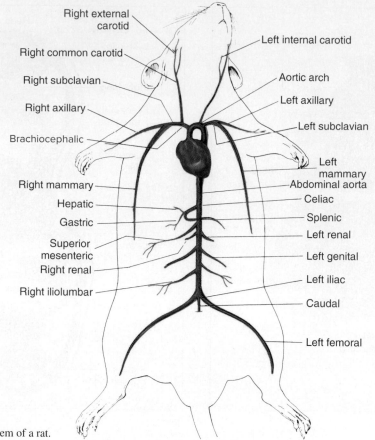

**Figure 49.11**    Arterial system of a rat.

## INQUIRY-BASED LEARNING

### *Do larger bodies have larger hearts than smaller bodies?*

Observations: Vertebrate hearts are similar in structure, but their hosts vary widely in size. The required output of a vertebrate heart likely varies with the total size of the animal.

Question: Is the percentage of total animal mass occupied by the heart consistent among different vertebrates?

a. Establish a working lab group and obtain Inquiry-Based Learning Worksheet 49 from your instructor.

b. Discuss with your group and instructor well-defined questions relevant to the preceding observations and question. Choose the best of your questions and record it.

c. Translate your question into a testable hypothesis and record it.

d. Outline on Worksheet 49 your experimental design and supplies needed to test your hypothesis. Ask your instructor to review your proposed investigation.

e. Conduct your procedures, record your data, answer your question, and make relevant comments.

f. Discuss with your instructor any revisions to your questions, hypotheses, or procedures. Repeat your work as needed.

### Question 3

*a.* How do the thicknesses of the atrial and ventricular walls differ? How does wall thickness relate to function?

*b.* Which structures compose the pulmonary circulation? Which compose the systemic circulation?

## Branches of the Aortic Arch and Descending Aorta

Locate and examine the structures noted in bold print in the following descriptions. **Coronary arteries** serve the heart muscle but are not easily seen because they branch from the aorta within the heart. The first visible artery branching from the aorta is the **brachiocephalic artery** (fig. 49.11), which originates from the aortic arch. The brachiocephalic artery is short and divides into the **right common carotid artery,** which supplies the right side of the neck, and the **right subclavian artery,** which supplies the right shoulder and arm. At the most anterior part of the bend in the aortic arch you should find the **left common carotid artery,** which conducts blood up the left side of the neck. Immediately to the left of the left common carotid artery is the left **subclavian artery.** This artery supplies blood to the left shoulder and arm.

## Branches of the Abdominal Aorta

Push the abdominal viscera to the left to locate the abdominal arteries. The first arterial branch from the abdominal aorta below the diaphragm is the **celiac artery.** This artery delivers blood toward the stomach, liver, spleen, and pancreas. The second artery arising from the abdominal artery is the **superior mesenteric artery,** which is larger than the celiac artery. The superior mesenteric artery delivers blood directly to the small intestine and intestinal mesenteries. The **renal arteries** are short and lead directly to the kidneys. Just posterior to the renal arteries are the **genital arteries** leading to the testes of a male and the ovaries of a female. Farther along the posterior end of the abdominal aorta you will find a pair of **iliolumbar arteries** leading to the dorsal muscles of the back. Next, the **inferior mesenteric artery** leads to intestinal mesenteries. The abdominal aorta also gives rise to the **caudal artery,** which passes into the dorsal body wall and supplies blood to the tail. Gently pull the posterior end of the abdominal aorta slightly to the side to see the caudal artery. The abdominal aorta finally divides posteriorly to form the **iliac arteries,** which deliver blood to the pelvic area and hind extremities.

### Question 4
*a.* Which abdominal arteries are paired? Which are singular?

*b.* Which artery is a significantly different length in mature males and females? Why?

## Secondary Branches of the Thoracic Arteries

Locate and examine the structures noted in bold print in the following descriptions. The **right subclavian artery** branches from the brachiocephalic artery. The right subclavian artery is short, passes under the clavicle, and branches into the **right internal mammary artery** (look along the inside of the ventral chest wall) and the **right axillary artery** leading toward the armpit. The **right common carotid** passes anteriorly along the neck after branching from the brachiocephalic artery and gives rise to the **right**

**external carotid artery,** servicing the face, and the **right internal carotid artery,** servicing the inner areas of the head. The left common carotid artery arises from the aortic arch and carries blood along the left side of the neck. This artery gives rise to the **left external** and **internal carotid arteries.** The left subclavian artery branches from the aortic arch and divides to form the arterial branches similar to those on the right side.

## Secondary Branches of the Abdominal Arteries

The celiac artery gives rise to (1) the **hepatic artery,** which supplies blood to the liver; (2) the **gastric artery,** which leads to the stomach; and (3) the **splenic artery,** which leads to the spleen and pancreas. Carefully trace the celiac artery from the abdominal aorta until you find these branches. If the arteries were cut during dissection, you will not be able to positively identify these branches. The **iliac arteries** give rise to many branches and continue on the inner thighs as the left and right **femoral arteries.**

## The Systemic Veins

Locate and examine the structures noted in bold print in the following descriptions. The **left** and **right superior venae cavae** conduct blood from the upper part of the body into the right atrium (fig. 49.12). Trace these veins from the atrium. As you follow a superior vena cava anteriorly, it produces a small **internal jugular vein** and continues as the **subclavian vein.** The subclavian vein then divides anteriorly into the **external jugular vein** and an **axillary vein.** The jugular veins drain blood from the head, and the axillary veins bring blood from the shoulders and arms.

The **inferior vena cava** carries blood from the lower body to the right atrium. The **hepatic vein** drains the liver and enters the inferior vena cava near the diaphragm. **Renal veins** drain the kidneys. **Genital veins** lead from the gonads either to the inferior vena cava or to the renal veins. The **iliac** and **femoral veins** drain the legs, and the **caudal vein** drains the tail.

### Question 5
*a.* Which veins are paired? Which are singular?

*b.* What is the adaptive significance of arteries in appendages being buried deeper than veins?

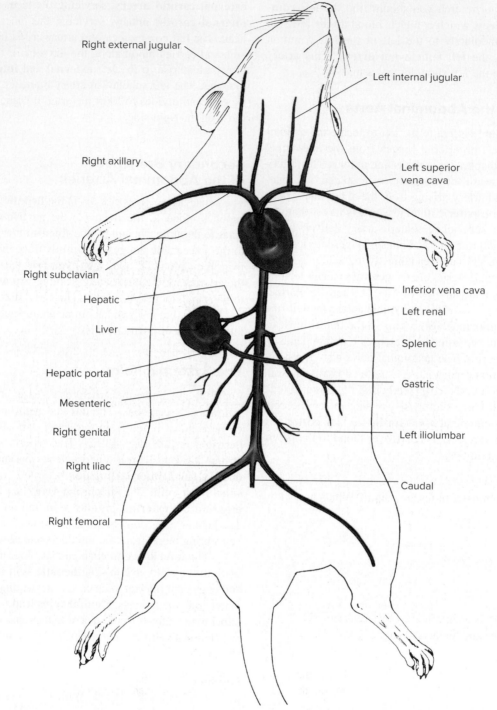

Right external jugular

Left internal jugular

Right axillary

Left superior
vena cava

Right subclavian

Inferior vena cava

Hepatic

Left renal

Liver

Splenic

Hepatic portal

Gastric

Mesenteric

Right genital

Left iliolumbar

Right iliac

Caudal

Right femoral

**Figure 49.12** Venous system of a rat.

## The Hepatic Portal System

Locate and examine the structures noted in bold print in the following descriptions. A **portal system** is a system of veins that carries blood from one bed of capillaries to another bed of capillaries (fig. 49.13). The **hepatic portal system** carries blood from capillaries in the mesenteries, small intestine, spleen, stomach, and pancreas to the liver. Specifically, the **gastric, splenic,** and **mesenteric veins** drain the digestive system and unite to form the **hepatic portal vein,** which carries blood to the liver. In the small intestine, digestive enzymes break down food, and the products pass across the intestinal wall into the bloodstream. Thus, the liver is strategically located to receive blood after nutrients have been absorbed in the intestinal tract. The liver cells can easily modify these nutrients and secrete materials into the circulatory system. In the liver, blood from the hepatic

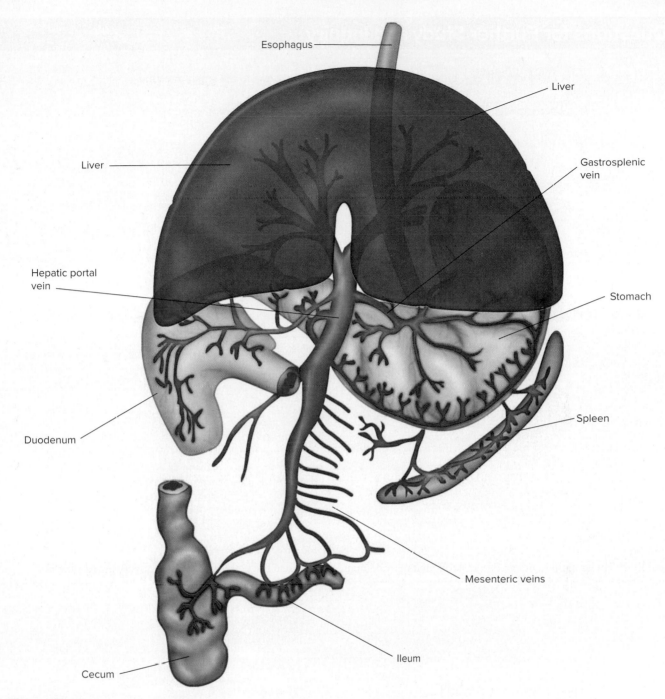

Esophagus

Liver

Gastrosplenic vein

Liver

Hepatic portal vein

Stomach

Duodenum

Spleen

Cecum

Mesenteric veins

Ileum

**Figure 49.13**  Hepatic portal system.

portal vein passes through another capillary system, which ultimately coalesces to form the hepatic vein leading to the inferior vena cava. In some preserved rats, the hepatic portal system has been injected with yellow latex. Ask your instructor to help you locate the portal system if your rat was not injected.

After you have located all of the arteries and veins on your specimen, test your knowledge by tracing the flow of blood from the kidney to the liver. Name all of the vessels and organs through which a blood cell might pass. You will be surprised at the number of vessels and structures you mention.

**Question 6**

Does the hepatic portal vein carry oxygenated or deoxygenated blood? Why should blood pass through the liver after leaving the digestive system and before entering the other organs?

1. Trace the path of a blood cell as it moves from the capillaries of the brain to the small intestine.

2. Do you recall any vertebrates with external reproductive organs on their dorsal surface? What advantages or disadvantages would this provide?

3. Body fluids include urine, blood, food, and reproductive fluids. Trace the path of each as it moves through the body.

### WRITING TO LEARN BIOLOGY

How does the circulatory system serve as a communication system as well as just a delivery system?

# Embryology
## Comparative Morphologies and Strategies of Development

---

### Learning Objectives

By the end of this exercise you should be able to:
1. Describe the early stages of embryological development common to advanced invertebrates and vertebrates.
2. Understand the formation of a three-layered embryo with ectoderm, endoderm, mesoderm, and a presumptive digestive cavity.
3. Relate the major structures of early embryos to the environment in which they develop.

---

Please visit **connect.mheducation.com** to review online resources tailored to this lab.

---

**A**ll sexually reproducing multicellular organisms begin life as single-celled zygotes and progress through stages of **growth, differentiation,** and **morphogenesis.** Growth is the irreversible increase in size of an organism. Differentiation is the structural and functional specialization of groups of cells. Morphogenesis is the development of pattern, shape, and form. **Embryology** is the study of these processes in the early development of an organism.

Early stages of embryology are similar in advanced invertebrates and vertebrates and include formation of body cavities, membranes, and multiple cell layers. During these stages, all embryos must overcome similar fundamental problems. For example, development always requires energy (i.e., food) and protection, as well as waste disposal. Although the specific evolutionary adaptations and strategies for successful development may vary among

groups of animals, all developing organisms must cope with the same problems (fig. 50.1).

During this exercise you will compare features and developmental strategies of the early stages of development of the sea star, frog, and chick. These developmental stages are:

- **Fertilization**—Male and female gametes fuse to form a zygote.

- **Cleavage**—Zygote divides into a larger and larger number of smaller and smaller cells that eventually form a hollow sphere of many cells. Cleavage ends when groups of cells begin to differentiate.

- **Gastrulation**—Cells of the sphere formed during cleavage continue to divide and move inward to form three cellular layers.

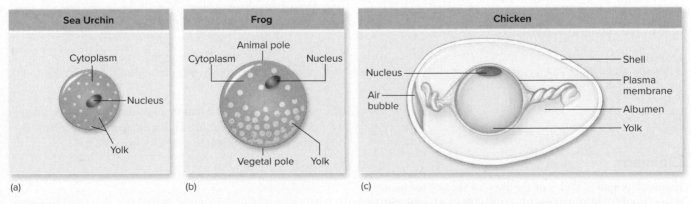

**Figure 50.1** Distribution of yolk in three kinds of eggs. (*a*) In an echinoderm egg, the cytoplasm contains a small amount of evenly distributed yolk and a centrally located nucleus. (*b*) In a frog egg, there is much more yolk, and the nucleus is displaced toward one pole. (*c*) Bird eggs are complex, with the nucleus contained in a small disc of cytoplasm that sits on top of a large, central yolk mass.

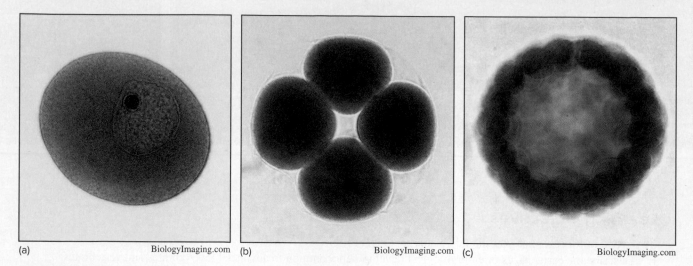

**Figure 50.2**  Early stages in sea star embryology. (*a*) Zygote (100×). (*b*) Early cleavage (100×). (*c*) Blastula (100×).

## SEA STAR DEVELOPMENT

Examine either a preserved or living *Asterias* on display and review in your textbook the general characteristics of echinoderms. *Asterias* is a common sea star belonging to the invertebrate phylum Echinodermata (see Exercise 40).

## Fertilization

If models are available in addition to living and prepared material, use them to examine sea star development. Examine a prepared slide of an unfertilized sea star egg and compare it with a slide of a zygote (fig. 50.2*a*). Eggs and sperm of a sea star are extruded from the body into seawater, and fertilization occurs externally. If living organisms are available, your instructor will prepare the organisms and stimulate them to shed eggs and sperm. These small eggs have only a small amount of yolk distributed evenly in the cytoplasm. Sea star eggs are **isolecithal,** meaning that their yolk is distributed evenly. This is important because the amount and distribution of yolk in an organism's egg strongly affect its pattern of development.

Fertilization of an egg gamete by a sperm gamete produces a diploid single-celled zygote. When the zygote cleaves (i.e., divides), it becomes an embryo.

## Question 1
Are both the egg and zygote of a sea star the same size?

## Cleavage

Examine prepared slides of the two- or four-celled cleavage stage (fig. 50.2*b*) and of the blastula of a sea star (fig. 50.2*c*). Approximately 1 h after fertilization, the zygote cleaves (divides) to form two cells. Cleavage of the entire zygote is termed **holoblastic,** meaning that only a small amount of yolk is present and easily divides evenly. Continued cleavage forms a ball of 16–32 cells called a **morula** and later forms a hollow sphere of many cells called a **blastula.** The fluid-filled cavity within the blastula is the **blastocoel,** and the cells are called **blastomeres.**

## Question 2
***a.***  Are the cells of the two-celled stage of a sea star similar in size?

***b.***  How does the size of an embryo during early cleavage compare with that of the zygote? Is the sea star embryo growing during cleavage?

***c.***  How many distinct layers of cells do developing sea stars have at the blastula stage?

***d.***  The blastomeres of a mature blastula eventually develop cilia and the entire blastula can rotate and move. How is movement important to a developing sea star?

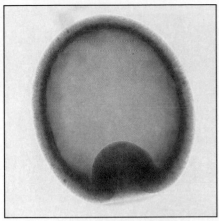
(a)
BiologyImaging.com

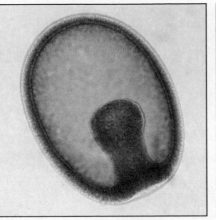
(b)
BiologyImaging.com

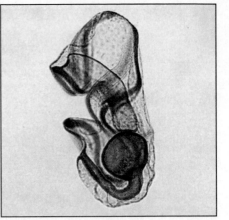
(c)
BiologyImaging.com

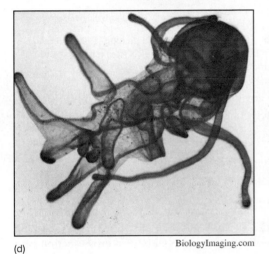
(d)
BiologyImaging.com

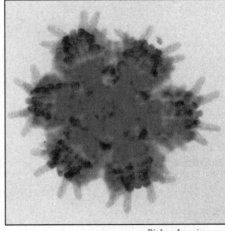
(e)
BiologyImaging.com

**Figure 50.3** Developmental stages of a sea star. (*a*) Early gastrula (70×). (*b*) Late gastrula (70×). (*c*) Bipinnaria larva (50×). (*d*) Brachiolaria larva (50×). (*e*) Juvenile sea star (35×).

## Gastrulation

Examine prepared slides of gastrulation stages and early feeding stages of sea star development (fig. 50.3). Gastrulation occurs after the blastula has matured and consists of many cells. During early gastrulation, cells invaginate into the blastocoel and form a **gastrula** and a new cavity, the **archenteron** (fig. 50.3*a, b*). After invagination the gastrula has two germ layers of cells, the endoderm and ectoderm. The blastocoel soon disappears, and the archenteron (sometimes called the gastrocoel) becomes the presumptive digestive tract.

The outer layer of gastrula cells is the **ectoderm** and will form the skin and nervous system of the mature organism. The inner layer is the **endoderm** and will form the digestive tract. Refer to your textbook for a list of other organs formed by these two germinal layers. The third and final germ layer is the **mesoderm** that forms between the ectoderm and endoderm. The mesoderm forms from cells that disassociate from the endoderm. The mesoderm produces muscular tissue and parts of the reproductive and circulatory systems.

The development of organs is called **organogenesis.** In sea stars, gastrulation is followed by organogenesis and the formation of the motile and feeding larval stage, the **bipinnaria larva** (fig. 50.3*c*). This stage is followed by the **brachiolaria larva** (fig. 50.3*d*). These larvae attach to the substrate and undergo considerable metamorphosis. Soon thereafter, sea star arms grow from the body with no relation to the original bilateral symmetry. Eventually, the maturing organism detaches from the substrate as a mobile, **juvenile** sea star (fig. 50.3*e*).

### Question 3

Why are the ectoderm and endoderm called "germ" (from the word *germinate*) layers?

## FROG DEVELOPMENT

Review in your textbook the general characteristics of chordates, vertebrates, and amphibians. Frogs are vertebrates in class Amphibia of phylum Chordata (Exercise 40). Although frogs often live on land, they require freshwater to lay eggs and reproduce.

UK City Images/Alamy Stock Photo

**Figure 50.4** External fertilization. When frogs mate, as these two are doing, the clasp of the male induces the female to release a large, gelatinous mass of mature eggs, over which the male discharges his sperm.

## Fertilization

Examine prepared slides of a frog egg and zygote. If models are available, use them to examine frog development.

Fertilization is external for frogs, but it coincides with contact between a male and female frog during **amplexus** (fig. 50.4). In this process a male clasps a female between his front legs and applies pressure to stimulate her to release eggs. The male then releases sperm cells into the water immediately surrounding the eggs. Fertilization occurs externally.

Penetration of an egg by a sperm cell stimulates the cytoplasm to become specialized in certain regions. A darkly pigmented hemisphere, the **animal pole,** and a lighter hemisphere, the **vegetal pole,** appear opposite each other. A gray crescent of interior cytoplasm will appear opposite the point of sperm entry (fig. 50.5). This gray crescent will determine orientation of the first cell division. The cells of the animal pole give rise to the body. The cells of the vegetal pole provide nutrition for the developing embryo. The vegetal pole contains a significant amount of yolk; thus, frog eggs are **telolecithal** (fig. 50.6).

### Question 4
***a.*** How large is a frog zygote compared to that of a sea star?

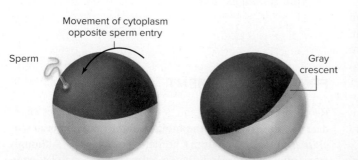

**Figure 50.5** Gray crescent formation in frog eggs. The gray crescent forms on the side of the egg opposite the point of penetration by the sperm.

Movement of cytoplasm opposite sperm entry

Sperm

Gray crescent

BiologyImaging.com

**Figure 50.6** Frog zygote (25×). A large diploid cell rich with nutritious cytoplasm and nucleic acids (mRNA) is ready to guide rapid protein production and development.

***b.*** Are any areas of the frog zygote more darkly pigmented than other areas?

## Cleavage

Examine slides of early and late cleavage of a developing frog embryo (fig. 50.7) and a slide of a cross section of a frog blastula (fig. 50.8).

Cleavage of frog zygotes is holoblastic, but the resulting cells are not equal in size. As they continue to divide, a morula develops. Fluid then collects in the center

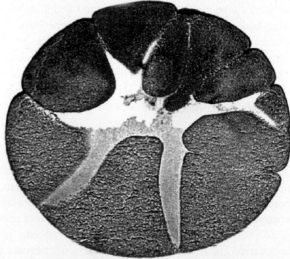

BiologyImaging.com

**Figure 50.7** Frog late cleavage, cross section (25×). The cytoplasm of the zygote is compartmentalized into 20–30 smaller cells. Not all of the cells are visible in this cross section.

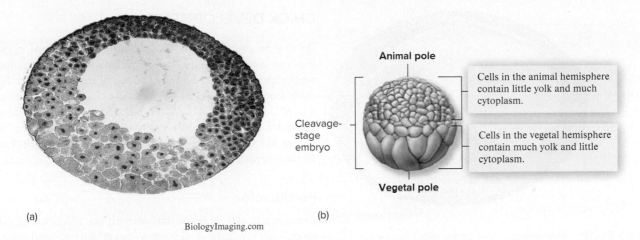

(a)

(b)

**Animal pole**

Cells in the animal hemisphere contain little yolk and much cytoplasm.

Cleavage-stage embryo

Cells in the vegetal hemisphere contain much yolk and little cytoplasm.

**Vegetal pole**

BiologyImaging.com

**Figure 50.8** (*a*) Frog blastula, cross section (25×). The first embryonic cavity becomes distinctive among more numerous but smaller cells. The cavity in the center is the blastocoel. (*b*) Animal and vegetal poles differentiate as cells continue to compartmentalize the cytoplasm by dividing but not growing in size.

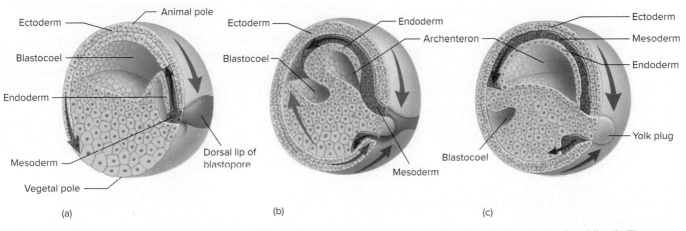

(a)

Ectoderm
Animal pole
Blastocoel
Endoderm
Mesoderm
Vegetal pole
Dorsal lip of blastopore

(b)

Ectoderm
Endoderm
Archenteron
Blastocoel
Mesoderm

(c)

Ectoderm
Mesoderm
Endoderm
Yolk plug
Blastocoel

**Figure 50.9** Frog gastrulation. (*a*) A layer of cells from the animal pole folds down over the yolk cells, forming the dorsal lip. (*b*) The dorsal lip zone then invaginates into the hollow interior, or blastocoel, eventually pressing against the far wall. The three principal tissues (ectoderm, endoderm, and mesoderm) become distinguished here. (*c*) The inward movement of the dorsal lip creates a new internal cavity, the archenteron, which opens to the outside through the yolk plug remaining at the point of invagination. The gastrula is a three-layered system with a cavity destined to become the digestive tract. In phyla such as protostomes, the blastopore will give rise to the mouth; in other phyla (deuterostomes), it gives rise to the anus. Humans and frogs (phylum Chordata) are deuterostomes.

of the morula and the hollow mass of cells forms a blastula. Vegetal pole cells of the blastula are laden with yolk and divide more slowly than do the animal pole cells. This produces an asymmetrical blastula.

**Question 5**

*a.* Which are larger, animal pole cells or vegetal pole cells?

*b.* Is the developing blastocoel visible in late cleavage?

*c.* How does orientation of the blastocoel within the blastula compare with that of a sea star?

## Gastrulation

Examine a cross section of a frog gastrula (fig. 50.9). Because yolk-laden cells of the vegetal pole of a frog blastula are larger than cells of the animal pole, the layer of blastomeres does not simply bend inward and invaginate as it does in sea stars. Instead, the layer of animal-pole cells moves, grows down over the vegetal cells, and moves inward at a depression called the **blastopore.** This growth and movement is called **involution.**

The yolk cells are enveloped during gastrulation. However, some of them protrude as a **yolk plug** through the **blastopore** (fig. 50.10). The mesodermal layer develops between the endoderm and ectoderm from cells proliferating within the dorsal lip of the blastopore.

Gastrulation in vertebrates is followed by formation of the **neural tube (neurulation)** and other organs (organogenesis), thereby producing a functional larva.

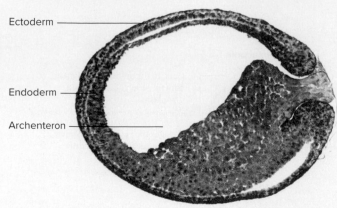

Ectoderm

Endoderm

Archenteron

BiologyImaging.com

**Figure 50.10** Frog gastrula, cross section (40×). Involution produces the first two germ layers (ectoderm and endoderm) and the primitive gut cavity (archenteron).

### Question 6

*a.* Is the degenerating blastocoel visible in the frog gastrula?

*b.* What is the relative size of the frog gastrula compared to the blastula and zygote?

*c.* How much yolk protrudes through the blastopore of a frog gastrula?

*d.* What is the major function of the larval stage in the overall life history of an organism?

*e.* Are all of the organ systems fully developed in a free-swimming larva? Which organ systems are unlikely to be fully developed?

*f.* How has the developing frog embryo dissipated its metabolic wastes?

## CHICK DEVELOPMENT

Review in your textbook the characteristics of bird eggs. Birds are vertebrates and have developmental structures and strategies adapted for a totally terrestrial existence. Specifically, birds (and reptiles) have eggs with protective shells, a large amount of yolk, and an intricate membrane system for various functions. Ostriches have the largest bird eggs at 15 cm diameter, while hummingbirds have the smallest at 10 mm diameter. A human egg is of only 0.1 mm diameter.

### Fertilization

Crack open and examine an unfertilized chicken egg. If models are available in addition to living and prepared material, use them to examine chick development.

Fertilization occurs internally before a hard shell is produced. The egg cell is a large yellow mass of yolk with a small area of cytoplasm on its surface called a **germinal disc,** which contains the egg nucleus. Cells formed from the germinal disc later form the embryo. Surrounding the egg cell is clear, watery **albumen** that supplies the embryo with water and food in addition to the yolk. Albumen is surrounded by two shell membranes and a hard shell porous to gases.

### Question 7

*a.* Is all of the albumen of a chicken egg the same consistency?

*b.* If you pull a probe through the albumen, it seems to cling to the probe and to itself. What macromolecules are abundant in albumen and make it viscous?

*c.* Why must an eggshell be porous?

*d.* What functions does the shell have?

### Cleavage

Yolk is so abundant in bird eggs that the entire zygote cannot divide; instead, cleavage is confined to the germinal disc. Such partial cleavage is termed **meroblastic** and produces a flat **blastodisc** (blastoderm) of developing cells lying atop the bulky yolk. As the blastodisc develops, it separates from the underlying yolk and then forms two layers: the **epiblast**

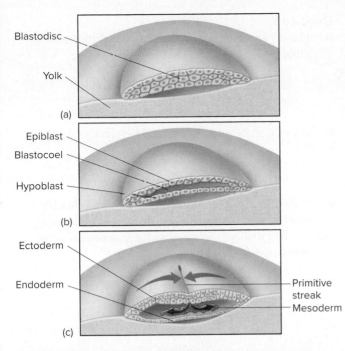

Blastodisc
Yolk

(a)

Epiblast
Blastocoel
Hypoblast

(b)

Ectoderm
Endoderm
Primitive streak
Mesoderm

(c)

**Figure 50.11** Gastrulation of the chick blastodisc. (*a*) The two sides of the chick blastodisc are not separated by yolk. (*b*) The upper layer of the blastodisc differentiates into ectoderm, the lower layer into endoderm. (*c*) Among the cells that migrate into the interior through the dorsal primitive streak are future mesodermal cells.

and **hypoblast** separated by a blastocoel (fig. 50.11*a*, *b*). The epiblast will become the ectoderm.

## Gastrulation

As in the sea star and frog, gastrulation involves movement of cells into the hollow blastula to establish a multilayered system. However, the opening of the blastodisc is not a round blastopore; rather, it is a linear furrow called the **primitive streak** (fig. 50.11*c*). Cells migrate across the surface of the blastodisc and into the primitive streak. Most of the migrating cells become the mesoderm, some of whose cells merge with the hypoblast to form the endoderm. The primitive streak becomes the midline of the developing chick.

Surrounding the embryo are networks of arteries and veins leading to extraembryonic membranes. These membranes in eggs developing in a terrestrial environment are important because they assume functions performed by the aquatic environment for other organisms such as frogs and sea stars (fig. 50.12). Notice that this homologous membrane system persists in mammalian development.

The **amnion** surrounds the embryo, the **yolk sac** surrounds the yolk, and the **allantois** forms a waste disposal sac. The **chorion** surrounds the entire embryo and yolk. Review in your textbook the function of each of these membranes and complete table 50.1.

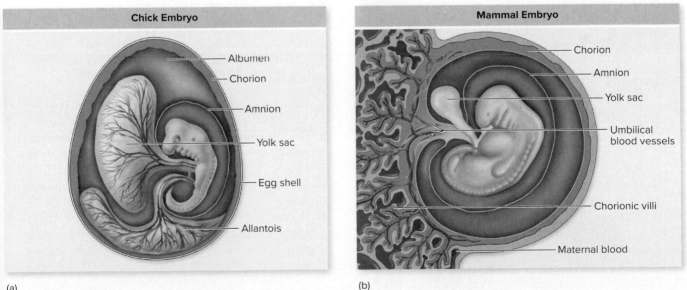

**Chick Embryo**

Albumen
Chorion
Amnion
Yolk sac
Egg shell
Allantois

(a)

**Mammal Embryo**

Chorion
Amnion
Yolk sac
Umbilical blood vessels
Chorionic villi
Maternal blood

(b)

**Figure 50.12** The amniotic egg and its membrane system are perhaps the most important features that allow vertebrates to live in a variety of terrestrial habitats. (*a*) An egg shell protects the chick embryo from drying out, nourishes it, and enables it to develop outside of water. An amniotic egg contains a large nourishing yolk and abundant albumen with nutrients and water. Nitrogenous wastes are excreted into the saclike allantois for storage. The amnion surrounds the developing embryo, and the chorion helps control diffusion of substances into and out of the embryo. The shell offers protection but is porous for diffusion of oxygen and carbon dioxide. (*b*) A mammalian embryo has the same membrane system. The allantois contributes blood vessels to the developing umbilical cord.

## Procedure 50.1  Examine chick embryos

1. Observe the demonstration specimens of living 33- and 72-h chick embryos.

2. Examine plastic mounts of 24-, 33-, and 72-h chick embryos, and follow the formation of the brain, heart, and eye.

3. Compare the living and mounted specimens with figures 50.13 and 50.14.

After 33 h of development the blastodisc is about 20 mm in diameter. The newly developed heart can be seen on the right side. Two veins bring blood to the heart, and a single ventral aorta drains blood from the heart. The developing brain continues as a neural tube the length of the body. After 72 h of development the anterior part of the embryo lies on its left side. The central nervous system and eye have developed considerably.

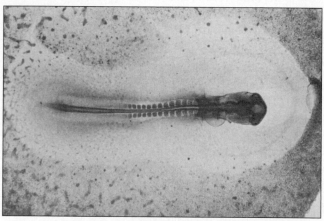

Carolina Biological Supply Company/Phototake

**Figure 50.13**   Chick embryo at 33 h, whole mount (25×).

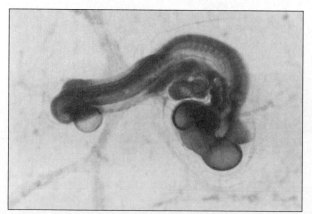

Carolina Biological Supply Company/Phototake

**Figure 50.14**   Chick embryo at 72 h, whole mount (10×).

### Question 8

*a.* Has organogenesis begun for 33- and 72-h chick embryos?

*b.* Is the heart beating in 33- and 72-h chick embryos?

*c.* Sea star and frog embryos do not have membranes similar to those of a chick embryo. How do sea star and frog embryos accomplish the functions of these membranes?

*d.* Chickens lack a larval stage. How do they compensate for this?

## INQUIRY-BASED LEARNING

### *How permeable are egg shells?*

Observations: The success of land animals requires eggs that can withstand desiccation while exchanging gases with the environment. Egg shells must be optimally permeable.

Question: How permeable are egg shells?

a. Establish a working lab group and obtain Inquiry-Based Learning Worksheet 50 from your instructor.

b. Discuss with your group well-defined questions relevant to the preceding observations and question. Choose the best of your questions and record it.

c. Translate your question into a testable hypothesis and record it.

d. Outline on Worksheet 50 your experimental design and supplies needed to test your hypothesis. Consider the effects of temperature, pH, and molecular size on the rate of movement through an egg shell. Ask your instructor to review your proposed investigation.

e. Conduct your procedures, record your data, answer your question, and make relevant comments.

f. Discuss with your instructor any revisions to your questions, hypotheses, or procedures. Repeat your work as needed.

# REVIEW

To review the comparative features of early development of the sea star, frog, and chick, complete tables 50.2 and 50.3 with appropriate descriptions and associated structures.

## Table 50.1

### Functions of the Major Membranes of a Developing Chick

| Membrane | Function |
|----------|----------|
| Amnion | |
| Allantois | |
| Chorion | |
| Yolk sac | |

## Table 50.2

### A Comparison of Embryological Features of a Developing Sea Star, Frog, and Chick

| Feature | Sea Star | Frog | Chick |
|---------|----------|------|-------|
| Relative egg size | | | |
| Amount of yolk | | | |
| Distribution of yolk | | | |
| Blastulation | | | |
| Gastrulation | | | |
| Larval stage | | | |

**Table 50.3**

| A Comparison of Ecological Aspects of Early Development of Sea Stars, Frogs, and Chicks | | | |
|---|---|---|---|
| Feature | Sea Star | Frog | Chick |
| Environment of development | | | |
| Fertilization | | | |
| Mechanism of waste disposal | | | |
| Physical protection | | | |
| Parental care | | | |

# Questions for Further Study and Inquiry

*1.*  Where does a sea star embryo get its nutrition during prelarval stages of development?

*2.*  More recently evolved organisms have modified the stages of embryological development. Would you expect early or late stages to be modified the most? Why?

*3.*  The basic stages of embryological development are remarkably similar for a wide range of organisms. How would you explain such consistency?

*4.*  As cells invaginate and move about, do they "know" where they are? If not, how is movement controlled? If so, how do they perceive their position?

## WRITING TO LEARN BIOLOGY

Is parental care of a species a factor that affects embryological development? If so, how?

# Animal Behavior
## Taxis, Kinesis, and Agonistic Behavior

## Learning Objectives

By the end of this exercise you should be able to:
1. Test how various stimuli influence the behavior of brine shrimp and pillbugs.
2. Classify a behavior as an example of an agonistic behavior, kinesis, or taxis.
3. Discuss the adaptive significance of kinesis, agonistic behavior, and taxis in the organisms you study.

Please visit **connect.mheducation.com** to review online resources tailored to this lab.

**B**iologists explain animal behavior by observing animals and by studying the physiology of behavior. Explanations of animal behavior are often complex. To minimize this complexity, biologists often study animals that have a limited range of behaviors. Complex behaviors are best understood after understanding simple and isolated behaviors.

**Orientations** are behaviors that position an animal in its most favorable environment. There are two types of orientations: taxis and kinesis. A **taxis** is a movement toward or away from a stimulus. Prefixes such as *photo-* and *chemo-* are usually added to describe the nature of the stimulus. For example, a fruit fly that flies toward light is positively **phototactic.** A **kinesis** is a random movement not oriented to the direction of the stimulus; that is, the stimulus initiates but does not necessarily orient the movement. The speed of the kinesis is determined by the intensity of the stimulus.

Some animals also exhibit **agonistic behaviors.** These behaviors usually occur when the animal is in a confrontation in which there may be an attack or withdrawal. An agonistic behavior that results in retreat or avoidance is a **submission,** whereas one that produces a more forceful response is an **aggression.** Agonistic behaviors seldom lead to death; rather, they usually help an animal maintain its territory. In many animals, agonistic behaviors make the animal look larger or more threatening.

In this exercise, you'll study three examples of animal behavior: agonistic behavior in Siamese fighting fish, kinesis in pillbugs, and taxis in brine shrimp. You'll also design and do your own experiments to better understand these behaviors.

## KINESIS IN PILLBUGS

Pillbugs (also called sow bugs and rolypolies) are terrestrial crustaceans that spend much of their time avoiding

BiologyImaging.com

**Figure 51.1** Pillbugs are terrestrial crustaceans whose behavior is strongly influenced by moisture.

dry environments (fig. 51.1). Pillbugs are easily collected in warm weather under logs, under flower pots, and in leaf litter. As their name suggests, rolypolies often respond to mechanical stimuli by rolling into a ball.

### Question 1
How do you think that pillbugs will respond when they're placed in an environment whose surface is moist? In one that is dry?

**Table 51.1**

| Observations of Kinesis in Pillbugs | | | |
|---|---|---|---|
| Dish | Number of Pillbugs Moving | Number of Turns per Minute | Rate of Movement |
| Wet | | | |
| Dry | | | |

## Procedure 51.1  Study kinesis in pillbugs

1. Place five pillbugs in each of two large petri dishes, one containing wet filter paper and the other containing dry filter paper. Place the dishes in a dark drawer.

2. Before you open the drawer, assign each of the following tasks to a member of your group. After 5 min, remove the dish from the drawer and observe the pillbugs in the dishes.

   - Count the number of pillbugs moving in each dish.
   - Count the number of turns (changes in direction) per minute for a single pillbug in each dish.
   - Choose a moving pillbug in each dish. Determine its rate of movement by counting the number of times that the animal circles the dish.

3. Record your observation in table 51.1.

**Question 2**

***a.*** What other stimuli might affect the behavior of pillbugs?

***b.*** What is the adaptive significance of a kinesis?

## AGONISTIC BEHAVIOR IN SIAMESE FIGHTING FISH

In male Siamese fighting fish (*Betta splendens*), the sight of another male triggers an innate, intraspecific, ritualized series of responses toward the intruder (figs. 51.2 and 51.3). These agonistic behaviors include broadside movements and facing movements. During broadside movements, the *Betta* turns its side toward the opponent. Typical aspects of broadside movements include tail flashing (closing and reopening the caudal fin), tail beating (sudden movement with its caudal fin toward an opponent), and pelvic fin flickering (moving its pelvic fin vertically opposite an opponent).

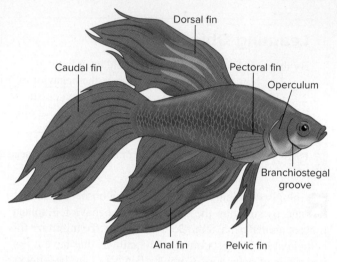

**Figure 51.2**  Male Siamese fighting fish.

After these broadside movements, a fish may perform a facing movement, at a right angle to the head of its opponent, while holding its fins erect. Simultaneously, the fish usually erects its gill covers (opercula); the branchiostegal membranes, under the gill covers, may protrude during the display. The fish may also perform other agonistic behaviors, such as darkening its skin, slightly arching its back, raising its dorsal fin, lowering its head, darting, or biting and nipping an opponent.

In this exercise, you will determine (1) what stimulus initiates agonistic behavior in *Betta* and (2) what parts of the stimulus are most important to trigger the response. Before you do this exercise, familiarize yourself with the anatomy of a Siamese fighting fish; be sure you can identify the dorsal fin, pectoral fin, anal fin, gill cover (operculum), and caudal fin (fig. 51.2). Also know how to recognize the agonistic behaviors of these fish. Figures 51.3*a* and *b* show *Betta* in nonaggressive poses, whereas figures 51.3*c* and *d* show *Betta* in aggressive poses.

## Procedure 51.2  Understand agonistic behavior in Siamese fighting fish

1. Review with your instructor the proper considerations for the ethical and humane treatment of animals.

2. The *Betta* you will study have been kept in separate aquaria such that the fish cannot see each other or their own reflections. Agonistic behaviors occur most frequently when the water temperature is 28°C.

**Nonaggressive postures**     **Aggressive postures**

**Figure 51.3**   Male Siamese fighting fish in (*a, b*) nonaggressive postures and (*c, d*) aggressive postures.

3. Because some fish are more aggressive than others, you'll have to do a preliminary test to determine which fish are most likely to respond to the stimuli. Do this by holding a small mirror against the side of each aquarium so that each fish can see its reflection. Observe the fish's reaction.

4. Select the four or five most aggressive males for your experiment. Assign an identification number to each fish.

5. Decide which stimulus you will test (e.g., color of the opposing fish, photocopy with modified anatomy, geometric shapes, frontal vs. lateral views). Be creative!

6. Form a hypothesis about the result you expect. Your instructor will advise you about how to write a testable hypothesis. Write your hypothesis here.

In your experimental design, use photocopies of figure 51.2 to discover what parts of the stimulus are sufficient to trigger agonistic behavior. For example, if your group decides to test whether color causes the agonistic response, color each of the images of the same shape a different color. Then attach each image to a stick or straw and present the images to the fish one at a time.

7. While doing your experiment, do not present the image to a fish for longer than 5 min. Also, be sure that you study only one variable at a time. For example, do not color the fish with different colors; if you do, you'll not be able to isolate which variable (i.e., shape or color) causes the response.

8. Count and record in table 51.2 the number of times each display occurs during a test.

**Question 3**
*a.* Are the displays "all or none"? That is, do fish ever exhibit partial responses?

## Table 51.2

**Agonistic Behaviors in Siamese Fighting Fish**

| Behavior | Test 1 | Test 2 | Test 3 | Test 4 | Test 5 |
|---|---|---|---|---|---|
| Pelvic fin flickering | | | | | |
| Tail flickering | | | | | |
| Broadside movements | | | | | |
| Tail beating | | | | | |
| Facing movements | | | | | |
| Dorsal fin erect | | | | | |
| Gill cover erect | | | | | |
| Back arching | | | | | |
| Head lowered | | | | | |
| Skin darkening | | | | | |
| Biting and nipping | | | | | |

*b.* What is the simplest stimulus that initiates a response?

*c.* Which factor is more important for triggering agonistic behavior?

*d.* After repeated identical stimuli, does a fish become "conditioned"? That is, does the duration of the behavior change or stop?

*e.* Will another species of fish trigger the behavior?

*f.* What are the differences between the aggressive and nonaggressive postures?

*g.* Will a female *Betta* trigger agonistic behavior in a male? If so, how does the response compare with that elicited by a male?

*h.* What is the adaptive significance of agonistic behavior?

*i.* What agonistic behaviors do humans display?

## TAXIS IN BRINE SHRIMP (ARTEMIA SALINA)

Brine shrimp (*Artemia*) are small crustaceans that sense many stimuli with two large compound eyes and two pairs of antennae (fig. 51.4). *Artemia* live in salt lakes and can be bought at most pet stores. In this exercise, you will study how various stimuli affect behavior of these shrimp.

## Procedure 51.3  Understand taxis in brine shrimp

1. Place brine shrimp in a test tube of salt water provided by your instructor. Then place the tubes against a black background to see the shrimp. Do not feed or disturb the shrimp.
2. Note their behavior in this relatively stimulus-free environment.

### Question 4

*a.* Are the shrimp near the top or bottom of the tube? Are they moving? Are they in groups or are they solitary?

*b.* What stimuli (e.g., light, food, acids) might initiate taxis in *Artemia?*

*c.* How do you predict the shrimp will respond to each of these stimuli?

Dan Olsen/Shutterstock

**Figure 51.4**  A young brine shrimp. This crustacean has segmented appendages for swimming and pigmented eyespots for detecting light (40×).

4. Do your experiment. What do you conclude? Do your data support your hypothesis? _____
_____
_____
_____
_____
_____
_____

## Procedure 51.4  Test how stimuli influence behavior of brine shrimp

1. Choose one of the stimuli listed in your answer to Question 4b.
2. Form a hypothesis to determine how this stimulus might influence the behavior of the brine shrimp. Write your hypothesis here. _____
_____
_____
_____
_____
_____

3. Decide how you will test your hypothesis. Describe your experimental design here. _____
_____
_____
_____
_____
_____
_____

### Question 5

*a.* How might the responses to the stimuli listed in Question 4b be adaptive?

*b.* Did the responses of the shrimp match your predicted responses?

# INQUIRY-BASED LEARNING

## *How close will a fish allow a threat to approach before responding?*

Observations: Fleeing from danger is a highly adaptive response for animals in the wild. The distance at which an animal senses danger and initiates flight would depend on many factors such as the current environment and the nature of the threat.

Question: What variables influence the flight initiation distance (FID) of squirrels sensing danger?

a. Establish a working lab group and obtain Inquiry-Based Learning Worksheet 51 from your instructor.

b. Discuss with your group well-defined questions relevant to the preceding observations and question. Choose the best of your questions and record it.

c. Translate your question into a testable hypothesis and record it.

d. Design an experiment to quantify FID in response to a chosen stimulus.

e. Outline on Worksheet 51 your experimental design and supplies needed to test your hypothesis. Ask your instructor to review your proposed investigation.

f. Conduct your procedures, record your data, answer your question, and make relevant comments.

g. Discuss with your instructor any revisions to your questions, hypotheses, or procedures. Repeat your work as needed.

# Questions for Further Study and Inquiry

1. Killdeer are ground-nesting birds common throughout North America. When a predator approaches, they often feign broken wings. What is the selective advantage of this type of innate behavior?

2. Mating (reproductive behavior) often involves a complex sequence of events that helps an animal find, court, and mate with a member of the same species. What are some examples of these behaviors? Do humans exhibit mating behaviors? If so, what are they?

3. What are the adaptive advantages of agonistic behaviors not followed by damaging fights? What animals have strong displays not followed by a damaging fight? What animals do engage in damaging fights?

4. How could each of the behaviors that you studied today have arisen by natural selection?

5. Are behaviors inherited or are they learned? If you believe behaviors are inherited, what is your evidence? If you believe behaviors are learned, what is your evidence?

6. In this exercise, you studied three aspects of animal behavior: taxis, kinesis, and agonistic behavior. Why is it important to study all aspects of animal behavior?

## WRITING TO LEARN BIOLOGY

Mating behaviors are species-specific. Why is that important?

# Dissection of a Fetal Pig

## Learning Objectives

By the end of this exercise you should be able to:
1. Perform a whole-body dissection of a vertebrate animal.
2. Identify the major anatomical features of the vertebrate body in a dissected specimen.

Please visit **connect.mheducation.com** to review online resources tailored to this lab.

**F**etal pig anatomy provides an excellent model of general mammalian anatomy. Although there are some significant differences, the body plan is the same, and the functional relationships within and between the anatomical systems model that of humans as well as other commonly studied mammals.

## MATERIALS NEEDED

- Preserved (plain or double-injected) fetal pig
- Dissection tools and trays
- Mounted fetal pig skeleton (optional)
- Storage container (if specimen is to be reused)

Read the directions and safety tips for this exercise carefully before starting any procedure. In a short course, a well-preserved specimen can be used from time to time throughout your studies. If you will be looking at your dissected specimen from time to time during the next several months, make sure that it is kept in the appropriate container under conditions suggested by your instructor.

## THE EXTERNAL ANATOMY

Observe the usual precautions when working with a preserved or fresh specimen. Heed the safety advice accompanying preservatives used with your specimen. Use protective gloves when you handle your specimen. Avoid injury with dissection tools and dispose of your specimen as instructed.

### Procedure A.1  Examine the external characteristics of your specimen

1. Determine the anatomical orientation of the specimen. Which direction is anterior? Posterior? Which direction is ventral? Dorsal? Identify sagittal, transverse, and frontal planes in your specimen.

2. Identify these externally visible features:
   pinna (auricle)
   external nares (nostrils)
   umbilicus (umbilical cord)
   forelimbs
   hind limbs
   thoracic region
   abdominal region
   nipples
   anus
   tail

3. Determine the sex of your specimen by examining the external genitals:

   **Female**—Immediately anterior to the anus, on the ventral surface, is the vulva with an opening to the vagina and the urethra. The vulva is also called the urogenital opening.

   **Male**—In the male fetal pig, there is a rather loose area of skin immediately posterior to the anus, perhaps even hiding the anus from view. Around the time of birth, each testis will descend from its position inside the body into the space under this skin. The skin will pouch out to form the outer wall of the scrotum. Just posterior to the umbilical cord is the distal end of the penis with its prepuce, or skin-fold covering. Locate the opening of the urethra in the penis.

# SKIN, BONES, AND MUSCLES

## *Procedure A.2*  **Remove the skin from the specimen**

1. Place the animal in a tray; make sure the animal's ventral surface is facing you.

2. Pull up on the skin over the neck and puncture it with the tip of a scissors. Slide the bottom tip of the scissors into the subcutaneous area under the skin.

3. Begin cutting along the lines indicated in figure A.1a. The posterior cuts are different for male and female specimens.

4. Be careful not to cut into the skeletal muscles under the skin or through the base of the umbilical cord.

5. With your forceps, pull the two flaps of skin over the neck away from the animal's body. Notice the areolar tissue under the skin that is pulled apart as you remove the skin. Sometimes it helps if you scrape at the loose connective tissue under the skin with your scalpel as you peel the skin away. The skinning process is difficult unless you have patience and proceed slowly. Pull the flaps of skin over the abdomen and over the groin area away in a similar fashion. If you have a male pig, leave the skin posterior to the umbilical cord in place for now.

6. Pin the flaps of skin to the floor of the dissection tray or cut them away from the body entirely. If you plan to study the musculature of the dorsum, you must similarly remove the skin from the animal's back.

7. Now examine the skin, identifying these features of the integument:

   • Dermis—the thick inner layer of the skin

   • Epidermis—the thinner outer layer of the skin

   • Hypodermis—the subcutaneous tissue under the skin proper, made of areolar and adipose tissue

Explore the shape of the skinned fetal pig body. How many bones of the fetal pig's skeleton can you see or feel? If you have a mounted fetal pig skeleton available, identify as many of the bones of the skeleton as you can. If you become stumped, refer to figure A.1a, b. Notice the similarity between the fetal pig's skeletal plan and that of a human. One difference easy to see is in the vertebral column: The pig has a different number of each type of vertebral bone compared with a human and has numerous caudal vertebrae instead of a single coccyx. Also, because the fetal pig's skeleton is just beginning its development, many of the bones are cartilaginous rather than bony.

Observe the fetal pig's musculature. Some of the external muscles of the torso can be separated from each other for easier viewing. Slide a probe into the loose connective tissue joining adjacent muscles and run the probe along their margins. Using figure A.2 as a guide, try to identify the major skeletal muscles of the fetal pig's body. Because the animal was not yet active when the specimen was prepared, the muscles appear underdeveloped and make muscle identification difficult.

# CARDIOVASCULAR STRUCTURES

## *Procedure A.3*  **Examine cardiovascular structures**

1. Cut flaps in the neck, abdomen, and groin areas as you did with the skin. Be careful not to damage any visceral organs with your scissors as you cut. Likewise, avoid injuring the internal structures associated with the umbilicus. Fold back the flaps and anchor them with pins or remove them.

2. Open the ventral body cavity by cutting into its muscular wall in a manner similar to your earlier cut into the skin over the abdomen.

3. Locate the heart near the middle of the thoracic cavity, in the mediastinum. Can you identify the four chambers? After identifying the major vessels (see steps 4 and 5), you may want to remove the heart and dissect it.

4. Locate the aorta, which is the large artery leaving the heart and arching posteriorly. If you have a double-injected preserved specimen, the systemic arteries are filled with red latex and the systemic veins are filled with blue latex. If not, the arteries usually can be distinguished from veins because they are stiffer and lighter in color than veins. Trace the major branches of the aorta, naming them if you can. Use figure A.3 if you need help.

5. Locate the anterior vena cava and note where it drains into the heart. It is sometimes called the precava and is analogous to the superior vena cava in a human. Follow its tributary veins and identify them with the help of figure A.3. Locate the posterior vena cava (postcava) and trace its tributaries. Once you have cut into the ventral body cavity, you may be tempted to cut and remove organs. It is important that you keep everything as intact as possible. You may pull organs to the side to view deeper structures, but avoid making cuts.

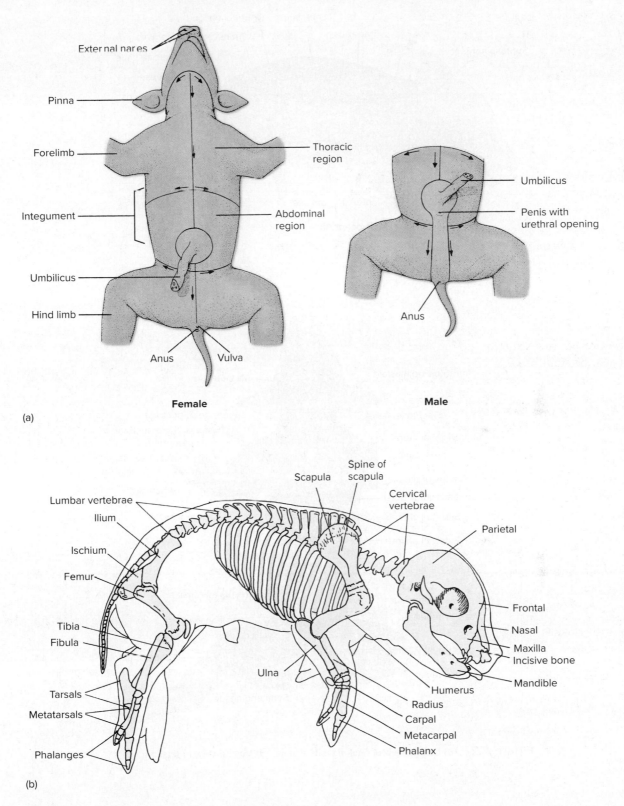

(a)

**Female**

**Male**

External nares

Pinna

Forelimb

Integument

Umbilicus

Hind limb

Anus

Vulva

Thoracic region

Abdominal region

Umbilicus

Penis with urethral opening

Anus

(b)

Lumbar vertebrae

Ilium

Ischium

Femur

Tibia

Fibula

Tarsals

Metatarsals

Phalanges

Scapula

Spine of scapula

Cervical vertebrae

Parietal

Frontal

Nasal

Maxilla
Incisive bone

Mandible

Humerus

Radius

Carpal

Metacarpal

Phalanx

Ulna

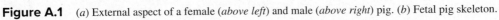

**Figure A.1** (*a*) External aspect of a female (*above left*) and male (*above right*) pig. (*b*) Fetal pig skeleton.

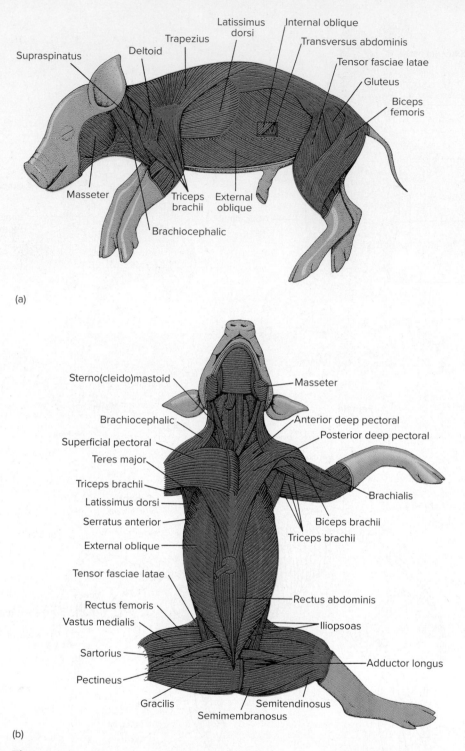

**Supraspinatus** — **Deltoid** — **Trapezius** — **Latissimus dorsi** — **Internal oblique** — **Transversus abdominis** — **Tensor fasciae latae** — **Gluteus** — **Biceps femoris**

**Masseter** — **Triceps brachii** — **External oblique** — **Brachiocephalic**

(a)

**Sterno(cleido)mastoid** — **Masseter**

**Brachiocephalic** — **Anterior deep pectoral** — **Posterior deep pectoral**

**Superficial pectoral**

**Teres major**

**Triceps brachii** — **Brachialis**

**Latissimus dorsi** — **Biceps brachii**

**Serratus anterior** — **Triceps brachii**

**External oblique**

**Tensor fasciae latae**

**Rectus femoris** — **Rectus abdominis**

**Vastus medialis** — **Iliopsoas**

**Sartorius** — **Adductor longus**

**Pectineus**

**Gracilis** — **Semitendinosus**

**Semimembranosus**

(b)

**Figure A.2**   (*a*) Lateral view of fetal pig musculature. (*b*) Ventral view of fetal pig musculature.

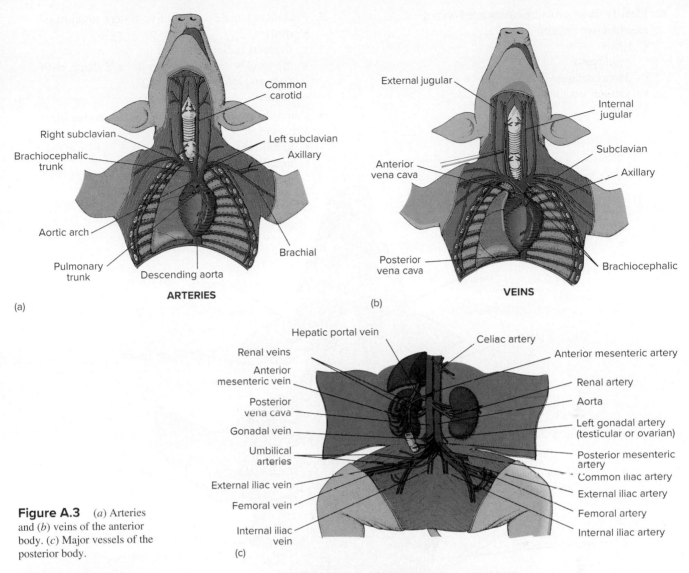

**Figure A.3** (a) Arteries and (b) veins of the anterior body. (c) Major vessels of the posterior body.

## THE VISCERA

The viscera, or major internal organs, can be seen within the ventral body cavity. Use figures A.4 and A.5 to guide you in locating the following:

1. Locate some of these features of the lower respiratory system:
   - larynx
   - trachea
   - primary bronchi
   - lungs (Can you distinguish the parietal and visceral pleurae?)
   - diaphragm

2. Locate these structures of the digestive system:
   - esophagus
   - stomach
   - liver (four separate lobes)
   - gallbladder
   - pancreas
   - small intestine
   - mesentery
   - large intestine (spiral colon)

3. Locate these lymphatic organs:
   - spleen
   - thymus

4. Locate these features of the urinary system:
   - kidney
   - renal cortex
   - renal pyramid
   - renal pelvis
   - renal calyx
   - ureter
   - urinary bladder
   - urethra

5. Try to locate these endocrine glands in your specimen:
   - thyroid gland
   - thymus gland
   - pancreas
   - adrenal glands
   - testes
   - ovaries

6. Identify these structures associated with the male reproductive system:
   - testes
   - epididymis
   - ductus deferens
   - seminal vesicle
   - penis

7. Find these female reproductive system structures:
   - ovaries
   - oviducts (fallopian tubes)
   - uterus (The fetal pig uterus has a Y shape, with right and left uterine horns.)
   - vagina

8. Carefully examine the umbilical cord, noting these structures:
   - umbilical vein
   - umbilical arteries

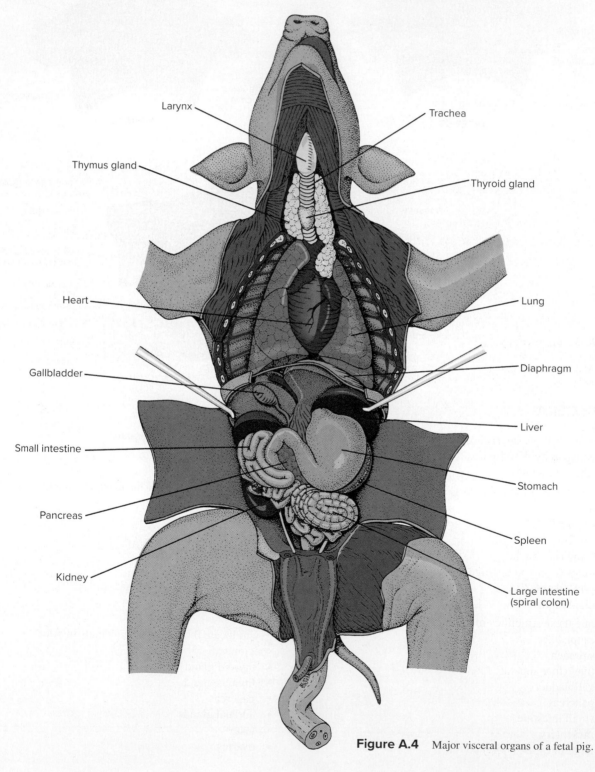

**Figure A.4** Major visceral organs of a fetal pig.

9. In the space to the right, sketch a cross section of the umbilical cord and label all identifiable structures. Unless your lab group has both a male and a female specimen, you may want to temporarily trade specimens with a group that has a pig of the opposite gender from yours. By doing so, you will be able to find the features of both reproductive systems.

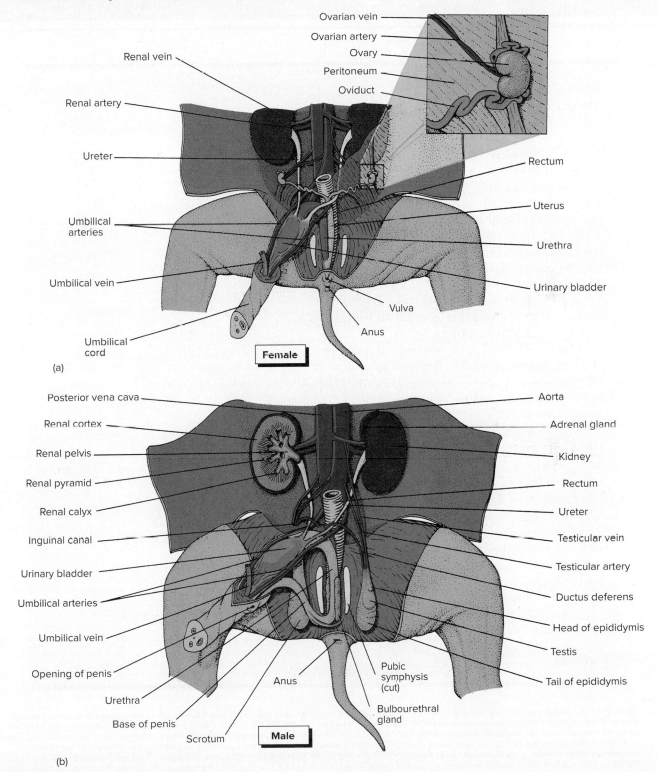

(a)

Ovarian vein
Ovarian artery
Ovary
Peritoneum
Oviduct

Renal vein
Renal artery
Ureter
Umbilical arteries
Umbilical vein
Umbilical cord

Rectum
Uterus
Urethra
Urinary bladder
Vulva
Anus

**Female**

(b)

Posterior vena cava
Renal cortex
Renal pelvis
Renal pyramid
Renal calyx
Inguinal canal
Urinary bladder
Umbilical arteries
Umbilical vein
Opening of penis
Urethra
Base of penis
Scrotum
Anus

Aorta
Adrenal gland
Kidney
Rectum
Ureter
Testicular vein
Testicular artery
Ductus deferens
Head of epididymis
Testis
Tail of epididymis

Pubic symphysis (cut)
Bulbourethral gland

**Male**

**Figure A.5** Urinary and reproductive organs of (*a*) a female and (*b*) a male fetal pig.

# Conversion of Metric Units to English Units

## Units of Length

The meter (m) is the basic unit of length.

1 m = 39.4 inches (in)
= 1.1 yards (yd)           1 in = 2.54 cm
= 3.28 feet (ft)

1 km = 1000 m = $10^3$ m     1 ft = 30.5 cm
= 0.62 mile (mi)                = 0.305 m

1 cm = 0.01 m = $10^{-2}$ m
= 0.39 in = 10 mm          1 yd = 0.91 m

1 nm = $10^{-9}$ m = $10^{-6}$ mm
= 10 angstroms (Å)         1 mi = 1.61 km

Units of area are squared (two-dimensional) units of length.

1 $m^2$ = 1.20 $yd^2$ = 1550 $in^2$ = $1.550 \times 10^3$ $in^2$

1 hectare = 10,000 square meters ($m^2$) = 2.47 acres

Measurements of area and volume can use the same units.

1 $m^3$ = 35.314 $ft^3$ = 1.31 $yd^3$

1 $cm^3$ (cc) = 0.000001 $m^3$ = 0.061 $in^3$

## Units of Mass

The kilogram (kg) is the basic unit of mass. The kilogram is the only base-unit of the International System of Units (SI) with a prefix ("kilo," symbol "k").*

1 kg = 1000 g = $10^3$ g = 2.2 lb

1 g = mass of 1 $cm^3$ of water at 4°C = 0.035 oz

## Units of Volume

The liter (L) is the basic unit of volume. Units of volume are cubed (three-dimensional) units of length.

1 liter = 1000 $cm^3$

1 liter = 2.1 pints = 1.06 qt        1 cup = 240 mL

1 liter = 0.26 gal = 1 $dm^3$

1 mL = 0.034 fl oz

## Units of Temperature

5 × degrees Fahrenheit = (9 × degrees Celsius) + 160

For example:

40°C = 104°F (a hot summer day)

75°C = 167°F (hot coffee)

−5°C = 23°F (coldest area of freezer)

37°C = 98.6°F (human body temperature)

−40°C = −40°F (extremely cold day)

---

* *Before 2019, the kilogram was the only base-unit in the International System of Units defined by a physical object (i.e., an artifact) rather than a universal constant of nature. (For example, a meter is defined as the distance that light travels in a vacuum in 1/299,792,458 second.) That centuries-old, rarely handled kilogram, which was called the "International Prototype of the Kilogram (IPK)" and "Big K," was a 3.9-cm cylinder of platinum/iridium alloy that had been stored by the International Bureau of Weights and Measures in a vault (under three domes of glass) since 1889 in Saint-Cloud, France. In the past 100 years, however, that cylinder had lost 50 millionths of a gram in weight because of reactions on the surface of the cylinder. On May 20, 2019, the definition of a kilogram changed; a kilogram is now defined with the use of Planck's constant, which relates mass to the energy of a photon. Planck's constant is believed to be a fundamental property of the universe that does not vary over time.*